T0269558

Soil behaviour and critical state soil mechanics

Soil behaviour and critical state soil mechanics

DAVID MUIR WOOD

CAMBRIDGE UNIVERSITY PRESS

Cambridge
New York Port Chester Melbourne Sydney

CAMBRIDGE UNIVERSITY PRESS
Cambridge, New York, Melbourne, Madrid, Cape Town, Singapore, São Paulo

Cambridge University Press
The Edinburgh Building, Cambridge CB2 8RU, UK

Published in the United States of America by Cambridge University Press, New York

www.cambridge.org
Information on this title: www.cambridge.org/9780521332491

First published 1990
Transferred to digital printing 1999
Digital reprint (with corrections) 2007

A catalogue record for this publication is available from the British Library

Library of Congress Cataloguing in Publication data

Wood, David Muir

Soil behaviour and critical state soil mechanics/David Muir
Wood.

p. cm.
Includes bibliographical references.
ISBN 0-521-33249-4. – ISBN 0-521-33782-8 (pbk.)
1. Soil mechanics. I. Title.
TA710.W598 1990
624.1′5136 – dc20 89-71189
CIP

ISBN 978-0-521-33249-1 hardback
ISBN 978-0-521-33782-3 paperback

To H, J, and A

Contents

Preface

It could be said that this book was conceived in December 1975 in Göteborg, Sweden, where Göran Sällfors had invited Peter Wroth and myself to Chalmers Tekniska Högskola to give a five-day course on Critical State Soil Mechanics. It was felt then that the material presented in that course ought to become a book. As the years passed outlines for such a book were sketched, but progress was slow in spite of good intentions. The eventual stimulus for labour to begin in earnest was provided in February 1985 by Yudhbir, who organised the Workshop on Critical State Models and Behaviour of Soils at the Indian Institute of Technology, Kanpur, as part of their Silver Jubilee celebrations. This book took shape in the course of discussions with Hideki Ohta in our smoke-filled room at Kanpur.

Since Peter Wroth and I (together or separately) had given many courses of lectures on Critical State Soil Mechanics, much of the material in this book has been aired previously in some form or other. I have endeavoured to give due acknowledgement for those ideas for which I cannot claim originality. Those who have attended any of these courses will discover that the sequence adopted here is quite different from that of the courses. The first lectures of the courses were devoted to the notion of critical states for soils (the latter part of Chapter 6), followed by simple applications of critical state lines to estimation of undrained strengths (early part of Chapter 7) and interpretation of index tests (Chapter 9). Only then was attention turned to the development and description of an elastic–plastic model for soil (Chapters 2, 3, 4, 5) and its application (Chapter 11), with some discussion of stress paths (Chapter 10) and strengths (latter part of Chapter 7), towards the end of the course.

Here the sequence has been reversed and the elastic–plastic models of soil behaviour are introduced first to provide a framework against which various aspects of soil behaviour, including the existence of critical states, can be studied. This seems to provide a more logical progression by providing a reason for looking for critical states, which are otherwise produced rather out of the blue. The aim here is to link the behaviour and modelling of soils to the prior knowledge that the reader may have of the behaviour and modelling of other engineering materials, represented by ideas of elasticity and plasticity. As a result the development of the numerical model comes first. However, those who wish to approach the subject by the route that was used in the courses can follow the sequence outlined in the previous paragraph.

In one way this book does not attempt to be a textbook on soil mechanics as traditionally taught, but in another way it does provide a new approach to the teaching of soil mechanics. The topics which are most obviously left out are seepage and consolidation. It can be argued on the one hand that there is nothing new to add to the large number of textbooks which treat these topics. On the other hand, seepage is merely an application to geotechnical problems of the solution of Laplace's equation; and similarly, consolidation is conventionally taught as the time-dependent one-dimensional deformation of soils resulting from transient flow of water and dissipation of excess pore pressures. This is merely an application of the solution of the one-dimensional diffusion equation. Both seepage and consolidation are, thus, topics that might be more appropriately placed in a course on engineering mathematics. Of course, there are many transient geotechnical situations involving the flow of water which cannot be described as one-dimensional. Proper analysis of these problems requires a coupling of the equations describing the flow of the water with the equations describing the behaviour of the soil, which require a properly formulated constitutive model for soil, and that is very much the subject of this book. Several of the applications of elastic–plastic models of soil behaviour described in Chapter 11 involve just such coupled consolidation analyses.

There is a blurring in the literature of the terms *consolidation* and *compression*. Whereas time-dependent deformation of soils (consolidation) is hardly mentioned here, the change in volume of soils resulting from changes in effective stress (compression) (which might be observed in the consolidometer or oedometer) is a central and vitally important theme running throughout the book. Here the term *consolidation* is reserved for the transient phenomenon, and the equilibrium relationship between volume and effective stress which is often called a 'normal consolidation line' is here called a 'normal compression' line to underline this distinction.

Some of the material for this book has been drawn from courses entitled

Critical State Soil Mechanics, and the phrase forms part of the title of this book. What is critical state soil mechanics?

The phrase was used by Andrew Schofield and Peter Wroth as the title of their 1968 book (Schofield and Wroth, 1968), from which this book has drawn much inspiration. Their purpose in that book 'is to focus attention on the critical state concept and demonstrate what [they] believe to be its importance in a proper understanding of the mechanical behaviour of soils'. To me, critical state soil mechanics is about the importance of considering volume changes as well as changes in effective stresses when trying to understand soil behaviour. Critical state soil mechanics is then concerned with describing various aspects of soil behaviour of which a clearer picture is obtained when differences in volume as well as differences in effective stresses are considered. Critical state soil mechanics is also concerned with building numerical models of soil behaviour in which a rational description of the link between volume change and effective stress history is a fundamental ingredient.

This is not to be taken to imply that critical state soil mechanics is about nothing more than one particular soil model, Cam clay. In this book, this model is introduced in Chapter 5 as a particular example of a general class of elastic–plastic models which happen to show critical states (the idea of critical states is discussed in detail in Chapter 6) and then used to illustrate various features of the observed experimental behaviour of real soils.

Some workers have decided that critical state soil mechanics is concerned only with one particular model of soil behaviour, and because that particular model does not reproduce all the features of their experimental observations, they conclude that neither that particular model nor, by extension, critical state soil mechanics has anything to offer, and hence they reject both. Some veer to the opposite extreme and suppose that everything said in the name of critical state soil mechanics represents a unique and complete description of Truth so that any experimental observations that appear to be at variance with this Truth must be in error.

Here a more tolerant, ecumenical line is taken. Critical state soil mechanics is not to be regarded as a campaign for a particular soil model but rather as providing a deeply running theme that volume changes in soils are at least as important as changes in effective stresses in trying to build a general picture of soil behaviour. This could probably be taken as the definition of critical state soil mechanics adopted for this book.

General and particular models of soil behaviour are described in Chapters 4 and 5, but it is certainly implicit throughout this book that Truth lies in experimental observations: models can at best be an aid to understanding and never a substitute for observation. It is hoped that the study of soil behaviour through the patterns predicted by a simple model may help to show that in many ways soil is not a particularly incomprehensible material, provided that the real possibility of major volumetric changes is accepted. The discovery that some observations do not fit the predictions of this simple model may lead one to reject it but should not lead to the rejection of the whole underlying framework.

It is necessary to defend the choice of symbols used in this text to represent specific volume, and the increments of volumetric strain and shear strain in the conditions of the triaxial test. Those who read the first draft will note that there has been a major change since that was prepared. Regular readers of books on critical state soil mechanics will be aware that the sets of symbols used in the books by Schofield and Wroth (1968), Atkinson and Bransby (1978), and Bolton (1979) are all different. So there is no consistent tradition to follow except one of variety. All the earlier books use v (rather than V) for specific volume, so I have reverted to this. The use of $\delta\varepsilon$ on its own for triaxial shear strain does not convey any information about its nature. I do not like $\delta\varepsilon_\mathrm{v}$ for the volumetric strain increment because I think the subscript v should be reserved for vertical strains. Once one starts trying to think of suitable subscripts to use, the only logical approach seems to be that proposed by Calladine (1963) according to which $\delta\varepsilon_p$ and $\delta\varepsilon_q$ are the increments of volumetric strain and triaxial shear strain and 'the subscripts suggest the association of the stress and incremental strain vectors in pairs'. The concordance between these symbols and those used in the earlier books is shown in the table.

Reference	Volumetric strain	Triaxial shear strain
Schofield and Wroth (1968)	$\dot{v}/v$ ($\dot{v} = -\delta v$)	$\dot{\varepsilon}$
Atkinson and Bransby (1978)	$\delta\varepsilon_\mathrm{v}$	$\delta\varepsilon_\mathrm{s}$
Bolton (1979)	$\delta\varepsilon_\mathrm{v}$	$\delta\varepsilon_q$
Present text	$\delta\varepsilon_p$	$\delta\varepsilon_q$

Acknowledgements

I should like to thank Jim Graham, Poul Lade, Serge Leroueil, and Neil Taylor for their very detailed comments on the first draft of this book. Steve Brown, Andrzej Drescher, Hon-Yim Ko, Steinar Nordal, Bob Schiffman, Andrew Schofield, Stein Sture, and Peter Wroth have also fed me suggestions for amendment and improvement. I have endeavoured to take note of all these comments, particularly if clarification of my text was required. I have given courses based around the content of this book in Boulder, Cambridge, Trondheim, Luleå, Catania, Glasgow, and Otaniemi over the past few years, and I have tried to incorporate improvements that were suggested by those who have been on the receiving end of these courses. This work was originally developed in the environment of the Cambridge Soil Mechanics Group and Cambridge University Engineering Department, and I am grateful to many colleagues for their discussions. Many of the exercises at the ends of chapters have been adapted from Cambridge University Engineering Department example sheets and examination papers.

The manuscript of the original draft of the book was typed by Reveria Wells and Margaret Ward. I am grateful to Les Brown, Peter Clarkson, Gloria Featherstone, and Ruth Thomas for their assistance in preparing some of the figures, most of which were drawn by Dennis Halls and Helen Todd. Diana Phillips and Clare Willsdon provided valuable last-minute assistance on picture research, and Hilary McOwat answered some bibliographical queries.

List of symbols

This list contains definitions of symbols and also an indication of the section in the book where they are first used. All symbols are defined in the text. Although there is obviously some duplication, it is hoped that this will not cause any confusion.

Symbol	Definition	Section
a	area of ram in triaxial cell	(1.4.1)
a	pore pressure parameter	(1.6)
a	exponent in variation of K_0 with overconsolidation	(10.3.2)
a	radius of loaded area	(11.2.1)
a	dimension of rectangular loaded area	(11.2.2)
A	cross-sectional area of triaxial sample	(1.4.1)
A	activity	(9.4.3)
A	slope of line in $w_L:I_P$ plot	(9.4.4)
b	pore pressure parameter	(1.6)
b	width of element in infinite slope	(7.6)
b	dimension of rectangular loaded area	(11.2.2)
B	intercept on line in $w_L:I_P$ plot	(9.4.4)
c	critical shear stress for yield criterion	(3.2)
c	one-dimensional compliance	(12.2)
c'	cohesion in Mohr–Coulomb failure	(7.1)
c_L	undrained strength of remoulded soil at liquid limit	(9.2)
c_P	undrained strength of remoulded soil at plastic limit	(9.4.2)
c'_{pe}	Hvorslev cohesion parameter for triaxial conditions	(7.4.1)
c_u	undrained shear strength	(7.2)

c_{ur}	remoulded undrained strength	(9.5)
c_v	coefficient of consolidation	(11.2.3)
c'_{ve}	Hvorslev cohesion parameter for shear box	(7.4.1)
c_α	coefficient of secondary consolidation	(12.2)
C	clay content	(9.4.3)
C'_c	compression index	(4.2)
C_k	permeability variation coefficient	(11.3.2)
C'_s	swelling index	(4.2)
d	depth of lake	(1.3)
d	diameter	(2.1)
d	penetration of fall-cone	(9.2)
d	depth to water table	(10.3.3)
D	cross-anisotropic elastic parameter = $3K^*G^* - J^2$	(2.3)
D	diameter of split-cylinder test specimen	(9.4.2)
e	void ratio	(1.2)
e_g	granular void ratio	(1.2)
E	Young's modulus	(2.1)
E	energy dissipated per unit volume	(8.4)
E'	Young's modulus in terms of effective stresses	(2.2)
E^*	cross-anisotropic elastic modulus	(2.3)
E_h	Young's modulus for horizontal direction	(2.3)
E_t	tangent stiffness	(12.3)
E_v	Young's modulus for vertical direction	(2.3)
f	yield locus	(4.5)
F	axial force in triaxial apparatus	(1.4.1)
g	plastic potential	(4.5)
g	Hvorslev strength parameter in $p':q$ plane	(7.4.1)
g	acceleration due to gravity	(9.2)
G	shear modulus	(2.1)
G'	shear modulus for soil (in terms of effective stresses)	(2.2)
G^*	shear modulus for cross-anisotropic soil	(2.3)
G_s	specific gravity of soil particles	(1.2)
G_t	tangent shear stiffness	(12.3)
G_{vh}	cross-anisotropic shear modulus	(2.3)
h	excess head of water	(1.3)
h	sample height in simple shear apparatus	(8.3)
h_c	Hvorslev strength parameter in $p':q$ plane (compression)	(7.4.1)

h_e	Hvorslev strength parameter in $p':q$ plane (extension)	(7.4.1)
H	slope height in Casagrande liquid limit device	(9.4.1)
I_D	relative density of sand	(7.4.2)
I_L	liquidity index	(9.3)
I_P	plasticity index	(7.2) (9.3)
I_ρ	settlement influence factor	(11.2.2)
J	cross-anisotropic elastic parameter	(2.3)
k	permeability	(1.2)
k	dummy variable	(4.4.1)
k	constant describing variation of sensitivity with liquidity	(9.5)
k	spring stiffness	(12.4)
k_h	horizontal permeability	(11.3.2)
k_{hh}	horizontal permeability from horizontal flow test	(11.3.3)
k_{hi}	horizontal permeability from in situ test	(11.3.3)
k_{hr}	horizontal permeability from radial flow test	(11.3.3)
k_v	vertical permeability	(11.3.2)
k_α	cone factor	(9.2)
K	bulk modulus	(2.1)
K	constant in Rowe's stress–dilatancy relation	(8.5)
K'	bulk modulus for soil (in terms of effective stresses)	(2.2)
K^*	bulk modulus for cross-anisotropic soil	(2.3)
K_0	earth pressure coefficient at rest	(9.4.5) (10.3.1)
K_{0nc}	value of K_0 for normally compressed soil	(7.4.1) (10.3.1)
l	length of sample	(1.4.1)
m	load factor in combined tension and torsion of tubes	(3.2)
m	mass of fall-cone	(9.2)
m_v	coefficient of volume compressibility	(11.2.3)
M	shape factor for Cam clay ellipse/slope of critical state line	(5.2)
M^*	value of M in triaxial extension	(7.1)
n	porosity	(1.2)

n	overconsolidation ratio (σ'_{vmax}/σ'_v)	(7.2)
n_p	isotropic overconsolidation ratio (p'_{max}/p')	(7.2)
N	location of isotropic normal compression line in $v:\ln p'$ plane	(5.2)
N	model scale	(11.3.3)
p	mean stress	(1.4.1)
p'_e	equivalent consolidation pressure	(6.2)
p'_n	mean effective stress on a normal compression line	(9.4.5)
p'_o	reference size of yield locus	(4.2)
P	normal load in simple shear apparatus/ shear box	(1.4.2) (8.3)
P	axial load on wire or tube	(2.1) (3.2)
P	diametral load in split cylinder test	(9.4.2)
P_0	preload value of axial load	(3.2)
q	deviator stress, generalised deviator stress	(1.4.1) (10.6.2)
q_m	cyclic deviator stress amplitude	(12.3)
q_o	reference deviator stress for size of shear yield loci	(12.4)
Q	shear load in simple shear apparatus/shear box	(1.4.2) (8.3)
Q	torque on tube	(3.2)
Q_x, Q_y	shear loads on sliding block	(4.4.1)
r	radius of tube	(3.2)
r	ratio of pressures on normal compression and critical state lines	(7.2) (9.4.5)
R	ratio of undrained strengths at plastic and liquid limits	(9.4.3)
s	mean stress in plane strain	(1.5)
s	length of stress path in $p':q$ plane	(3.3)
S_r	degree of saturation	(1.2)
S_t	sensitivity	(9.5)
t	maximum shear stress in plane strain	(1.5)
t	wall thickness of tube	(3.2)
t	time	(12.2)
t_1	reference time	(12.2)
u	pore pressure	(1.3)
u_0	back pressure	(1.4.1)
v	specific volume	(1.2)

v_c	intercept on normal compression line in v:$\log_{10}\sigma'_v$ plane	(4.2)
v_c, v_d	reference specific volumes on unloading–reloading line	(10.3.2)
v_g	granular specific volume	(1.2)
v_{max}	maximum specific volume of a sand	(7.4.2)
v_{min}	minimum specific volume of a sand	(7.4.2)
v_o	specific volume as prepared	(6.5)
v_s	intercept on unloading–reloading line in v:$\log_{10}\sigma'_v$ plane	(4.2)
v_s	reference value of specific volume	(9.3)
v_κ	intercept on unloading–reloading line	(4.2)
v_λ	intercept on normal compression line	(4.2)
v'_λ	reference specific volume on one-dimensional normal compression line	(10.3.1)
v_1	reference value of specific volume	(11.3.3)
V	volume of sample	(1.4.1)
w	water content	(1.2)
w_L	liquid limit	(7.6) (9.2)
w_P	plastic limit	(7.6) (9.3)
W	work input per unit volume	(1.4.1)
W	weight of element in infinite slope	(7.6)
W_d	distortional work input per unit volume	(1.4.1)
W_T	total work input to shear box sample	(8.3)
W_v	volumetric work input per unit volume	(1.4.1)
x, y	shearing and normal displacement in shear box or simple shear apparatus	(6.5) (7.4) (8.3)
x, y	movement along and perpendicular to failure plane	(7.4)
x, y, z	coordinates	(1.3) (1.4.1)
x^s, y^s, z^s	sliding movements for frictional block	(4.4.1)
Y, y	sliding loads	(12.4)
α	cross-anisotropic elastic parameter	(2.3)
α	angle of fall-cone	(9.2)
$\alpha_p, \alpha_q, \alpha_r, \alpha_z$	coefficients of elastic total stress change	(11.2.1)

β	slope angle	(7.6)
β	dilatancy parameter $= \tan^{-1} \delta\varepsilon_q^p/\delta\varepsilon_p^p$	(8.2)
β	slope of failure line in $t{:}s'$ plane	(10.4.1)
γ	shear strain	(1.1)
γ	total unit weight of soil	(1.3)
γ'	buoyant unit weight of soil	(1.3)
γ_w	unit weight of water	(1.3)
$\gamma_{yz}, \gamma_{zx}, \gamma_{xy}$	shear strains	(1.4.1)
Γ	location of critical state line in compression plane	(6.1)
δ	small increment	(1.2)
δ	axial displacement	(12.4)
Δ	large increment	(4.2)
Δw	water content shift in fall-cone tests	(9.2)
Δw_{100}	water content shift for 100-fold change in strength	(9.2)
ε	normal strain	(1.1)
ε_a	axial strain	(1.4.1)
ε_h	horizontal strain	(11.3.3)
ε_p	volumetric strain	(1.2)
ε_q	triaxial shear strain	(1.4.1)
ε_r	radial strain	(1.4.1)
ε_s	volumetric strain in plane strain	(1.5)
ε_t	maximum shear strain in plane strain	(1.5)
ε_v	vertical strain	(11.3.3)
$\varepsilon_{xx}, \varepsilon_{yy}, \varepsilon_{zz}$	normal strains	(1.4.1)
$\varepsilon_1, \varepsilon_2, \varepsilon_3$	principal strains	(1.4.1)
ζ	dummy parameter to describe size of plastic potential	(4.5)
ζ	pressure applied at ground surface	(11.2.1)
η	stress ratio $= q/p'$	(3.3)
η_K	value of η for one-dimensional conditions	(9.4.5)
η_{Knc}	value of η_K for one-dimensional normal compression	(9.4.5) (10.3.1)
θ	coordinate, twist of tube	(3.2)
θ	dilatancy angle for triaxial conditions	(8.3)
θ	inclination of axis of elliptical yield loci in $s'{:}t$ plane	(11.3.3)
κ	slope of unloading–reloading line in $v{:}\ln p'$ plane	(4.2)

κ^*	unloading index	(9.3)
λ	slope of normal compression line in v:ln p' plane	(4.2)
λ_{r}	λ for remoulded clay	(9.5)
λ_{u}	λ for undisturbed clay	(9.5)
Λ	$(\lambda - \kappa)/\lambda$	(5.4)
Λ^*	$(\lambda - \kappa^*)/\lambda$	(9.3)
μ	friction coefficient	(4.4.1)
μ	exponent in expression linking strength with overconsolidation	(7.2)
μ	frictional constant	(8.3)
μ	Bjerrum's correction factor for vane strength	(9.6)
μ	shape factor for elliptical yield loci in s':t plane	(11.3.3)
ν	Poisson's ratio	(2.1)
ν'	Poisson's ratio for soil in terms of effective stresses	(2.2)
$\nu_{\mathrm{hh}}, \nu_{\mathrm{vh}}, \nu^*$	Poisson's ratios for cross-anisotropic soil	(2.3)
ρ	settlement	(11.2.2)
σ	normal stress	(1.1)
σ'	normal stress on failure plane	(7.1)
σ_{a}	axial stress	(1.4.1)
σ_{c}	compressive stress in split cylinder test	(9.4.2)
σ'_{h}	horizontal effective stress	(9.4.5)
σ'_{hc}	horizontal preconsolidation pressure	(11.2.1)
σ_{r}	cell pressure, radial stress	(1.4.1)
σ_r	radial stress	(3.2)
σ_{t}	tensile stress in split cylinder test	(9.4.2)
σ_{v}	vertical stress	(1.3)
σ'_{v}	vertical normal effective stress in shear box	(7.4.1)
σ'_{v}	vertical effective stress	(9.4.5)
σ'_{vc}	vertical preconsolidation pressure	(3.3)
σ'_{ve}	equivalent one-dimensional consolidation pressure	(7.4.1)
$\sigma_{xx}, \sigma_{yy}, \sigma_{zz}$	normal stresses	(1.4.1)
σ_z	axial stress	(3.2)
σ_θ	circumferential stress	(3.2)
σ_0	preload value of axial stress	(3.2)
$\sigma_1, \sigma_2, \sigma_3$	principal stresses	(1.4.1)
$\sigma'_{\mathrm{I}}, \sigma'_{\mathrm{II}}, \sigma'_{\mathrm{III}}$	major, intermediate, and minor principal effective stresses	(7.1)

τ	shear stress	(1.1)
τ	shear stress on failure plane	(7.1)
τ_h	shear stress on horizontal plane in shear box	(7.4.1)
$\tau_{yz}, \tau_{zx}, \tau_{xy}$	shear stresses	(1.4.1)
$\tau_{z\theta}, \tau_{\theta z}$	shear stresses on radial planes	(3.2)
ϕ'	Mohr–Coulomb friction angle	(7.1)
ϕ'_{cs}	critical state angle of shearing resistance	(7.4.1)
ϕ'_e	Hvorslev angle of shearing resistance	(7.4.1)
ϕ'_f	angle of shearing resistance in Rowe's stress–dilatancy relation	(8.5)
ϕ'_m	mobilised angle of shearing resistance	(8.3)
ϕ'_r	residual angle of shearing resistance	(7.7)
ϕ'_μ	interparticle angle of friction	(8.5)
χ	scalar multiplier	(4.4.1) (4.5)
χ	fall-cone parameter	(9.2)
ψ	angle of dilation in plane strain $= \sin^{-1}(-\delta\varepsilon_s/\delta\varepsilon_t)$	(8.3)

Superscripts

$'$	effective stress quantity	(1.3)
e	elastic component	(4.2)
p	plastic component	(4.2)

Subscripts

c	preconsolidation value	(11.2.1)
cs	critical state value	(6.1)
f	failure value	(6.3)
i	initial value	(5.4) (6.3)
L	value at liquid limit	(9.3)
P	value at plastic limit	(9.3)
u	undrained	(2.2)
y	value at yield	(11.2.1)
Ω	value at Ω-point	(9.4.4)

1

Introduction: models and soil mechanics

1.1 Use of models in engineering

Scientific understanding proceeds by way of constructing and analysing models of the segments or aspects of reality under study. The purpose of these models is not to give a mirror image of reality, not to include all its elements in their exact sizes and proportions, but rather to single out and make available for intensive investigation those elements which are decisive. We abstract from non-essentials, we blot out the unimportant to get an unobstructed view of the important, we magnify in order to improve the range and accuracy of our observation. A model is, and must be, unrealistic in the sense in which the word is most commonly used. Nevertheless, and in a sense, paradoxically, if it is a good model it provides the key to understanding reality. (Baran and Sweezy, 1968)

Engineering is concerned with understanding, analysing, and predicting the way in which real devices, structures, and pieces of equipment will behave in use. It is rarely possible to perform an analysis in which full knowledge of the object being analysed permits a complete and accurate description of the object to be incorporated in the analysis. This is particularly true for geotechnical engineering. The soil conditions under a foundation or embankment can be discovered only at discrete locations by retrieving samples of soil from boreholes or performing in situ tests; soil conditions between such discrete locations can be deduced only by informed interpolation. (This is a major difference between geotechnical engineering and structural or mechanical engineering, in which it is feasible to specify and control the properties of the steel, concrete, or other material from which a structural member or mechanical component is to be manufactured.)

Not only is it rarely possible to perform such an analysis, it is rarely desirable. Understanding of the behaviour of real objects is improved if intelligent simplifications of reality are made and analyses are performed using simplified models of the real objects. The models considered here are conceptual models. Predictions can also be based on physical models in which, for example, small prototype structures are placed on small blocks of soil. Such physical models are also simplified versions of reality because it is not usually feasible to reproduce at a small scale all the in situ variability of natural soils.

The objective of using conceptual models is to focus attention on the important features of a problem and to leave aside features which are irrelevant. The choice of model depends on the application. For example, the orbit of a spacecraft can be analysed by considering the spacecraft as a point mass concentrated at its centre of gravity. However, to calculate how to operate the engines to get the spacecraft into orbit, it is necessary to know about the distribution of mass in the spacecraft, its moments of inertia about various axes, and the way in which its shape influences its motion. To plan the distribution of seats and fixtures, the spacecraft might be considered as a box of a certain internal shape and size. These are three conceptual models of the same object.

Similarly, an architect might model a steel-framed building as a series of spaces in which to place furniture, partitions, services, and so on, whereas the structural engineer might model the steel frame in two different ways: (1) to calculate bending moments at various points in the frame, the steel beams and columns may be represented by line members for which the dimensions of the cross section are irrelevant; (2) to design the connections between the beams and columns of the frame, the cross-sectional details are all important.

Point masses and line members are convenient idealisations of real objects and structures; with such simplified representations, analyses can be readily performed and patterns of response deduced. Idealisation can

Fig. 1.1 Observed behaviour of mild steel in pure tension.

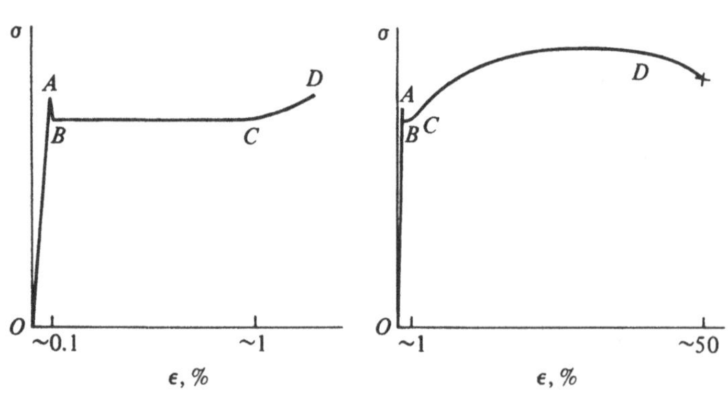

extend also to the characterisation of the material behaviour: for example, a stress:strain relationship for a mild steel specimen loaded in simple tension (Fig. 1.1). This figure shows an initial linear climb OA to a so-called upper yield point at A, a drop AB with almost no further strain to a lower yield point at B, an extension at essentially constant stress BC, followed by strain hardening CD with increase of stress to ultimate rupture. (There is a small drop in stress shortly before the specimen actually parts.) However, to perform analyses of the behaviour of steel structures (Baker and Heyman, 1969), this actual stress:strain curve is replaced by the idealised stress:strain curve (Fig. 1.2). In this figure, the distinction between upper and lower yield points has been removed so that there is a direct transition from the initial linear elastic section OM to a plastic plateau MN. Subsequent strain hardening is also ignored so that the plastic deformation MN can be assumed to continue at constant stress to indefinitely large strains. The whole body of plastic design of steel structures has been successfully based on this idealised stress:strain relationship.

Classical soil mechanics makes much implicit use of idealised stress: strain relationships. A typical shear stress (τ):shear strain (γ) curve for a soil specimen might be OXY in Fig. 1.3a. Two groups of calculations are regularly performed in geotechnical engineering: stability calculations and settlement calculations. *Settlement calculations* (Fig. 1.3b) are concerned with the stiffness of soil masses under applied loads. An obvious idealisation of the stress:strain curve is to assume that over the range of stresses applied under working loads, the stress:strain behaviour is linear and elastic, represented by OA in Fig. 1.3a. *Stability calculations* (Fig. 1.3c) are concerned with complete failure of soil masses, with large deformations occurring on rupture planes, accompanied by collapse of geotechnical structures. If the deformations are large, the precise shape of the early stages of the stress:strain curve is of little importance, and the stress:strain

Fig. 1.2 Idealised behaviour of mild steel in pure tension.

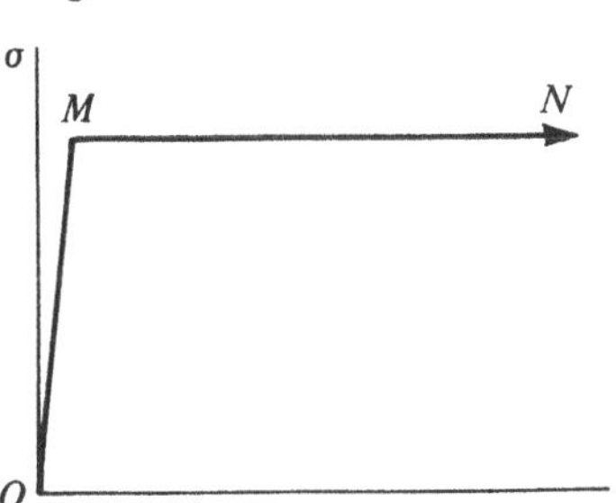

behaviour can be idealised as rigid: perfectly plastic, represented by OBC in Fig. 1.3a. These simple elastic and rigid plastic models lie behind much of classical theoretical soil mechanics and lead to a concentration of site investigation effort on seeking *the* stiffness of the soil (the slope of OA) and *the* strength of the soil (the level of BC).

Vermeer and de Borst (1984) call such elementary simple elastic and rigid plastic models of soil behaviour 'student's models'. However, this book suggests that the general picture of soil behaviour is better understood from more realistic models. Students should be interested in more than just the perfectly elastic and perfectly plastic idealisations (which could perhaps be called 'children's models'). Here, these more realistic models are called student's models; they too are idealisations and simplifications of real soil behaviour, but less radical idealisations than those in Fig. 1.3. There are two reasons for wanting to proceed to the more realistic student's models. The first is that such models bring together many of the apparently unrelated aspects of soil behaviour – strength, compression, dilatancy (volume change on shearing), and the existence of critical states (in which unlimited deformations can occur without changes of stresses or volume) – and they provide a background against which data of actual soil behaviour can be studied. Learning about soil behaviour then becomes more coherent, and models of soil behaviour can be seen as extensions of the concepts of plasticity and yielding which have become familiar from the descriptions of the mechanical behaviour of metals.

The second reason is that the simple children's models are inadequate for the description of real soil response. There is an advantage in supposing that the states of stress in soil elements in a geotechnical structure are

Fig. 1.3 (a) Observed and idealised shearing behaviour of soil for (b) settlement and (c) stability calculations.

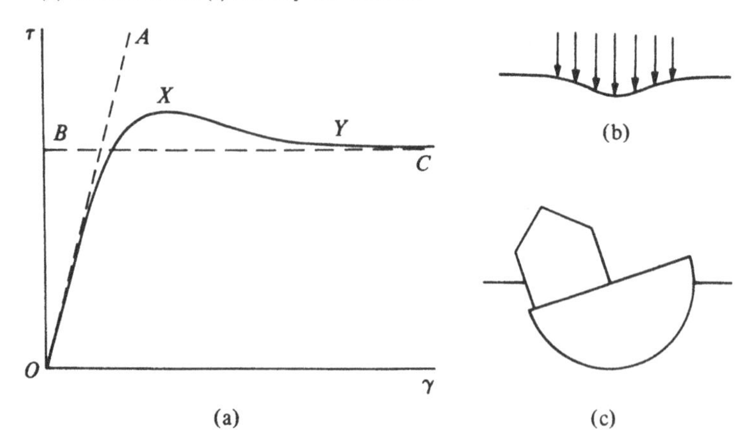

sufficiently remote from a failure state that their response can be assumed to be linear and elastic: elastic analysis of the distribution of stresses and deformations in an elastic material is comparatively straightforward, and for many problems exact results have been or can be obtained. However, the real non-linearities of soil response exercise an important influence on actual distributions of stresses and deformations; and with the increased availability of computers of various sizes, it is becoming more common to predict the responses of geotechnical structures using numerical analyses which incorporate more realistic models of soil behaviour. These analyses can be no better than the models and idealisations on which they are based, and a geotechnical engineer needs to understand the capabilities of the models to be able to assess the relevance of the analyses.

The models discussed in this book go beyond the elementary elastic and plastic models illustrated in Fig. 1.3a. Some hints at ways in which the models can be developed further are suggested in Chapter 12.

1.2 Soil: volumetric variables

The stress:strain behaviour of soils can be described by models which are essentially similar to those that might be used to describe the stress:strain behaviour of metals; in other words, at one level there is nothing particularly special about soils as compared with other materials except, of course, that a large proportion of the volume occupied by a mass of soil is made up of voids (Fig. 1.4). The voids may be filled with two (or more) pore fluids, usually water and air (Fig. 1.5) but possibly oil or gas instead (or in addition). When a soil is deformed, significant and often irreversible changes in volume can occur as the relative positions of the soil particles change. By contrast, irrecoverable deformation of metals occurs at essentially constant volume. Any successful description of soil response must obviously incorporate the possibility of large volumetric changes.

It might seem unlikely that the behaviour of a material that is clearly so heterogeneous at the particulate level could be described in terms of stresses and strains, which are more obviously useful for continuous materials. Most geotechnical structures are large by comparison with the size of individual soil particles, and stresses and strains must be thought of as quantities which are averaged over volumes of soil containing many particles.

Soil particles are usually considered to be rigid, but each one is in physical contact with some of its neighbours to form a highly redundant skeletal, cellular framework. If the volume occupied by this particle structure is to change, then the fluid in the voids must flow through the soil. Different characters of response are obtained depending on the extent to which the soil structure impedes the movement of the pore fluid. The permeability k is the quantity usually measured; it indicates the ease with which water (or other fluid) can move through the soil. The permeability is an inverse indication of the drag exerted on the viscous flowing pore fluid by the tortuous passages through the structure of the soil. The drag increases as the proportion of boundary layer (where the viscous effects of the soil particles are large) to total volume of flow increases. The ratio of surface area to the volume of particles increases as the typical dimension of the particles falls, and hence a soil composed of very small particles is likely to have a low permeability, while a soil composed of large particles is likely to have a high permeability.

The range of possible particle sizes is enormous (Fig. 1.6): clay particles have a typical dimension of less than 2 micrometres (μm) (compare Fig. 1.4)

Fig. 1.4 Scanning electron micrograph of Leda clay (picture width, 13 μm) (micrograph supplied by A. Balodis).

whereas coarse gravel has a typical dimension up to 60 millimetres (mm). (This range of sizes is roughly equivalent to the length of a cricket pitch compared with the distance from Cambridge to Glasgow, or the side of a baseball diamond compared with the distance from Washington to Detroit.) Any soil may contain a great range of particle sizes, depending on the way in which the soil was formed: some typical grading curves are shown in Fig. 1.6.

The range of particle shapes is also great: clay particles are often flat and platelike (Fig. 1.4) whereas sand and gravel particles are more likely to be sub-spherical (Fig. 1.7). Here immediately is an example of the use of a simple model to describe soils. The plotting of typical particle dimensions on the grading chart of Fig. 1.6 models real particle shapes and sizes with equivalent spheres. For coarser soils, the size of the equivalent sphere is determined from the size of the square sieve mesh through which the particles can just pass (Fig. 1.8a). For finer soils, the

Fig. 1.5 Assembly of soil particles with voids filled with mixture of air and water.

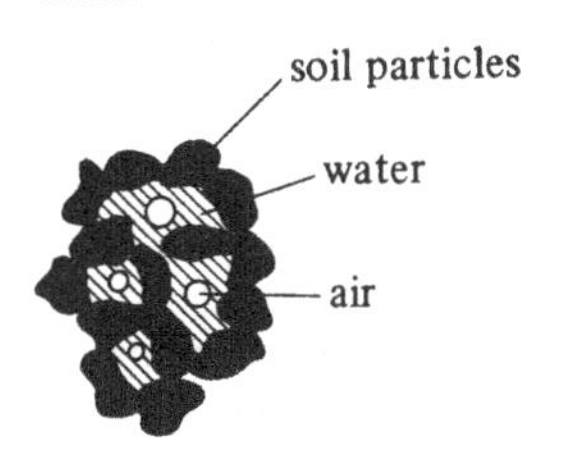

Fig. 1.6 Grading curves for (1) kaolin, (2) London clay, (3) Vienna clay, (4) estuarine silt, (5) medium sand, (6) coarse sand, (7) glacial till (after Atkinson and Bransby, 1978), and (8) residual granitic soil (data from Yudhbir, 1982).

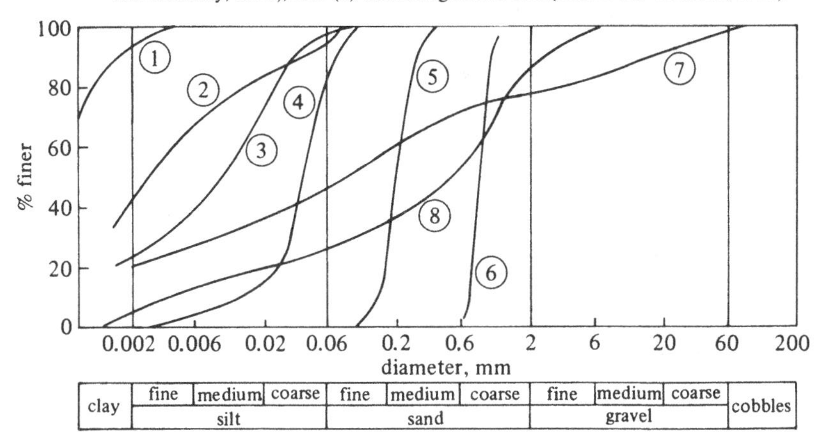

Fig. 1.7 Particles of coarse Leighton Buzzard sand (picture width, 37 mm).

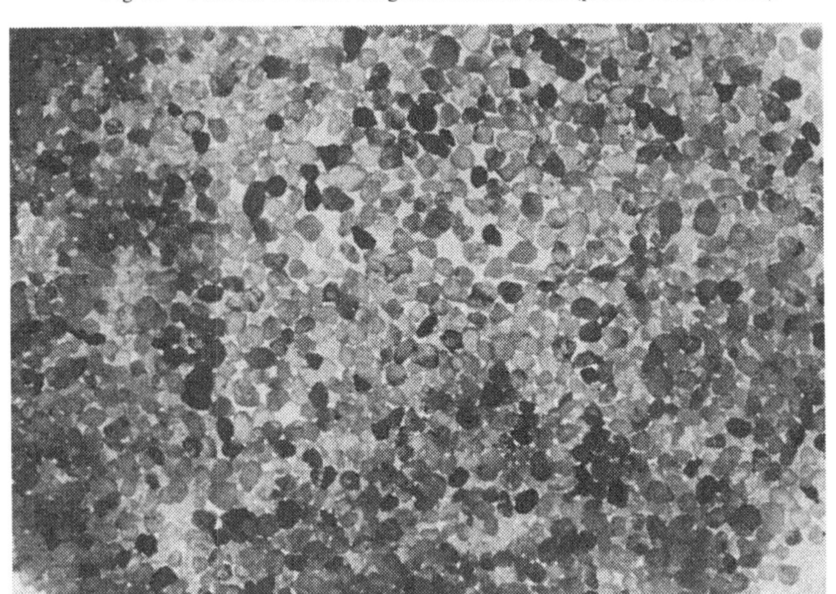

Fig. 1.8 (a) Particles of various shapes passing through uniform sieve apertures; (b) soil particles treated as equivalent falling spheres.

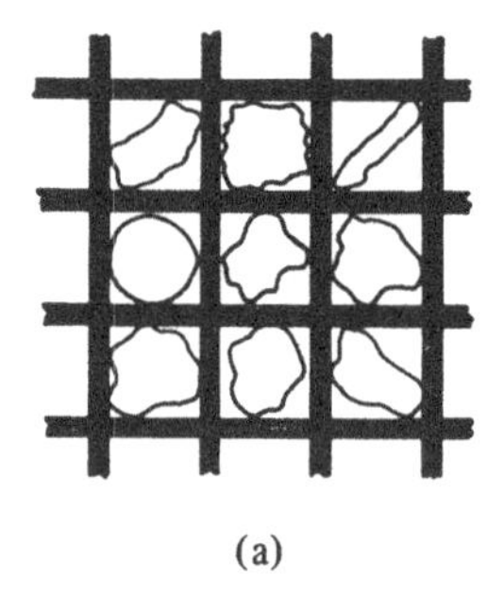

(a)

(b)

determination of particle size distribution relies on Stokes' law describing the terminal velocity of spherical bodies falling through a viscous fluid (Fig. 1.8b). As indicated by Wroth and Houlsby (1985), this 'is a convenient way of providing an index of particle sizes, but clearly would be useless for understanding the details of clay particle interactions and van der Waals forces'.

For a given soil type, for which the shapes and sizes of particles are fixed, the permeability will depend primarily on the proportions of the volume of the soil that are occupied by particles and by voids. This volumetric packing of the soil is one part of the description of the current structure of the soil. It can be expected that the past history of the soil deposit will be reflected in its present structure and that the present structure will control the future response, whether this is judged in terms of permeability or in terms of the mechanical response to changes in stresses. This book is largely concerned with models of soil behaviour for describing this latter mechanical behaviour.

The description of the present structure of a soil is not straightforward. The soil may appear homogeneous at the macroscopic scale but, as has been seen, may be extremely inhomogeneous at the microscopic scale (e.g. Fig. 1.4). Because of the redundancy of the particulate structure formed by the soil particles, the arrangement of the particles can be expected to depend not only on the current stresses but also on the history of stresses that has brought the soil to its current stress state. A complete description of the structure requires knowledge of the packing, arrangement, and shape of all the particles of the soil. Such a description could be attempted. Perhaps the average particle shape could be described as approximately ellipsoidal and the dimensions and orientations of the ellipsoids, together with the distribution of the directions of the contacts between the ellipsoids could be ascertained. However, even if the particles were all essentially the same size and shape, the assembly of such a structural description would be complex (though possible; e.g. Oda, Konishi, and Nemat-Nasser, 1980), and the manipulation of such a structural description to take account of the changes that occur as the stresses imposed on the soil change would be cumbersome.

Simple models are easier to describe and comprehend, and extra complexities can always be incorporated subsequently if it is then felt that they cannot be dispensed with. Here, then, we prefer to use the simplest possible partial description of the structure of a soil and to allow macroscopic and phenomenological observation of the effects of changing microscopic structure. The simplest partial description of the structure of the soil can be made by using a volumetric variable to describe the proportions of space in the soil occupied by solid material and by pores. Such a variable, which might be thought of as a first (scalar) invariant of a particle structure tensor, is relatively easy to determine (particularly by comparison with all the variables that might be needed to describe the three-dimensional orientations of the particles), but it may take various forms.

For an intact soil with the voids saturated with water, the simplest quantity to measure is the water content w, which is the ratio of the masses of water and of dry soil particles in any volume of soil. With a knowledge of the specific gravity of the material of the soil particles G_s (typically in the range 2.6–2.75), the void ratio e (the ratio of the volumes of void and soil particles, Fig. 1.9a) can be determined:

$$e = G_s w \tag{1.1}$$

This link between water content and void ratio applies only for saturated soil, but the void ratio may be regarded as the more fundamental variable (even though it may be less easy to measure), representing as it does the relative disposition of particles and voids for any soil whether saturated or not. For unsaturated soil, the link between water content and void ratio requires additional knowledge of the degree of saturation S_r, the proportion of the volume of the voids occupied by water (Fig. 1.9b); then (1.1) is replaced by

$$e = \frac{G_s w}{S_r} \tag{1.2}$$

In general, however, we prefer the specific volume v as a volumetric variable for reasons of mathematical convenience which will become clear. This is the volume composed of unit volume of soil particles with their surrounding voids (Figs. 1.9a, b):

$$v = 1 + e \tag{1.3}$$

so for saturated soil,

$$v = 1 + G_s w \tag{1.4}$$

(Another volumetric quantity is porosity n, which is the ratio of volume of void to total soil volume:

$$n = \frac{e}{1+e} = \frac{v-1}{v} \tag{1.5}$$

Porosity is mentioned here for completeness, but for the purposes of model building it offers no advantage over specific volume v.)

A central theme of critical state soil mechanics is that it is not sufficient to consider the mechanical behaviour of soils in terms of stresses alone;

some structural variable such as specific volume is also required. It may be remarked that the expectation that the behaviour of a soil can be linked with a single volumetric variable is likely to be valuable for soils that have a reasonably narrow range of particle sizes; for example, reasonably pure clays, silts, sands, or gravels. Grading curves for typical examples of these soils were shown in Fig. 1.6. Many natural soils – particularly glacial tills, boulder clays, and residual soils – may have a very wide grading, with substantial contents both of clay-size particles and of larger, sand- or gravel-size particles (Fig. 1.6, curves 7 and 8). It appears that the clay particles in such soils may form packets surrounding or bridging between

Fig. 1.9 Divisions of volume of soil with (a) one pore fluid, (b) two pore fluids, and (c) rotund and platy particles.

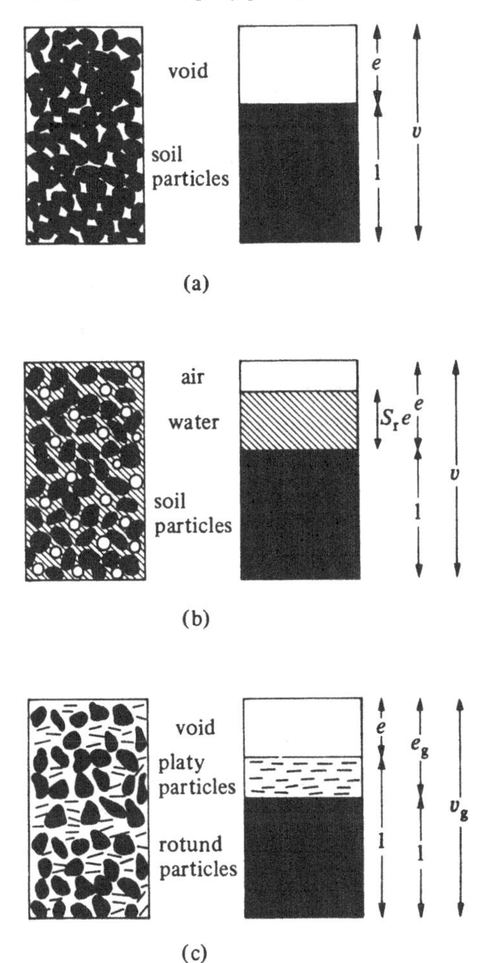

larger particles (e.g. see Collins and McGown, 1974). The mechanical response of such soils can be expected to reflect a mixture of the responses of the coarser granular structure and of the finer clay bridges. In such a case, two volumetric variables might be relevant: one describing the packing of the clay particles within their packets at the microscopic level and one describing the density of packing of the coarser particles at the macroscopic level. A variable of the second type is the granular specific volume v_g (Fig. 1.9c), which is the volume composed of unit volume of rotund particles together with surrounding platy particles and voids [related to the granular void ratio e_g introduced by Lupini, Skinner, and Vaughan (1981)]:

$$v_g = 1 + \frac{\text{volume of voids} + \text{volume of platy particles}}{\text{volume of rotund particles}} \tag{1.6}$$

$$v_g = 1 + e_g \tag{1.7}$$

A volumetric variable has been chosen. However, in analysing soil response it is often useful to work in terms of volumetric strains. An increment of volumetric strain $\delta\varepsilon_p$ can be defined as

$$\delta\varepsilon_p = \frac{-\delta v}{v} \tag{1.8}$$

where the negative sign is required because, by convention in soil mechanics, compressive strains (and stresses) are taken as positive. This is a logarithmic strain increment defined as the ratio of the change of the quantity to its current value. (The reason for using the symbol $\delta\varepsilon_p$ for the volumetric strain increment will be explained subsequently.) Economy and elegance are achieved in this expression by using specific volume v as the volumetric variable, compared with the corresponding expressions using void ratio e or porosity n, for which

$$\delta\varepsilon_p = \frac{-\delta e}{1+e} = \frac{-\delta n}{1-n} \tag{1.9}$$

1.3 Effective stresses: pore pressures

It has been mentioned that the pores of a soil may be filled with one or more pore fluids. Each of these fluids may be under pressure, and it is probable that if more than one fluid is present, the pressures in the different fluids will be different. Here attention will be concentrated on saturated soils containing only one pore fluid, which is usually water or air. There will then be just one pore pressure in the pore water or air that fills the pores of the soil, which may, in general, be assumed to be continuous.

Consider first a soil saturated with water. In a swimming pool filled with water (Fig. 1.10a), the absolute pressure varies with depth in the pool. Referred to any horizontal datum, such as the surface of the pool, the relative water pressure is everywhere zero. In the absence of thermal gradients, the water is motionless.

If the pool is filled with dry soil and the voids of the soil are allowed to fill with water, with the surface of the soil at the previous surface of the water (Fig. 1.10b), some water of course has been displaced, but the statements in the last paragraph remain true: the absolute or total pore water pressure varies with depth, but referred to the surface of the pool, the relative or excess pore water pressure is everywhere zero, and the water in the pores of the soil is motionless.

If the equilibrium of the pore water is somehow disturbed, then, in some elements, pore pressures will be observed which are different from the pore pressures arising from the positions of the elements relative to the datum. Such non-equilibrium pore pressures are called *excess* pore pressures. The pore water will flow to reestablish equilibrium. Pore water flows as a result of *gradients* of excess pore water pressure. At two adjacent points A and B, at depths z_A and z_B below a datum level (Fig. 1.11), the total pore pressures can be made up of two parts:

$$u_A = \gamma_w(z_A + h_A) \tag{1.10a}$$

and

$$u_B = \gamma_w(z_B + h_B) \tag{1.10b}$$

where γ_w is the unit weight of water, and h_A and h_B represent the excess heads of water at A and B referred to the datum $z = 0$. Then there is a component of flow of water from A towards B if $h_A > h_B$, irrespective of the relative depths z_A and z_B.

Although flow of water in soils is controlled by excess pore pressures,

Fig. 1.10 (a) Standpipes at different depths in a pool of water; (b) standpipes at different depths in a pool filled with water-saturated soil.

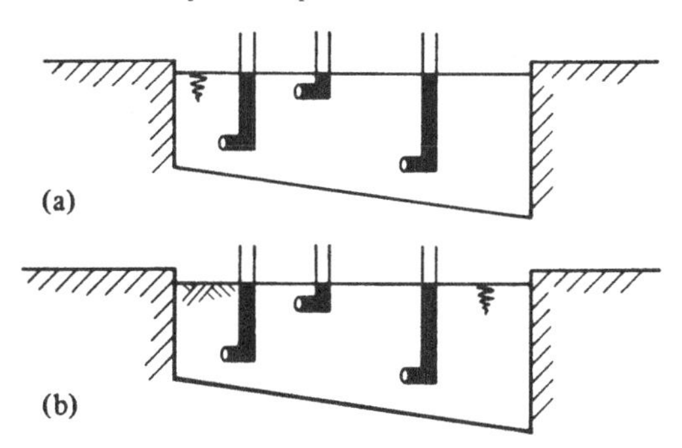

the presence of any pore pressure in a soil will have an influence on its mechanical response to changes in stresses. The effects of pore pressures can be approached in several ways: only one, an empirical or phenomenological approach, will be considered in this chapter, two other approaches, in terms of particle contact stresses and in terms of components of work input to a deforming soil element, have been described by Bishop (1959) and by Houlsby (1979), respectively.

The response of a soil element to certain changes in stress depends on a number of factors, such as

total stresses acting on the element (calculated as total force divided by total area),

total pore water pressure acting in the pores of the element,

specific volume of the element,

history of loading of the element,

temperature of the element, and

magnetic and electric fields at the element.

However, it is observed experimentally that in most situations of engineering interest, so far as the first two quantities are concerned, only the differences between total stresses and total pore water pressures are of importance. If the pore pressure is changed by an amount x and the normal stresses acting on the element are also changed by the same amount x, then there is no noticeable change in the element, and the response of the element to subsequent changes in stress is unchanged.

The total stress, the ratio of total force to total area, is typically denoted by σ. The difference between total stress σ and total pore water pressure u is called the *effective stress* and is denoted by σ' (the prime signifies effective stress), where

$$\sigma' = \sigma - u \tag{1.11}$$

The existence of pore water pressure does not affect the shear stresses

Fig. 1.11 Standpipes at neighbouring points in soil through which seepage is occurring.

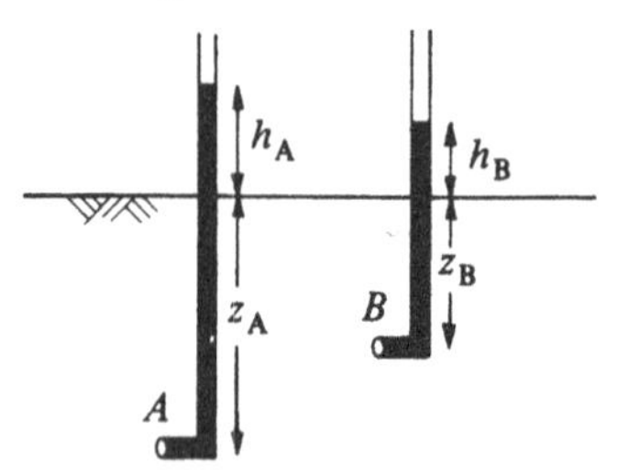

experienced by a soil element. The shearing resistance of water is assumed to be negligible.

An example of the distinction between total and effective stresses is illustrated in Fig. 1.12. Element A (Fig. 1.12a) is at depth z below the ground surface, and the static, equilibrium water table is also at the ground surface. If the total unit weight of the soil is γ, then the total vertical stress σ_{vA} at element A is

$$\sigma_{vA} = \gamma z \tag{1.12}$$

and the total pore pressure u_A is

$$u_A = \gamma_w z \tag{1.13}$$

so that the effective vertical stress σ'_{vA} at element A is

$$\sigma'_{vA} = \sigma_{vA} - u_A = (\gamma - \gamma_w)z = \gamma' z \tag{1.14}$$

where $\gamma' = \gamma - \gamma_w$ is the submerged or buoyant unit weight of the soil.

Element B (Fig. 1.12b) is also at depth z below the soil surface, but this soil is at the bottom of a lake or ocean of depth d. The total vertical stress σ_{vB} at element B is

$$\sigma_{VB} = \gamma z + \gamma_w d \tag{1.15}$$

and the total pore pressure u_B is

$$u_B = \gamma_w(z + d) \tag{1.16}$$

Hence, the effective vertical stress σ'_{vB} at element B is

$$\sigma'_{vB} = \sigma_{vB} - u_B = (\gamma - \gamma_w)z = \gamma' z = \sigma'_{vA} \tag{1.17}$$

and although the total stresses at A and B are quite different, the effective stresses are identical; thus, in the absence of other differences, the responses of elements A and B to subsequent changes in effective stress are identical.

Fig. 1.12 (a) Element A at depth z below ground level; (b) element B at depth z below floor of lake of depth d.

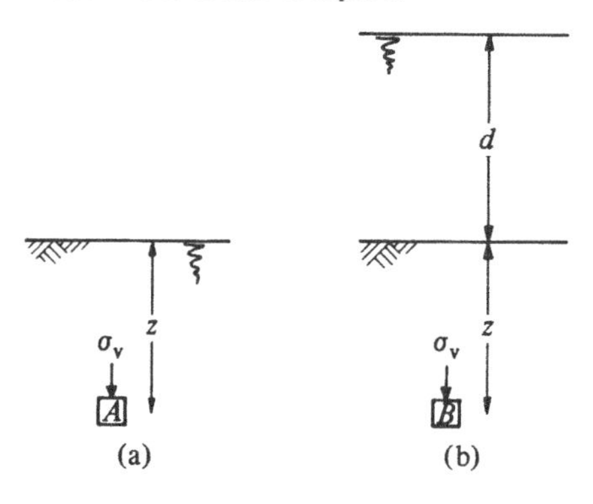

All descriptions of soil behaviour in this book start from the hypothesis, well supported by experiment, that the effective stresses determine the response of a soil. We concentrate on saturated soils, those in which the pores are filled with one pore fluid, usually air or water. In many parts of the world, geotechnical engineers are concerned with soils which are not saturated: the voids contain mixtures of air and water. It is not clear whether simple ideas of effective stresses are useful for unsaturated soils, in which different pressures in the different pore fluids can coexist. Coleman (1962) suggests that changes of stress and pore water pressure relative to pore air pressure are important; and Fredlund (1979), introducing a 'contractile skin' as an independent phase (representing the air–water interface) in addition to the soil particles and the two pore fluids, appears to arrive at an equivalent result. It seems clear that, for unsaturated soils, the idea of a single notional definition of effective stress as simple as (1.11) (a basic combination of stress and pore pressure which controls the mechanical behaviour of soils) no longer exists.

1.4 Soil testing: stress and strain variables

1.4.1 Triaxial apparatus

If the stress:strain properties of a metal are required for engineering purposes, then the metal is typically tested in simple tension by pulling a specimen between end grips (Fig. 1.13a) with no lateral stress being applied. When the stress:strain properties of a concrete are required, then the concrete is typically cast into the form of a cubical or cylindrical specimen and tested in simple compression between parallel loading

Fig. 1.13 (a) Metal wire specimen being tested in simple (unconfined) tension; (b) concrete cube specimen being tested in simple (unconfined) compression.

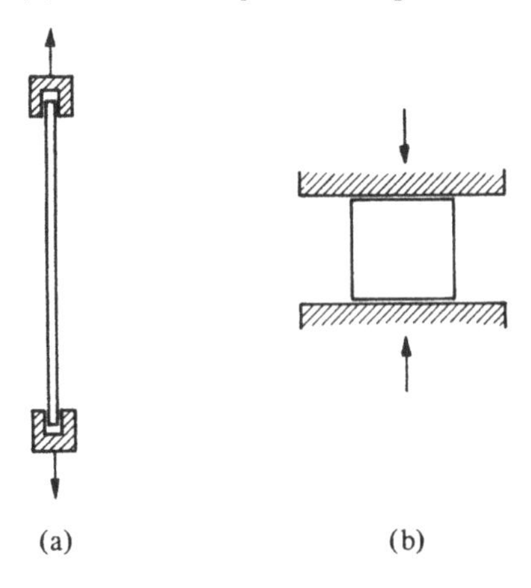

platens (Fig. 1.13b), again with no lateral stress being applied. These are uniaxial extension and compression tests.

For many soils, such unconfined tests are not possible; without confinement, a sample of an uncemented sandy soil just falls apart. Samples of clayey soils can stand without lateral support (the potter's task would be impossible if they did not) because negative pore pressures provide an effective confinement. The emphasis in Section 1.2 on the influence of the present structure of the soil on its subsequent response suggests that soil testing should start with the soil sample in a condition approximating as closely as possible to its condition in situ in the ground; and this also requires that the testing apparatus should be able to impose some lateral stress on the soil sample.

The most widely used testing apparatus for investigating the stress:strain behaviour of soils is, in origin, a 'confined uniaxial' testing device known (misleadingly) as the *triaxial apparatus.* Its capabilities are wider than those of confined uniaxial testing, as will be seen, and it can be described as an axially symmetric testing device having two degrees of freedom. Many variants of triaxial apparatus are commercially available, but the basic principles are illustrated in Fig. 1.14. The capabilities of the triaxial test have been fully described by Bishop and Henkel (1957) and in the volume edited by Donaghe, Chaney, and Silver (1988).

A cylindrical soil sample is contained in a rubber membrane to isolate it from direct contact with the surrounding fluid (typically water) with which the testing cell is filled. The cell fluid can be pressurised. The sample

Fig. 1.14 Schematic diagram of triaxial apparatus.

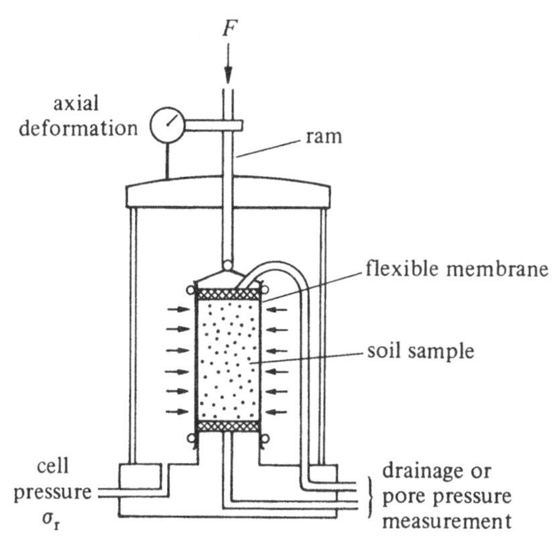

sits in the cell between a rigid base and a rigid top cap which is loaded by means of a ram passing out of the cell. Loads may be applied to this ram either by placing dead loads on a hanger or by reaction against a loading frame. Part or all of the rigid base and/or the rigid top cap is porous so that pore fluid can drain from the soil. Alternatively, if drainage is prevented, then the pressure in the pore fluid can be measured.

The quantities that are usually measured during any triaxial test are the pressure in the cell fluid, the cell pressure σ_r, which provides an all-round pressure on the soil sample; the axial force F applied to the loading ram; the change in length δl of the sample; and, for a test in which drainage is permitted, the change in volume of the sample δV (actually measured as the volume of pore water moving into or out of the sample, with an implicit assumption of incompressibility of soil particles); or, for a test in which drainage is not permitted, the pore pressure u in the pore fluid. The force transmitted through the loading ram, of area a, to the rigid top cap of the sample, of current cross-sectional area A, provides an axial stress of

$$\sigma_a = \sigma_r\left(1 - \frac{a}{A}\right) + \frac{F}{A} \tag{1.18}$$

The area correction term $(1 - a/A)$ can be ignored in conventional triaxial tests in which the cell pressure is held constant; the change in axial stress is then

$$\delta\sigma_a = \frac{\delta F}{A} \tag{1.19}$$

The effect of the area correction may be small but can readily be allowed for, particularly when test results are being received through an intelligent data acquisition unit, or microcomputer.

The stress quantity

$$q = \sigma_a - \sigma_r \simeq \frac{F}{A} \tag{1.20}$$

is known as the *deviator stress*. The *axial stress* can be written

$$\sigma_a = \sigma_r + q \simeq \sigma_r + \frac{F}{A} \tag{1.21}$$

Stress variables can be chosen in various ways in order to express the two degrees of freedom of the triaxial apparatus: different pairs of stress variables have been adopted at different times and in different parts of the world. An obvious choice might be the cell pressure σ_r and the deviator stress q because these stress variables are directly measured and controlled externally to the triaxial cell. Alternatively, axial stress σ_a and radial stress σ_r might be chosen as the principal stresses which the soil sample experiences. Here, however, a different line of argument is followed (see also Wood, 1984b).*

First, note that with a pressure u in the pore fluid, if total axial and radial stresses σ_a and σ_r are being experienced externally by the sample through its end platens and enveloping membrane, then there are also effective axial and radial stresses

$$\sigma'_a = \sigma_a - u \tag{1.22a}$$

and

$$\sigma'_r = \sigma_r - u \tag{1.22b}$$

According to experiment and hypothesis, these effective stresses play a more fundamental role than total stresses in controlling soil behaviour. Clearly, since the deviator stress q (1.20) is defined as a difference of two stresses, its value is not affected whether it is calculated as a difference of total stresses or a difference of effective stresses:

$$q = \sigma_a - \sigma_r = \sigma'_a - \sigma'_r \tag{1.23}$$

Second, it was noted in Section 1.2 that volumetric deformations play an important role in influencing the mechanical behaviour of soils because of the changing relative positions of soil particles that volumetric deformations imply. In discussions of numerical models of soil behaviour, it is important to be able to write expressions for the work input to a soil sample subjected to certain stresses when it undergoes certain deformations. It is helpful to be able to separate deformations of soils and this input of work into the components associated with change of volume and with change of shape. Here discussion is limited to the conditions obtaining in the triaxial test.

When a cubical soil element supporting effective normal stresses $\sigma'_{xx}, \sigma'_{yy}, \sigma'_{zz}$ and shear stresses $\tau_{yz}, \tau_{zx}, \tau_{xy}$ (referred to rectangular axes x, y, z) experiences corresponding normal strain increments $\delta\varepsilon_{xx}, \delta\varepsilon_{yy}, \delta\varepsilon_{zz}$

*It is usually implicitly assumed that radial and circumferential stresses in a cylindrical triaxial sample are principal stresses and are equal. Consideration of the equation of radial equilibrium shows that this is true only if there are no variations of the radial stress in the radial direction. In some situations, the provision of radial drainage to speed flow of water into or out of the sample may inadvertently provoke variations of radial stress and hence inequalities of radial and circumferential stresses; see, for example, Gibson, Knight, and Taylor (1963) and Atkinson, Evans, and Ho (1985). Also, it is usually implicitly assumed that the radial and circumferential strains are principal strains and are equal. This places certain constraints on the properties of a sample and requires it to be either isotropic or cross-anisotropic with an axis of symmetry aligned with the axis of the cylindrical sample (e.g. see Saada and Bianchini, 1975).

and shear strain increments $\delta\gamma_{yz}, \delta\gamma_{zx}, \delta\gamma_{xy}$, then the work input per unit volume of the element is

$$\delta W = \sigma'_{xx}\,\delta\varepsilon_{xx} + \sigma'_{yy}\,\delta\varepsilon_{yy} + \sigma'_{zz}\,\delta\varepsilon_{zz} + \tau_{yz}\,\delta\gamma_{yz} + \tau_{zx}\,\delta\gamma_{zx} + \tau_{xy}\,\delta\gamma_{xy} \tag{1.24}$$

When the stresses on the soil element are principal effective stresses $\sigma'_1, \sigma'_2, \sigma'_3$, and the principal strain increments $\delta\varepsilon_1, \delta\varepsilon_2, \delta\varepsilon_3$ have the same principal axes, then this expression becomes

$$\delta W = \sigma'_1\,\delta\varepsilon_1 + \sigma'_2\,\delta\varepsilon_2 + \sigma'_3\,\delta\varepsilon_3 \tag{1.25}$$

For a cylindrical triaxial sample supporting effective axial and radial stresses $\sigma'_a = \sigma'_1$ and $\sigma'_r = \sigma'_2 = \sigma'_3$ (by assumption, as discussed), undergoing increments of axial and radial strain $\delta\varepsilon_a = \delta\varepsilon_1$ and $\delta\varepsilon_r = \delta\varepsilon_2 = \delta\varepsilon_3$, the work input to the sample is

$$\delta W = \sigma'_a\,\delta\varepsilon_a + 2\sigma'_r\,\delta\varepsilon_r \tag{1.26}$$

The increment of volumetric strain $\delta\varepsilon_p$ is

$$\delta\varepsilon_p = \delta\varepsilon_a + 2\,\delta\varepsilon_r \tag{1.27}$$

The increment of volumetric work δW_v can be written as a simple product of an effective stress quantity with this volumetric strain increment

$$\delta W_v = p'\,\delta\varepsilon_p \tag{1.28}$$

only if the effective stress quantity p' is chosen to be the mean effective stress

$$p' = \frac{\sigma'_a + 2\sigma'_r}{3} \tag{1.29}$$

This mean effective stress can be thought of as indicating the extent to which all the principal stresses are the same, just as volumetric strain is about uniform change in all dimensions of an element without change of shape. If the axial and radial stresses are equal, $\sigma'_a = \sigma'_r$, then $p' = \sigma'_r$.

Change of shape of an element results from different changes in the different dimensions of the element, and strain increment and stress quantities are required to indicate the extent to which all the principal strain increments or principal stresses are not the same. Evidently, the deviator stress q is one such stress quantity because if the axial and radial stresses are not different ($\sigma'_a = \sigma'_r$), then the deviator stress is zero. It is also a convenient variable because, as seen in (1.20), it is directly measured in triaxial tests.

The increment of distortional work δW_d associated with change of shape of the triaxial element can be written as a simple product of this deviator stress with a strain increment quantity

$$\delta W_d = q\,\delta\varepsilon_q \tag{1.30}$$

only if the strain increment quantity $\delta\varepsilon_q$ is chosen to be

$$\delta\varepsilon_q = \frac{2(\delta\varepsilon_a - \delta\varepsilon_r)}{3} \tag{1.31}$$

which can be called the *triaxial shear strain increment.* It is straightforward to demonstrate from (1.26), (1.27), (1.29), (1.31), and (1.20) that

$$\delta W = \sigma'_a \,\delta\varepsilon_a + 2\sigma'_r \,\delta\varepsilon_r$$

$$\delta W = \frac{(\sigma'_a + 2\sigma'_r)(\delta\varepsilon_a + 2\,\delta\varepsilon_r)}{3} + \frac{(\sigma'_a - \sigma'_r)2(\delta\varepsilon_a - \delta\varepsilon_r)}{3}$$

$$\delta W = p'\,\delta\varepsilon_p + q\,\delta\varepsilon_q \tag{1.32}$$

$$\delta W = \delta W_v + \delta W_d \tag{1.33}$$

The subscripts p and q on the symbols for the volumetric strain and triaxial shear strain increment quantities have been chosen as in Calladine (1963) so that they suggest the correct association of stress and incremental strain variables in work-conjugate pairs.

The definitions (1.27) and (1.20) are particular forms of general expressions for volumetric and distortional stresses

$$p' = \frac{\sigma'_{xx} + \sigma'_{yy} + \sigma'_{zz}}{3} \tag{1.34}$$

and

$$q = \left[\frac{(\sigma'_{yy} - \sigma'_{zz})^2 + (\sigma'_{zz} - \sigma'_{xx})^2 + (\sigma'_{xx} - \sigma'_{yy})^2}{2} + 3(\tau^2_{yz} + \tau^2_{zx} + \tau^2_{xy})\right]^{1/2} \tag{1.35}$$

given the conditions that the principal axes coincide with the reference axes x, y, z and that two of the principal stresses are equal to the radial effective stress σ'_r. Similarly, the definitions (1.27) and (1.31) are particular forms of expressions for general volumetric and distortional strain increments

$$\delta\varepsilon_p = \delta\varepsilon_{xx} + \delta\varepsilon_{yy} + \delta\varepsilon_{zz} \tag{1.36}$$

and

$$\delta\varepsilon_q = \tfrac{1}{3}\{2[(\delta\varepsilon_{yy} - \delta\varepsilon_{zz})^2 + (\delta\varepsilon_{zz} - \delta\varepsilon_{xx})^2 + (\delta\varepsilon_{xx} - \delta\varepsilon_{yy})^2] + 3(\delta\gamma^2_{yz} + \delta\gamma^2_{zx} + \delta\gamma^2_{xy})\}^{1/2} \tag{1.37}$$

These quantities are related to the first and second invariants of the stress tensor and the strain increment tensor (e.g. see Spencer, 1980).

The factor $\frac{2}{3}$ in the expression (1.31) for the triaxial shear strain increment makes the definition appear awkward, but it is clearly necessary in order to give the correct product for distortional work. For the particular case of deformation at constant volume, from (1.27), $\delta\varepsilon_p = 0$ implies

$\delta\varepsilon_a = -2\,\delta\varepsilon_r$, and then, from (1.31), the expression for $\delta\varepsilon_q$ degenerates to

$$\delta\varepsilon_q = \delta\varepsilon_a \tag{1.38}$$

The strain increments $\delta\varepsilon_a, \delta\varepsilon_r, \delta\varepsilon_p$, and $\delta\varepsilon_q$ can be related directly or indirectly to the quantities usually measured during a triaxial test: change of length of sample δl and change in volume of sample δV:

$$\delta\varepsilon_a = \frac{-\delta l}{l} \tag{1.39}$$

$$\delta\varepsilon_r = \frac{-\delta V/V + \delta l/l}{2} \tag{1.40}$$

$$\delta\varepsilon_p = \frac{-\delta V}{V} \tag{1.41}$$

$$\delta\varepsilon_q = \frac{-\delta l}{l} + \frac{\delta V/V}{3} \tag{1.42}$$

With p' defined as the mean effective stress (1.29), a mean total stress p can be defined as

$$p = \frac{\sigma_a + 2\sigma_r}{3} \tag{1.43}$$

$$p = p' + u \tag{1.44}$$

This total mean stress can be calculated from the measured cell pressure σ_r and deviator stress $q \simeq F/A$ (1.20):

$$p = \sigma_r + \frac{q}{3} \tag{1.45}$$

Although the triaxial apparatus does indeed have two degrees of freedom, the triaxial test that is most commonly performed is one in which the cell pressure is held constant, $\delta\sigma_r = 0$, and the axial load is increased. This test is known as a *conventional triaxial compression test*. From the differential form of (1.45), if $\delta\sigma_r = 0$, then

$$\delta p = \frac{\delta q}{3} \tag{1.46}$$

and the total stress path followed in this conventional triaxial compression test rises at gradient 3 from an initial stress condition (*AB* in Fig. 1.15). In the typical initial stress condition *A* in Fig. 1.15, the soil sample has been compressed isotropically (*0A*) by simply increasing the cell pressure ($\Delta\sigma_r > 0$) without applying any deviator stress ($q = 0$) so that the initial total mean stress is equal to the cell pressure at the end of this initial isotropic compression.

If some provision is made for pulling (instead of pushing) the top cap, which requires a positive connection between the loading ram and the top cap, then it is possible for both the ram force F and the deviator stress q [as defined by (1.20)] to be negative. Triaxial tests in which $q < 0$ are called triaxial extension tests, and a *conventional triaxial extension test* is one in which the deviator stress is decreased while the cell pressure is held constant. The relationship (1.46) between changes in p and changes in q still holds, and the path of a conventional triaxial extension test falls at gradient 3 from the initial stress condition (AC in Fig. 1.15).

Nothing, apart from convenience in performing the tests, necessitates a restriction of triaxial testing to conventional compression and extension tests; and particularly with the advent of more subtly computer controlled triaxial testing apparatus, no one stress path is in principle any more difficult to apply than any other. The term *conventional* may soon be regarded as having purely historical significance. Among other specific stress paths that might be considered purely for the purposes of familiarisation with plotting in this particular p:q stress plane are paths for which the total axial stress is held constant and paths for which the total mean stress is held constant.

To keep the total axial stress constant while changing the cell pressure requires, from (1.18), that the ram force (or deviator stress) and cell pressure must be changed simultaneously. For $\delta\sigma_a = 0, \delta\sigma_r = -\delta q$. Then, from the differential form of (1.43),

$$\delta p = \frac{2\,\delta\sigma_r}{3} = \frac{-2\,\delta q}{3} \tag{1.47}$$

and the total-stress path followed in such a compression test (or extension test if $\delta q < 0$) has a slope $-\frac{3}{2}$ (AD and AE in Fig. 1.15).

Fig. 1.15 Total stress paths for triaxial compression and extension tests.

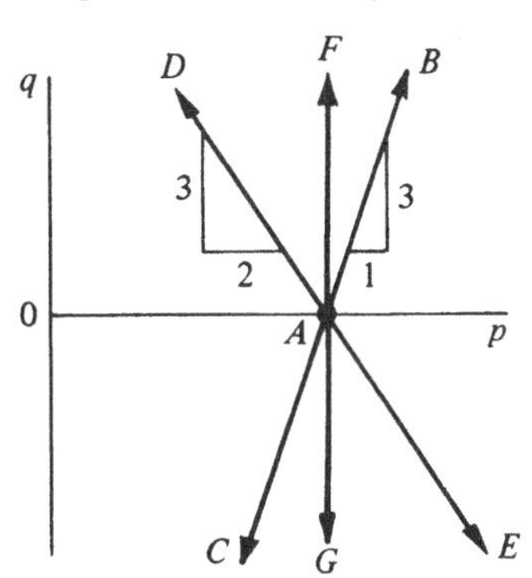

To keep the total mean stress constant requires that any change in deviator stress is accompanied by a change in cell pressure such that, from (1.43),

$$\delta\sigma_a = -2\,\delta\sigma_r \tag{1.48}$$

or, from (1.45),

$$\delta\sigma_r = \frac{-\delta q}{3} \tag{1.49}$$

Of course, a compression (or extension) test with constant total mean stress will rise (or fall) vertically in the p:q stress plane (AF and AG in Fig. 1.15). Such tests may be especially important if separation of volumetric and distortional aspects of soil response is desired.

So far, the discussion has been restricted to total stress paths. If drainage can occur freely from the soil sample to atmospheric pressure, then the pore pressure will be zero, and total and effective stresses will be identical. In such drained tests, the effective stress paths will be the same as the total stress paths shown, for example, in Fig. 1.15. It was mentioned in Section 1.2 that the pores of a soil might not be saturated with a single pore fluid: the presence of a small proportion of air in pore water is quite typical. The measurement of volume changes of triaxial samples requires the measurement of the volume of pore fluid flowing into or out of the sample. If part of this pore fluid is emerging as gas (air), then part of the volume change of the sample will not be observed as a change in the level of liquid in a burette. It is often desirable to ensure that the pore fluid is indeed saturated by subjecting the whole pore fluid system to a pressure, called a *back pressure*, which is held constant at a value u_0 during the test in order to ensure that the gaseous phase remains in solution. Drainage can occur freely, but against this back pressure. In such tests, the total

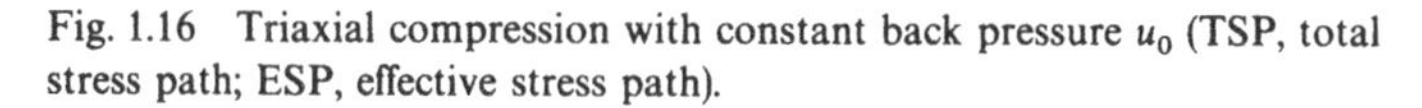
Fig. 1.16 Triaxial compression with constant back pressure u_0 (TSP, total stress path; ESP, effective stress path).

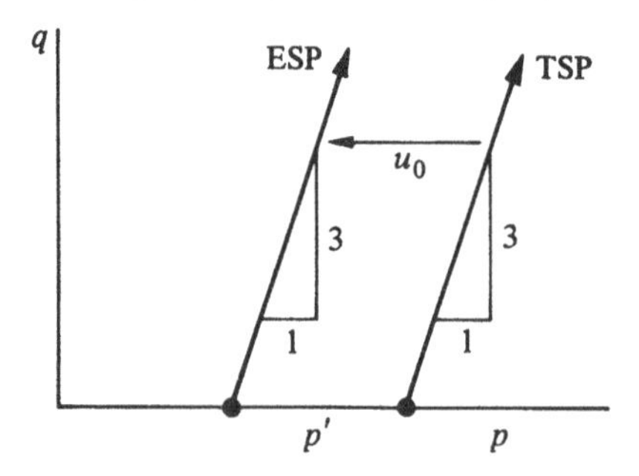

and effective stress paths will be separated by a constant distance u_0 parallel to the mean stress axis (Fig. 1.16).

The next important feature of soil behaviour is that unlike metals, which only change in volume when the mean stress p' is changed, soils usually change in volume when they are sheared. This phenomenon of *dilatancy*, which will be discussed in detail in Chapter 8 and will be a recurrent theme in other chapters, can simplistically be visualised with the aid of Fig. 1.17, which shows two layers of discs, one on top of the other. If a shear stress is applied to the upper layer, then each disc in this layer has

Fig. 1.17 Layers of circular discs dilating as they are sheared.

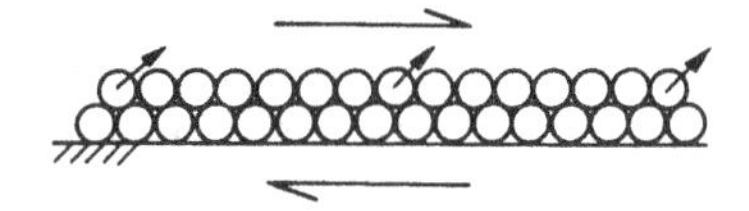

Fig. 1.18 Total and effective stress paths for undrained triaxial test: (a) on soil that wishes to contract as it is sheared, (b) on soil that wishes to expand as it is sheared, and (c) on soil that wishes to expand as it is sheared performed with back pressure u_0.

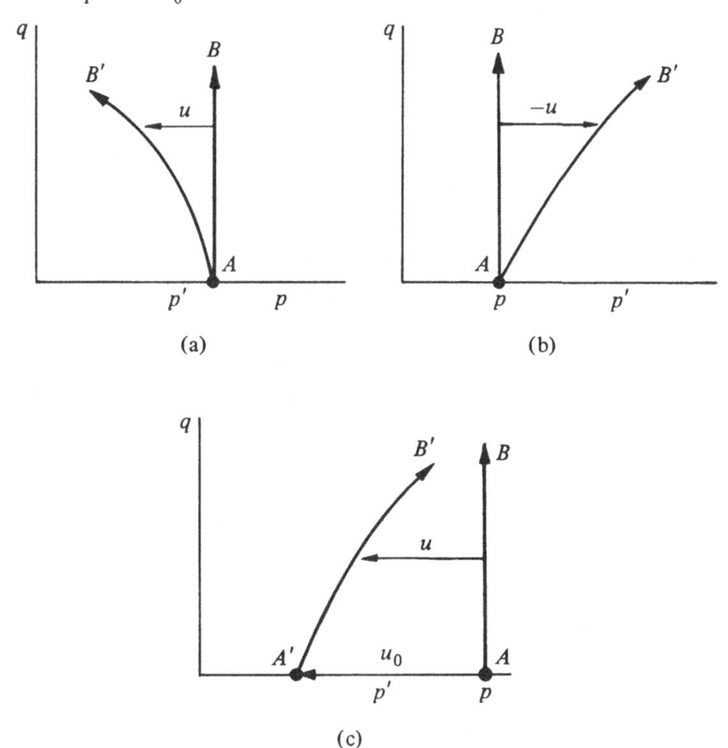

to rise (increasing the volume occupied by this proto-granular material) for the 'sample' to undergo any shear deformation. In general, soils may either compress or expand as they are sheared; but dilatancy is the primary reason for the significant difference between drained and undrained testing.

In drained tests, drainage can occur freely from the sample, and the volume occupied by the soil structure can change freely as it deforms. In undrained tests, drainage is prevented; the pore fluid is not permitted to flow into or out of the sample, and the soil itself is not able to do what it wants to do. Positive or negative pore pressures are generated in order that the soil may nevertheless be able to cope. If the soil wishes to contract as it is sheared, a positive pore pressure will develop because the pore fluid is prevented from flowing out of the sample. In an undrained compression test on such a soil, conducted at constant mean stress, the effective stress path will lie to the left of the total stress path ($u > 0$ implies $p' < p$; AB' in Fig. 1.18a). If the soil wishes to expand as it is sheared, it will need to suck in pore fluid. However, because pore fluid is prevented from flowing into the sample, a negative pore pressure will develop. In an undrained compression test on such a soil, performed with constant mean stress, the effective stress path lies to the right of the total stress path ($u < 0$ implies $p' > p$; AB' in Fig. 1.18b).

An undrained test is usually assumed to be a constant volume test (the term *isochoric* is sometimes used), but it is more strictly a constant mass test (*isomassic*) because closure of the drainage tap merely prevents any material from leaving the sample. It will be a truly constant volume test only if all the constituents of the soil sample are incompressible. For most soils this is a reasonable assumption provided the pore fluid is a saturated liquid. Air is certainly compressible, and if the pore fluid is a mixture of water and air not in solution, then it too will be significantly compressible. Hence, it may be important to perform undrained tests on soils which might develop negative pore pressures with an extra back pressure, to prevent the absolute pressure in the pore water from becoming so low that cavitation occurs and air comes out of solution. The limiting negative pressure that can be sustained without cavitation is of the order of the atmospheric pressure, in other words about -100 kilopascals (kPa). For an undrained test on a dilating soil performed with a back pressure, the relative positions of total stress path AB and effective stress path $A'B'$ are shown in Fig. 1.18c; though the pore pressure falls during the test, it never becomes negative, and cavitation is never in prospect.

Data from conventional triaxial compression tests performed with constant cell pressure ($\delta\sigma_r = 0$) are usually presented in terms of the quantities that are most readily determined: plots of the variation with (nominal) axial strain ε_a of deviator stress q and (nominal) volumetric strain ε_p (for drained tests) or pore pressure u (for undrained tests) (Fig. 1.19). These are unlimited plots in the sense that the axial strain can in principle increase indefinitely. It seems unlikely, however, that the stresses can increase indefinitely – for someone familiar with the mechanical behaviour of metals, a limit to the shear stress (or deviator stress) that can be supported might be anticipated – and it is clear that the volume of the sample cannot either decrease below the point at which the voids have disappeared (specific volume $v = 1$, void ratio $e = 0$) or increase above the point at which the particles are no longer in contact with each other. (Of course, during the process of deposition and formation of a soil, the particles may initially not be in contact as they settle through water towards the bottom of a lake or sea; but once a soil has formed, it can be assumed for engineering purposes that no particle will again lose contact with all its neighbours.) Stresses and specific volumes are thus limited quantities. In this book, although familiar stress:strain curves will be presented, much of the discussion will be conducted with the aid of limited plots involving stresses (in particular, effective stresses, given the importance of these in determining soil response) and specific volumes. The particular plots to be used are the effective stress plane $p':q$, which

Fig. 1.19 Standard plots of results of conventional triaxial compression tests: (a) drained tests and (b) undrained tests.

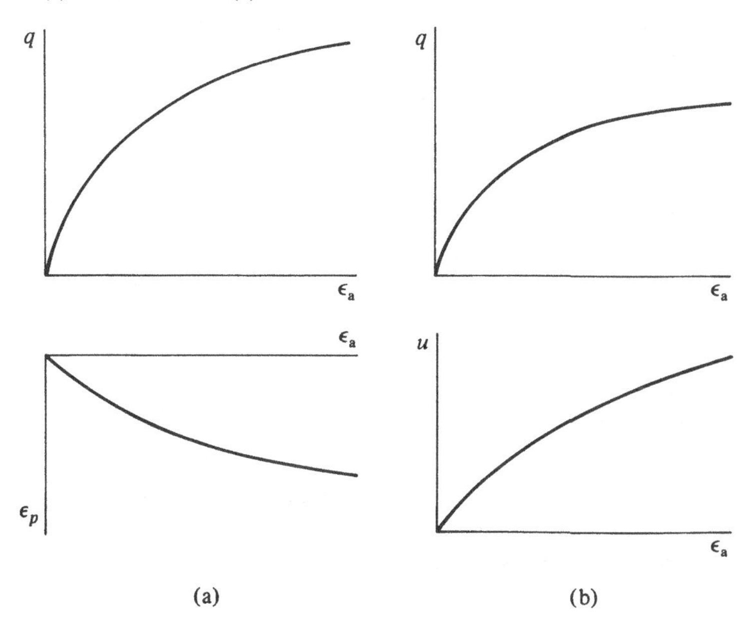

has already been used in Figs. 1.15, 1.16, and 1.18 to illustrate the capabilities of the triaxial apparatus, and also a plot of specific volume and mean effective stress $p':v$, which will be called the *compression plane*. The paths of the typical drained and undrained conventional triaxial compression tests, for which data were shown in Fig. 1.19a, b, are replotted in the $p':q$ effective stress plane and the $p':v$ compression plane in Fig. 1.20. AB is the drained test for which the effective stress path is forced to have slope $\delta q/\delta p' = 3$, and AC is the undrained test for which the volume remains constant, $\delta v = 0$. The way in which plots of test data in the stress plane and the compression plane can lead to clearer insights into patterns of soil behaviour will become evident in Chapter 6.

1.4.2 *Other testing apparatus*

Some of the practical field stress paths that one might attempt to emulate in the triaxial apparatus are described in Section 10.3. One will be mentioned here because it has led to the development of its own testing device, the *oedometer*. Many soils are deposited under seas or lakes, over areas of large lateral extent. Symmetry dictates that the depositional

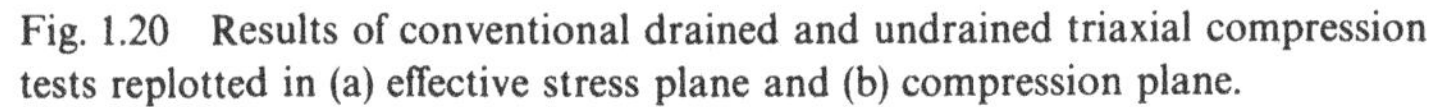

Fig. 1.20 Results of conventional drained and undrained triaxial compression tests replotted in (a) effective stress plane and (b) compression plane.

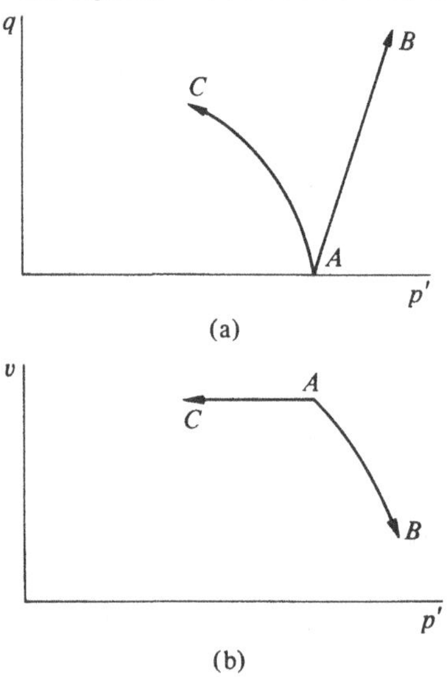

Fig. 1.21 Purely vertical movement of soil particles during one-dimensional deposition.

history of such soils must have been entirely one-dimensional (Fig. 1.21); there is no possibility for a soil element to have moved or deformed to one side or another. If a large surface load, such as an embankment or large spread foundation, is then placed on such a soil deposit, conditions beneath the centre of the loaded area will be similarly one-dimensional.

One-dimensional deformation is a special case of axisymmetric deformations, which can be applied in a triaxial apparatus by suitable control of the cell pressure as the deviator stress is changed in such a way that the radial strain of the sample is always zero ($\delta\varepsilon_r = 0$), which implies from (1.40) that

$$\frac{\delta V}{V} = \frac{\delta l}{l} \tag{1.50}$$

Continual application of this equation to check that volume and length changes are in step is possible though tedious to do manually but is perfectly feasible with computer controlled triaxial apparatus. However, if all that is required is one-dimensional deformation of soils, then this complexity can be avoided by using a device intended for just this purpose, the oedometer.

In the oedometer, the cylindrical soil sample is contained in a stiff metal ring which prevents any lateral deformation of the soil (Fig. 1.22). The use of oedometers to study time-dependent dissipation of excess pore water pressures in soils – the process of consolidation – is described in many textbooks of soil mechanics (e.g. Atkinson and Bransby, 1978). This

Fig. 1.22 Schematic section through oedometer.

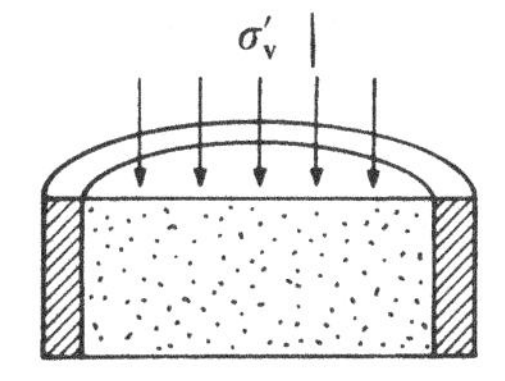

Fig. 1.23 Typical results of oedometer test.

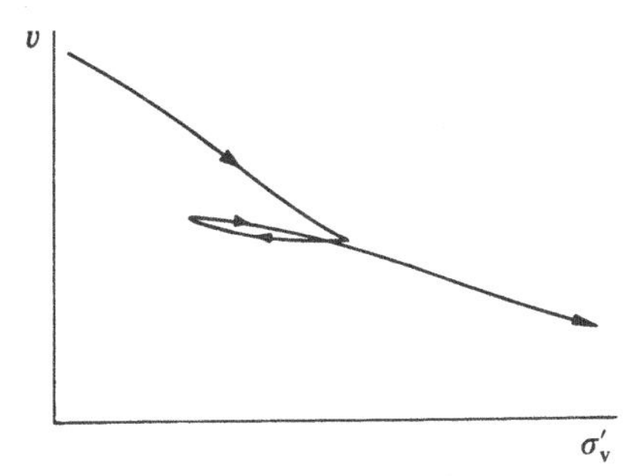

is a straightforward application of the solution to the one-dimensional diffusion equation in order to deduce the diffusion coefficient, known in soil mechanics as the *coefficient of consolidation*; so it will not be repeated here. The oedometer also produces information about the stress:strain behaviour of soils during one-dimensional deformation. Experimental data are usually plotted in a compression plane of specific volume v and vertical effective stress σ'_v (Fig. 1.23). The horizontal effective stress acting on the soil is not usually measured, so it is not usually possible to calculate values of mean effective stress p' or deviator stress q.

Historically, the first soil-testing apparatus to be developed was the direct shear box (Fig. 1.24). It was originally developed (e.g. see Collin, 1846) because it was observed in many landslides in clay that failure occurred on a thin, well-defined failure surface. In the shear box, failure of the soil sample on just such a thin, well-defined failure surface is produced by moving the top half of the box laterally relative to the bottom half of the box by means of a shear load Q (Fig. 1.24). The soil is loaded by a normal load P through a piston which can move up or down if the soil wishes to change in volume as it is sheared. This apparatus is still widely used, primarily for estimating strengths of soils. Its use for studying the stress:strain behaviour of soils before failure is limited because there can be no pretence that the stress and strain conditions in the soil sample are uniform. Although knowledge of P and Q might be sufficient to estimate the average normal and shear stress acting on a horizontal plane through the sample, there is no information about the stresses on other planes (in particular, the vertical plane); hence, there is insufficient information to

Fig. 1.24 Schematic section through direct shear box.

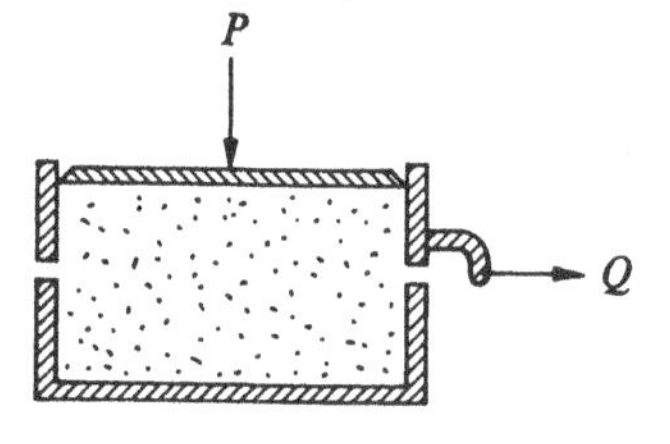

Fig. 1.25 Schematic diagram of simple shear apparatus.

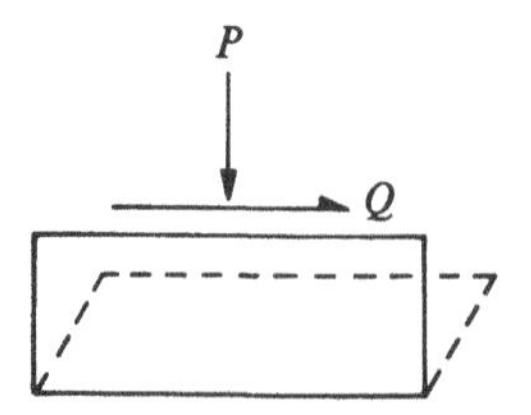

specify the position and size of the Mohr circle of stress for the plane of shearing.

A slight improvement on the direct shear box is provided by the simple shear apparatus (Roscoe, 1953; Bjerrum and Landva, 1966), shown schematically in section in Fig. 1.25. In this apparatus, there is more guarantee that the deformation of the soil sample will be reasonably uniform as the cross section changes from rectangle to parallelogram, but it is still not possible to guarantee that the stress state in the sample will be uniform (e.g. see Wood, Drescher, and Budhu, 1980). The simple shear apparatus is also commercially available and quite widely used.

Some data from direct shear box tests and simple shear tests on soils will be presented in Chapters 6–8 to illustrate ideas connected with the strength and dilatancy of soils.

1.5 Plane strain

Although most of the discussion in this book will be concerned with states of stress that can be reached in a conventional axially symmetric triaxial apparatus, there will be some discussion of stress paths appropriate to plane problems. The direct shear and the simple shear apparatus test soil samples under conditions of plane strain; such conditions are of more frequent practical occurrence than conditions of axial symmetry since they are relevant for any long geotechnical structure – embankment, wall, or footing – whose length is large compared with its cross section (Fig. 1.26a). For such structures, the displacement and strain along the length of the structure (the y axis in Fig. 1.26a) are zero:

$$\varepsilon_{yy} = 0 \tag{1.51}$$

Discussion of plane-strain stress paths can usefully be conducted in terms of the stresses in the plane of shearing – $\sigma'_{xx}, \sigma'_{zz}, \tau_{xz}$ (Fig. 1.26a) – leaving the effective stress in the y direction, σ'_{yy}, which from symmetry must be a principal stress, to be considered as a dependent rather than an independent variable because it is forced to adopt whatever value is necessary to comply with the imposed condition of zero strain in the y direction.

Under such conditions of plane strain, expression (1.24) for the increment of work per unit volume when an increment of strain occurs under certain imposed stresses becomes

$$\delta W = \sigma'_{xx}\delta\varepsilon_{xx} + \sigma'_{zz}\delta\varepsilon_{zz} + \sigma_{zx}\delta\gamma_{zx} \tag{1.52}$$

since $\delta\varepsilon_{yy} = 0$ and, from symmetry,

$$\tau_{xy} = \tau_{yz} = 0 \tag{1.53}$$

and

$$\delta\gamma_{xy} = \delta\gamma_{yz} = 0 \tag{1.54}$$

The principal effective stresses in the plane of shearing can be calculated from the Mohr circle construction (Fig. 1.26b):

$$\left.\begin{matrix}\sigma'_1\\ \sigma'_3\end{matrix}\right\} = \frac{\sigma'_{xx}+\sigma'_{zz}}{2} \pm \sqrt{\left(\frac{\sigma'_{xx}-\sigma'_{zz}}{2}\right)^2 + \tau_{zx}^2} \tag{1.55}$$

For the special case in which the principal axes of strain increment coincide with the principal axes of stress, the expression for the work increment (1.52) can be written in terms of σ'_1, σ'_3, and the corresponding principal

Fig. 1.26 Stress and strain variables for plane strain.

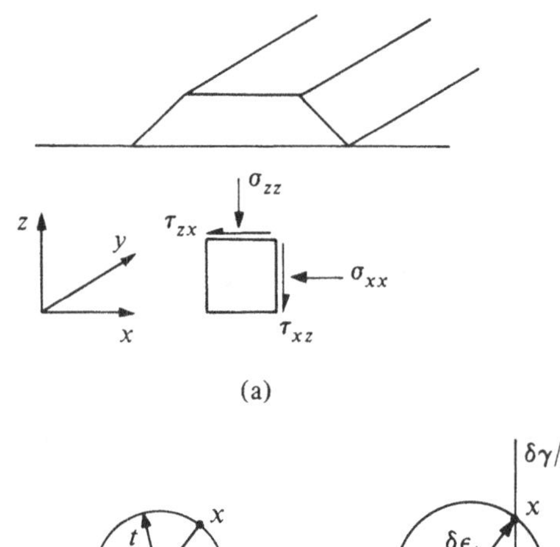

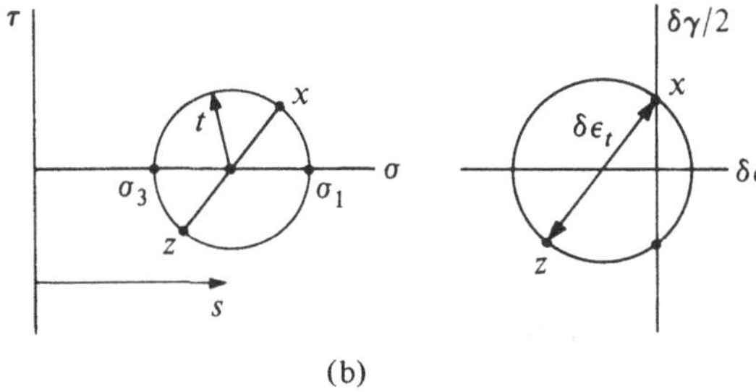

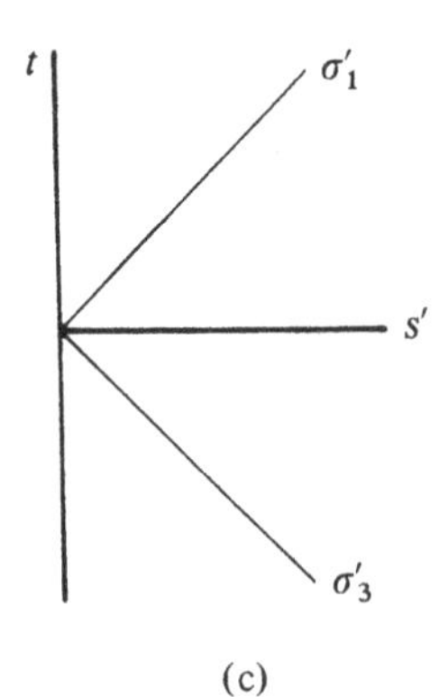

strain increments $\delta\varepsilon_1$ and $\delta\varepsilon_3$:

$$\delta W = \sigma'_1\,\delta\varepsilon_1 + \sigma'_3\,\delta\varepsilon_3 \tag{1.56}$$

For this case, with coaxiality of principal axes, it is again convenient to separate the strain increment and stress into volumetric and distortional parts. Appropriate pairs of variables are then the volumetric strain increment

$$\delta\varepsilon_s = \delta\varepsilon_1 + \delta\varepsilon_3 \tag{1.57}$$

the mean effective stress *in the plane of shearing*

$$s' = \frac{\sigma'_1 + \sigma'_3}{2} \tag{1.58}$$

the shear strain increment (diameter of the Mohr circle of strain increment, Fig. 1.26b)

$$\delta\varepsilon_t = \delta\varepsilon_1 - \delta\varepsilon_3 \tag{1.59}$$

and the maximum shear stress in the plane of shearing

$$t = \frac{\sigma'_1 - \sigma'_3}{2} \tag{1.60}$$

The variables s' and t will be used in Section 10.4 to present stress paths for a number of geotechnical situations of plane strain. The stress plane $s':t$ is a rotation through $\pi/4$ of the stress plane $\sigma'_1:\sigma'_3$ (Fig. 1.26c), the shear stress t is the radius of the Mohr circle of stress, and the mean stress s' is the centre of this circle (Fig. 1.26b). With these variables, the expression (1.56) for the work increment can be written as

$$\delta W = s'\,\delta\varepsilon_s + t\,\delta\varepsilon_t \tag{1.61}$$

Again, the subscripts on the symbols for the strain increment variables have been chosen to suggest the correct pairing of work-conjugate stress and strain increment quantities.

Although the validity of (1.56) and (1.61) is limited to situations in which the principal axes of strain increment and of stress coincide, this limitation may not be unduly restrictive. In many of the situations considered in Section 10.4, the principal effect of construction is expected to be a change of the horizontal and vertical stresses from an initial state in which the soil is one-dimensionally compressed, and the horizontal and vertical stresses are principal stresses. Thus, rotation of principal axes may not be of primary significance.

1.6 Pore pressure parameters

The response of soils depends on the effective stresses acting on them, but in many geotechnical situations, it is much easier to estimate, from rough considerations of equilibrium, the changes in total stresses that may occur as a result of the progress of construction. Thus, an analysis of the expected behaviour of the ground (such as whether it will fail or how much it will deform) requires an estimate of the pore pressures that are likely to develop as a result of the applied changes in total stresses so that the changes in effective stresses may in turn be estimated. The detailed linking of pore pressure and stress changes requires a consistent model of soil response: for example, an elastic model (Chapter 2) or an elastic–plastic model (such as those in Chapters 4 and 5). It is often useful to describe pore pressure development with the aid of so-called pore pressure parameters. The basic idea of pore pressure parameters will be introduced here, and the relationship between this basic idea and the models of soil response will be discussed in later chapters.

The principle of effective stress indicates that pore pressure is the difference between the total and effective mean stress:

$$u = p - p' \tag{1.62}$$

In incremental form this can be written as

$$\delta u = \delta p - \delta p' \tag{1.63}$$

which says that the pore pressure increase is due in part to the increase in total mean stress and in part to the decrease in effective mean stress (Fig. 1.27). For a sample sitting in a triaxial apparatus with the drainage connections closed, the change in total stress is the result of actions taken by the operator, who can vary the total stresses at will. The change in effective stress is not under the control of the operator; it merely assumes whatever value is necessary for the volume of the soil to remain constant. The volume of the soil may want to change as it is sheared and as a result of the changes in deviator stress which the operator has chosen to apply. In a typical soil which wants to contract as it is sheared (Fig. 1.18a), the

Fig. 1.27 Generation of pore pressure from changes in total and effective mean stresses.

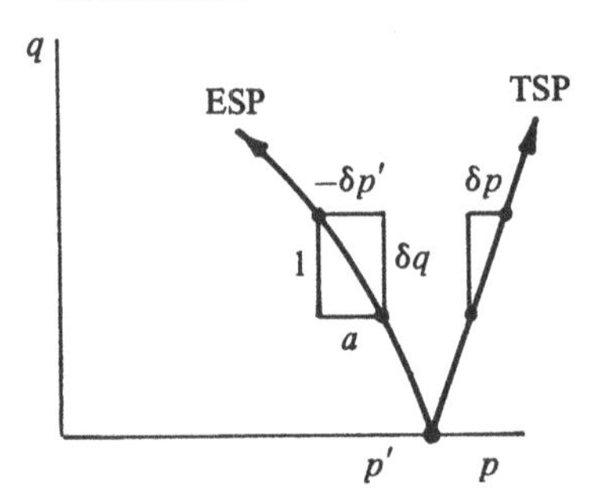

mean effective stress has to drop as the deviator stress is increased, to keep the volume constant. Suppose that at a particular stage of an undrained test, the changes of deviator stress and mean effective stress (Fig. 1.27) are linked by

$$\delta p' = -a\,\delta q \tag{1.64}$$

Then, (1.62) becomes

$$\delta u = \delta p + a\,\delta q \tag{1.65}$$

and if the pore pressure parameter a is known, the pore pressure change can be deduced from the changes in total stresses. It is important to appreciate that it is only the second term $a\,\delta q$ which says anything about the soil; the first part of the pore pressure change δp is entirely the result of arbitrary external actions over which the soil has no control. The pore pressure parameter a is not usually a soil constant but will depend on the past history of the soil and on the nature of the loading.

A slightly modified form of (1.65),

$$\delta u = b(\delta p + a\,\delta q) \tag{1.66}$$

has rather wider application. The extra pore pressure parameter b should be unity for saturated soils but will be lower than unity for soils with compressible pore fluids, or for porous materials in which the compressibility of the matrix is of the same order as the compressibility of the pore fluid. This will be discussed in Chapter 2.

1.7 Conclusion

The scene has now been set for the development of numerical models for the description of soil behaviour, which will be presented in later chapters. These models will concentrate on the response of soil in the axially symmetric conditions imposed by the conventional triaxial apparatus. It is necessary to choose correct pairs of strain increment and stress variables to study soil response in the triaxial apparatus and to develop numerical models. These variables separate volumetric and distortional effects and are the mean effective stress $p' = (\sigma'_a + 2\sigma'_r)/3$ and volumetric strain increment $\delta\varepsilon_p = \delta\varepsilon_a + 2\delta\varepsilon_r$ on the one hand, and the deviator stress $q = \sigma'_a - \sigma'_r$ and triaxial shear strain increment $\delta\varepsilon_q = 2(\delta\varepsilon_a - \delta\varepsilon_r)/3$ on the other.

Exercises

E1.1. A triaxial sample supporting total axial and radial stresses σ_a and σ_r undergoes increments of axial and radial strain $\delta\varepsilon_a$ and $\delta\varepsilon_r$. The pressure in the pore water remains constant. Show that the work input to the soil skeleton is given by $p'\,\delta\varepsilon_p + q\,\delta\varepsilon_q$.

E1.2. Data of triaxial tests are often presented in terms of the so-called MIT stress variables $\{p'\} = (\sigma'_a + \sigma'_r)/2$ and $\{q\} = (\sigma'_a - \sigma'_r)/2$. Find, as functions of the strain increments $\delta\varepsilon_a$ and $\delta\varepsilon_r$, the corresponding work conjugate strain increment quantities $\delta\alpha$ and $\delta\beta$, which should be associated with these stress variables so that the work increment can be written as $\delta W = \{p'\}\,\delta\alpha + \{q\}\,\delta\beta$.

E1.3. The form of the pore pressure expression that has achieved widest use is not (1.66) but is the expression proposed by Skempton (1954a),

$$\Delta u = B[\Delta\sigma_3 + A(\Delta\sigma_1 - \Delta\sigma_3)]$$

which was produced with *exclusively triaxial compression* conditions in mind, so that $\Delta\sigma_3 = \Delta\sigma_r$ and $\Delta\sigma_1 = \Delta\sigma_a$. Show that the pore-pressure parameter B is identical to b in (1.66) but the parameter $A = a + \frac{1}{3}$ (with a assumed constant over the large stress changes $\Delta\sigma_a$ and $\Delta\sigma_r$); so, for non-dilatant materials, $a = 0$ but $A = \frac{1}{3}$.

2

Elasticity

2.1 Isotropic elasticity

Elastic, recoverable, material response is easier to describe and comprehend than plastic, irrecoverable response. Some essential elements of the theory of elasticity will be presented in this chapter.

A familiar introduction to the elastic properties of materials is obtained by the simple experimental procedure of hanging weights on a wire and measuring the elongation (Fig. 2.1a). For many materials, there is a range of loads for which the elongation varies linearly with the applied load P (Fig. 2.1b) and is recovered when the load is removed. There is no permanent deformation of the wire.

Such a procedure provides a direct indication of the validity of Hooke's law as originally stated (1675); *ut tensio sic vis* (as the extension so the force). Hooke's law was originally published as an anagram *ceiiinosssttuv* (Fig. 2.2) to fill space in his book on helioscopes, an enigmatic way of attempting to guarantee scientific precedence (Heyman, 1972).

The slope of the linear relationship is related to the unconfined uniaxial stiffness of the material of the wire. Young's modulus E is expressed as

$$E = \frac{P/A}{\delta l/l} \tag{2.1}$$

where P is the load on the wire of area A and length l, and δl is the extension of the wire.

The experimental introduction to elasticity is usually restricted to observation of the load:extension properties of tensile specimens. However, if a micrometer were available to measure the changing diameter d of the wire, then one would observe that as the wire becomes longer, its diameter becomes smaller (Figs. 2.1c and 2.3a). The ratio of the magnitude of the induced diametral strain to the imposed longitudinal strain is Poisson's ratio ν:

$$\nu = \frac{-\delta d/d}{\delta l/l} \tag{2.2}$$

It is well known that the pair of constants E and ν is sufficient to describe the elastic response of isotropic materials. These two constants are probably the most easily understood elastic constants because direct experimental observation of them is so straightforward. However, in many ways it is more fundamental to use an alternative pair of elastic constants: the bulk modulus K and shear modulus G, which divide the elastic deformation into a volumetric part (change of size at constant shape, Fig. 2.3b) and a distortional part (change of shape at constant volume, Fig. 2.3c), respectively.

The benefit of using K and G is particularly great when elasticity of soils is considered. Undrained deformation of soils is specifically concerned with deformation of soil at constant volume, that is, pure distortion of soil, a change of shape without change in size. The distinction between undrained and drained processes is only relevant because there are some processes (in general, inelastic ones) during which soils express a desire to change in size as well as shape as they are sheared.

The relationships between the two sets of constants can be deduced by considering the change in volume of the wire as it is extended, which shows that the bulk modulus K is

$$K = \frac{E}{3(1-2\nu)} \tag{2.3}$$

Fig. 2.1 Tensile test on metal wire: (a) test arrangement; (b) load, extension relationship; (c) changes in diameter and length.

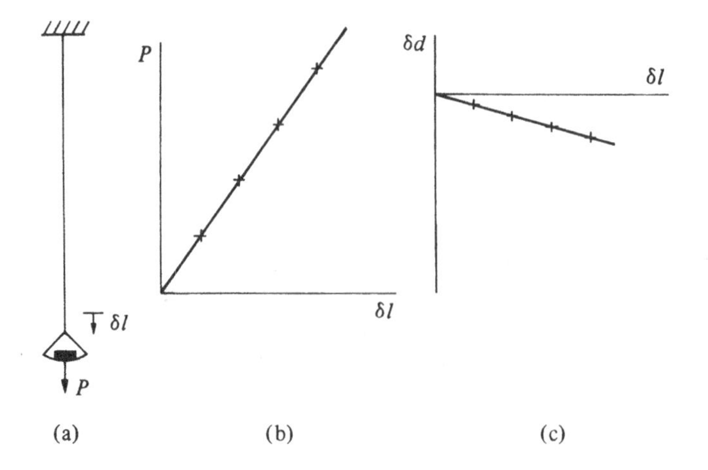

Fig. 2.2 Hooke's original statement of the law *ut tensio sic vis.*

A

DESCRIPTION

OF

HELIOSCOPES,

And ſome other

INSTRUMENTS

MADE BY

ROBERT HOOKE,

Fellow of the *Royal Society.*

Hos ego, &c.

Sic vos non vobis——.

LONDON,

Printed by *T. R.* for *John Martyn* Printer to the *Royal Society,* at the *Bell* in St. *Pauls* Church-yard, 1676.

To fill the vacancy of the enſuing page, I have here added a *decimate* of the *centeſme* of the Inventions I intend to publiſh, though poſſibly not in the ſame order, but as I can get opportunity and leaſure; moſt of which, I hope, will be as uſeful to Mankind, as they are yet unknown and new.

1. *A way of Regulating all ſorts of* Watches *or* Time-keepers, *ſo as to make any way to equalize, if not exceed the* Pendulum-Clocks *now uſed.*

2. *The true Mathematical and Mechanichal form of all manner of* Arches *for building, with the true butment neceſſary to each of them.* A Problem which no *Architectonick* Writer hath ever yet attempted, much leſs performed. abccc ddeeeee f gg iiiiiii llmmmmmnnnnnooprr sssttttttuuuuuuuux.

3. *The true Theory of* Elaſticity *or* Springineſs, *and a particular Explication thereof in ſeveral Subjects in which it is to be found: And the way of computing the velocity of Bodies moved by them.* c e i i i n o s s s t t u u.

and by considering the change of the right angle between rays 'drawn' in the material of the wire at $\pm\pi/4$ to the axis of the wire (Fig. 2.3c), which shows that the shear modulus G is

$$G = \frac{E}{2(1+\nu)} \tag{2.4}$$

2.2 Soil elasticity

Unconfined tensile tests on soils are not usually feasible. Compression tests are more commonly performed using the triaxial apparatus and with some lateral confinement provided by the cell pressure. The results of a typical drained test on a soil sample might resemble those shown in Fig. 2.4. This is a conventional triaxial compression test in which the axial stress (or deviator stress) is increased while the lateral stress (or cell pressure) is held constant. The initial linear sections of the stress:strain curve (Fig. 2.4a) (deviator stress q plotted against triaxial shear strain ε_q) and of the volume-change curve (Fig. 2.4b) (volumetric strain ε_p plotted

Fig. 2.3 (a) Young's modulus describing change in length and Poisson's ratio describing change in width; (b) bulk modulus describing change in size at constant shape; (c) shear modulus describing change in shape at constant volume.

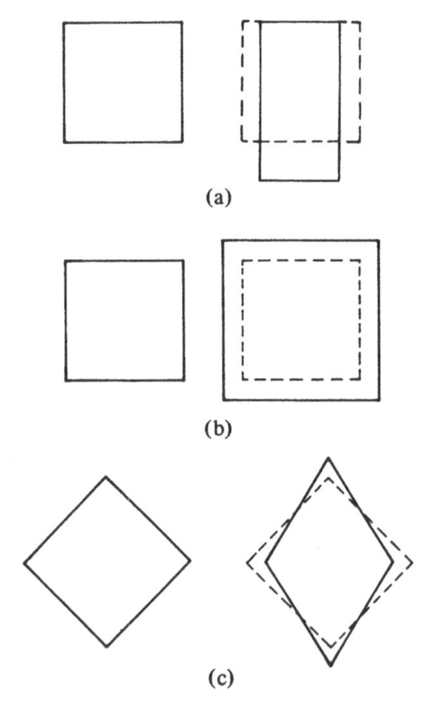

against triaxial shear strain ε_q) might be interpreted as the elastic response of the soil to the imposed changes of stress. Thus, the values of the elastic constants could be deduced.

It is a presumption throughout this book that soil response is governed by changes in effective stresses and that elastic response is no exception. Elastic properties of the soil skeleton will be written here with a prime to emphasise this point.

Working in terms of Young's modulus and Poisson's ratio, one can describe the response of a soil specimen to a general triaxial change of effective stress by these equations:

$$\begin{bmatrix} \delta\varepsilon_a \\ \delta\varepsilon_r \end{bmatrix} = \frac{1}{E'} \begin{bmatrix} 1 & -2\nu' \\ -\nu' & 1-\nu' \end{bmatrix} \begin{bmatrix} \delta\sigma'_a \\ \delta\sigma'_r \end{bmatrix} \tag{2.5}$$

Matrices are introduced here simply as a convenient shorthand means of writing sets of equations. Thus, (2.5) implies the following pair of equations:

$$\delta\varepsilon_a = \frac{1}{E'}(\delta\sigma'_a - 2\nu'\,\delta\sigma'_r) \tag{2.5a}$$

$$\delta\varepsilon_r = \frac{1}{E'}[-\nu'\,\delta\sigma'_a - (1-\nu')\,\delta\sigma'_r] \tag{2.5b}$$

Fig. 2.4 Elastic constants deduced from conventional drained triaxial compression test: (a) deviator stress q and triaxial shear strain ε_q; (b) volumetric strain ε_p and triaxial shear strain ε_q; (c) deviator stress q and axial strain ε_a.

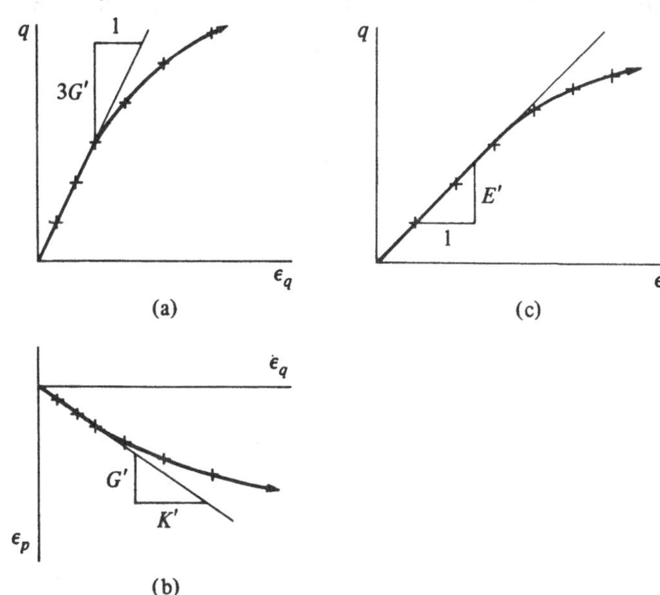

Writing equations in terms of matrix products is extremely helpful when trying to write compact, efficient computer programs. However, no aptitude for manipulation of combinations of matrices is either expected or demanded in this book.

The preferred strain increment and effective stress variables that were introduced in Section 1.4.1 for description and analysis of triaxial tests were the mean effective stress p' and deviator stress q,

$$\begin{bmatrix} p' \\ q \end{bmatrix} = \begin{bmatrix} \frac{1}{3} & \frac{2}{3} \\ 1 & -1 \end{bmatrix} \begin{bmatrix} \sigma'_a \\ \sigma'_r \end{bmatrix} \tag{2.6}$$

and the increments of volumetric strain $\delta\varepsilon_p$ and triaxial shear strain $\delta\varepsilon_q$,

$$\begin{bmatrix} \delta\varepsilon_p \\ \delta\varepsilon_q \end{bmatrix} = \begin{bmatrix} 1 & 2 \\ \frac{2}{3} & -\frac{2}{3} \end{bmatrix} \begin{bmatrix} \delta\varepsilon_a \\ \delta\varepsilon_r \end{bmatrix} \tag{2.7}$$

The elastic response in (2.5) can then be written more elegantly using bulk modulus and shear modulus to separate effects of changing size and changing shape:

$$\begin{bmatrix} \delta\varepsilon_p \\ \delta\varepsilon_q \end{bmatrix} = \begin{bmatrix} 1/K' & 0 \\ 0 & 1/3G' \end{bmatrix} \begin{bmatrix} \delta p' \\ \delta q \end{bmatrix} \tag{2.8}$$

The off-diagonal zeroes in (2.8) indicate the absence of coupling between volumetric and distortional effects for this isotropic elastic material. Change in mean stress p' produces no distortion $\delta\varepsilon_q$, and change in the distortional deviator stress q produces no change in volume.

The initial gradient of the stress:strain curve in Fig. 2.4a is then $3G'$. The initial gradient of the volume change curve in Fig. 2.4b is

$$\frac{\delta\varepsilon_p}{\delta\varepsilon_q} = \frac{3G'}{K'}\frac{\delta p'}{\delta q} \tag{2.9}$$

For a conventional drained triaxial compression test on a soil specimen, in which

$$\delta q = 3\,\delta p' \tag{2.10}$$

this becomes

$$\frac{\delta\varepsilon_p}{\delta\varepsilon_q} = \frac{G'}{K'} \tag{2.11}$$

and the elastic properties have been recovered. Evidently, values of Young's modulus and Poisson's ratio could be deduced using (2.3) and (2.4).

If, alternatively, drainage from the triaxial sample is prevented, then undrained constant volume response is observed. The initial response of the soil specimen may still be elastic, but now, with volume change prevented, pore pressures develop. The response of the soil can be depicted in plots of deviator stress and pore pressure against triaxial shear strain (Fig. 2.5).

The imposition of a condition of constant volume on (2.8) implies that

$$\frac{\delta p'}{K'} = 0 \tag{2.12}$$

which requires either that

$$K' = \infty$$

or that

$$\delta p' = 0$$

There is no reason why the bulk modulus of the soil skeleton should be infinite; certainly the act of closing the drainage tap on the triaxial apparatus can have no influence on the elastic properties of the soil skeleton. Consequently, the only reasonable solution to (2.12) is

$$\delta p' = 0 \tag{2.13}$$

The pore pressure changes reflect directly the imposed changes in total mean stress

$$\delta u = \delta p \tag{2.14}$$

Fig. 2.5 Elastic constants deduced from conventional undrained triaxial compression test: (a) deviator stress q and triaxial shear strain ε_q; (b) pore pressure u and triaxial shear strain ε_q; (c) deviator stress q and axial strain ε_a.

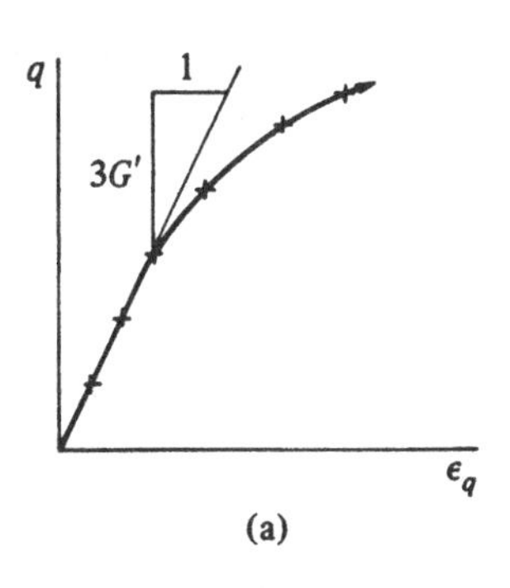

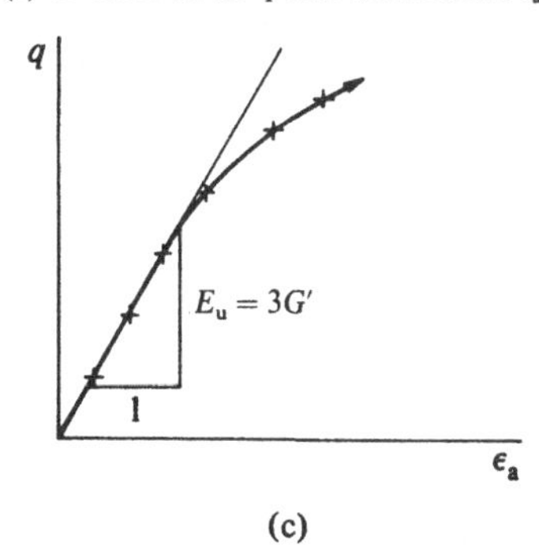

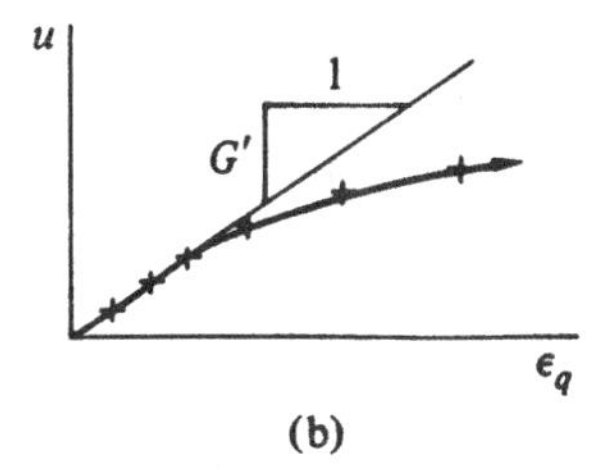

and, irrespective of the total stress path, the effective stress path is vertical in the $p':q$ plane (AB in Fig. 2.6a). Evidently, *any* total stress path could be imposed; if one of these paths had no applied change in total mean stress, $\delta p = 0$, then the soil would have no desire to change in volume under the purely distortional stress changes and hence no tendency to generate any pore pressure. This result implies that for this isotropic elastic material the pore pressure parameter a in (1.65) is zero, which is just another way of saying that there is no coupling between volumetric and distortional effects, as illustrated in (2.8).

The constant volume condition imposes no constraint on the change in *shape* of the soil sample, and (2.8) makes it clear that the slope of the deviator stress:triaxial shear strain plot (Fig. 2.5a) will again be $3G'$, as in the drained test. For a conventional undrained triaxial compression test in which the cell pressure is held constant while the axial stress is increased,

$$\delta p = \frac{\delta q}{3} = \delta u \tag{2.15}$$

and the slope of the plot of pore pressure against triaxial shear strain (Fig. 2.5b) is G'.

If at some stage of the undrained loading the drainage tap is opened and the pore pressure is allowed to dissipate at constant total stress, then some deformation of the soil will occur. With the total stresses constant, the deviator stress $q = \sigma'_a - \sigma'_r = \sigma_a - \sigma_r$ cannot change, and the effective stress path for this dissipation process is parallel to the p' axis (BC in Fig. 2.6a). It is then apparent from (2.8) that the accompanying deformation will involve only change of size ($\delta\varepsilon_p > 0$) with no change in shape ($\delta\varepsilon_q = 0$).

Fig. 2.6 Undrained shearing AB and subsequent pore pressure dissipation BC in conventional triaxial compression test: (a) total and effective stress paths; (b) deviator stress q and axial strain ε_a.

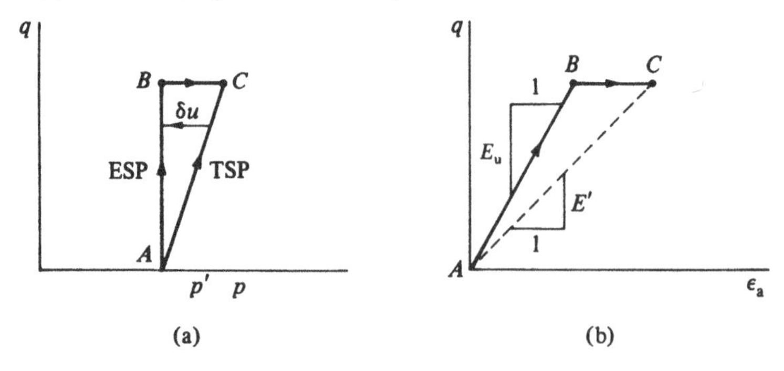

Although the behaviour of soil elements is controlled by changes in effective stresses, it is often useful to describe the elastic response of soil in terms of changes in total stresses. Equilibrium equations for a soil continuum can be written in terms of total stresses without having to introduce pore pressures, and analytical procedures may often lead more readily to distributions of total stresses than to distributions of effective stresses. The observed response of a soil element must, however, be identical whether it is treated in terms of total stresses or effective stresses.

The total stress equivalent of (2.8) is

$$\begin{bmatrix} \delta\varepsilon_q \\ \delta\varepsilon_q \end{bmatrix} = \begin{bmatrix} 1/K_u & 0 \\ 0 & 1/3G_u \end{bmatrix} \begin{bmatrix} \delta p \\ \delta q \end{bmatrix} \tag{2.16}$$

The distinction between elastic properties in terms of total or effective stresses is only helpful for constant volume undrained conditions; this is the reason for the subscript u on the bulk modulus and shear modulus in (2.16).

For an undrained constant volume condition, $\delta\varepsilon_p = 0$ implies that

$$\frac{\delta p}{K_u} = 0 \tag{2.17}$$

Whereas the effective stress description of (2.8) and (2.12) looks at the triaxial soil sample from inside the membrane, the total stress description of (2.16) and (2.17) looks at the sample from outside the membrane. There can now be no constraint on the total stress path that is imposed. The condition of no volume change must emerge whatever the externally applied changes in total stress. Hence, the total stress undrained bulk modulus K_u must be infinite,

$$K_u = \infty \tag{2.18}$$

which, from the general equation (2.3), implies that the undrained Poisson's ratio is

$$\nu_u = \tfrac{1}{2} \tag{2.19}$$

The deviator stress q is not affected by drainage conditions because, as a difference of two stresses, it is independent of pore pressure. The shearing, or change of shape, of the soil $\delta\varepsilon_q$ calculated from (2.8) and (2.16) must be identical and hence

$$G_u = G' \tag{2.20}$$

and the shear modulus is independent of the drainage conditions.

Given the link between shear modulus, Young's modulus, and Poisson's ratio implied by (2.4), the undrained and drained values of Young's modulus, E_u and E', respectively, are not independent. For from (2.4) and (2.20),

$$\frac{E_u}{2(1+\nu_u)} = \frac{E'}{2(1+\nu')}$$

and then with (2.19),

$$E_u = \frac{3E'}{2(1+\nu')} \tag{2.21}$$

Young's modulus describes the slope of the axial stress:axial strain relationship. In conventional triaxial compression tests $\delta\sigma_a = \delta q$, and Young's modulus is the slope of the deviator stress:axial strain relationship. Different slopes, in the ratio given by (2.21), will be seen in drained and undrained tests (Figs. 2.4c and 2.5c). The effect of allowing drainage to occur after some increments of undrained loading (effective stress path *BC* after *AB* in Fig. 2.6a) is to take the deviator stress: axial strain state from the undrained to the drained line, as shown in Fig. 2.6b.

In summary, for isotropic elastic soil there are only two independent elastic soil constants. Elastic constants to describe the behaviour of soil under special conditions (e.g. in terms of total stresses for undrained conditions) can be deduced from the more fundamental effective stress constants and cannot be chosen independently.

2.3 Anisotropic elasticity

The discussion in previous sections has been restricted to the ideal case of isotropic elasticity. Real soil may not fit into this simple picture. Deviations from this picture may result from inelasticity, but they can also occur if the soil is elastic but anisotropic.

A completely general description of an anisotropic elastic material requires the specification of 21 elastic constants (e.g. see Heyman, 1982; Love, 1927), but analyses using such general material characteristics are rarely practicable. Besides, the depositional history of many soils introduces symmetries which may reduce considerably the number of independent elastic constants.

Many soils have been deposited over areas of large lateral extent, and the deformations they have experienced during and after deposition have been essentially one-dimensional. Soil particles have moved vertically downwards (and possibly also upwards) with, from symmetry, no tendency to move laterally (Fig. 1.21). The anisotropic elastic properties of the soil reflect this history. The soil may respond differently if it is pushed in vertical or horizontal directions, but it will respond in the same way if it is pushed in any horizontal direction. For example, cylindrical sample A in Fig. 2.7, taken from the ground with its axis vertical, behaves differently from samples B, C, D, and E, which have been taken from the ground with their axes in various horizontal directions; but samples B, C, D, and E all behave identically.

This special form of anisotropy, known as *transverse isotropy* or *cross anisotropy*, requires only five elastic constants for its specification The form of the relationship between stress increments and strain increments takes the form

$$\begin{bmatrix} \delta\varepsilon_{xx} \\ \delta\varepsilon_{yy} \\ \delta\varepsilon_{zz} \\ \delta\gamma_{yz} \\ \delta\gamma_{zx} \\ \delta\gamma_{xy} \end{bmatrix} = \begin{bmatrix} 1/E_\text{h} & -\nu_\text{hh}/E_\text{h} & -\nu_\text{vh}/E_\text{v} & 0 & 0 & 0 \\ -\nu_\text{hh}/E_\text{h} & 1/E_\text{h} & -\nu_\text{vh}/E_\text{v} & 0 & 0 & 0 \\ -\nu_\text{vh}/E_\text{v} & -\nu_\text{vh}/E_\text{v} & 1/E_\text{v} & 0 & 0 & 0 \\ 0 & 0 & 0 & 1/2G_\text{vh} & 0 & 0 \\ 0 & 0 & 0 & 0 & 1/2G_\text{vh} & 0 \\ 0 & 0 & 0 & 0 & 0 & 2(1+\nu_\text{hh})/E_\text{h} \end{bmatrix} \begin{bmatrix} \delta\sigma'_{xx} \\ \delta\sigma'_{yy} \\ \delta\sigma'_{zz} \\ \delta\tau_{yz} \\ \delta\tau_{zx} \\ \delta\tau_{xy} \end{bmatrix} \tag{2.22}$$

where the stress and strain increments are referred to rectangular Cartesian axes x, y, and z with the z axis vertical (Fig. 2.7).

Most of the routine soil tests that are performed in practice are triaxial compression tests on samples such as A in Fig. 2.7, taken out of the ground with their axes vertical, for example, from some sort of borehole. Graham and Houlsby (1983) show that it is not possible from such tests to recover more than three elastic constants for the soil; since two constants are needed for the description of isotropic elastic response, that leaves only one constant through which some anisotropy can be incorporated. They propose a particular form of one-parameter anisotropy which allows certain analytical advantages and leads to a particular form of stiffness

Fig. 2.7 Cylindrical soil samples taken out of the ground with their axes vertical (A) and horizontal (B, C, D, E).

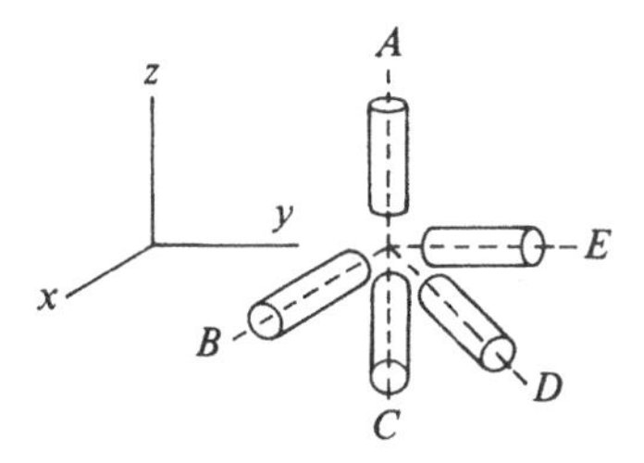

matrix relating stress increments and strain increments:

$$\begin{bmatrix} \delta\sigma'_{xx} \\ \delta\sigma'_{yy} \\ \delta\sigma'_{zz} \\ \delta\tau_{yz} \\ \delta\tau_{zx} \\ \delta\tau_{xy} \end{bmatrix} = \frac{E^*}{(1+\nu^*)(1-2\nu^*)}$$

$$\times \begin{bmatrix} \alpha^2(1-\nu^*) & \alpha^2\nu^* & \alpha\nu^* & 0 & 0 & 0 \\ \alpha^2\nu^* & \alpha^2(1-\nu^*) & \alpha\nu^* & 0 & 0 & 0 \\ \alpha\nu^* & \alpha\nu^* & (1-\nu^*) & 0 & 0 & 0 \\ 0 & 0 & 0 & \alpha(1-2\nu^*)/2 & 0 & 0 \\ 0 & 0 & 0 & 0 & \alpha(1-2\nu^*)/2 & 0 \\ 0 & 0 & 0 & 0 & 0 & \alpha^2(1-2\nu^*)/2 \end{bmatrix} \begin{bmatrix} \delta\varepsilon_{xx} \\ \delta\varepsilon_{yy} \\ \delta\varepsilon_{zz} \\ \delta\gamma_{yz} \\ \delta\gamma_{zx} \\ \delta\gamma_{xy} \end{bmatrix} \quad (2.23)$$

This expression can be inverted so that the stiffness matrix becomes a compliance matrix:

$$\begin{bmatrix} \delta\varepsilon_{xx} \\ \delta\varepsilon_{yy} \\ \delta\varepsilon_{zz} \\ \delta\gamma_{yz} \\ \delta\gamma_{zx} \\ \delta\gamma_{xy} \end{bmatrix} = 1/E^* \begin{bmatrix} 1/\alpha^2 & -\nu^*/\alpha^2 & -\nu^*/\alpha & 0 & 0 & 0 \\ -\nu^*/\alpha^2 & 1/\alpha^2 & -\nu^*/\alpha & 0 & 0 & 0 \\ -\nu^*/\alpha & -\nu^*/\alpha & 1 & 0 & 0 & 0 \\ 0 & 0 & 0 & 2(1+\nu^*)/\alpha & 0 & 0 \\ 0 & 0 & 0 & 0 & 2(1+\nu^*)/\alpha & 0 \\ 0 & 0 & 0 & 0 & 0 & 2(1+\nu^*)/\alpha^2 \end{bmatrix} \begin{bmatrix} \delta\sigma'_{xx} \\ \delta\sigma'_{yy} \\ \delta\sigma'_{zz} \\ \delta\tau_{yz} \\ \delta\tau_{zx} \\ \delta\tau_{xy} \end{bmatrix} \quad (2.24)$$

In these expressions, E^* and ν^* represent modified values of Young's modulus and Poisson's ratio for the soil, and α is the anisotropy parameter. Equation (2.24) can be compared with the completely general five-constant description of transverse isotropy (2.22).

Expressions (2.22)–(2.24) give the complete stiffness and compliance matrices which are necessary for any analysis of transversely isotropic elastic soil. However, to interpret the results of triaxial tests, it is helpful once again to look at description of changes in size and in shape. Still following Graham and Houlsby (1983), one finds the stiffness equation to be

$$\begin{bmatrix} \delta p' \\ \delta q \end{bmatrix} = \begin{bmatrix} K^* & J \\ J & 3G^* \end{bmatrix} \begin{bmatrix} \delta\varepsilon_p \\ \delta\varepsilon_q \end{bmatrix} \quad (2.25)$$

where K^* and G^* are modified values of bulk modulus and shear modulus, and the presence of the two off-diagonal terms J shows that there is now some cross-coupling between volumetric and distortional effects. The quantities K^*, G^*, and J can be expressed in terms of the quantities E^*, ν^*, and α from (2.23):

$$K^* = \frac{E^*(1 - \nu^* + 4\alpha\nu^* + 2\alpha^2)}{9(1 + \nu^*)(1 - 2\nu^*)} \tag{2.26}$$

$$G^* = \frac{E^*(2 - 2\nu^* - 4\alpha\nu^* + \alpha^2)}{6(1 + \nu^*)(1 - 2\nu^*)} \tag{2.27}$$

$$J = \frac{E^*(1 - \nu^* + \alpha\nu^* - \alpha^2)}{3(1 + \nu^*)(1 - 2\nu^*)} \tag{2.28}$$

It may be confirmed that for $\alpha = 1$, these expressions for K^* and G^* reduce to (2.3) and (2.4) and that $J = 0$. The compliance form of (2.25) is

$$\begin{bmatrix} \delta\varepsilon_p \\ \delta\varepsilon_q \end{bmatrix} = \frac{1}{D}\begin{bmatrix} 3G^* & -J \\ -J & K^* \end{bmatrix}\begin{bmatrix} \delta p' \\ \delta q \end{bmatrix} \tag{2.29}$$

with

$$D = 3K^*G^* - J^2 \tag{2.30}$$

The coupling between volumetric and distortional effects implies that constant volume effective stress paths are no longer vertical constant p' paths in the $p':q$ plane [(2.13) and Fig. 2.6a]. The direction of the path will depend on the value of α. With $\alpha > 1$ the soil is stiffer horizontally than vertically, and the undrained effective stress path shows a decrease in p' (Fig. 2.8). With $\alpha < 1$ the soil is stiffer vertically, and the undrained effective stress path shows an increase in p' (Fig. 2.8). The pore pressure that is observed in an undrained test will be different from the change in total mean stress [see (2.14)], with the difference depending on α and ν^*. The direction of the effective stress path is, from (2.29):

$$\frac{\delta q}{\delta p'} = \frac{3G^*}{J} \tag{2.31}$$

Fig. 2.8 Effective stress paths for constant volume deformation of cross-anisotropic elastic soil.

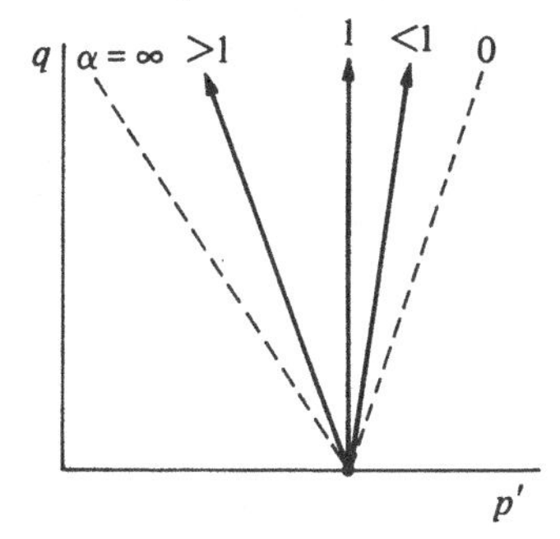

or

$$\frac{\delta q}{\delta p'} = \frac{3(2 - 2v^* - 4\alpha v^* + \alpha^2)}{2(1 - v^* + \alpha v^* - \alpha^2)} \tag{2.32}$$

This ratio has limiting values $-\frac{3}{2}$ and $+3$ for α very large and very small, which imply effective stress paths with constant axial stress and constant radial stress, respectively. Comparison with (1.65) shows that for this cross-anisotropic elastic soil, the pore pressure parameter a is given by

$$a = \frac{-J}{3G^*} \tag{2.33}$$

Fig. 2.9 Changes in effective stress in undrained triaxial compression of Winnipeg clay (data from Graham and Houlsby, 1983).

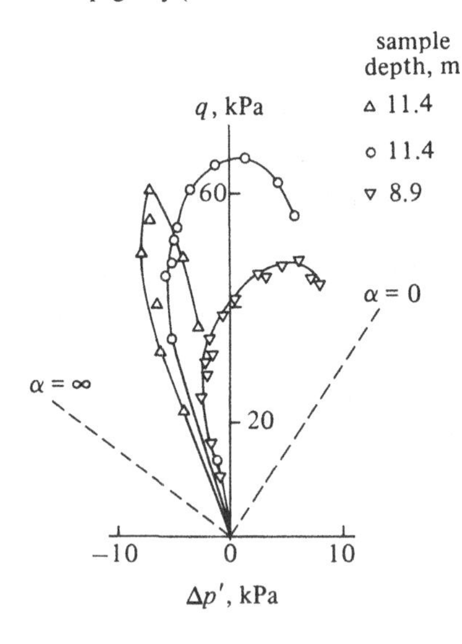

Fig. 2.10 Volumetric strain:triaxial shear strain paths for compression of cross-anisotropic elastic soil under isotropic stresses.

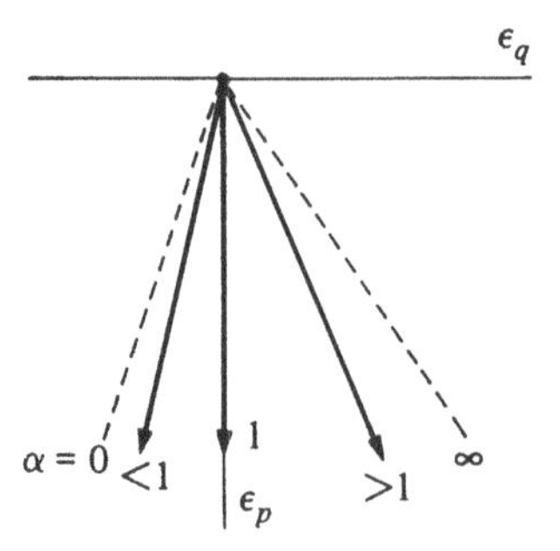

Some typical data for a natural clay from Winnipeg, Manitoba are shown in Fig. 2.9 (after Graham and Houlsby, 1983). For this clay, Graham and Houlsby estimate a value of $\nu^* = 0.2$; they quote a range of slopes of effective stress paths $\delta q/\delta p'$ between -15.8 and -4.45, and they deduce an average value for α^2 of 1.52, which is the ratio of horizontal to vertical stiffness, E_h/E_v in (2.22).

The coupling between change of size and change of shape also implies that shear strains will occur along stress paths in which q is held constant, for example, isotropic compression $q = 0$ (Fig. 2.10). For such a path, from (2.25),

$$\frac{\delta\varepsilon_q}{\delta\varepsilon_p} = \frac{-J}{3G^*} \tag{2.34}$$

or

$$\frac{\delta\varepsilon_q}{\delta\varepsilon_p} = \frac{-2(1 - \nu^* + \alpha\nu^* - \alpha^2)}{3(2 - 2\nu^* - 4\alpha\nu^* + \alpha^2)} \tag{2.35}$$

and the path of this isotropic compression will not in general lie along the $\delta\varepsilon_p$ axis ($\varepsilon_q = 0$) in the $\varepsilon_p:\varepsilon_q$ strain plane. This ratio (2.35) has limiting values $\frac{2}{3}$ and $-\frac{1}{3}$ for α very large and very small. These limiting values correspond to compression at constant radial strain ($\delta\varepsilon_r = 0$) and constant axial strain ($\delta\varepsilon_a = 0$), respectively.

Typical data for the Winnipeg clay are shown in Fig. 2.11 (after Graham and Houlsby, 1983). Graham and Houlsby quote a range of ratios $\delta\varepsilon_q/\delta\varepsilon_p$ between 0.1 and 0.32 and deduce an average value of α^2 between 1.8 and 1.9.

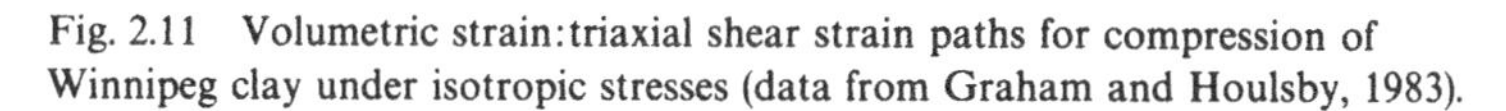

Fig. 2.11 Volumetric strain:triaxial shear strain paths for compression of Winnipeg clay under isotropic stresses (data from Graham and Houlsby, 1983).

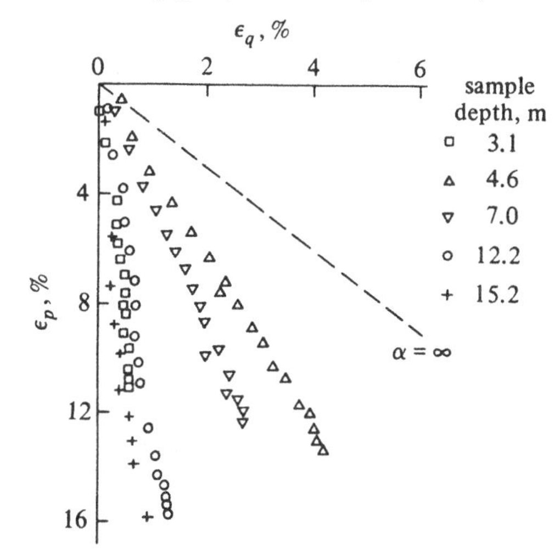

2.4 The role of elasticity in soil mechanics

Because many subsequent sections of this book are devoted to discussion of the inelastic or plastic behaviour of soils, it may be wondered what real role elasticity plays in soil mechanics. Two applications are briefly presented here: calculation of deformations of geotechnical structures under working loads and selection of stress paths to guide appropriate laboratory testing.

Soil mechanics has traditionally been concerned firstly with ensuring that geotechnical structures do not actually collapse and secondly with ensuring that the working deformations of these structures are acceptable. If it can be assumed that the soil will respond elastically to applied loads, then the whole body of elastic theory becomes available to analyse the deformations of any particular problem. Settlements of foundations and deformations of piles are frequently estimated by using charts computed from elastic analyses of more or less standard situations. Much site investigation, whether with triaxial tests in the laboratory or with pressuremeter tests or plate loading tests in the field, is devoted to determining accurate values of moduli for soils for subsequent use in deformation analyses. The results of such analyses will of course be as good as the quality of the determination of the moduli and the quality of the assumption of elasticity.

The behaviour of an isotropic elastic soil is encapsulated in just two elastic constants, which can be obtained from a simple programme of soil testing. Most soils cannot satisfactorily be described as isotropic and elastic, so a more elaborate model will be required to describe soil response. Such models are used to extrapolate from available experimental data (typically obtained under the rather restrictive stress conditions imposed in conventional laboratory tests) to the complex states of stress and strain which develop around a prototype structure. The quality of the prediction of soil response will depend on the extent of this extrapolation. If the soil response is very stress path dependent, then it is helpful if the laboratory testing can bear *some* relation to the stress paths to which soil elements in the ground around a geotechnical structure may be subjected.

This argument is circular since the stress paths which are predicted to develop in the ground depend on the details of the stress:strain response which the laboratory testing is trying to evaluate. This circle has to be broken. Elastic stress distributions are available for many loading situations (e.g. see the comprehensive collection of Poulos and Davis, 1974); these are frequently a useful starting point in assessing plausible stress changes for soil elements. Stress paths for certain simple geotechnical problems will be considered in more detail in Chapter 10.

Paradoxically, then, this major application of elasticity in soil mechanics is guiding the study of the inelastic stress:strain behaviour of soils.

Exercises

E2.1. Use Hooke's law to deduce relationships between stress increments for samples of isotropic elastic soil which are being (i) deformed in plane strain and (ii) compressed one-dimensionally.

E2.2. A conventional undrained triaxial compression test, with the cell pressure σ_r held constant, is carried out on a sample of stiff overconsolidated clay. The stress:strain relationship is found to be linear up to failure, so it is deduced that the clay behaves as an isotropic perfectly elastic material.

i. After an axial strain $\Delta\varepsilon_a = 0.9$ per cent, the deviator stress is measured to be 90 kPa. For this stage of the test, calculate the values of Δu, Δp, $\Delta p'$, $\Delta\sigma'_a$, $\Delta\sigma'_r$, $\Delta\sigma_a$, $\Delta\varepsilon_r$, and E_u.

ii. At this time, the axial stress and cell pressure are kept constant, and the sample is allowed to drain so that the pore pressures dissipate and the sample undergoes a volumetric strain $\Delta\varepsilon_p = 0.3$ per cent. What are the values of $\Delta\sigma'_a, \Delta\sigma'_r, \Delta u, \Delta p', \Delta q$, $\Delta\varepsilon_a$, and $\Delta\varepsilon_r$ for this stage of the test, and what are the values of the elastic constants K', G', E', and ν'?

E2.3. The effective stress elastic behaviour of a cross-anisotropic soil is characterised by a Young's modulus E^* for the vertical direction, a Young's modulus $\alpha^2 E^*$ for the horizontal direction, a Poisson's ratio ν^*/α indicating the strain in a horizontal direction due to a strain in a vertical direction, and a Poisson's ratio ν^* indicating the strain in a horizontal direction due to a strain in an orthogonal horizontal direction [see (2.24)].

A cuboidal specimen of this soil is compressed by a total normal stress increment $\Delta\sigma_1$ in the vertical direction, with no change in total stress in one horizontal direction, $\Delta\sigma_3 = 0$, and with no strain in the other horizontal direction, $\Delta\varepsilon_2 = 0$.

Find an expression for the pore pressure change Δu in the soil if drainage is not permitted. Check that your expression gives $\Delta u = \Delta\sigma_1/2$ for an isotropic elastic soil.

E2.4. Conventional drained and undrained triaxial compression tests are performed on samples of cross-anisotropic soil taken from the ground with their axes horizontal. Show that the cross section of the samples does not remain circular as they are compressed. Determine the slope of the effective stress path in the p':q plane for the undrained test in terms of appropriate elastic constants.

Assuming that measurements are made only of axial and volumetric strain in the usual way, determine the slope of the strain path in the ε_p:ε_q plane that would be deduced for the drained test.

Check that these expressions reduce to the correct values for isotropic elastic soil.

E2.5. Consider the strains that occur on the stress cycle $(p', q) = (p'_a, 0)$; $(p'_b, 0)$; (p'_b, q_b); (p'_a, q_b); and $(p'_a, 0)$ applied to a sample of elastic soil which has a bulk modulus dependent on mean stress ($K' = \alpha p'$). Show that if the shear modulus G' is also dependent on p', then energy can be created or lost on this closed cycle, and hence that it would not be thermodynamically admissible to assume a constant value of Poisson's ratio for this soil.

3

Plasticity and yielding

3.1 Introduction

The behaviour of an elastic material can be described by generalisations of Hooke's original statement, *ut tensio sic vis*: the stresses are uniquely determined by the strains; that is, there is a one-to-one relationship between stress and strain. Such a relationship may be linear or non-linear (Fig. 3.1), but an essential feature is that the application and removal of a stress leaves the material in pristine condition and no nett energy is dissipated.

For many materials the overall stress:strain response cannot be condensed into such a unique relationship; many states of strain can correspond to one state of stress and vice versa. For example, the first loading of an annealed copper wire in simple tension may follow a curved load:deformation path which is not retraced when the load is removed, but the wire is left with a permanent extension under zero load (AA_1B_1 in Fig. 3.2, from Taylor and Quinney, 1931). If the wire is reloaded to loads less than the previous maximum load, then an essentially elastic response is observed (B_1C_1 in Fig. 3.2, though there is a slight departure from the unloading path as the previous maximum load is approached), that is, there is a one-to-one relation between load and deformation. As soon as the previous maximum load is exceeded, the elastic description of the response ceases to apply and unloading from a higher load leaves the wire with a further permanent extension ($B_1C_1A_2B_2$ in Fig. 3.2).

In principle, the reloading of the copper wire up to and beyond the previous maximum load could be modelled with a non-linear, elastic description of the behaviour. Such a description would be of extremely limited application, however, because it would not be able to cope with the observation that subsequent unloading does not retrace the same path. Such a pattern of behaviour can, however, be described using an elastic–plastic model. The irrecoverable, permanent extensions that remain under zero load are plastic deformations and can be regarded as defining new reference states from which subsequent elastic response can be measured, provided the past maximum load is not exceeded. The departure from stiff elastic response that occurs as reloading proceeds beyond the past maximum load may be called *yielding*, and the past maximum load becomes a current *yield point* for the copper wire being loaded in simple tension.

In general, in this chapter, yielding is associated with a transition from stiff to less stiff response, and the more or less well-defined kinks in stress:strain curves that can be established as marking such transitions are termed yield points. This is a convenient, though not particularly rigorous, definition of yield.

Fig. 3.1 (a) Linear and (b) non-linear elastic stress:strain relationships.

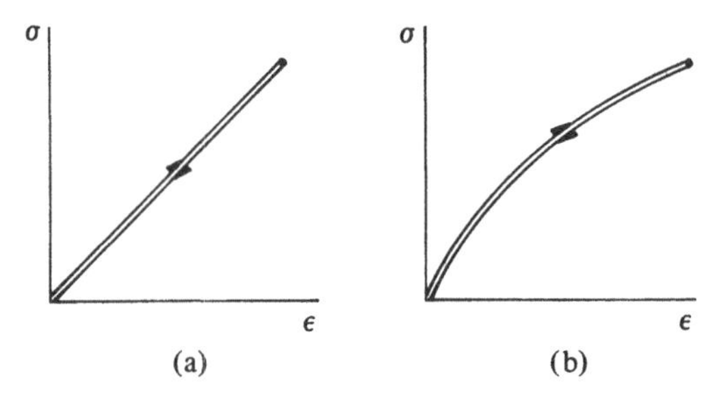

Fig. 3.2 Repeated tensile test on annealed copper wire (initial length 251.5 mm, diameter 3.23 mm) (after Taylor and Quinney, 1931).

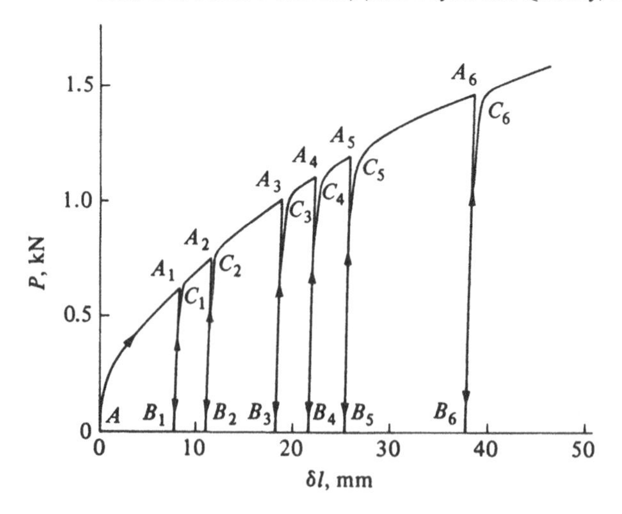

3.2 Yielding of metal tubes in combined tension and torsion

Figure 3.2 shows the developing plastic deformation and steadily increasing yield load or elastic limit for an annealed copper wire under simple uniaxial tension. We expect a thin-walled tube of the same material to show the same sort of response under pure tensile loading and unloading. Essentially similar behaviour would also be seen if the thin-walled tube were subjected to increasing cycles of loading and unloading in pure torsion instead of pure tension. It is of interest to investigate the effect of applying combinations of tension and torsion on the yielding and plastic deformation of the tube.

The data of uniaxial tension of annealed copper wire shown in Fig. 3.2 were taken from the classic paper of Taylor and Quinney (1931). They followed these simple tensile tests with combined tension and torsion tests on thin-walled tubes of annealed copper (and other materials). The schematic arrangement for their tests is shown in Fig. 3.3. Copper tubes (of external diameter 6.3 mm, internal diameter 4.5 mm, and length 292 mm) were initially loaded in tension with a load P_0. This load was then reduced to a value $P = mP_0$, and a torque Q was applied until plastic deformations were observed. The path of a typical test is plotted in a load plane (P:Q) in Fig. 3.4.

Taylor and Quinney performed tests with eight different values of m, from 0.025 to 0.95. Through the eight yield points thus established, a yield curve could be drawn defining the combinations of tension P and torsion Q for which plastic deformations would begin to occur. This is the current

Fig. 3.3 Combined tension and torsion test of thin-walled metal tubes.

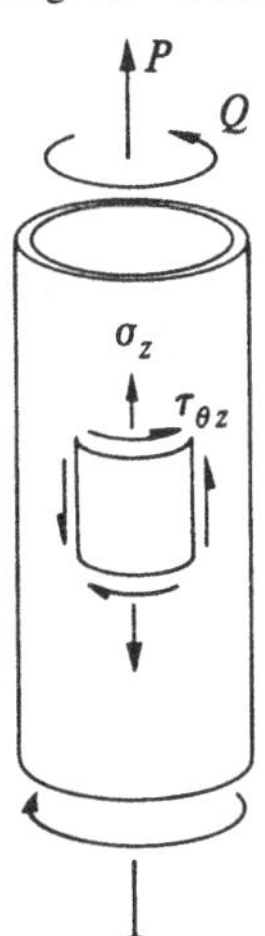

yield curve for a set of copper tubes with one particular history of preloading.

Evidently, with $m = 1$ the point A ($P = P_0$, $Q = 0$) is expected to lie on the yield curve, and a yield point B ($P = 0$, $Q = Q_1$) could in principle be found in a test with $m = 0$ (Fig. 3.5). If there were no interaction between the effects of tension and torsion, then the onset of plastic deformations would be associated with combinations of loads lying on the rectangle ACB, implying that the torque required to produce yield would be Q_1, irrespective of the value of the tension. In fact, the experimental data lie on a curve fitting inside this rectangle.

The shape of this curve can be predicted from a theoretical assumption about the yield criterion for the copper. Yield criteria are more satisfactorily discussed in terms of stress components rather than components of load P and Q. The stresses acting on an element of metal

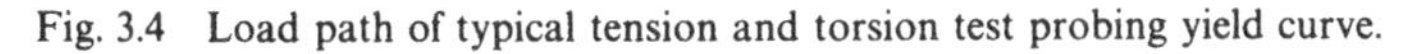
Fig. 3.4 Load path of typical tension and torsion test probing yield curve.

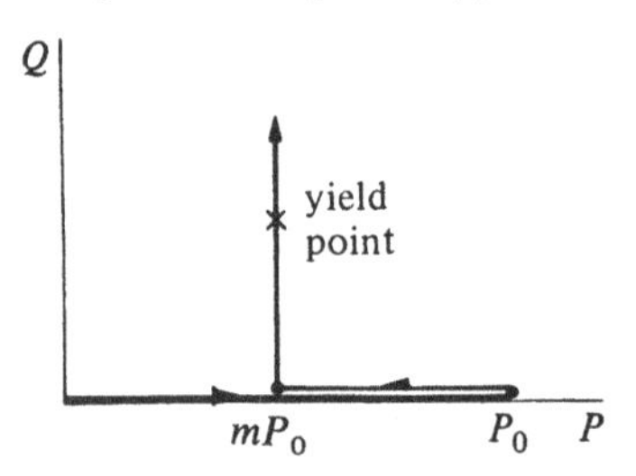

Fig. 3.5 Yield points observed in combined tension and torsion of annealed copper compared with yield curves of Tresca and von Mises (data from Taylor and Quinney, 1931).

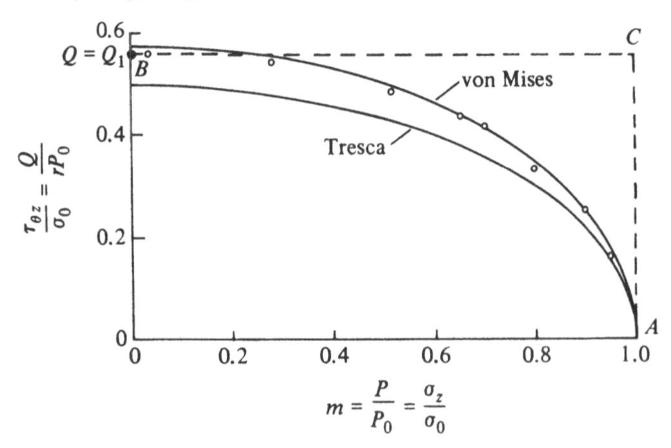

from the thin-walled tube (Fig. 3.3) are an axial tension,

$$\sigma_z = \frac{P}{2\pi r t} \tag{3.1}$$

and a shear stress on transverse and longitudinal planes,

$$\tau_{z\theta} = \tau_{\theta z} = \frac{Q}{2\pi r^2 t} \tag{3.2}$$

where r and t are the radius and wall thickness of the tube. In the absence of internal and external pressures the radial normal stress through the tube (which is a principal stress) and the tangential normal stress in the tube are both zero,

$$\sigma_r = \sigma_2 = \sigma_\theta = 0 \tag{3.3}$$

and the other two principal stresses are (Fig. 3.6)

$$\begin{matrix}\sigma_1 \\ \sigma_3\end{matrix} = \frac{\sigma_z}{2} \pm \sqrt{\left(\frac{\sigma_z}{2}\right)^2 + \tau_{\theta z}^2} \tag{3.4}$$

Part of the object of the experiments of Taylor and Quinney (1931) was to discover which of the yield criteria due to Tresca and to von Mises best fitted the data. These two yield criteria have found widespread acceptance and use in the theory of plasticity (Hill, 1950). According to Tresca (1869), yielding occurs when the maximum shear stress in the material reaches a critical value. This can be written in terms of principal stresses σ_1, σ_2, and σ_3 as

$$\max(\sigma_i - \sigma_j) = 2c \qquad (i, j = 1, 2, 3) \tag{3.5}$$

where $2c$ is the yield stress in uniaxial tension and where any of $\sigma_1, \sigma_2, \sigma_3$ may be major, minor, or intermediate principal stresses. The three principal stress $\sigma_1, \sigma_2, \sigma_3$ can be used as three orthogonal cartesian coordinate axes to define a 'principal stress space'. Equation (3.5) then describes a regular hexagonal prism in principal stress space. This prism is centred on the 'space diagonal' of principal stress space, the line on which all three principal stresses are equal, that is, $\sigma_1 = \sigma_2 = \sigma_3$ (Fig. 3.7a). The surface

Fig. 3.6 Mohr's circle of stress for an element in wall of tube.

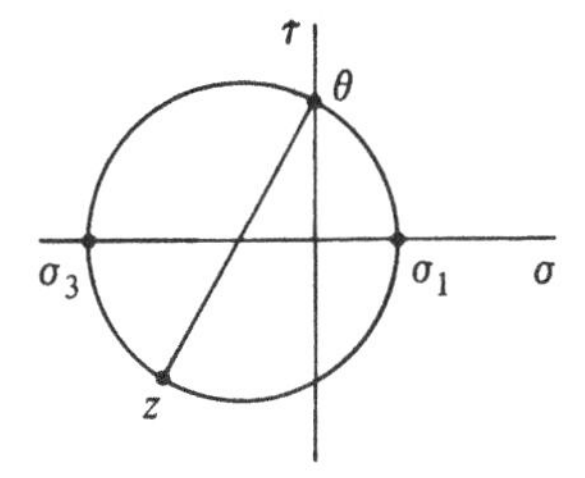

of this prism, which links combinations of principal stress at which yield occurs, can be called a *yield surface*. For the thin-walled tubes, Tresca's criterion implies

$$2\sqrt{\left(\frac{\sigma_z}{2}\right)^2 + \tau_{\theta z}^2} = 2c$$

or

$$\sigma_z^2 + 4\tau_{\theta z}^2 = 4c^2 \tag{3.6}$$

which is an ellipse in the $\sigma_z:\tau_{\theta z}$ plane (Fig. 3.5) and implies that in pure torsion ($\sigma_z = P = 0$) yielding occurs for $\tau_{\theta z} = c$.

According to von Mises (1913), yielding occurs when the second invariant of the stress tensor reaches a critical value. The statement of von Mises' yield criterion in this form can be better understood when read in conjunction with Section 1.4.1 on stress and strain variables. However, it can be more simply interpreted as implying that yielding occurs when the principal stress state reaches a critical distance from the space diagonal of principal stress space, that is, the line $\sigma_1 = \sigma_2 = \sigma_3$. It

Fig. 3.7 Yield surface according to yield criterion of (a) Tresca and (b) von Mises.

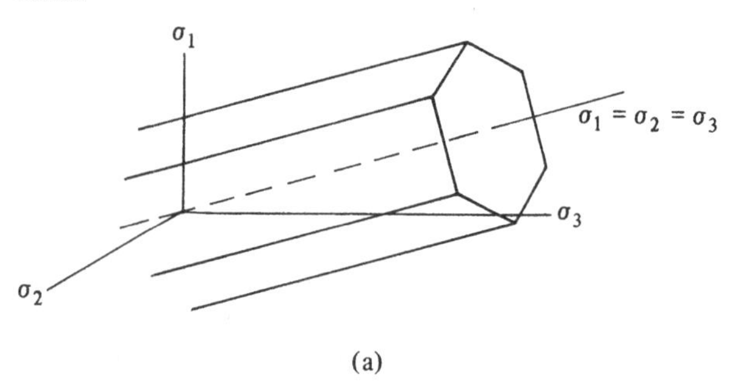

(a)

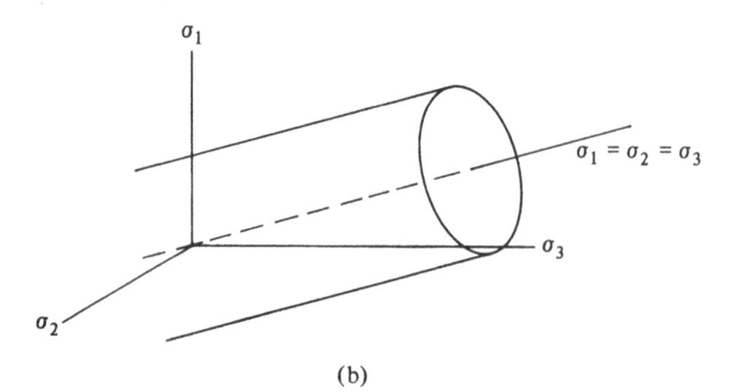

(b)

thus defines a right circular cylindrical yield surface centred on this line (Fig. 3.7b). It can be written as

$$(\sigma_2 - \sigma_3)^2 + (\sigma_3 - \sigma_1)^2 + (\sigma_1 - \sigma_2)^2 = 8c^2 \tag{3.7}$$

where $2c$ is the yield stress in uniaxial tension. The distance from the space diagonal to the principal stress state gives an indication of the magnitude of the distortional stress which is tending to change the shape of the material. Therefore, an alternative interpretation of von Mises' yield criterion is that yielding occurs when the elastic strain energy of distortion reaches a critical value.

For the thin-walled tubes under combined tension and torsion, von Mises' yield criterion becomes

$$\sigma_z^2 + 3\tau_{\theta z}^2 = 4c^2 \tag{3.8}$$

which is again an ellipse in the $\sigma_z{:}\tau_{\theta z}$ plane (Fig. 3.5), but this ellipse implies that in pure torsion ($\sigma_z = P = 0$) yielding occurs for $\tau_{\theta z} = 2c/\sqrt{3}$.

Plotting the experimental data of yielding and setting $2c = \sigma_0 = P_0/2\pi rt$, we find (Fig. 3.5) that the points lie much closer to the von Mises ellipse than to the Tresca ellipse. The same result can be seen when the data are plotted in the deviatoric view of principal stress space (Fig. 3.8) seen down the space diagonal $\sigma_1 = \sigma_2 = \sigma_3$. Positions in this deviatoric view are controlled only by differences of principal stresses, because a uniform increase in all three principal stresses merely moves a stress point along a line parallel to the space diagonal, and this movement cannot be seen in a view orthogonal to this line.

Fig. 3.8 Observed yield points in combined tension and torsion of annealed copper compared with yield curves of Tresca and von Mises in deviatoric view of principal stress space, and vectors of plastic deformation (data from Taylor and Quinney, 1931).

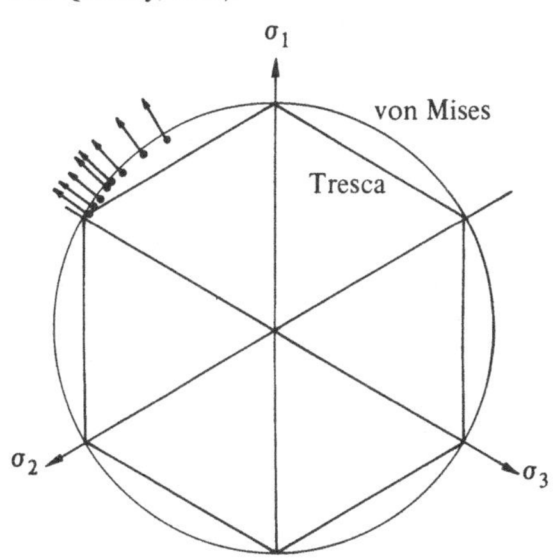

In this deviatoric view, the Tresca yield criterion becomes a regular hexagon and the von Mises criterion a circle. Because expressions (3.5) and (3.7) for the two yield criteria involve only differences of principal stresses, the size of the hexagon or circle is independent of the value of the mean stress p (compare Fig. 3.7), and the deviatoric view of either of these yield criteria is unique and not dependent on the current value of the mean stress. It is a familiar observation of metal plasticity that the presence of isotropic pressures has a negligible effect on the occurrence of yielding.

Although the experimental data show that the von Mises yield criterion provides the better description of metal plasticity, the difference between the two criteria is not particularly great, and it is often mathematically more convenient to make use of the Tresca criterion in analytical work. Indeed, the Tresca criterion is implicitly invoked in all calculations of plastic failure including calculations of bearing capacity of soils.

The loading paths that Taylor and Quinney used to probe the yield curve consisted of changes of torque with no change of tension (Fig. 3.4). They were careful to ensure that the specimens that they tested behaved closely isotropically. For a material behaving isotropically and elastically, the application of torque to a thin tube should produce twist but no change in length of the tube because this is a purely distortional process (Section 2.1).

Typical plots of shear stress $\tau_{\theta z} = Q/2\pi r^2 t$ against axial extension δl (Fig. 3.9) provide a good indication that inelastic effects are occurring; such plots were used by Taylor and Quinney to identify the yield points. As the torque is increased there is initially only twist and no change in

Fig. 3.9 Yielding detected from plots of shear stress and extension for combined tension and torsion of annealed copper tubes (after Taylor and Quinney, 1931).

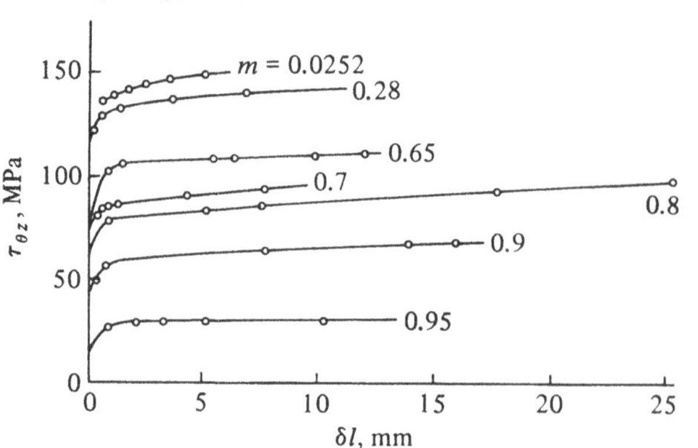

length; when yielding occurs, further twist is associated with significant extension of the tubes. The twist $\delta\theta$ plotted against the extension δl for these plastic deformations (Fig. 3.10) shows an essentially linear relationship between twist and extension for each value of $m = P/P_0$ and a clear dependence of the ratio of twist to extension on the value of m.

These deformation data can be presented on the stress diagram in which the yield data were plotted, provided they are converted to appropriate work-conjugate strain quantities. The work input to a tube per unit volume of material, when an extension δl and a twist $\delta\theta$ occur under a tension P and a torque Q, is

$$\delta W = \frac{P\,\delta l + Q\,\delta\theta}{2\pi r t l} \tag{3.9}$$

and this can be written in terms of the stress quantities σ_z and $\tau_{\theta z}$ [(3.1) and (3.2)] as

$$\delta W = \sigma_z \frac{\delta l}{l} + \tau_{\theta z} \frac{r\,\delta\theta}{l} \tag{3.10}$$

It is not surprising then that the appropriate work-conjugate strain components are the longitudinal strain $\delta l/l$ and the shear strain $r\,\delta\theta/l$.

The plastic deformation data of Fig. 3.10 can then be plotted as vectors

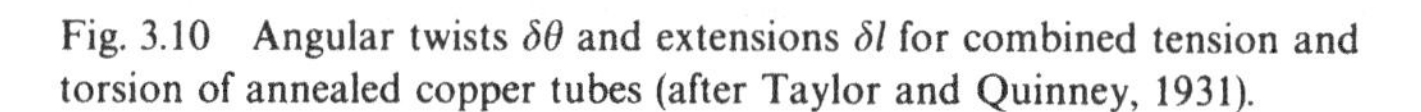
Fig. 3.10 Angular twists $\delta\theta$ and extensions δl for combined tension and torsion of annealed copper tubes (after Taylor and Quinney, 1931).

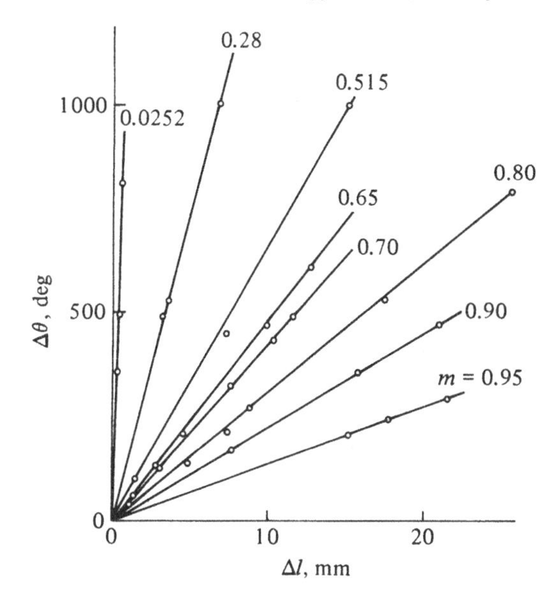

of plastic strain increment at each of the yield points as shown in Fig. 3.11. The direction of each vector indicates the relative amounts of plastic twist and extension that occur when the yield curve is reached. It is apparent that these vectors are approximately orthogonal to the von Mises ellipse which was found to fit the yield points.

In principal stress space the correct strain parameters to associate with the principal stresses are the principal strain increments (Taylor and Quinney observed that, as expected for an isotropic material, the principal axes of strain increment and of stress were coincident after yield, within their experimental accuracy). The plastic strain increment vectors are plotted in the deviatoric view of principal stress space in Fig. 3.8. The maximum deviation from the radial direction, the direction of the normal to the von Mises circle, is about 3.5°.

The link between mechanisms of plastic deformation and the yield curve that appears to have emerged in Figs. 3.11 and 3.8 seems to imply that the directions of the plastic strain increment vectors are governed not by the route through stress space that was followed to reach the yield surface, but by the particular combination of stresses at the particular point at which the yield surface was reached. This is a key feature of the behaviour of plastic materials, distinguishing them from elastic materials, and illustrated in Fig. 3.9. For elastic materials the mechanism of elastic deformation depends on the stress increments; for plastic materials which are yielding, the mechanism of plastic deformation depends on the stresses.

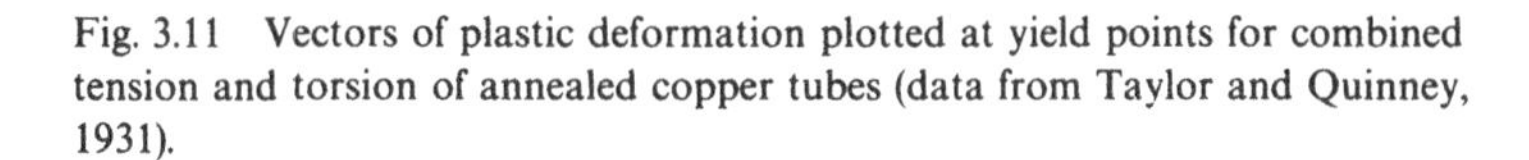

Fig. 3.11 Vectors of plastic deformation plotted at yield points for combined tension and torsion of annealed copper tubes (data from Taylor and Quinney, 1931).

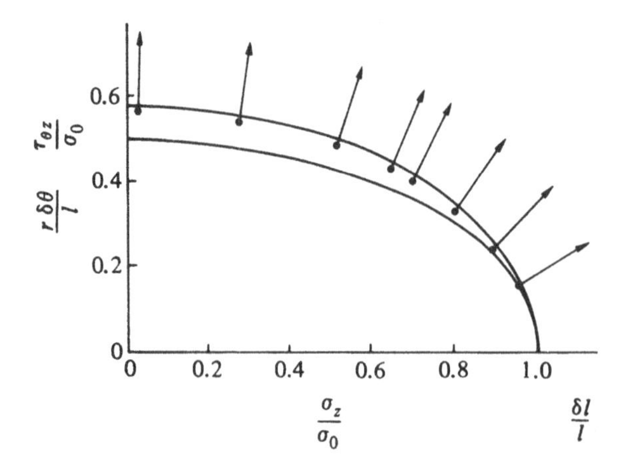

3.3 Yielding of clays

Although a single qualitative picture of the yielding of soils is being created, it is convenient to separate discussion of the yielding of clays from the yielding of sands because of the different experimental procedures that have been used to probe the yielding of these different soil types.

The discussion of the yielding of annealed copper in Section 3.2 began

Fig. 3.12 One-dimensional compression and unloading of speswhite kaolin in oedometer: (a) specific volume v and vertical effective stress σ'_v; (b) specific volume v and vertical effective stress σ_v' (logarithmic scale); (c) vertical effective stress σ'_v and specific volume v; (d) vertical effective stress σ'_v (logarithmic scale) and specific volume v (data from Al-Tabbaa, 1987).

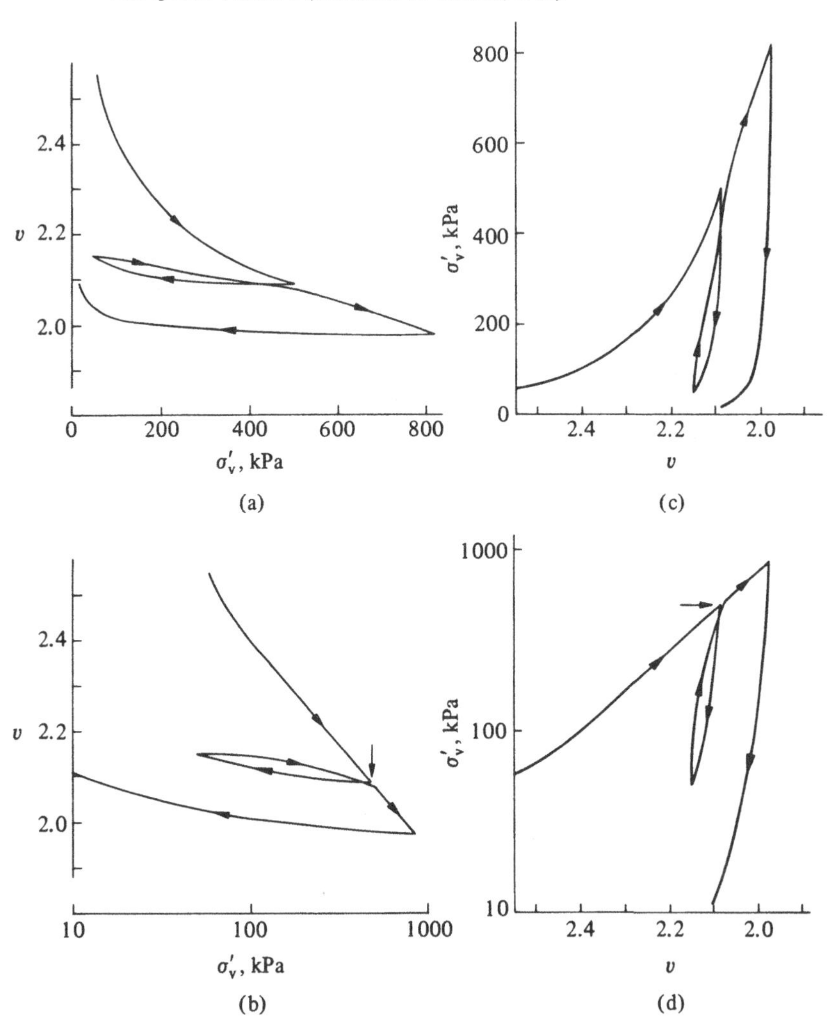

with consideration of a simple, one-degree-of-freedom, uniaxial tension test on copper wire. A simple, one-degree-of-freedom test that is familiar in geotechnical engineering is the oedometer test used to study the one-dimensional compression characteristics of soils. By convention, the results of oedometer tests are usually plotted with the height of the sample or a volumetric parameter as ordinate and the applied effective stress as abscissa, often plotted on a logarithmic scale (Figs. 3.12a, b). However, if the axes are interchanged (Figs. 3.12c, d), then similarity with the behaviour

Fig. 3.13 Spestone kaolin: (a) isotropic compression and unloading and (b) undrained triaxial compression and unloading (after Roscoe and Burland, 1968); (c) cycles of compression and unloading at constant mean effective stress p' (after Wood, 1974).

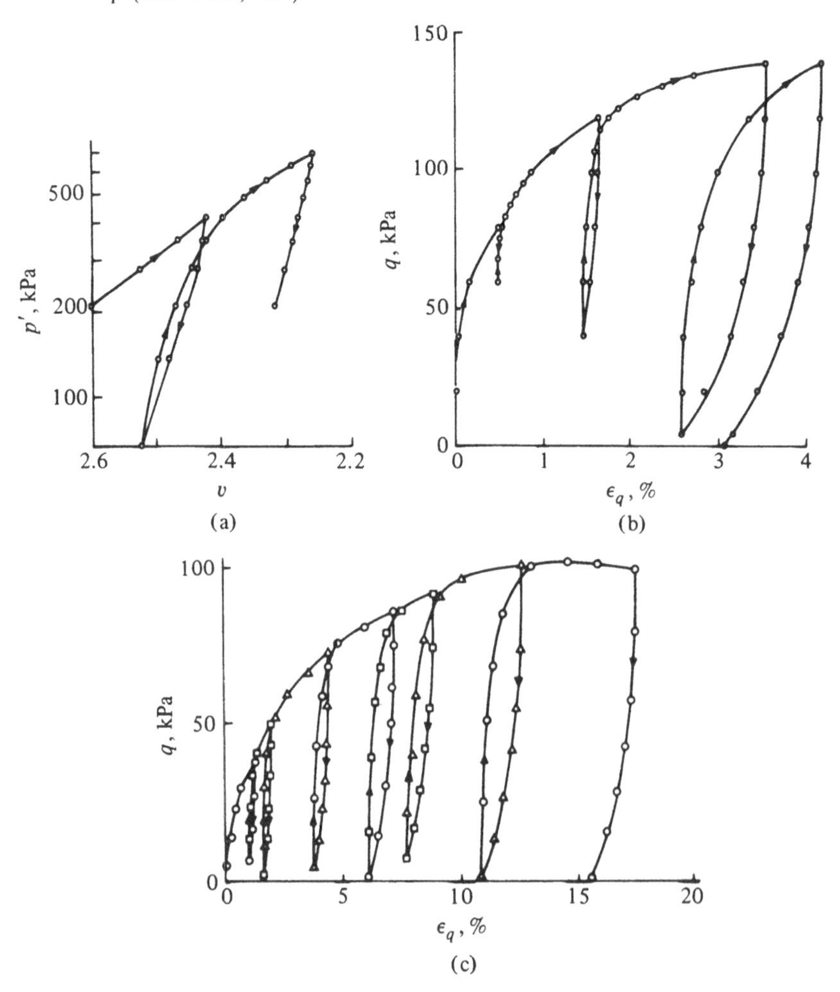

of the copper wire (Fig. 3.2) is immediately apparent. A *preconsolidation pressure* is often sought in such tests; that is, the pressure at which the stiffness of the soil in the oedometer falls rapidly, and the slope of the $v{:}\log\sigma'_v$ curve shows a sudden change (Figs. 3.12b, d). It is clear that this can be thought of as a yield point for the soil. For stresses below the preconsolidation pressure σ'_{vc}, the response in the oedometer test is stiff and essentially 'elastic'. If, after the preconsolidation pressure has been exceeded, the stresses are again reduced, then stiff elastic response can again be found. The hysteresis in these elastic regions may often be regarded as negligible particularly if the intention is to construct simple models of soil behaviour.

The preconsolidation pressure observed in oedometer tests is the most familiar example of yielding of soils, but a similar pattern can be found in isotropic compression tests (Fig. 3.13a), conventional undrained compression tests (Fig. 3.13b), or drained compression tests (Fig. 3.13c). In each case a stiff response is observed when the load is reduced below a previous maximum value, and the stiffness falls again when the load is increased beyond this past maximum value, which acts as a current yield point for the soil. Again, some hysteresis is observed in cycles of unloading and reloading. This may actually be insignificant if the excursion from the past maximum stress is not great, but even when hysteresis appears to be too large to ignore, it is usually small by comparison with the irrecoverable deformations that have already occurred in the test.

Note that plotting the results of oedometer tests or isotropic compression tests with a logarithmic stress axis (Figs. 3.12d and 3.13a) tends to mask the progressive increase in stiffness that occurs as the load is increased after yielding, as compared with the progressive decrease in stiffness that is seen after yielding in the uniaxial tension of the copper wire (compare Fig. 3.12c and Fig. 3.2). The copper wire is unconfined, and there is no change in lateral stress as deformation proceeds; the cycles of uniaxial tension apply successively higher and higher shear stresses to the material of the wire. The oedometer sample is confined laterally by a rigid ring, and the lateral stress builds up as one-dimensional compression proceeds. Though shear stresses are being imposed, the dominant effect is one of increasing general mean stress level and hence of increasing stiffness. This is, of course, the only effect in the isotropic compression test, where no shear stresses at all are applied. A pattern essentially identical to that seen for the copper wire is found in repeated *shearing* (as opposed to compression) of soil, as shown in Figs. 3.13b, c.

The tests illustrated in Figs. 3.12 and 3.13 have been thought of as independent tests akin to the uniaxial tension test on copper wire, each loading a soil sample in one particular way. The combined tension and torsion tests on copper tubes studied the response of a number of tubes, which had all been given the same preloading history, to different combinations of tension and torsion. The soil mechanics equivalent of this is a series of tests in which soil samples with the same preloading history are subjected to different modes of loading, such as one-dimensional compression, isotropic compression, and undrained shearing.

The ground provides a convenient source of soil samples with a single history of preloading. Many soil deposits have been laid down fairly uniformly over areas of fairly large lateral extent, so that a series of samples taken from the same depth can be considered essentially identical. Suppose that three such samples [(1), (2), (3)] have been set up in triaxial apparatus and are in equilibrium under the same cell pressure, so that they have effective stress state A in Fig. 3.14a. Sample (1) is subjected to isotropic

Fig. 3.14 Three tests probing yield curve for undisturbed soil samples: (a) effective stress paths in $p':q$ plane; (b) isotropic compression test (1), specific volume v and mean effective stress p'; (c) one-dimensional compression test (2), specific volume v and vertical effective stress σ'_v; (d) undrained compression test (3), deviator stress q and triaxial shear strain ε_q.

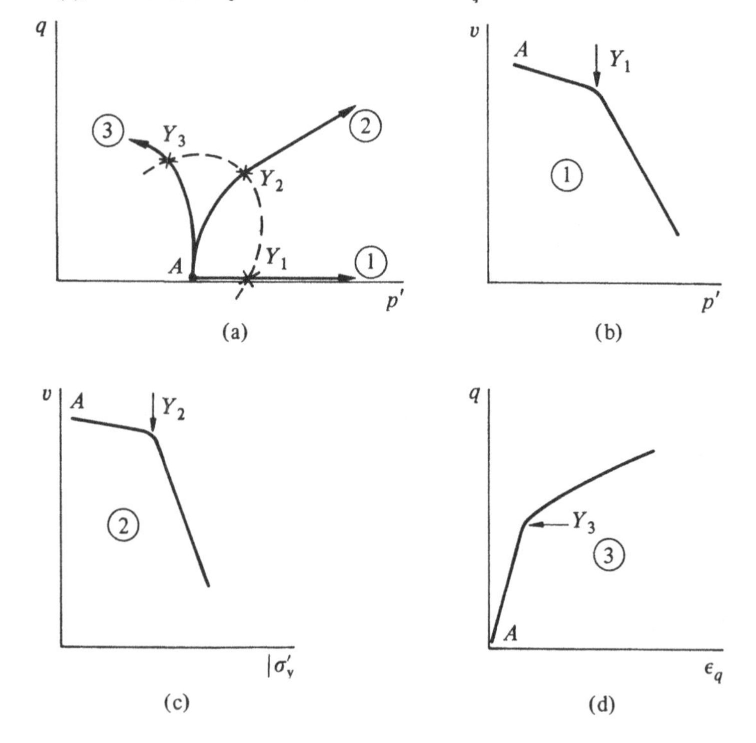

compression by increasing the cell pressure (Fig. 3.14b), and a yield point Y_1 is observed where the stiffness of the sample changes markedly. Sample (2) is subjected to one-dimensional compression (the sort of loading that could be imposed in an oedometer, though lateral stress is not usually measured in oedometer tests) by controlling the cell pressure as the axial stress is increased in such a way that lateral strain of the sample does not occur. In this way the effective stress path for one-dimensional loading can be followed and plotted in the $p':q$ effective stress plane (Fig. 3.14a). A yield point Y_2 is observed where the stiffness of the sample changes sharply (Figs. 3.14a, c). Sample (3) is subjected to a conventional undrained compression test with pore pressure measurement. The effective stress path is shown in Fig. 3.14a. Yielding is observed at Y_3, where the stiffness of the sample in a plot of deviator stress against triaxial shear strain changes sharply (Figs. 3.14a, d).

The three tests have probed in three different ways the boundary of the elastic region for the soil with one particular history. Already a yield curve could be sketched, linking the yield points observed in these tests (Fig. 3.14a). Other probing paths could be devised for the triaxial apparatus to confirm the position of this yield curve, or yield locus, in other parts of the $p':q$ effective stress plane. With testing apparatus more complex than the conventional triaxial apparatus, a series of yield points could be established forming a yield surface in principal effective stress space $(\sigma'_1:\sigma'_2:\sigma'_3)$ or in a general effective stress space $(\sigma'_{xx}:\sigma'_{yy}:\sigma'_{zz}:\tau_{yz}:\tau_{zx}:\tau_{xy})$. The yield locus in the triaxial or $p':q$ plane is just a particular section through this current yield surface bounding all elastically attainable states for the soil with one particular history. The yield surface can be regarded as a generalised preconsolidation pressure; the preconsolidation pressure observed in an oedometer test corresponds to just one point on this yield surface.

The test data that have been used in Figs. 3.12 and 3.13 to illustrate the yielding of clay have been obtained from triaxial and oedometer tests on laboratory prepared kaolin, reconstituted from powder and compressed one-dimensionally from a slurry. The phenomenon of yielding in insensitive soils such as this is often considerably less marked than the change in stiffness observed in the reloading of the annealed copper wire (Fig. 3.2). For such soils it is clear that considerable subjectivity may be involved in selecting precise yield points.

Yielding is often more readily observed in natural clays, which may have developed a certain structure over the millenia since their deposition. Probing these clays can disturb this structure: for very sensitive clays, the drop in stiffness associated with 'destructuration' may be extreme. Various strategies have been used to probe the yield surfaces of natural clays. For both the examples shown here, undisturbed samples of clay have been recompressed to a common initial effective stress state and then subjected to rosettes of stress probes.

The location of the yield locus established by Tavenas, des Rosiers, Leroueil, LaRochelle and Roy (1979) using undisturbed samples of clay from St. Louis, Canada is shown in Fig. 3.15. The initial effective stress state used in their tests corresponded to their estimate of the in situ effective stress state. On each radial path in Fig. 3.15, estimates of the yield point were obtained by plotting the experimental data in three ways.

Yielding of the copper wire or tubes, and of clay samples just discussed, has been deduced from an increase in the rate at which a strain parameter increases with continuing increase in stress. Since different loading paths generate different modes of deformation or straining, different strain variables may provide a more sensitive indication of the occurrence of yield on particular stress paths. Tavenas et al. have used plots of p' against volumetric strain ε_p, or q against axial strain ε_a (Figs. 3.16a, b) to provide alternative estimates of a yield point in Fig. 3.15.

A further estimate is possible from consideration of the energy required to deform a sample. A simple uniaxial (unconfined) loading test leads to the axial stress:axial strain curve shown in Fig. 3.17a. The work done in straining the sample can be calculated at any stage from the area underneath the stress:strain curve,

$$W = \int \sigma_a \, d\varepsilon_a \tag{3.11}$$

Fig. 3.15 Yield curve deduced from triaxial tests on undisturbed St. Louis clay (after Tavenas, des Rosiers, Leroueil, LaRochelle, and Roy, 1979).

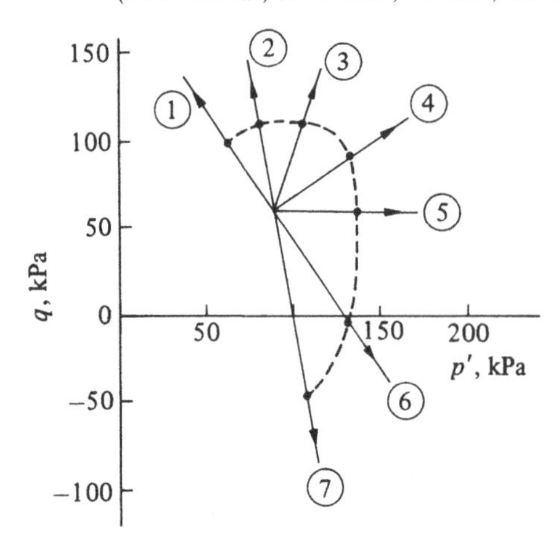

Plotting this cumulative work against the stress (Fig. 3.17b) shows that beyond the yield point B an increasing amount of energy is required to produce a given increment in stress, and a yield point could be deduced from the change in slope of this stress:work curve.

For such a one-dimensional system, the substitution of work for strain may not appear to provide much benefit. For more general states of stress

Fig. 3.16 Determination of yield points in triaxial tests on St. Louis clay: (a) mean effective stress p' and volumetric strain ε_p; (b) deviator stress q and axial strain ε_a; (c) mean effective stress p' and work input per unit volume W (after Tavenas, des Rosiers, Leroueil, LaRochelle, and Roy, 1979).

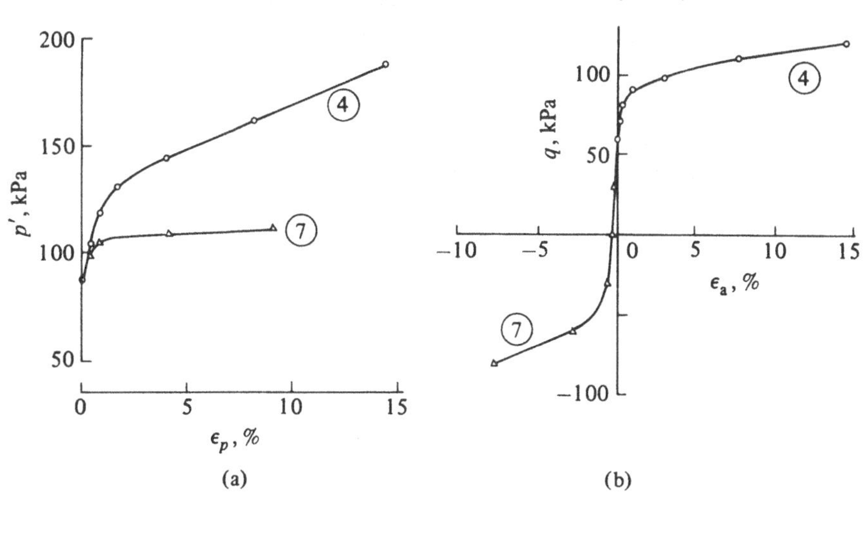

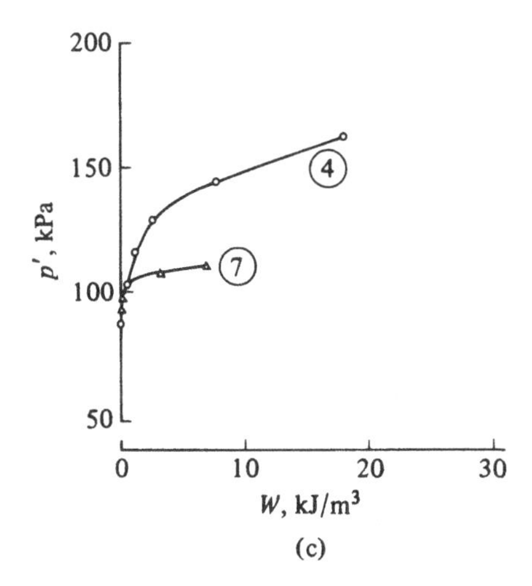

or strain the advantages can be greater. In the triaxial apparatus,

$$W = \int (p'\,d\varepsilon_p + q\,d\varepsilon_q) \tag{3.12}$$

and this expression is valid whatever changes in strain may occur – compression and distortion of the sample have both been incorporated. A third estimate of the position of the yield point was obtained by Tavenas et al. from plots of p' against W (Fig. 3.16c). The yield points deduced from these three different procedures were very similar, and the points plotted in Fig. 3.15 are the average of the three estimates.

In some of the probing tests performed by Tavenas et al., p' or q is held constant. No matter what strain or energy variable is used, if the stress axis shows an unchanging quantity, then there is no possibility of detecting a yield point as a kink in the curve. In general, the yielding can be sought in triaxial tests on a variety of paths in the $p'{:}q$ plane, on some of which any one of the simple effective stress variables $\sigma'_a, \sigma'_r, p', q$, or $\eta = q/p'$ may be constant (Fig. 3.18). No single plot is likely to be suitable for detection

Fig. 3.17 (a) Work W as area underneath stress:strain curve; (b) yielding deduced from variation of work done with applied stress.

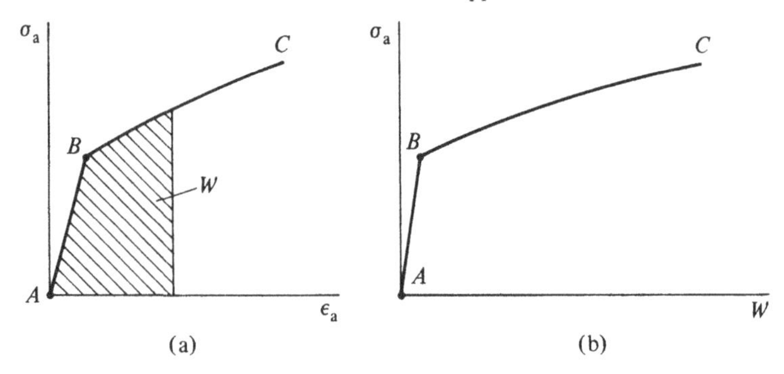

Fig. 3.18 Paths in effective stress plane on which different stress variables remain constant.

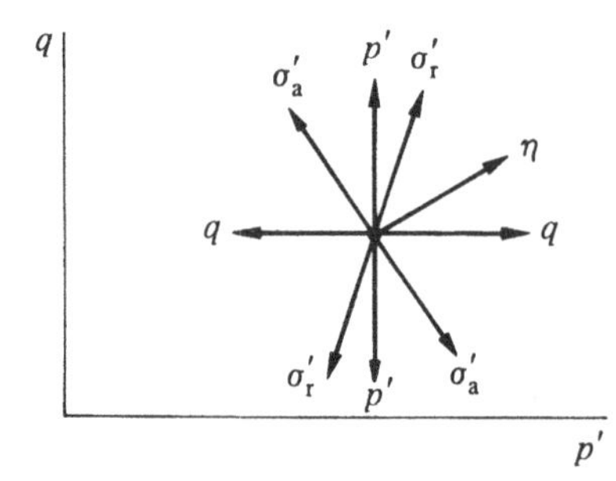

of yielding in all tests. Indeed, since yield points in tests on soils tend to be rather less marked than those seen in repeated tension, or combined tension and torsion of annealed copper, the best approach may be to use as many different plots as possible so that a number of independent estimates of yield points can be made.

There is, however, one derived plot which Graham, Noonan, and Lew (1983) have shown can be used in general: this uses the cumulative work input (3.12) as a quantity which incorporates all components of strain increment and uses as the stress variable a scalar quantity, the length s of the stress path (Fig. 3.19), where

$$\delta s = \sqrt{\delta p'^2 + \delta q^2} \tag{3.13}$$

This is a stress variable which increases monotonically whatever the direction of the stress probe.

In general, even this plot should not be used in isolation, but in combination with other possibilities, as shown in Fig. 3.20 adapted from Graham, Noonan, and Lew (1983). Here yield points have been sought in plots of $\sigma'_a:\varepsilon_a$, $\sigma'_r:\varepsilon_r$, $p':\varepsilon_p$, $q:\varepsilon_q$, and $s:W$ for an anisotropic compression probe on an undisturbed sample of Winnipeg clay, with q/p' constant at 0.46. The yield points obtained from a series of plots such as these are similar but by no means identical.

Graham and his co-workers have performed a large number of probing triaxial tests on Winnipeg clay in order to discover the shapes of the yield loci that are appropriate to samples at different depths in the soil deposit. Samples of soil taken from different depths (Fig. 3.21a) are expected to have different yield surfaces. However, the past history of loading of soil elements at various depths can be expected to be similar – for example, one-dimensional compression and unloading and possible secondary effects such as cementation or ageing. So it is reasonable to suppose that the general shape of the yield surface is the same for all depths and that

Fig. 3.19 Length of stress probe s.

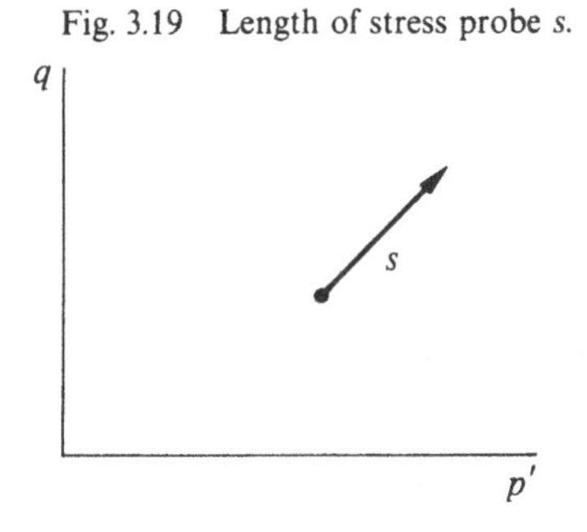

Fig. 3.20 Determination of yield points in triaxial tests on sample of undisturbed Winnipeg clay: (a) axial effective stress σ'_a and axial strain ε_a; (b) mean effective stress p' and volumetric strain ε_p; (c) deviator stress q and triaxial shear strain ε_q; (d) radial effective stress σ'_r and radial strain ε_r; (e) length of stress path s, (3.13) and Fig. 3.19, and work input per unit volume W (after Graham, Noonan, and Lew, 1983).

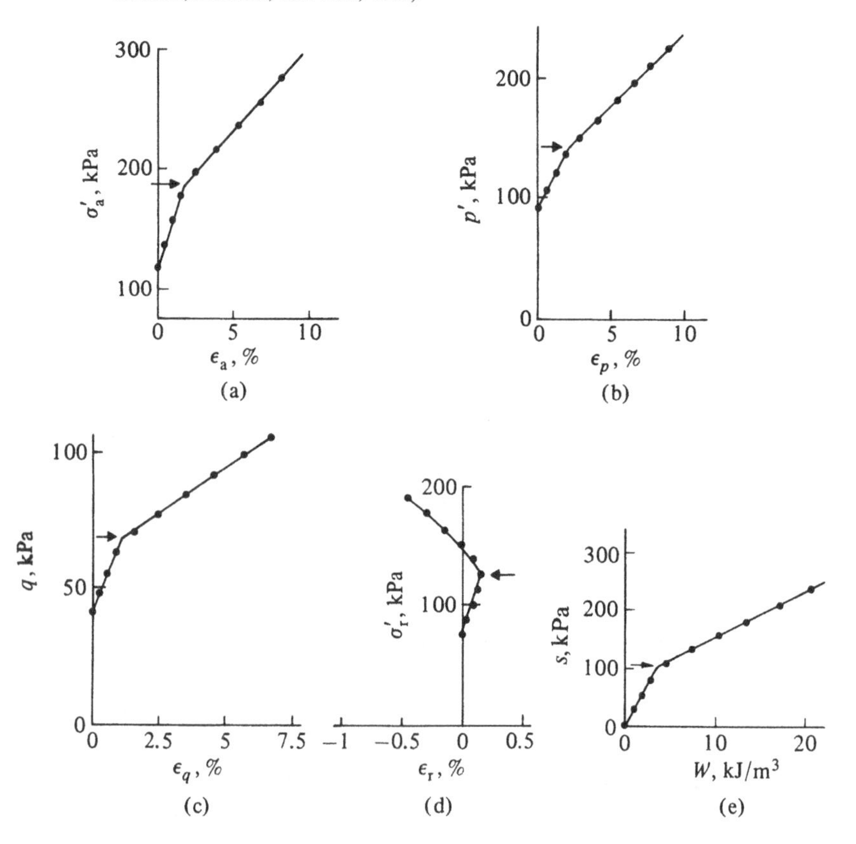

Fig. 3.21 (a) Soil elements at different depths; (b) yield curve and line $\sigma'_a = \sigma'_{vc}$.

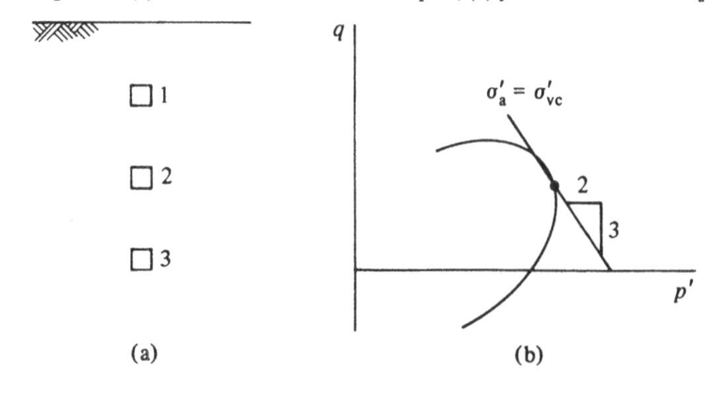

the size may vary because the past stresses are expected to have been greater at greater depths.

A simple indicator of the size of the yield surface at any particular depth is provided by the preconsolidation pressure σ'_{vc}, which, as previously mentioned, is the yield point observed in an oedometer or one-dimensional compression test. The vertical effective stress measured in a conventional oedometer test does not provide sufficient information to plot an effective stress state in the $p':q$ plane because the value of the horizontal stress is not known. Knowledge of a preconsolidation pressure σ'_{vc} thus only defines

Fig. 3.22 Yield curves deduced from triaxial tests on samples of undisturbed Winnipeg clay taken from four different depths (after Graham, Noonan, and Lew, 1983): (a) yield curves in $p':q$ effective stress plane; (b) yield curves normalised with preconsolidation pressure σ'_{vc}.

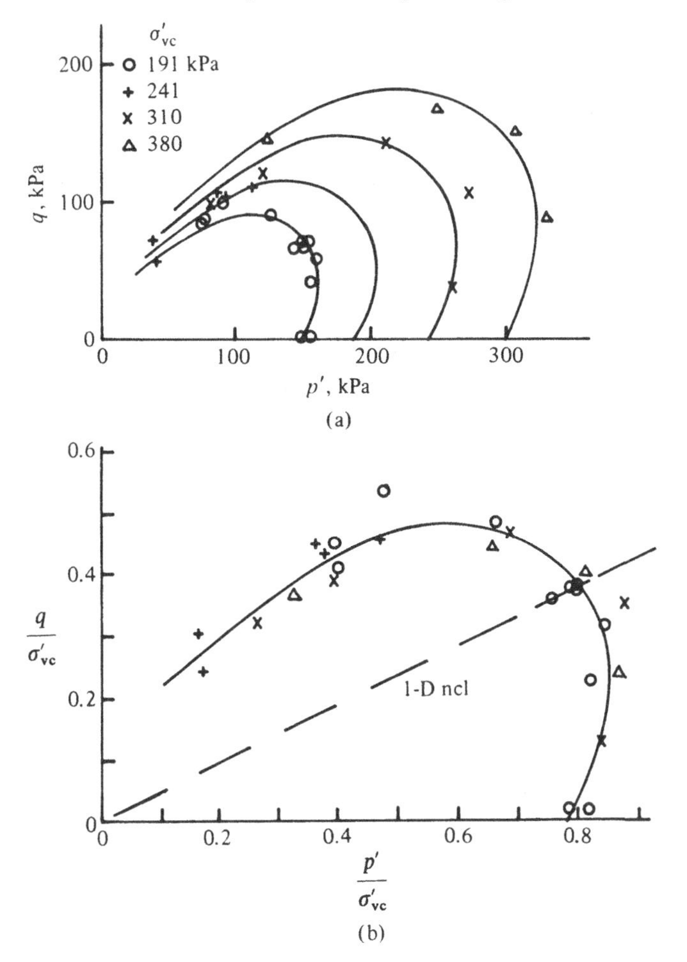

a line in the $p':q$ plane (Fig. 3.21b),

$$\sigma'_{a} = \sigma'_{vc} \tag{3.14}$$

or

$$p' + \frac{2q}{3} = \sigma'_{vc} \tag{3.15}$$

but the changing position of this line, which cuts the yield locus at the effective stress state corresponding to yielding in the oedometer, is sufficient to indicate the changing size of the yield locus in the $p':q$ plane.

Yield loci obtained by Graham, Noonan, and Lew (1983) for samples of Winnipeg clay from different depths are shown in Fig. 3.22a in the $p':q$ plane, and again in Fig. 3.22b normalised with respect to the appropriate value of preconsolidation pressure. A single non-dimensionalised yield curve can be sketched through the data points of Fig. 3.22b with reasonable confidence.

The examples shown here are concerned with Canadian clays. However, there is nothing exclusively transatlantic about the yielding of clays, and other studies of the yielding of natural clays include Larsson (1981), clay from Bäckebol, Sweden; Bell (1977), clay from Belfast, Northern Ireland; Berre (1975) and Ramanatha Iyer (1975), clay from Drammen, Norway; as well as Wong and Mitchell (1975), clay from Ottawa, Canada.

3.4 Yielding of sands

Samples of clay can be taken from the ground with relatively minor disturbance to the samples. A yield locus deduced from probing tests in the triaxial apparatus on a series of field samples should represent correctly the current yield locus for the clay in its in situ condition at a particular depth in the ground. Sampling sand, unless it is strongly cemented, inevitably leads to serious disturbance of the particle structure. It is not usually feasible to reestablish field conditions of particle arrangements in the triaxial apparatus, and a more fundamental route has to be taken to study the yielding of sands.

The determination of the entire shape of a current yield surface for a given soil sample is not feasible. The detection of yielding requires that the stress path that is being used to probe the yield surface be taken well beyond the yield point. As soon as the yield point has been passed, the yield surface starts to change. It is axiomatic that a stress state can lie on or inside but never outside a current yield surface. The passing of the yield point requires the current yield surface to change size and possibly shape to accommodate the new current stress state. Any subsequent probing investigates the shape, not of the original yield surface but of this new current yield surface, which may be called a subsequent yield surface. Thus, in the triaxial stress plane $p':q$, a stress path AB (Fig. 3.23a) might be used to discover one point Y on the current yield locus yl 1. The yield point might be detected in a plot of deviator stress q against triaxial shear strain ε_q (Fig. 3.23b), but by the time enough of the stress:strain curve has been recorded to confirm the position of the yield point Y, the current stress state is at B, and a new yield locus yl 2 must exist passing through B.

A major series of triaxial tests investigating the yielding of sand is reported by Tatsuoka (1972) and summarised by Tatsuoka and Ishihara (1974a). Tatsuoka subjected individual samples of Fuji River sand to elaborate triaxial stress paths in order to locate the position of small segments of developing yield loci for this sand. His procedure is illustrated in Fig. 3.24. The applied stress paths consisted of sections at constant cell pressure ($\delta q/\delta p' = 3$) and sections at constant deviator stress ($\delta q = 0$). A typical path might consist of isotropic compression from O to A followed by conventional, constant cell pressure compression from A to B. The yielding of the sand is now governed by a yield locus passing through B, and its local form is investigated with the stress path $BCDE$. On this path

Fig. 3.23 Expansion of yield locus from yl 1 to yl 2 on path AB.

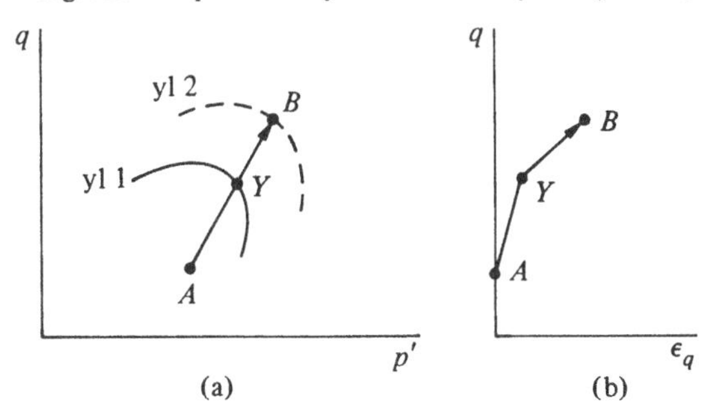

Fig. 3.24 Stress path to probe segment BY of yield locus passing through B.

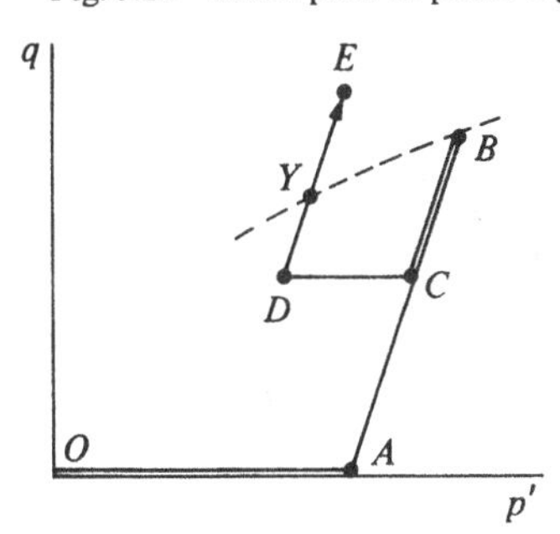

the deviator stress is reduced at constant cell pressure (B to C), then the cell pressure is reduced at constant deviator stress (C to D), and finally, the deviator stress is increased again at constant cell pressure (D to E). The stress:strain response on section DE is studied to establish the position of a yield point Y and thus deduce the local shape BY of a segment of the yield locus through B.

Evidently, a number of stress paths such as $BCDE$ can be combined to deduce the positions of a series of segments of developing yield loci. A typical complete test path applied to an initially dense sample is shown in Fig. 3.25a, and the corresponding stress ratio ($\eta = q/p'$): strain curves are shown in Figs. 3.25b, c. A series of segments of yield curves are marked on Fig. 3.25a, and all the yield segments for sand of this density are shown in Fig. 3.25d. This last figure contains sufficient experimental observations so that the shape of the developing yield curves can be sketched and a mathematical description of a general yield curve can be generated. Such a mathematical description contains a size parameter, equivalent to preconsolidation pressure, which can be used to normalise the segments in Fig. 3.25d onto a single curve.

Tatsuoka was primarily interested in probing the positions of yield loci with paths on which the deviator stress q was increasing and on which the ratio $\eta = q/p'$ was being steadily increased. Plastic deformations are

Fig. 3.25 Yielding in triaxial probing tests on dense Fuji River sand: (a) stress path in p':q plane used to examine yielding (●); (b) yielding observed in plot of stress ratio q/p' and volumetric strain ε_p; (c) yielding observed in plot of stress ratio q/p' and triaxial shear strain ε_q; (d) segments of yield curves in p':q plane (after Tatsuoka, 1972).

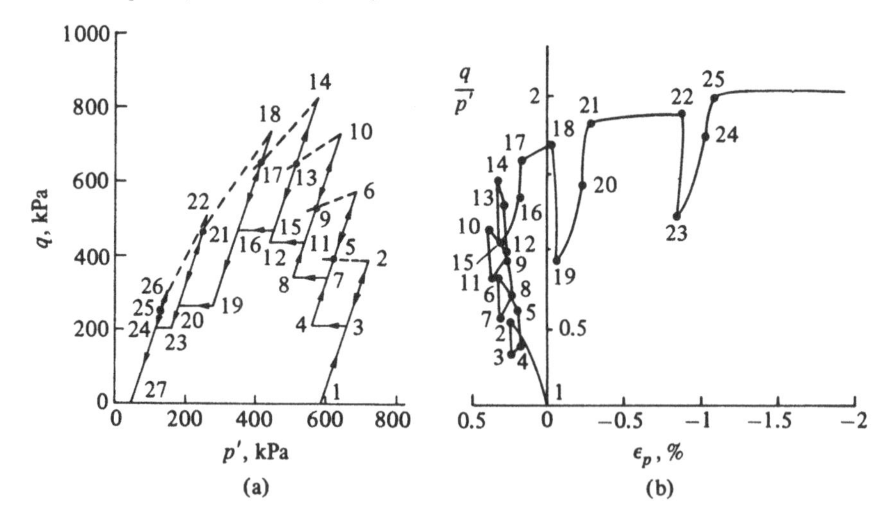

also a feature of paths in which q or η is held constant (as shown earlier by the work of El-Sohby, 1969, and others). A particular example is provided by isotropic compression: isotropic compression and unloading of Fuji River sand is shown in Fig. 3.26. The response is qualitatively the same as that seen earlier in isotropic compression and unloading of clays (Fig. 3.13a). It is to be expected, then, that the yield loci will in fact be closed in the direction of increasing p', as suggested by the dotted curves in Fig. 3.25d, though Tatsuoka's own data are rather sparse in this respect. Data of the yielding of another Japanese sand in triaxial compression and extension, presented by Miura, Murata, and Yasufuku (1984), go a little further towards the closure of the yield locus across the mean effective stress axis.

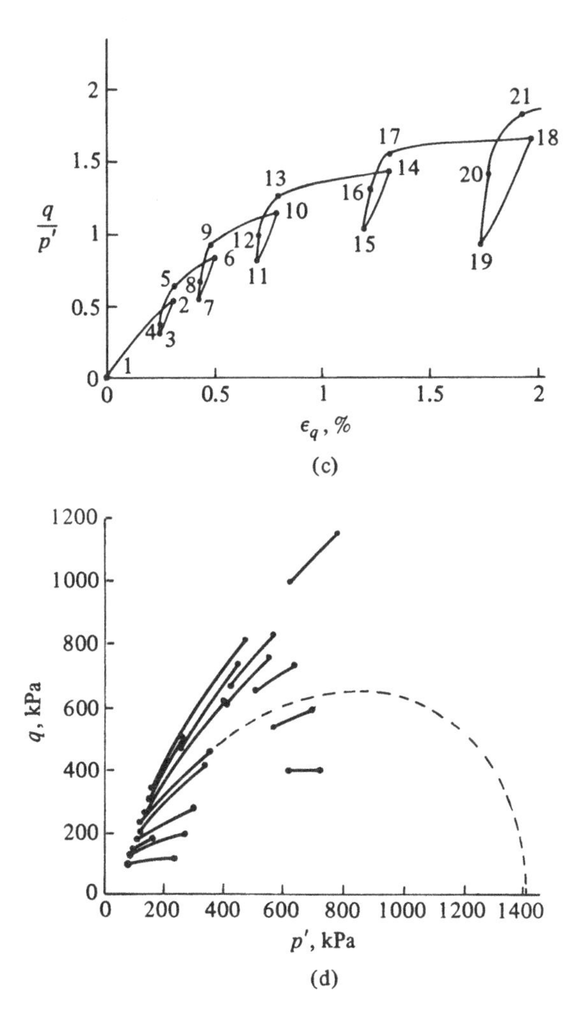

In an earlier set of tests reported by Poorooshasb, Holubec, and Sherbourne (1967), a procedure similar to that of Tatsuoka was used to study the yielding of Ottawa sand. Whereas the yield segments found by Tatsuoka (Fig. 3.25d) show some curvature with increasing mean effective stress, Poorooshasb et al. suggested that an approximate description of the yielding of their sand could be obtained by assuming that the yield loci were lines of constant stress ratio $q/p' = \eta =$ constant. Here, too, some extra statement is needed to describe the yielding that occurs under increase of mean effective stress. The sand data seem to be supporting a slightly different picture of yielding from that seen for clays. For sands, the dominant effect leading to irrecoverable changes in particle arrangement is the stress ratio or mobilised friction. High mean stress levels are required to produce significant irrecoverable deformations in purely isotropic compression: this can be related to the hard, somewhat rotund shape of typical sand particles. The quite different character of clay particles and their interactions leads to much more significant

Fig. 3.26 Isotropic compression and unloading of loose and dense Fuji River sand (after Tatsuoka, 1972).

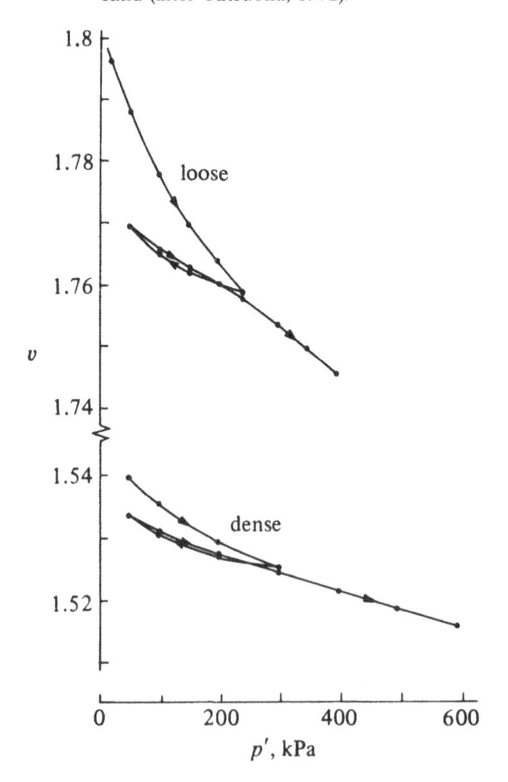

irrecoverable deformations under isotropic loading at the mean stress levels encountered in routine geotechnical structures.

3.5 Yielding of metals and soils

This chapter shows that the concepts of yielding and development of irrecoverable plastic strains are just as relevant for soils as for metals. Just as combinations of tension and torsion were applied by Taylor and Quinney to thin metal tubes to obtain information about the shape of the current yield locus or yield surface for different metals, so combinations of mean normal stress and deviator stress can be applied to samples of soil in the triaxial apparatus to obtain information about the shape of the current yield locus or yield surface for the soil. There is no difference in principle between these results: in either case, changes of stress which remain inside the current yield surface are associated with stiff response and essentially recoverable deformations, whereas changes of stress that push through the current yield surface are associated with less stiff response and the development of irrecoverable deformations.

Of course, important differences exist between the yielding of soils and the yielding of metals. One has already been noted: yield points seen in tests on soils are in general less marked than yield points that are commonly observed in tests on metals. As a result, yield surfaces for soils are rather less precisely defined than yield surfaces for metals.

There is a major difference between the shapes of yield surfaces for metals and those for soils. The tests of Taylor and Quinney were used to investigate whether the Tresca or the von Mises yield criterion was more relevant for metals. The yield surfaces that emerge from the application of these two yield criteria differ only in their shape in the deviatoric plane (Figs. 3.7 and 3.8) where, with p' constant, the Tresca yield surface has a hexagonal section, and the von Mises yield surface has a circular section. Both yield criteria are written in terms of differences of principal stresses; absolute values are of no significance, and the size of both yield surfaces is independent of the mean normal stress p'. It is a well-known experimental observation that the presence of mean stress p' has no effect on the yielding of metals. In a diagram such as Fig. 3.22b, the yield criteria of both Tresca and von Mises would plot as straight lines, $q/\sigma'_{vc} = \text{constant}$.

For soils, on the other hand, the nature of the deviatoric sections of the yield surface has yet to be explored, but the mean normal effective stress p' is of primary importance. Both for sands and for clays the deviatoric size of the yield surface, that is, the range of values of q for stiff elastic response, is markedly dependent on p'. Indeed, particularly for clays, yielding occurs with increase of p' even in the complete absence of deviator stress. For comparison with Figs. 3.7 and 3.8, Fig. 3.27 shows, in three-dimensional principal effective stress space, a view of the yield surface proposed by Clausen, Graham, and Wood (1984) for a natural clay from Mastemyr, Norway.

This marked difference of shape of yield surface should, however, be regarded as a difference in detail, and not a difference in principle, between the plastic behaviour of metals and soils. Though soils are in many ways more complex materials than metals to describe with numerical models, we will show in subsequent chapters that it is perfectly possible to develop simple elastic–plastic models for soils (helped by knowledge of the elastic–plastic behaviour of metals) which fit readily into the same general theoretical framework.

Exercises

E3.1. A block of material which yields according to Tresca's yield criterion, with yield stress $2c$ in uniaxial tension, is subjected to an initial principal stress state $(\sigma_1, \sigma_2, \sigma_3) = (3c/4, 3c/8, 3c/8)$. The stresses are then changed steadily, with fixed ratios of stress increments. Calculate the stress states at yield if the stress increment

Fig. 3.27 Yield surface in principal effective stress space for undisturbed Mastemyr clay (proposed by Clausen, Graham, and Wood, 1984).

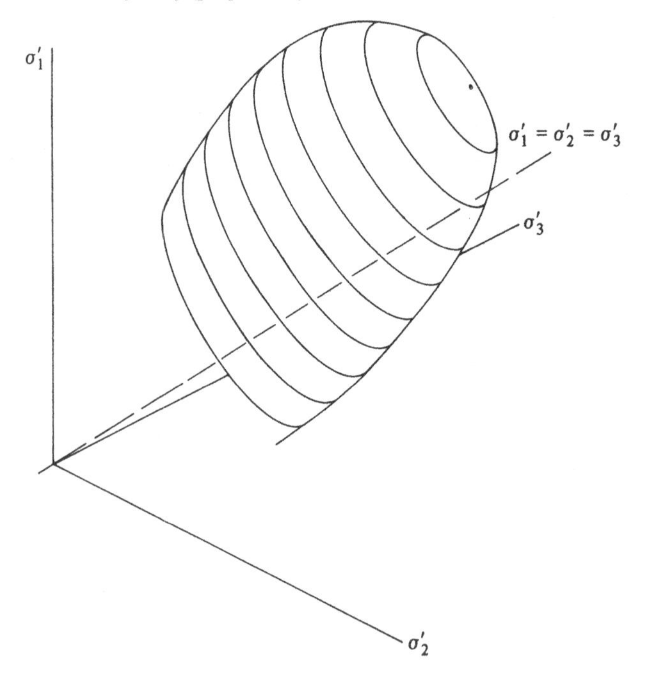

ratios $(\delta\sigma_1, \delta\sigma_2, \delta\sigma_3)$ are (a) $(1, -\frac{1}{2}, -\frac{1}{2})$; (b) $(-\frac{1}{2}, 1, -\frac{1}{2})$; (c) $(-\frac{1}{2}, -\frac{1}{2}, 1)$; (d) $(0, 1, -1)$; (e) $(-1, 0, 1)$; (f) $(1, -1, 0)$.

E3.2. Repeat E3.1 for initial principal stress states $(\sigma_1, \sigma_2, \sigma_3)$ equal to (a) $(5c/4, c/8, c/8)$ and (b) $(c/6, 2c/3, 2c/3)$, and note the different regions of principal stress space in which the yielding occurs.

E3.3. Repeat E3.1 for a material which yields according to the yield criterion of von Mises, with yield stress $2c$ in uniaxial tension.

E3.4. A block of material which yields according to the yield criterion of von Mises, with yield stress $2c$ in uniaxial tension, and which behaves isotropically and elastically before yield, is subjected to an initial principal stress state $(\sigma_1, \sigma_2, \sigma_3) = (2c/3, c/3, c/3)$, with the principal axes 1, 2, 3 coincident with Cartesian reference axes x, y, z. The block is subjected to a simple shear test, with shear stresses τ_{xz} being increased from zero in the $x{:}z$ plane without change in the normal stresses σ_x and σ_z, and with no strain in the y direction. Calculate the value of τ_{xz} at which yield occurs, and calculate the corresponding set of principal stresses and the directions of the principal axes referred to the x, y, z reference axes.

E3.5. A sample of Winnipeg clay from a depth of 11 metres is set up in a triaxial cell and allowed to come into equilibrium under effective stresses $(p', q) = (150, 0)$ kilopascals (kPa). At this depth, the preconsolidation pressure is found from oedometer tests to be $\sigma'_{vc} = 250$ kPa. Use the data of yielding of Winnipeg clay shown in Fig. 3.22 to estimate the effective stresses at which the yield locus is reached in a conventional drained compression test.

4

Elastic–plastic model for soil

4.1 Introduction

In this chapter we build a general but simple elastic–plastic model of soil behaviour, starting with the experimental observation of the existence of yield loci that was discussed in Chapter 3. Other features are added as necessary, and their selection is aided sometimes by our knowledge of well-known characteristics of soil response and at other times by knowledge of the elastic–plastic behaviour of metals.

Broadly, having established that yield surfaces exist for soils, it follows that, for stress changes inside a current yield surface, the response is elastic. As soon as a stress change engages a current yield surface, a combination of elastic and plastic responses occurs. It is necessary to decide on the nature of the plastic deformations: the magnitudes and relative magnitudes of various components of plastic deformation and the link between these magnitudes and the changing size of the yield surface.

It must be emphasised again that we are attempting to produce a simple broad-brush description of soil modelling which cannot hope to match all aspects of soil behaviour. Some of the shortcomings of such models are discussed in Chapter 12. For convenience of presentation, the discussion is largely restricted to combinations of stress and strain that can be applied in the triaxial apparatus, and the model is described in terms of triaxial stress variables p' and q and strain variables ε_p and ε_q. For convenience, it is assumed that changes in the size of the current yield locus are related to changes in volume, which permits the compression and shearing of clays to be brought simply into a single picture and leads to a class of what can be called volumetric hardening models. The possibility that changes in size of yield loci are related to distortional as well as volumetric effects is included in Section 4.5.

4.2 Elastic volumetric strains

A yield surface marks the boundary of the region of elastically attainable states of stress. Changes of stress within the yield surface are accompanied by purely elastic or recoverable deformations. The relationship between strain increments and stress increments can be written if the elastic properties of the soil are known. It might for convenience be assumed that the soil behaves isotropically and elastically within the yield surface; then the elastic stress:strain relationship becomes (from Section 2.2)

$$\begin{bmatrix} \delta\varepsilon_p \\ \delta\varepsilon_q \end{bmatrix} = \begin{bmatrix} 1/K' & 0 \\ 0 & 1/3G' \end{bmatrix} \begin{bmatrix} \delta p' \\ \delta q \end{bmatrix} \tag{4.1}$$

and recoverable changes in volume are associated only with changes in mean effective stress p'. (There would be no difficulty in incorporating an anisotropic elastic description of soil response within the yield surface, but it would make the presentation of this discussion rather less clear since the possibility of elastic volume changes accompanying changes of

Fig. 4.1 Normal compression line (ncl), yield locus (yl), and associated unloading–reloading line (url).

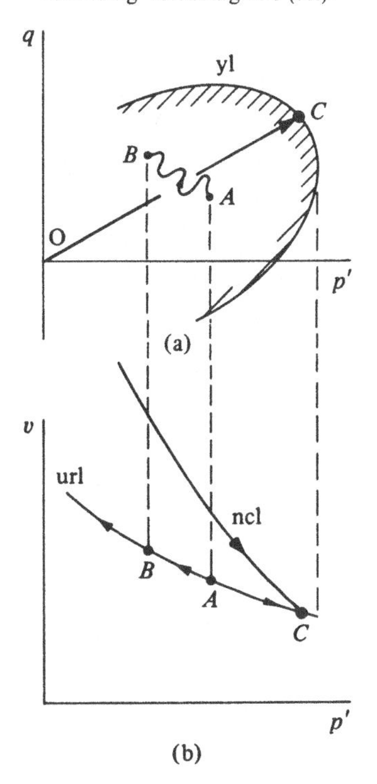

deviator stress q at constant mean effective stress p' would have to be admitted.)

Suppose that a particular soil sample has the yield locus (yl) in the $p':q$ plane shown in Fig. 4.1a. The specific volume v of this sample could be determined for some effective stress state such as A within the yield locus, and a point corresponding to A could be plotted in the compression plane $p':v$ (Fig. 4.1b). A change in stress which involves a change in mean stress p', such as from A to B in Fig. 4.1a, leads to a change in volume, from (4.1). A new point B can then be plotted in the compression plane (Fig. 4.1b). Because the response is elastic, the route taken in the stress plane from A to B is immaterial. As all stress states within the yield locus are visited, a series of points in the compression plane is obtained, forming a single unloading–reloading line (url) (Fig. 4.1b), which represents elastically attainable combinations of specific volume v and mean effective stress p'.

The position, shape, and size of the yield locus for the soil shown in Fig. 4.1a have resulted from the past history of loading of the soil. A likely history could be one-dimensional compression (and unloading). The stress path associated with one-dimensional or other anisotropic normal compression is a straight line such as OC in the $p':q$ stress plane (Fig. 4.1a), and the yield locus (yl) passes through the point C of maximum compression. The combinations of specific volume v and mean effective stress p' at various stages of normal compression form a normal compression line (ncl) to point C in the compression plane (Fig. 4.1b).

The two statements concerning the elastic behaviour within the yield locus and the history of normal compression which created the yield locus are combined in Fig. 4.1 to illustrate the predominantly irrecoverable and plastic nature of the volume changes occurring during normal compression, while the yield locus is being pushed out to its present position.

The compression plane diagram of Fig. 4.1b has been drawn with a linear scale for the mean stress p' axis. It is often found that the linearity of normal compression lines and unloading–reloading lines in the compression plane is improved if data are plotted with a logarithmic scale for the mean stress axis (Fig. 4.2). The equation for the normal compression line (ncl) then takes the form

$$v = v_\lambda - \lambda \ln p' \tag{4.2}$$

and the equation of the unloading–reloading line (url) takes the form

$$v = v_\kappa - \kappa \ln p' \tag{4.3}$$

where λ and κ are the slopes of the two lines and v_λ and v_κ the intercepts on the lines at $p' = 1$. Evidently (but unfortunately), the values of v_λ and

v_κ depend on the units chosen for the measurement of stress. Throughout this book it is assumed that the unit of stress is 1 kilopascal (kPa).

Equations (4.2) and (4.3) have been written in terms of natural logarithms because natural logarithms emerge automatically from mathematical manipulations. Results of oedometer tests are often plotted on a semilogarithmic basis, with a logarithmic stress axis, and the response in compression and unloading is described using a compression index C'_c and a swelling index C'_s (Fig. 4.3), such that the equation of the normal

Fig. 4.2 Normal compression line (ncl) and unloading–reloading line (url) in ($\ln p':v$) compression plane.

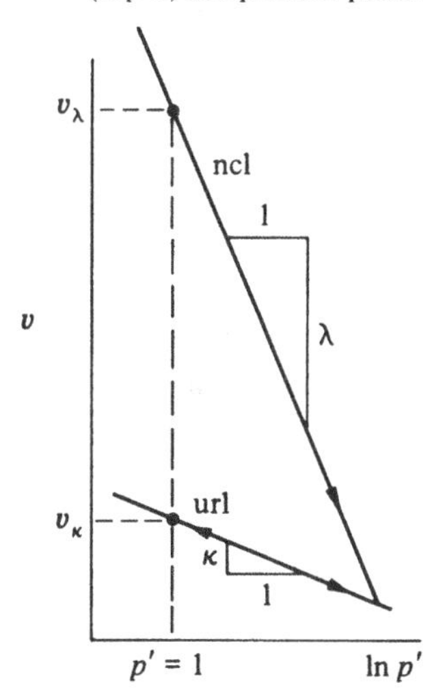

Fig. 4.3 Results of one-dimensional compression test in oedometer interpreted in terms of compression index C'_c and swelling index C'_s.

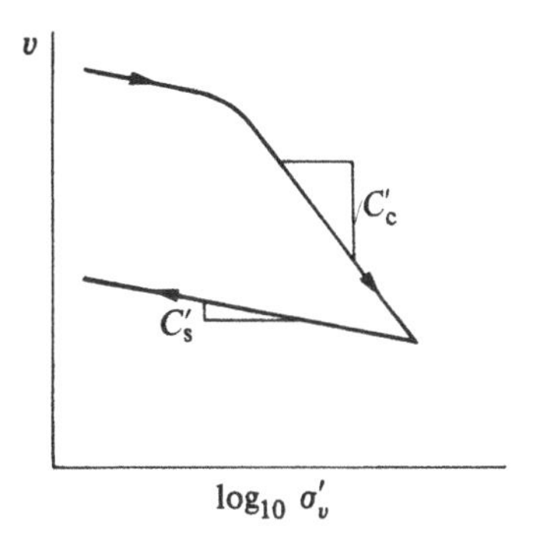

compression phase is

$$v = v_c - C'_c \log_{10} \sigma'_v \tag{4.4}$$

and the equation of the unloading or swelling phase is

$$v = v_s - C'_s \log_{10} \sigma'_v \tag{4.5}$$

where v_c and v_s are intercepts for $\sigma'_v = 1$.

The principal difference between equations (4.2) and (4.3) and equations (4.4) and (4.5) is the (conventional) use of logarithms to base ten in the latter pair. The slopes of the two sets of lines are simply related:

$$C'_c = \lambda \ln 10 \simeq 2.3\lambda \tag{4.6}$$

and

$$C'_s \simeq \kappa \ln 10 \simeq 2.3\kappa \tag{4.7}$$

(The reason for the approximate relation between C'_s and κ will emerge in Section 10.3.2.)

It is obviously helpful to have some expression describing the shape of normal compression and unloading–reloading lines in the compression plane. Expressions (4.2) and (4.3) provide a simple description of this shape, but the development of the elastic–plastic model for soil in this chapter does not depend on this particular shape. Assumption of a particular shape permits the discussion to be focused, but it should become clear how freedom to include more general relationships can be retained.

Equation (4.3) for the unloading–reloading line can be written in an incremental form,

$$\delta v^e = -\kappa \frac{\delta p'}{p'} \tag{4.8}$$

where the superscript e indicates that these are elastic recoverable changes in volume. Since an increment in specific volume δv produces an increment of volumetric strain

$$\delta \varepsilon_p = \frac{-\delta v}{v} \tag{4.9}$$

(Section 1.2), expression (4.8) can be rewritten as

$$\delta \varepsilon_p^e = \kappa \frac{\delta p'}{v p'} \tag{4.10}$$

where again the superscript e denotes an elastic strain increment.

By comparison with (4.1), this implies that

$$K' = \frac{v p'}{\kappa} \tag{4.11}$$

and a constant slope κ of the unloading–reloading line in the semi-logarithmic compression plane (Fig. 4.2) implies a bulk modulus K' that increases with mean stress p'. [The effect in (4.11) of the small decrease in volume v that occurs as the soil is reloaded is likely to be very much smaller than the effect of the increase in p'.]

Changes in deviator stress q within the yield locus, for isotropic elastic soil, cause no changes in volume but do produce elastic deviatoric, or triaxial shear strains $\delta\varepsilon_q^e$, which can be calculated from (4.1) with an appropriate value of shear modulus G'. With a bulk modulus dependent on mean stress p' (4.11), there are, strictly, certain limitations on the choice of a variable or constant shear modulus (Zytynski, Randolph, Nova, and Wroth, 1978; and see Exercise E2.5). In practice, a value for shear modulus might be deduced from the bulk modulus, an assumed value of Poisson's ratio ν', and a combination of (2.3) and (2.4):

$$G' = \frac{3(1 - 2\nu')K'}{2(1 + \nu')} \tag{4.12}$$

This would lead to a shear modulus that was dependent on mean stress in the same way as the bulk modulus. Alternatively, a constant value of shear modulus might be assumed, in which case the variation of bulk modulus with mean stress implies a variation of Poisson's ratio; rearranging (4.12) gives

$$\nu' = \frac{3K' - 2G'}{2G' + 6K'} \tag{4.13}$$

4.3 Plastic volumetric strains and plastic hardening

The previous section has considered changes in stress which lie within the current yield locus. Consider now a change in stress which causes the soil to yield (Fig. 4.4a), a short stress probe from a point K on the current yield locus (yl 1) to a point L outside the current yield locus. The stress state L must lie on a new yield locus (yl 2), and an assumption has to be made about the shape of this new yield locus.

One philosophy that might be followed has already been hinted at in the discussion of the yielding of Winnipeg clay in Section 3.3. In Fig. 3.22 it was shown that the preconsolidation pressure σ'_{vc} could be used to normalise the current yield loci for samples of Winnipeg clay taken from various depths in the ground. The assumption underlying that discussion was that no matter what the value of the preconsolidation pressure, the *shape* of the current yield surface would be the same, with only its size changing. The important assumption is now made that, irrespective of the stress path by which a new yield surface is created, its shape remains the same; in other words, subsequent yield surfaces always have the same shape. Thus, the yield locus (yl 2) passing through point L in Fig. 4.4a is assumed to have the same shape as the yield locus (yl 1) passing through point K. Again, it must be emphasised that this is a convenient assumption but not a necessary one, though the complexities that arise when yield surfaces are allowed to change shape (as well as size) as yielding proceeds are great (and are touched on in Section 12.4).

The yield locus (yl 1) through K in Fig. 4.4a is that which the soil possessed by virtue of having been (one-dimensionally) normally compressed to point A. Point K can be established in the compression plane (Fig. 4.4b) lying on the unloading–reloading line (url 1) through point A on the (one-dimensional) normal compression line. The yield locus (yl 2) through L in Fig. 4.4a is one which, as a result of this most recent assumption, could have been obtained by (one-dimensional) normal compression of the soil to B, followed by a stress path from B to L lying within this yield locus. Consequently, point L can be established in the

Fig. 4.4 (a) Expansion of yield locus from yl 1 to yl 2 and (b) corresponding change in unloading–reloading line from url 1 to url 2.

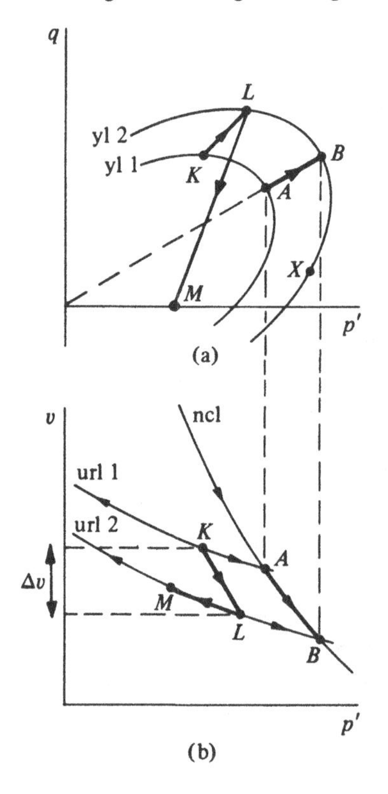

compression plane (Fig. 4.4b) lying on the unloading–reloading line (url 2) through point B on the (one-dimensional) normal compression line. Point L lies on yield locus yl 2 and hence lies on the boundary of the elastic region contained by yl 2.

The soil for which the yield locus is now yl 2 is in fact opaque to attempts to elucidate details of its history. If the sample were unloaded from L to the isotropic stress state M (Fig. 4.4) and left in a triaxial apparatus for an unsuspecting research worker to find, the research worker could discover with judicious probing that the soil had a yield locus of a certain size but could not deduce whether the soil had that particular yield locus because it had been previously loaded to stress state B, L, or any other point X lying on the same curve (Fig. 4.4a). Its past must remain hidden.

Fig. 4.5 Data of yielding deduced from triaxial tests on undisturbed Winnipeg clay in (a) $p':q$ effective stress plane and (b) $v:p'$ compression plane (after Graham, Noonan, and Lew, 1983).

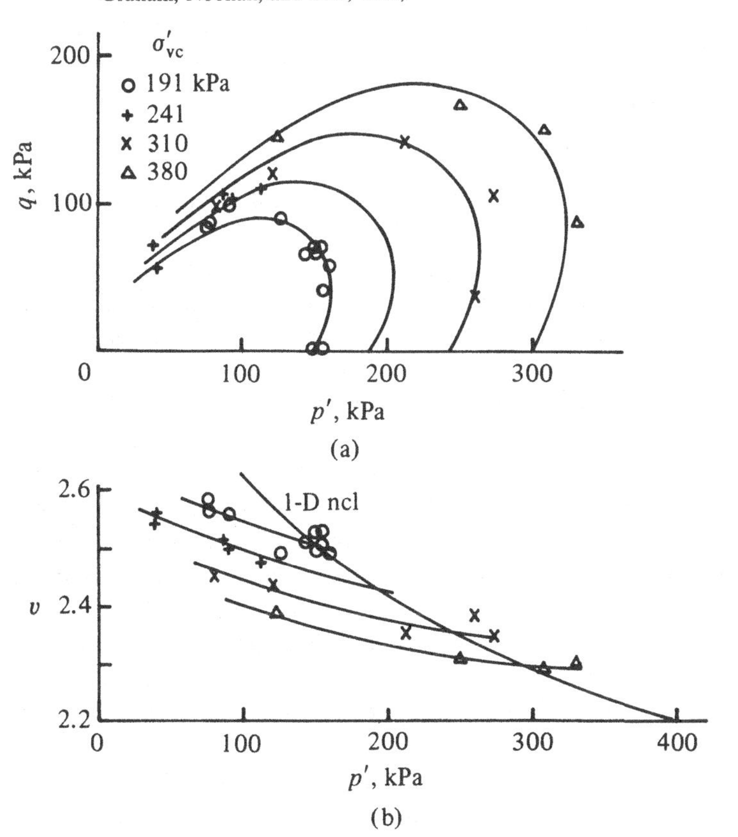

The data of yielding of Winnipeg clay (Graham, Noonan, and Lew, 1983) which were presented in Section 3.3 fit in with this general picture. Graham et al. not only determined the effective stresses at yield for soil specimens taken from various depths (Fig. 4.5a) but also recorded the corresponding values of specific volume. Consequently, their data can be presented in the compression plane (Fig. 4.5b) as well as in the effective stress plane. For each value of preconsolidation pressure σ'_{vc}, that is, for soil samples from each depth, the points in the compression plane lie around a fairly well-defined curve. It is now being assumed that the yield loci that Graham et al. deduced as current yield loci for samples taken from various depths can be treated as identical to the subsequent yield loci that would be observed if, for example, samples from the depth with the lowest value of σ'_{vc} were subjected to stress paths which explored regions of the $p':q$ effective stress plane lying outside the initial yield locus for soil from this depth.

The total change in volume that occurs as the stress state changes from K to L in Fig. 4.4a is given by the vertical Δv separation of K and L in the compression plane in Fig. 4.4b. It is necessary, for the construction of the elastic–plastic model for soil, to separate this total volume change Δv into recoverable, elastic, and irrecoverable, plastic, parts

$$\Delta v = \Delta v^{e} + \Delta v^{p} \tag{4.14}$$

where the superscripts e and p refer to elastic and plastic deformations, respectively.

The response of annealed copper wire to increasing cycles of loading and unloading in uniaxial tension was discussed in Section 3.2. When, having been loaded to A_1 (Fig. 3.2) and unloaded to B_1, the wire is reloaded from B_1 (Fig. 4.6a), irrecoverable extensions of the wire begin when the past maximum load is exceeded; and at a point such as A_2, the deformation consists of irrecoverable and recoverable parts. The recoverable part is found by unloading the wire to B_2, and the irrecoverable part is then seen to be the separation, on the $P = 0$ axis, of points B_1 and B_2. However, the elastic properties of the copper do not change as the irrecoverable deformation increases, so the increase in irrecoverable deformation on the loading cycle $B_1C_1A_2B_2$ is given by the horizontal, $P = \text{constant}$, separation of the unloading–reloading lines B_1C_1 and A_2B_2 at any value of load P lower than the yield point for this cycle C_1. It is not necessary to remove the load completely to deduce the magnitude of the irrecoverable deformation.

A direct analogy can be drawn between the response of the copper wire and the response of the soil, which has been redrawn in Fig. 4.6b turned on its side for ease of comparison. For the wire, B_1C_1 in Fig. 4.6a is an unloading–reloading line (url 1), a line joining a set of elastically attainable combinations of tension P and extension δl. For the soil, the unloading–reloading line url 1 (Fig. 4.6b) is a line joining a set of elastically attainable combinations of mean stress p' and specific volume v. For each material, after some yielding has occurred, there is a new unloading–reloading line (url 2 in Figs. 4.6a, b) joining a new set of elastically attainable combinations of tension or mean stress, and extension or specific volume. For the soil, then, by analogy with the copper, the irrecoverable change of volume Δv^p is the separation of the two unloading–reloading lines at constant mean stress (Fig. 4.6b).

Fig. 4.6 (a) Elastic and plastic behaviour in cycle of reloading and unloading of annealed copper wire (after Taylor and Quinney, 1931); (b) normal compression line (ncl) and unloading–reloading lines (url 1 and url 2) for soil; mean effective stress p' and specific volume v plotted with linear scale; (c) mean effective stress p' plotted with logarithmic scale.

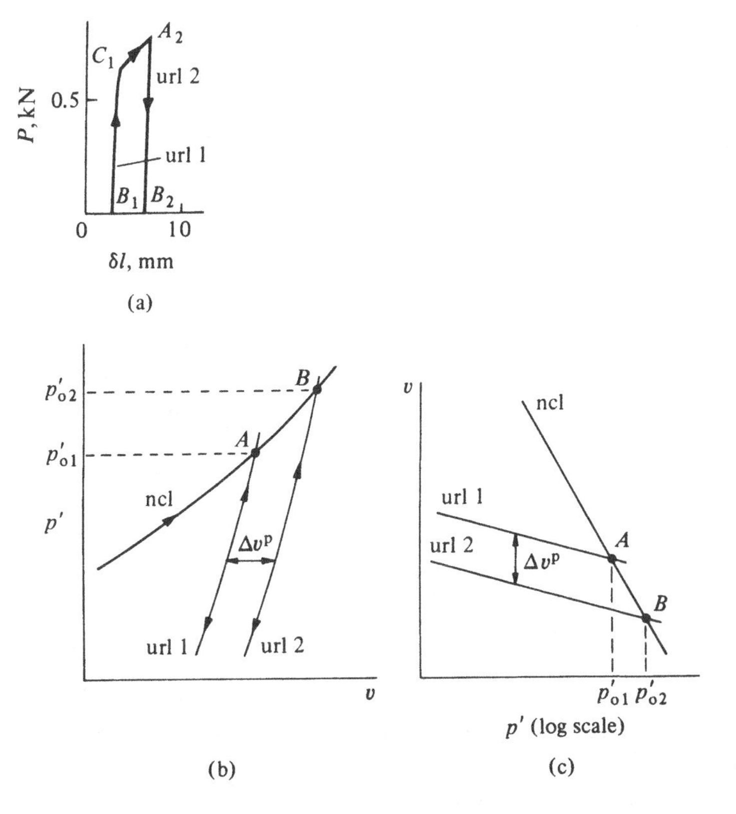

The equation of unloading–reloading line url 1 is, from (4.3),

$$v = v_{\kappa 1} - \kappa \ln p' \tag{4.15}$$

and the equation of unloading–reloading line url 2 is

$$v = v_{\kappa 2} - \kappa \ln p' \tag{4.16}$$

The irrecoverable change in volume is simply the change in the intercept v_κ in the equation for the unloading–reloading line:

$$\Delta v^{\mathrm{p}} = \Delta v_\kappa = v_{\kappa 2} - v_{\kappa 1} \tag{4.17}$$

An alternative expression for this irrecoverable volume change can be obtained by looking in greater detail at the region of the compression plane around the points at which the unloading–reloading lines meet the (one-dimensional) normal compression line (Figs. 4.6b, c). Point A in Figs. 4.4 and 4.6b, c is the point on the normal compression line with $p' = p'_{o1}$, the mean stress which in normal compression leaves the soil with yield locus yl 1 through A in the stress plane (Fig. 4.4a) and with unloading–reloading line url 1 in the compression plane (Fig. 4.4b). Point B is, correspondingly, the point on the normal compression line with $p' = p'_{o2}$. The irrecoverable change in specific volume between the unloading–reloading lines url 1 and url 2 is then seen to be the volume change remaining when the normal compression stress is increased from p'_{o1} to p'_{o2} and then reduced to p'_{o1} again. From (4.2) and (4.3), this is

$$\Delta v^{\mathrm{p}} = -\lambda \ln\left(\frac{p'_{o2}}{p'_{o1}}\right) + \kappa \ln\left(\frac{p'_{o2}}{p'_{o1}}\right) \tag{4.18}$$

$$= -(\lambda - \kappa) \ln\left(\frac{p'_{o2}}{p'_{o1}}\right) \tag{4.19}$$

where the first term of (4.18) represents the *total* change in volume occurring as the mean stress is increased from A to B along the normal compression line, and the second term is the part of this volume change which is recovered when the mean stress is reduced again. In the limit, (4.19) becomes

$$\delta v^{\mathrm{p}} = -(\lambda - \kappa) \frac{\delta p'_o}{p'_o} \tag{4.20}$$

and in terms of strains,

$$\delta \varepsilon^{\mathrm{p}}_p = (\lambda - \kappa) \frac{\delta p'_o}{v p'_o} \tag{4.21}$$

These two expressions, (4.20) and (4.21), for plastic volumetric changes are very similar to the two expressions (4.8) and (4.10) for elastic volumetric changes. The multipliers are different [$(\lambda - \kappa)$, which controls the plastic deformations, is typically about four times as big as κ, which controls the elastic deformations (see Section 9.4)] and whereas elastic changes in volume occur whenever the mean effective stress p' changes, plastic changes of volume occur only when the size of the yield locus changes, specified by a normal compression stress p'_o.

The total volumetric strain increment is the sum of its elastic and plastic components [compare (4.14)]:

$$\delta\varepsilon_p = \delta\varepsilon_p^e + \delta\varepsilon_p^p \tag{4.22}$$

or

$$\delta\varepsilon_p = \kappa\frac{\delta p'}{vp'} + (\lambda - \kappa)\frac{\delta p'_o}{vp'_o} \tag{4.23}$$

Similarly for the total change in specific volume v

$$\delta v = \delta v^e + \delta v^p \tag{4.24}$$

or

$$\delta v = -\kappa\frac{\delta p'}{p'} - (\lambda - \kappa)\frac{\delta p'_o}{p'_o} \tag{4.25}$$

When the soil is being (one-dimensionally) normally compressed, the stress state is always at the tip of the current yield locus – always in the same geometrical position (A, B, and C in Fig. 4.7) – and $p' = p'_o$. Then,

Fig. 4.7 Successive yield loci and unloading–reloading lines resulting from normal compression.

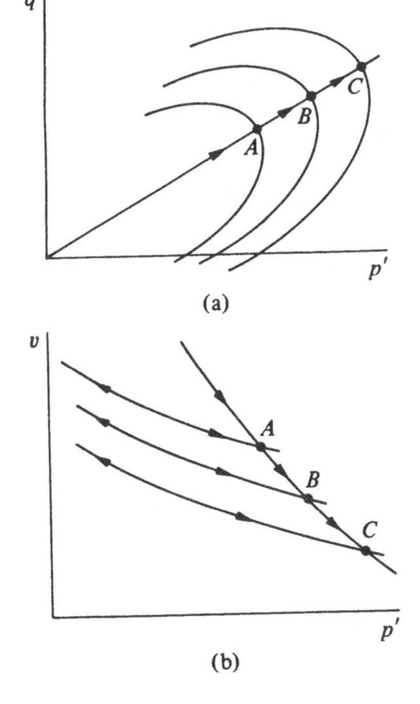

from (4.23), the total volumetric strain increment is

$$\delta\varepsilon_p = \lambda \frac{\delta p'}{vp'} \tag{4.26}$$

and the total change in specific volume is

$$\delta v = -\lambda \frac{\delta p'}{p'} \tag{4.27}$$

which, being integrated, recovers the equation of the normal compression line:

$$v = v_\lambda - \lambda \ln p' \tag{4.2bis}$$

As an illustration of the separation of volumetric strains into elastic and plastic components, consider the response of the clay to three stress probes PQ, PR, and PS starting from the same point P on the current yield locus (yl 1) (Fig. 4.8). The path PQ is directed towards the interior of the yield locus yl 1 and consequently produces purely elastic response. In the compression plane, the state of the soil moves up the current unloading–reloading line (url 1). The elastic change in volume is given by

$$\delta v^e = -\kappa \frac{\delta p'}{p'} \tag{4.8bis}$$

Fig. 4.8 Stress probes producing recoverable and irrecoverable changes in volume.

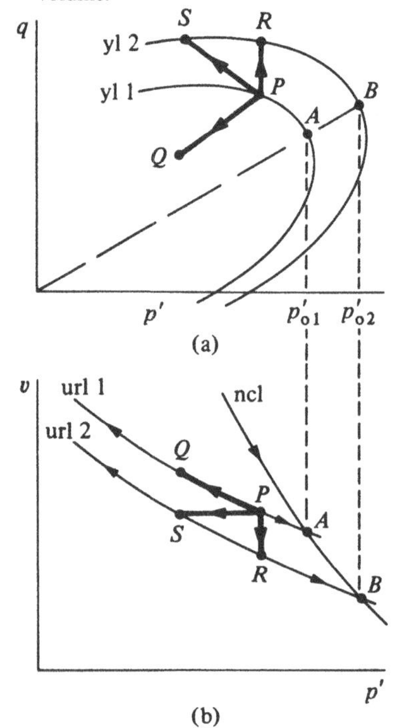

and the elastic volumetric strain by

$$\delta\varepsilon_p^{\mathrm{e}} = \kappa \frac{\delta p'}{vp'} \qquad (4.10\text{bis})$$

This elastic process occurs with no change of p'_{o}.

The path PR is directed vertically upwards in the $p':q$ plane, at constant p' (Fig. 4.8a). The new effective stress state R lies on a new larger yield locus (yl 2) which could have been obtained by further (one-dimensional) normal compression of the soil from $p'_{\mathrm{o}1}$ to $p'_{\mathrm{o}2}$ (A to B in Fig. 4.8). This new yield locus is associated with a new unloading–reloading line (url 2) meeting at $p'_{\mathrm{o}2}$ (B in Fig. 4.8b) with the (one-dimensional) normal compression line in the compression plane. Since there is no change of p' from P to R, there is no elastic volumetric strain (i.e. there is no change in the amount of elastic volumetric strain energy stored in the soil), and the volume change resulting from the change in p'_{o} is purely plastic, given by (4.20), and the plastic volumetric strain is given by (4.21):

$$\delta\varepsilon_p^{\mathrm{p}} = (\lambda - \kappa) \frac{\delta p'_{\mathrm{o}}}{vp'_{\mathrm{o}}} \qquad (4.21\text{bis})$$

Path PS is a stress path with the same change in p' as path PQ (Fig. 4.8), but the points have been chosen in such a way that S not only lies on the same new expanded yield locus as R (yl 2) but also has the same specific volume as the initial point P (Fig. 4.8b). This is clearly a path which involves both elastic and plastic changes in volume. Since the overall change in volume is zero, the sum of the elastic and plastic changes must be zero,

$$\delta v = \delta v^{\mathrm{e}} + \delta v^{\mathrm{p}} = 0 \qquad (4.28)$$

Similarly, in terms of volumetric strains

$$\delta\varepsilon_p = \delta\varepsilon_p^{\mathrm{e}} + \delta\varepsilon_p^{\mathrm{p}} = 0 \qquad (4.29)$$

The elastic volumetric strain is given by (4.10),

$$\delta\varepsilon_p^{\mathrm{e}} = \kappa \frac{\delta p'}{vp'} \qquad (4.10\text{bis})$$

Since, as drawn in Fig. 4.8, $\delta p' < 0$ therefore, $\delta\varepsilon_p^{\mathrm{e}} < 0$. The plastic volumetric strain is given by (4.21),

$$\delta\varepsilon_p^{\mathrm{p}} = (\lambda - \kappa) \frac{\delta p'_{\mathrm{o}}}{vp'_{\mathrm{o}}} \qquad (4.21\text{bis})$$

and to give a zero total volumetric strain $\delta\varepsilon_p^{\mathrm{p}} > 0$ and $\delta p'_{\mathrm{o}} > 0$, as drawn in Fig. 4.8.

The stress path PS is associated with zero volume change and is thus part of the effective stress path that would be followed in a constant volume or undrained test. The important conclusion is that a condition of no overall change in volume places no restrictions on the individual elastic and plastic components that make up that overall change. The implication is then that yielding can, and in general does, occur in an undrained test, with elastic and plastic contributions exactly balancing to give zero resultant total volumetric strain. In general, therefore, the effective stress path followed in an undrained test does not have the same shape as the yield locus. It is necessary for the yield curve to be expanding and producing plastic volumetric compression in order to balance the elastic volumetric expansion that occurs as a result of the reduction of mean effective stress as the pore pressure builds up in an undrained test. (It is shown in Section 5.4 that there may be some undrained paths on which these effects are reversed and plastic volumetric expansion occurs with a counterbalancing elastic volumetric compression.)

4.4 Plastic shear strains

In the previous section, plastic volumetric strain has been associated with change in size of the yield locus. A certain irrecoverable

Fig. 4.9 Stress probes between pair of yield loci.

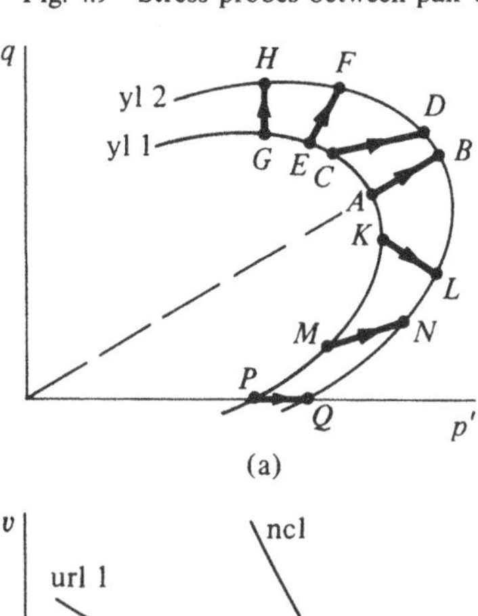

(a)

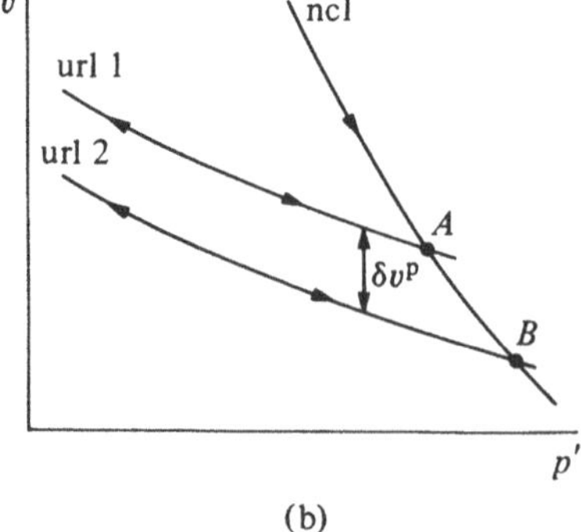

(b)

change in volume δv^p (Fig. 4.9b) is associated with the expansion of the yield locus from yl 1 to yl 2 (Fig. 4.9a), but this expansion could have been achieved with any one of the stress paths *AB*, *CD*, *EF*, *GH*, *KL*, *MN*, or *PQ* (Fig. 4.9a). The recoverable change in volume associated with each of those paths would be different since the changes in mean stress p' are different, but because each stress path forms a link between the same two yield loci, the change in p'_o (which indicates the size of the yield locus) caused by each path is the same; hence, from (4.20), the irrecoverable change in volume is the same. There is clearly a basic difference in the way in which elastic and plastic volume changes are generated.

Description of plastic volumetric strains provides only a partial description of the plastic deformation: it is necessary also to calculate the magnitude of any plastic shear strains that may occur. When the plastic deformation of thin-walled metal tubes was discussed in Section 3.2, it was inferred from the pattern that emerged in Figs. 3.8 and 3.11 that the directions of the plastic strain increment vectors are governed, not by the route through stress space that was followed to reach the yield surface, *but by the particular combination of stresses at the particular point at which the yield surface was reached.* This idea of linking relative magnitudes of plastic strain increments with stresses rather than stress increments is a distinguishing feature of plastic as opposed to elastic response. Before the application of this idea to soils is considered in detail, an example will be given of its application in another mechanical system.

4.4.1 Frictional block

The behaviour of a block sliding on a rough surface provides an analogy for plastic behaviour of materials which may be helpful to those unfamiliar with the theory of plasticity.

A block sitting under a normal load P and subjected to an increasing

Fig. 4.10 Block sliding on frictional interface (a) side view; (b) plan.

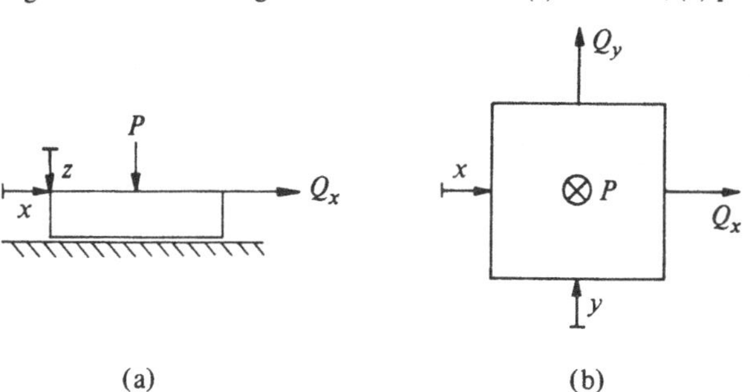

shear load Q_x in one direction (Fig. 4.10a) slides off in that direction when

$$Q_x = \mu P \tag{4.30}$$

where μ is the coefficient of friction for the rough interface (and the distinction between static and dynamic friction is being ignored). If the block is subjected to shear loads Q_x and Q_y, orthogonal to each other and to the normal load P (Fig. 4.10b), then sliding occurs when the resultant shear load is equal to μP, that is, when

$$\sqrt{Q_x^2 + Q_y^2} = \mu P \tag{4.31}$$

or

$$f = Q_x^2 + Q_y^2 - \mu^2 P^2 = 0 \tag{4.32}$$

Equations (4.31) and (4.32) describe a sliding surface for the block in $P{:}Q_x{:}Q_y$ space, which has the shape of a right circular cone centred on the P axis. A section through this sliding surface with $Q_y = 0$ produces a straight line (4.30) (Fig. 4.11a), and a section at constant P produces a circle of radius μP (Fig. 4.11b). The second equation (4.32) of the sliding surface introduces a sliding function f which can be computed for any set of values of P, Q_x, Q_y. If $f < 0$, then the block remains still. If $f = 0$, then the block slides. Values of $f > 0$ are not admissible.

Consider the block loaded only with the shear load Q_x applied at the upper surface of the block. There may be some elastic shear deformation of the block before it slides away, and the load : displacement relationship is as shown in Fig. 4.12a. In the $P{:}Q_x$ plane, the sliding motion can be indicated by a vector of sliding components of displacement $\delta z^s{:}\delta x^s$, where the superscript s indicates that the elastic contribution to the total

Fig. 4.11 Loads and displacements for sliding of frictional block: (a) sections through sliding surface (—) and sliding potentials (– –) for $Q_y = 0$; (b) section through sliding surface and sliding potential for constant P, with loading path $OABC$.

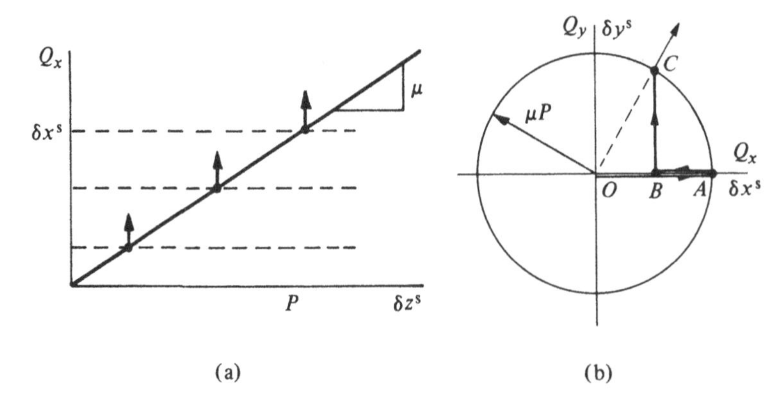

displacement has been subtracted, and where the z displacement is in the direction of the normal load P. Since the block does not lift off the rough surface, sliding always occurs with $\delta z^s = 0$, and these sliding vectors are directed parallel to the Q_x axis (Fig. 4.11a).

If, after some sliding has occurred in the x direction, the shear load Q_x is reduced to $\mu P/2$, without changing P (AB in Fig. 4.11b), and an increasing orthogonal shear load Q_y is applied (BC), then, from (4.31) or (4.32), sliding occurs when $Q_y = \sqrt{3}\mu P/2$. Even though sliding is induced by increments in Q_y, it occurs in the direction of the resultant shear load, so that a vector of the components of sliding displacement $\delta x^s:\delta y^s$ is always normal to the circular sliding locus in the $Q_x:Q_y$ plane (Fig. 4.11b).

The resulting load:deformation relationship for the increasing force Q_y and the displacement in the x direction is shown in Fig. 4.12b. Just as for the thin metal tubes, in which increments of torque started to cause axial extension when the material of the tubes began to yield (Fig. 3.9), so, for the frictional block, irrecoverable movements occur in the x direction even though sliding is induced by the changing shear load in the y direction. This is another example of the dependence of the components of irrecoverable deformation on the state of loading at which irrecoverable deformation is occurring and not on the path by which that state of loading was reached.

A second function g can be introduced,

$$g = Q_x^2 + Q_y^2 - k^2 = 0 \tag{4.33}$$

where k is a dummy variable. This function can be described as a sliding potential since the relative amounts of sliding displacement that occur in each of the x, y, z directions can be found by differentiating g with respect

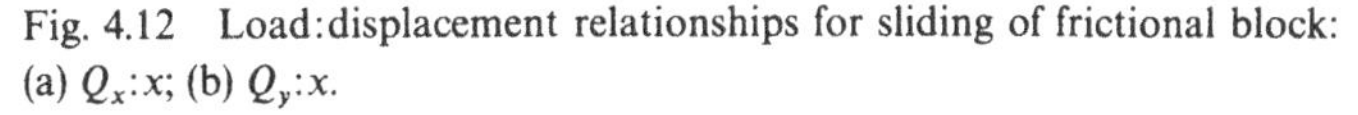

Fig. 4.12 Load:displacement relationships for sliding of frictional block: (a) $Q_x{:}x$; (b) $Q_y{:}x$.

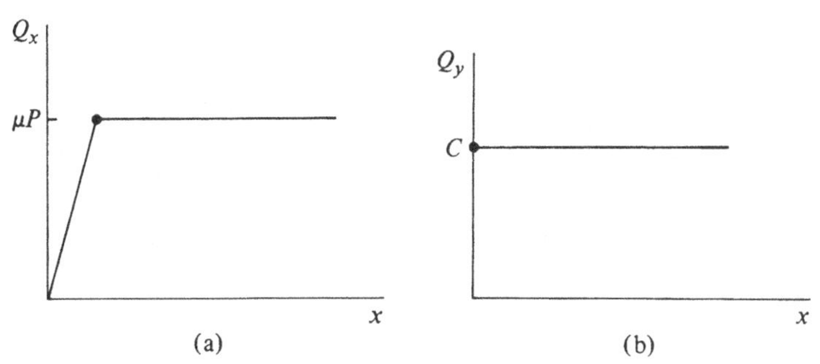

to each of the corresponding loads:

$$\delta x^s = \chi \frac{\partial g}{\partial Q_x} = \chi 2Q_x \tag{4.34a}$$

$$\delta y^s = \chi \frac{\partial g}{\partial Q_y} = \chi 2Q_y \tag{4.34b}$$

$$\delta z^s = \chi \frac{\partial g}{\partial P} = 0 \tag{4.34c}$$

The quantity χ is a scalar multiplier which does not enter into any considerations of the relative magnitudes of the three components of sliding displacement. For any value of k, (4.33) describes a right-circular cylinder in $P:Q_x:Q_y$ space, centred on the P axis.

4.4.2 Plastic potentials

The next idea to be absorbed into a simple model of soil behaviour, from these observations of the response of a simple mechanical system, is the dependence of plastic deformations on the stress state at which yielding of the soil occurs rather than on the route by which that stress state is reached.

Suppose that yielding is occurring at a stress state Y in the $p':q$ plane (Fig. 4.13a). Yielding is associated with the occurrence of, in general, some plastic, irrecoverable volumetric strain $\delta\varepsilon_p^p$ and some plastic shear strain $\delta\varepsilon_q^p$. The magnitudes of these two components of strain can be plotted at Y with axes parallel to p' and q to form a plastic strain increment vector YS (Fig. 4.13a). A short line AB can be drawn through Y orthogonal to this plastic strain increment vector.

Fig. 4.13 (a) Plastic strain increment vectors normal to family of plastic potential curves; (b) families of plastic potentials (– –) and yield loci (—).

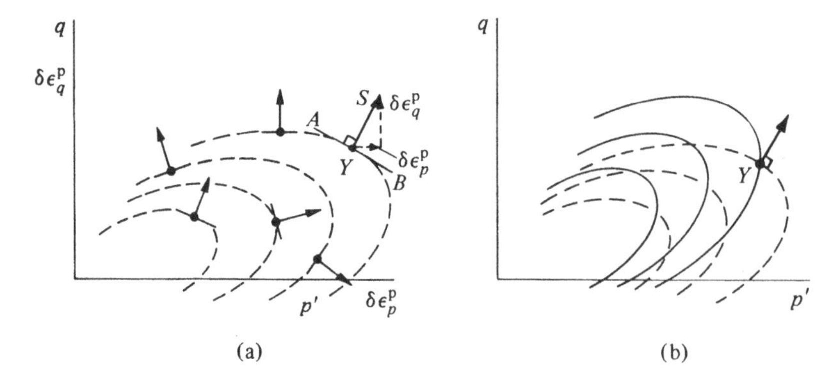

Yielding can occur under many combinations of stresses in the history of a soil, but for each combination, a vector of plastic strains can be drawn (Fig. 4.13a), and a short line can be drawn through each yield point orthogonal to the corresponding strain increment vector. As more data become available, these lines can be joined to form a family of curves to which the plastic strain increment vectors are orthogonal (Fig. 4.13a). These curves are called plastic potentials.

Given a general expression for the family of plastic potential curves, a member of the family can be drawn through any stress state Y at which yielding is occurring (Fig. 4.13b), and the direction of the outward normal then defines the direction of the plastic strain increment; or, in general, the direction of the outward normal to the plastic potential surface defines the *ratio* (or relative magnitudes) of the various components of plastic deformation that may be occurring. A direct similarity can thus be seen with the sliding potential introduced for the frictional block in Section 4.4.1.

The *magnitude* of the plastic volumetric strain increment was discussed in Section 4.3 and linked with the changing size of the yield locus. With knowledge of the plastic potentials, the mechanism of plastic deformation is defined. In particular, the *relative* magnitudes of plastic distortion (or change in shape or shear strain) and of plastic change of size (or volumetric strain) are specified. The magnitude of the plastic shear strain can then be calculated, and the description of the plastic behaviour of the soil is complete.

4.4.3 *Normality or associated flow*

The plastic potentials of Fig. 4.13a form a family of curves in the $p':q$ stress plane. The yield loci of Fig. 4.7 also form a family of curves in the $p':q$ stress plane. Complete specification of the soil model requires information about each of these sets of curves (Fig. 4.13b). Recall that, in studying the behaviour of the copper tubes (Figs. 3.8 and 3.11), the vectors indicating the relative magnitudes of the various components of irrecoverable deformation were orthogonal to the curve in the stress plane which specified the onset of these irrecoverable deformations – the yield curve. In other words, for this example from metal plasticity, the plastic potentials and yield loci coincide, and the two sets of curves are identical. For the frictional block of Section 4.4.1, however, the sliding surfaces were cones, but the sliding potentials were cylinders and therefore not coincident, though the sections at constant P were both circles.

When it comes to inventing models of soil behaviour or to attempting to describe observed patterns of response, it is clearly a great advantage if, for a given material, the shapes of yield loci and plastic potentials can be assumed to be the same; then the number of functions that has to be generated to describe the plastic response is reduced by one. It also turns out to be advantageous to have coincident yield loci and plastic potentials when numerical predictions are to be made using an elastic–plastic soil model in a finite element program: the solution of the equations that emerge in the analyses is faster and the validity of the numerical predictions can be more easily guaranteed.

If the yield surfaces and plastic potential surfaces for a material are identical, then the material is said to obey the postulate of *normality*: the plastic strain increment vector is in the direction of the outward normal to the yield surface. Alternatively, the material can be said to follow a law of *associated flow*: the nature of the plastic deformations, or flow, is associated with the yield surface of the material. Normality and associated flow are two terms for the same pattern of material response.

It is important not to confuse the postulate of normality with Drucker's (1954, 1966) postulate of stability, which states that for a material under a certain state of stress, subjected by an external agency to a perturbation of stress, the work done by the external agency on the displacements it produces must be positive or zero. It essentially rules out the possibility of strain softening and forces normality; that is, it forces the identity of yield surfaces and plastic potentials. However, whereas stability implies normality, normality does not imply stability, (see Section 5.3). Also, as has been seen for the frictional block, it is clear that, for many materials, plastic potentials and yield surfaces are not identical. While normality can be regarded as a convenient assumption, stability can hardly be proposed as a necessary one.

The yield data found by Graham, Noonan, and Lew (1983) for Winnipeg clay were shown in a non-dimensional stress plane $p'/\sigma'_{vc}:q/\sigma'_{vc}$ in Fig. 3.22b. Graham et al. also measured the strains occurring after yield, were able to separate pre-yield and post-yield strains to estimate the plastic contribution to the total strains, and plotted the directions of the plastic strain increment vectors $\delta\varepsilon_p^p:\delta\varepsilon_q^p$ at the appropriate yield points with local axes parallel to the p'/σ'_{vc} and q/σ'_{vc} axes (Fig. 4.14a).

The immediate impression is that these plastic strain increment vectors are roughly normal to the average yield locus. Closer examination shows that the deviation from normality does vary between $\pm 20°$ with an average value of about $1°$ (Fig. 4.14b). There is perhaps a tendency for the ratio of components of plastic shear strain to plastic volumetric strain to be greater than that expected from normality, particularly around the top of the yield locus ($0.3 < q/p' < 1.0$), but for this soil the error in assuming normality to be valid would not appear to be very large. Further, since the yield curve was itself drawn as an average curve, passing close to many somewhat scattered data points, there would be scope for adjusting its shape slightly to improve the agreement with the postulate of normality.

Data have been shown for a natural clay illustrating normality (or relatively minor deviations from normality) of plastic strain increment vectors to yield loci. For sands, the proposition of normality is much less acceptable, and the elastic–plastic models that have been most successfully used for matching the stress:strain response of sands have incorporated separate shapes for yield loci and plastic potentials (e.g. Lade, 1977; Vermeer 1984).

In Section 3.4 it was noted that Poorooshasb, Holubec, and Sherbourne (1967) had found experimentally that yield loci for Ottawa sand in triaxial

Fig. 4.14 (a) Vectors of plastic strain increment plotted at yield points deduced from triaxial tests on undisturbed Winnipeg clay; (b) departure from normality for undisturbed Winnipeg clay (data from Graham, Noonan, and Lew, 1983).

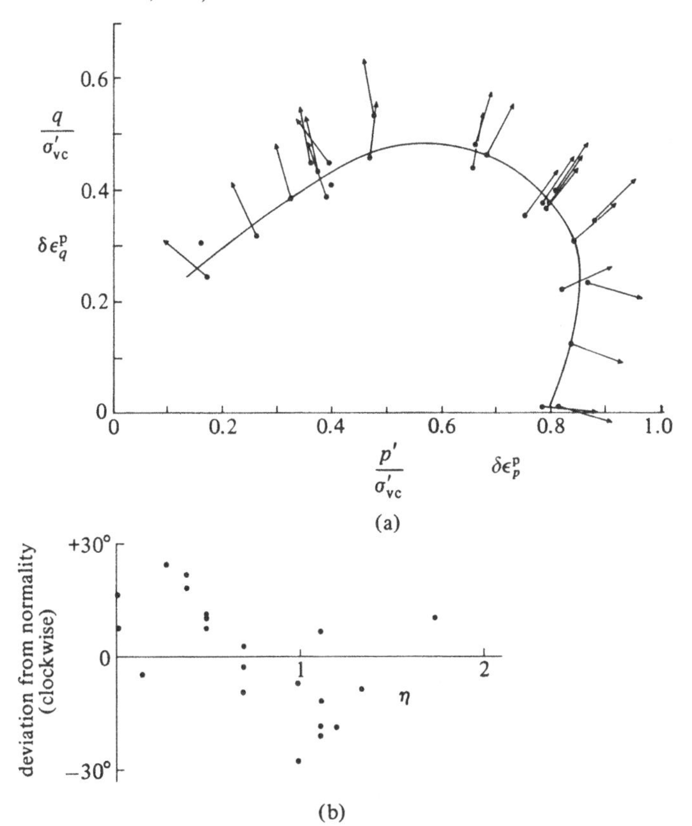

compression could be approximately described by lines at constant stress ratio, $q/p' = \eta = \text{constant}$. In a previous paper (Poorooshasb, Holubec, and Sherbourne, 1966), data of experimental plastic strain increment vectors are used to define a set of plastic potentials. The results for dense Ottawa sand are shown in Fig. 4.15. It is apparent that the assumption of normality of plastic strain increment vectors to the yield loci would result in much greater *negative* plastic volumetric strains (plastic volumetric expansion or dilation) than are actually observed. The frictional block analogy was introduced in Section 4.4.1 specifically to reassure the reader that there is no need to be alarmed by non-associated flow. The plastic flow of sands is discussed further in Chapter 8.

4.5 General plastic stress:strain relationship

The plastic compliance matrix relating increments of effective stress $\delta p'$ and δq in a triaxial apparatus and corresponding increments of irrecoverable strain $\delta\varepsilon_p^p$ and $\delta\varepsilon_q^p$ is derived here. Suppose that the soil has yield loci

$$f(p', q, p'_o) = 0 \tag{4.35}$$

defining the boundary of the region of elastically attainable combinations of effective stress in the $p':q$ plane. The parameter p'_o indicates the size of any particular member of the family of yield loci.

Suppose that the soil has plastic potentials

$$g(p', q, \zeta) = 0 \tag{4.36}$$

where ζ is a parameter controlling the size of the plastic potential which passes through the effective stress state $p':q$. The plastic strain increments form a mechanism of plastic deformation related to the normal to the

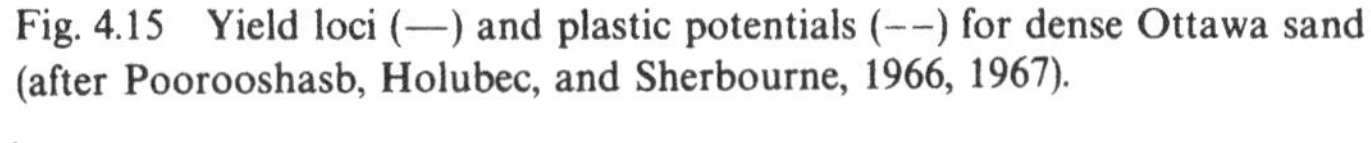

Fig. 4.15 Yield loci (—) and plastic potentials (--) for dense Ottawa sand (after Poorooshasb, Holubec, and Sherbourne, 1966, 1967).

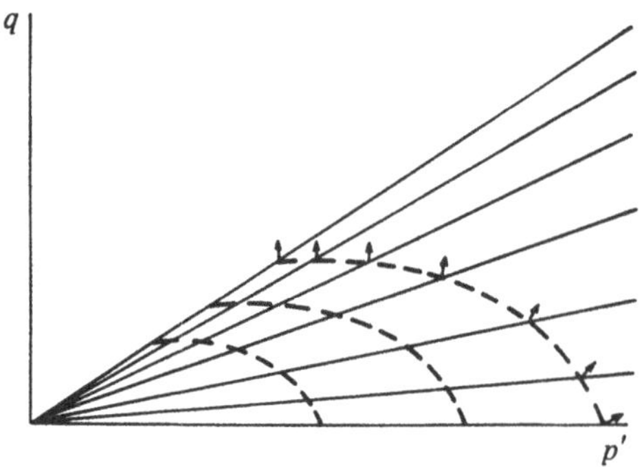

plastic potential at the current effective stress state so that

$$\delta\varepsilon_p^{\mathrm{p}} = \chi \frac{\partial g}{\partial p'} \tag{4.37}$$

$$\delta\varepsilon_q^{\mathrm{p}} = \chi \frac{\partial g}{\partial q} \tag{4.38}$$

where χ is a scalar multiplier whose value will be derived subsequently from the assumed hardening characteristics of the soil.

Suppose that change in size of the yield loci, that is change in p'_{o}, is linked with increments of both plastic volumetric strain and plastic shear strain according to a hardening rule:

$$\delta p'_{\mathrm{o}} = \frac{\partial p'_{\mathrm{o}}}{\partial \varepsilon_p^{\mathrm{p}}} \delta\varepsilon_p^{\mathrm{p}} + \frac{\partial p'_{\mathrm{o}}}{\partial \varepsilon_q^{\mathrm{p}}} \delta\varepsilon_q^{\mathrm{p}} \tag{4.39}$$

The differential form of the yield loci is

$$\frac{\partial f}{\partial p'} \delta p' + \frac{\partial f}{\partial q} \delta q + \frac{\partial f}{\partial p'_{\mathrm{o}}} \delta p'_{\mathrm{o}} = 0 \tag{4.40}$$

Combining (4.37)–(4.40) yields an expression for the scalar multiplier χ:

$$\chi = -\left(\frac{\partial f}{\partial p'} \delta p' + \frac{\partial f}{\partial q} \delta q \right) \Bigg/ \frac{\partial f}{\partial p'_{\mathrm{o}}} \left(\frac{\partial p'_{\mathrm{o}}}{\partial \varepsilon_p^{\mathrm{p}}} \frac{\partial g}{\partial p'} + \frac{\partial p'_{\mathrm{o}}}{\partial \varepsilon_q^{\mathrm{p}}} \frac{\partial g}{\partial q} \right) \tag{4.41}$$

Substituting this expression for χ back into (4.37) and (4.38) gives the general plastic stress:strain relationship:

$$\begin{bmatrix} \delta\varepsilon_p^{\mathrm{p}} \\ \\ \delta\varepsilon_q^{\mathrm{p}} \end{bmatrix} = \frac{-1}{\left[\dfrac{\partial f}{\partial p'_{\mathrm{o}}} \left[\dfrac{\partial p'_{\mathrm{o}}}{\partial \varepsilon_p^{\mathrm{p}}} \dfrac{\partial g}{\partial p'} + \dfrac{\partial p'_{\mathrm{o}}}{\partial \varepsilon_q^{\mathrm{p}}} \dfrac{\partial g}{\partial q} \right] \right]} \begin{bmatrix} \dfrac{\partial f}{\partial p'} \dfrac{\partial g}{\partial p'} & \dfrac{\partial f}{\partial q} \dfrac{\partial g}{\partial p'} \\ \dfrac{\partial f}{\partial p'} \dfrac{\partial g}{\partial q} & \dfrac{\partial f}{\partial q} \dfrac{\partial g}{\partial q} \end{bmatrix} \begin{bmatrix} \delta p' \\ \\ \delta q \end{bmatrix} \tag{4.42}$$

From (4.42), it is clear that if plastic potentials and yield loci coincide so that the soil obeys the principle of associated flow, and if

$$f = g \tag{4.43}$$

then the compliance matrix of (4.42) is symmetric.

4.6 Summary: ingredients of elastic–plastic model

In the next chapter we discuss the response of a particularly simple elastic–plastic model which fits into the more general volumetric hardening framework which has just been presented. The generality of this framework may make the whole process of model building seem a rather arbitrary affair, but the intention has been to indicate that the framework of elastic–

plastic models is a very broad framework, so that there is nothing mutually exclusive about particular models which sit in different, possibly overlapping, parts of that framework. The general principles are of very wide application.

In general, assumptions, which can to some extent be guided by experimental observations, are needed about four aspects of the elastic–plastic response of the soil:

About the way in which elastic, recoverable deformations of the soil are to be described.

About the boundary in a general stress space of the region within which it is reasonable to describe the deformations as elastic and recoverable: a yield surface is needed.

About the mode of plastic deformation that occurs when the soil is yielding: a plastic potential is needed to specify the relative magnitudes of various components of plastic deformation.

About the way in which the absolute magnitude of the plastic deformation is linked with the changing size of the yield locus; this link is known as a hardening rule describing the expansion of the yield locus (hardening of the soil) with a sort of generalised plastic tangent modulus.

(In some cases it is also necessary to make some additional assumption about the conditions under which failure occurs, in other words, to define a limiting surface in effective stress space beyond which the stress state can in no circumstances pass.)

Each of these assumptions may contain other assumptions. For example, to make the discussion a little less general, it has been assumed here that elastic volumetric response is described by (4.10) and that this equation remains valid, with a constant value of κ, when the soil yields. It has been assumed that the elastic response is isotropic. It has been assumed (except in Section 4.5) that the expansion of the yield locus is controlled only by plastic volumetric strain, and it has been assumed that this can be linked with the normal compression of the soil: plastic volumetric strain describes irrecoverable changes in volumetric packing of the soil particles (changes in specific volume or void ratio) but does not describe major geometrical rearrangements or realignment of particles. It has been assumed that the yield loci always have the same shape, though their sizes may change. Finally, it has been suggested that it may often be convenient to assume that plastic potentials and yield loci have the same shape so that only one (instead of two) sets of curves has to be defined. Each of these assumptions is *convenient*; none of them is *necessary*.

In summary, the ingredients of the elastic–plastic model are these:

1. Elastic properties – how much recoverable deformation?
2. Yield surface – is plastic deformation occurring?
3. Plastic potential – what is the mechanism of plastic deformations?
4. Hardening rule – what are the magnitudes of the plastic deformations, and how much has the yield surface changed in size?

Exercises

E4.1. A block of material which yields according to the criterion of Tresca, with yield stress $2c$ in uniaxial tension, is subjected to an initial principal stress state $(\sigma_1, \sigma_2, \sigma_3) = (3c/4, 3c/8, 3c/8)$. The stresses are then changed steadily, with fixed ratios of stress increments. Calculate the ratios of principal plastic strain increments at yield if the stress increment ratios $(\delta\sigma_1, \delta\sigma_2, \delta\sigma_3)$ are (a) $(1, -\frac{1}{2}, -\frac{1}{2})$; (b) $(-\frac{1}{2}, 1, -\frac{1}{2})$; (c) $(-\frac{1}{2}, -\frac{1}{2}, 1)$; (d) $(0, 1, -1)$; (e) $(-1, 0, 1)$; (f) $(1, -1, 0)$. It can be assumed that the normality rule holds for this material.

E4.2. Repeat Exercise E4.1 assuming that the material yields according to the criterion of von Mises, with yield stress $2c$ in uniaxial tension. Assume that the normality rule holds for this material.

E4.3. It is found that the stress:strain behaviour of a particular soil in triaxial compression tests can be represented by an elastic–plastic soil model with yield loci and coincident plastic potentials

$$q = Mp' \ln\left(\frac{p'_o}{p'}\right)$$

where p'_o indicates the size of the current yield locus and M is a soil constant. Hardening of the yield loci is described by

$$\frac{\delta p'_o}{p'_o} = \frac{v}{\lambda - \kappa}(\delta\varepsilon^p_p + A\,\delta\varepsilon^p_q)$$

where A, λ, and κ are other soil constants. The elastic behaviour of the soil is represented by a bulk modulus $K' = vp'/\kappa$ and a shear modulus G'.

Generate the elastic and plastic compliance matrices linking changes in volumetric strain $\delta\varepsilon_p$ and triaxial shear strain $\delta\varepsilon_q$ with changes in p' and q. Find expressions for the limiting values of stress ratio $\eta = q/p'$ reached in triaxial compression tests performed on initially normally compressed soil with (i) constant mean effective stress p' and (ii) constant volume v.

E4.4. A soil is found to have yield loci

$$f = q - 2.5M[\sqrt{p'_o(p'_o - p')} + (p'_o - p')] = 0$$

of size governed by p'_o, with M a soil constant, and plastic potentials

$$g = q - Mp' \ln\left(\frac{\alpha}{p'}\right) = 0$$

where α is a size parameter.

Sketch these curves in the $p':q$ effective stress plane and discuss the effective stress response to be expected in (i) a strain controlled and (ii) a stress controlled undrained triaxial compression test on a normally consolidated sample of this soil.

It can be assumed that p'_o varies only with plastic volumetric strain $\delta\varepsilon_p^p$ and is not affected by plastic shear strain $\delta\varepsilon_q^p$.

E4.5. A sample of dense dry sand is set up in a plane strain biaxial apparatus with initial stresses $\sigma'_x = \sigma'_y = \sigma'_z = 100\,\text{kPa}$. A test is carried out in which σ'_y is kept constant at 100 kPa and σ'_x is steadily increased, with no strain in the z direction. It is assumed that σ'_x and σ'_y remain principal stresses throughout the test. Appropriate stress and strain variables for presentation of results of plane strain tests were introduced in Section 1.5.

The experimental results in Fig. 4.E1 suggest an idealisation using a perfectly elastic, perfectly plastic model. Yielding of the ideal material in these plane-strain tests is governed by the function

$$f(s', t) = t - s' \sin\phi' = 0$$

and plastic flow is governed by the plastic potential

$$g(s', t) = t - s' \sin\psi = 0$$

where ϕ' and ψ are angles of friction and dilation, respectively.

The elastic behaviour is represented in matrix form by

$$\begin{bmatrix} \delta s' \\ \delta t \end{bmatrix} = \begin{bmatrix} A & B \\ B & C \end{bmatrix} \begin{bmatrix} \delta\varepsilon_s \\ \delta\varepsilon_t \end{bmatrix}$$

Show that $A = K' + G'/3$, $B = 0$, and $C = G'$, where K' is the bulk

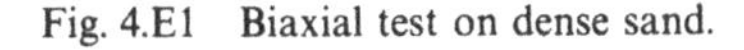
Fig. 4.E1 Biaxial test on dense sand.

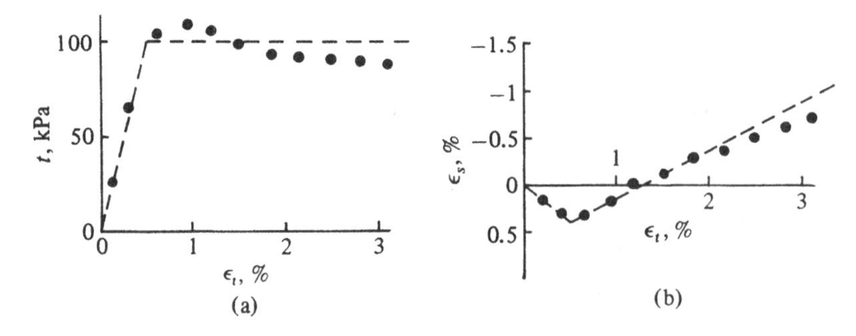

modulus, and G' is the shear modulus. Estimate values of K', G', and Poisson's ratio ν', and the friction angle ϕ' for the sand.

Show that after yield the ideal material experiences no elastic strains, and hence derive a value for the dilation angle ψ.

E4.6. Predictions are to be made of the behaviour of normally compressed clay in triaxial compression tests using a work-hardening elastoplastic model. The elastic strain increments are related to the stress increments by the expressions

$$\delta\varepsilon_p^{\mathrm{e}} = \frac{\delta p'}{K'}, \qquad \delta\varepsilon_q^{\mathrm{e}} = \frac{\delta q}{3G'}$$

where the bulk modulus K' and the shear modulus G' are independent of stress level.

The yield function is given by

$$f = q^2 - M^2\{p'(p'_{\mathrm{o}} - p')\} = 0$$

where p'_{o} is the value of p' on the yield surface during isotropic compression. The plastic potential is identical to f, and the hardening law gives the increment of plastic volumetric strain as

$$\delta\varepsilon_p^{\mathrm{p}} = \frac{\delta p'_{\mathrm{o}}}{H}$$

where H is a constant, also independent of stress level.

Determine expressions for the total strain increments $\delta\varepsilon_p$ and $\delta\varepsilon_q$ in terms of the current stress ratio $\eta = q/p'$ and the stress increments $\delta p'$ and δq when yielding occurs. Hence, show that for a test in which p' is kept constant

$$\varepsilon_q = \left(\frac{1}{3G'} - \frac{4}{HM^2}\right)q + \frac{2p'_{\mathrm{c}}}{HM}\ln\left(\frac{Mp'_{\mathrm{c}} + q}{Mp'_{\mathrm{c}} - q}\right)$$

where the initial conditions are $q = \varepsilon_q = 0$ and $p' = p'_{\mathrm{o}} = p'_{\mathrm{c}}$.

E4.7. A soil is to be described with a volumetric hardening elastic–plastic soil model within the framework presented in this chapter, with

$$\delta p'_{\mathrm{o}} = vp'_{\mathrm{o}}\frac{\delta\varepsilon_p^{\mathrm{p}}}{\lambda - k}$$

and elastic volumetric strains

$$\delta\varepsilon_p^{\mathrm{e}} = \kappa\frac{\delta p'}{vp'}$$

Show that if the yield loci and undrained effective stress paths are plotted in terms of p' and $\eta = q/p'$, then at any chosen value of η, the ratio of the slopes of the two sets of curves is $\Lambda = 1 - \kappa/\lambda$.

5

A particular elastic–plastic model: Cam clay

5.1 Introduction

In the previous chapter, elastic–plastic models for soil were discussed in a general way. Yield loci and plastic potentials were sketched, but no attempt was made to suggest possible mathematical expressions for these curves. In this chapter, a particular model of soil behaviour is described and used to predict the response of soil specimens in standard triaxial tests. In subsequent chapters, this model is used to illustrate a number of features of soil behaviour. This particular model can be regarded as one of the set of volumetric hardening models covered by the discussion in Chapter 4. When the model was originally described in the literature by Roscoe and Burland (1968), it was called 'modified' Cam clay to distinguish it from an earlier model called Cam clay (Roscoe and Schofield, 1963). The qualifier *modified* is dropped here because the modified Cam clay model has probably been more widely used for numerical predictions. The 'original' Cam clay model is mentioned in Section 8.4.

The model is described here in terms of the effective stress quantities p' and q which are relevant to the discussion of soil response in conventional triaxial tests. Most of the examples given are for triaxial compression, though it is tacitly assumed that triaxial extension can be accommodated merely by allowing the deviator stress q to take negative values. The extension of the model to other more general states of stress is described briefly in Section 10.6.2.

It may be noted that the shape of yield locus assumed in the $p':q$ effective stress plane for this Cam clay model bears very little resemblance to the shapes of yield loci that were discovered experimentally for natural soils (Chapter 3). There are three reasons for choosing, apparently perversely, to discuss this particular model in detail. Firstly, on the grounds of simplicity: the description of the shapes of the yield loci introduces just one shape parameter.

Secondly, in many of the experimental investigations of soil behaviour that have been carried out with programmes of triaxial tests in the laboratory, the soil samples have been initially isotropically compressed by subjecting them to a cell pressure with no deviator stress being applied. Such a starting state of isotropic compression may not be very much like the stress state that soils experience in the ground, but it is easy to apply. Also, to avoid some of the variability of undisturbed natural soils, many of the experimental investigations of fundamental soil behaviour have deliberately concentrated on reconstituted soils because of the greater ease of obtaining reproducible and comparable results from many essentially identical samples. Consequently, many of the patterns of soil response that have been observed have been observed in tests on initially isotropically compressed reconstituted soils, and it is these patterns that have particularly guided the development of models such as Cam clay, even though these soils may appear to be somewhat remote from reality.

Thirdly, as is shown in Chapters 10 and 11, for many practical applications the differences between this simple model, Cam clay, and apparently more realistic models may not be important.

The most successful applications of this Cam clay model have been to problems involving the loading of samples of clay or geotechnical constructions on clay. Nevertheless as is shown in subsequent chapters, many of the patterns of response that can be illustrated with this particular model are relevant to other, non-claylike soils. It must be emphasised that this is a pedagogic model, deliberately presented at the simplest possible level. If the presentation of this model can be successfully followed, then it will become apparent that it is relatively straightforward to make the changes that are required to incorporate more realistic features of soil response.

5.2 Cam clay

The four ingredients of an elastic–plastic model listed in Section 4.6 are

1. Elastic properties
2. Yield surface
3. Plastic potential
4. Hardening rule

It is assumed that recoverable changes in volume accompany any changes in mean effective stress p' according to the expression

$$\delta\varepsilon_p^e = \kappa \frac{\delta p'}{vp'} \qquad (5.1)(4.10\text{bis})$$

implying a linear relation, in the compression plane, between specific volume v and logarithm of mean stress p' for elastic unloading–reloading of the soil (url in Fig. 5.1c).

It is assumed that recoverable shear strains accompany any changes in deviator stress q according to the expression

$$\delta\varepsilon_q^e = \frac{\delta q}{3G'} \tag{5.2}$$

with constant shear modulus G'.

As shown in Section 4.2, the combination of (5.1) and (5.2) implies a variation of Poisson's ratio with mean effective stress but to assume instead a constant value of Poisson's ratio would be equally acceptable.

The simplest shape that could be imagined for the yield locus in the $p':q$ stress plane would be a circle (or perhaps a sphere in principal stress space). The next most simple shape is an ellipse (an ellipsoid in principal stress space) which offers some extra freedom by comparison with the circle since the ratio of major to minor axes can be retained as a shape parameter of the model. For this isotropic model, the ellipse is centred

Fig. 5.1 (a) Elliptical yield locus for Cam clay model in $p':q$ plane; (b), (c) normal compression line and unloading–reloading line in compression plane.

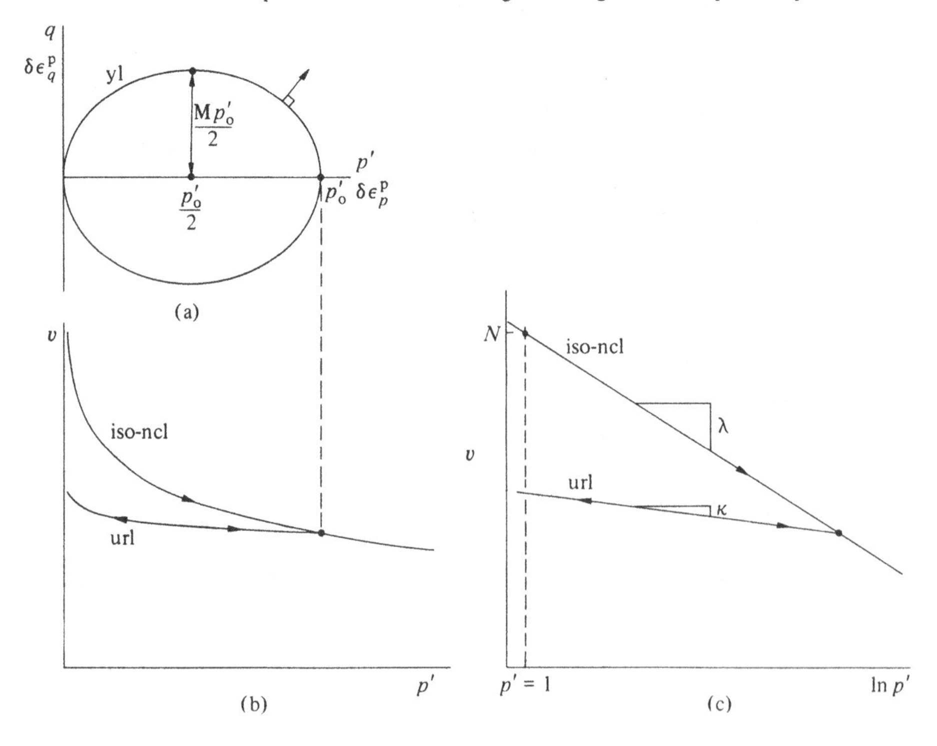

on the p' axis (yl in Fig. 5.1). It is convenient to make it always pass through the origin of effective stress space, though this is not essential: It seems reasonable to propose that unless the soil particles are cemented together, a soil sample will not be able to support an all-round tensile effective stress and that irrecoverable volumetric deformations would develop if an attempt were made to apply such tensile effective stresses.

A convenient way in which to write the equation of the ellipse in Fig. 5.1a is

$$\frac{p'}{p'_o} = \frac{M^2}{M^2 + \eta^2} \tag{5.3}$$

where $\eta = q/p'$. This equation describes a set of ellipses, all having the same shape (controlled by M), all passing through the origin, and having sizes controlled by p'_o. When the soil is yielding, the change in size p'_o of the yield locus is linked with the changes in effective stresses p' and $q = \eta p'$, through the differential form of (5.3):

$$\frac{\delta p'}{p'} + \frac{2\eta\,\delta\eta}{M^2 + \eta^2} - \frac{\delta p'_o}{p'_o} = 0 \tag{5.4a}$$

or

$$\left(\frac{M^2 - \eta^2}{M^2 + \eta^2}\right)\frac{\delta p'}{p'} + \left(\frac{2\eta}{M^2 + \eta^2}\right)\frac{\delta q}{p'} - \frac{\delta p'_o}{p'_o} = 0 \tag{5.4b}$$

To incorporate this particular form of yield locus into the general framework presented in Section 4.5, (5.3) can be rewritten as

$$f = q^2 - M^2[p'(p'_o - p')] = 0 \tag{5.5}$$

It is assumed that the soil obeys the normality condition; so that having assumed an equation for the family of yield loci (5.5), we find that the plastic potentials are automatically given by the same family of curves in the $p':q$ plane:

$$g = f = q^2 - M^2[p'(p'_o - p')] = 0 \tag{5.6}$$

Then the vector of plastic strain increments $\delta\varepsilon^p_p:\delta\varepsilon^p_q$ is in the direction of the outward normal to the yield locus (Fig. 5.1). This implies, from (4.37) and (4.38), that

$$\frac{\delta\varepsilon^p_p}{\delta\varepsilon^p_q} = \frac{\partial g/\partial p'}{\partial g/\partial q} = \frac{M^2(2p' - p'_o)}{2q} = \frac{M^2 - \eta^2}{2\eta} \tag{5.7}$$

when plastic deformations are occurring.

It is assumed that yield loci expand at constant shape, the size being controlled by the tip stress p'_o, and that the expansion of the yield loci, the hardening of the soil, is linked with the normal compression of the soil. Following Section 4.3, we assume a linear relationship between specific volume v and logarithm of mean effective stress p'_o during isotropic normal compression of the soil (iso-ncl in Fig. 5.1c),

$$v = N - \lambda \ln p'_o \tag{5.8}$$

where N is a soil constant specifying the position of the isotropic normal compression line in the compression plane $p':v$ (Fig. 5.1c). Then the magnitude of plastic volumetric strains is given by

$$\delta\varepsilon^p_p = [(\lambda - \kappa)/v] \frac{\delta p'_o}{p'_o} \qquad (5.9)(4.21\text{bis})$$

and the elements of the hardening relationship (4.39) become

$$\frac{\partial p'_o}{\partial \varepsilon^p_p} = \frac{v p'_o}{\lambda - \kappa} \tag{5.10a}$$

$$\frac{\partial p'_o}{\partial \varepsilon^p_q} = 0 \tag{5.10b}$$

The description of the model is now complete. Combining (5.1) and (5.2), we can summarise the elastic stress:strain response in the matrix equation

$$\begin{bmatrix} \delta\varepsilon^e_p \\ \delta\varepsilon^e_q \end{bmatrix} = \begin{bmatrix} \kappa/vp' & 0 \\ 0 & 1/3G' \end{bmatrix} \begin{bmatrix} \delta p' \\ \delta q \end{bmatrix} \tag{5.11}$$

Comparing (5.4b) with (4.40) and substituting in (4.42), we can summarise the plastic stress:strain response in the matrix equation

$$\begin{bmatrix} \delta\varepsilon^p_p \\ \delta\varepsilon^p_q \end{bmatrix} = \frac{(\lambda - \kappa)}{vp'(M^2 + \eta^2)} \begin{bmatrix} (M^2 - \eta^2) & 2\eta \\ 2\eta & 4\eta^2/(M^2 - \eta^2) \end{bmatrix} \begin{bmatrix} \delta p' \\ \delta q \end{bmatrix} \tag{5.12}$$

Equation (5.12) operates only if plastic strains are occurring. The compliance matrix in (5.12) is symmetric because of the assumption of associated flow (normality). Its determinant is zero because, from (5.7), the ratio of plastic volumetric strain to plastic shear strain is dependent on the stress state – in fact, the stress ratio, at which yielding occurs – and not on the stress increments; hence, the two rows of the plastic compliance matrix are multiples of each other.*

*Although the equations that have been presented and the following discussion relate to the restricted axially symmetric conditions of the triaxial apparatus, the model can be applied more generally. It was shown in Section 1.4.1 that the stress variables p' and q and strain increment variables $\delta\varepsilon_p$ and $\delta\varepsilon_q$ used to describe conditions in the triaxial apparatus are special cases of more general variables which can be defined for all strain increment and stress states, expressions (1.34)–(1.37). There is no difficulty in principle in interpreting (5.11) and (5.12) as though they were written in terms of these more general variables.

Equations (5.11) and (5.12) are completely general and are relevant for *all* effective stress paths that might be followed in the $p':q$ plane. The use of this Cam clay model is most easily understood by deducing the strain increment response to a given increment of effective stresses. The steps described are those which would be needed in, for example, the implementation of this Cam clay model in a computer program.

The elastic component of the strain increment can be calculated from (5.11). It is then necessary to determine whether or not plastic strains are occurring. Given an initial stress state $A(p':q)$ and a current yield locus defined by a current value of $p'_o = p'_{oA}$, increments of effective stress $AB(\delta p', \delta q)$ are imposed (Fig. 5.2). From (5.3), a value of $p'_o = p'_{oB}$ for the member of the family of yield loci passing through the new effective stress

Fig. 5.2 (a) Stress increment expanding current yield locus; (b) stress increment inside current yield locus.

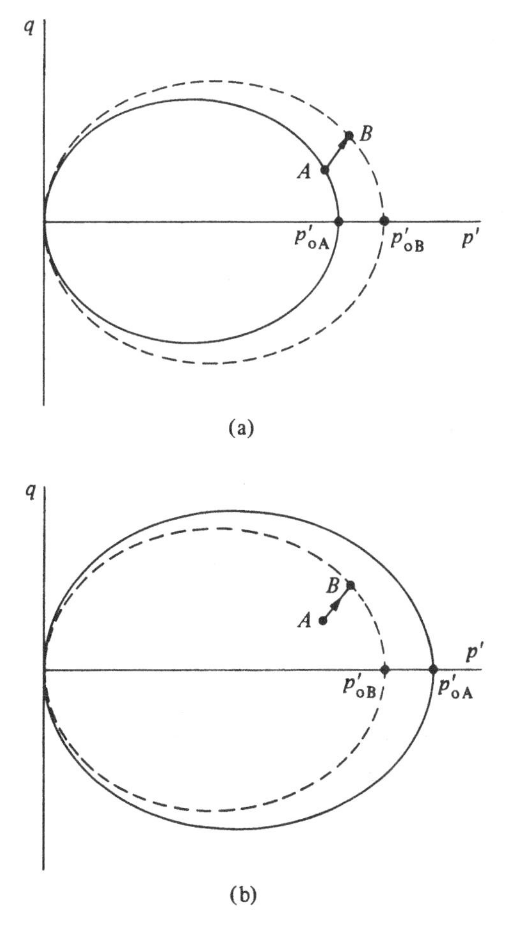

state $B(p' + \delta p' : q + \delta q)$ (Fig. 5.2) can be calculated. If this value $p'_{oB} > p'_{oA}$ (Fig. 5.2a), then the yield locus has had to expand to accommodate the new stress state, and plastic strains are occurring. If this value $p'_{oB} < p'_{oA}$ (Fig. 5.2b), then the new stress state lies inside the current yield locus; so plastic strains do not occur, and the yield locus does not change in size. The total strain increments can be calculated from the combination of (5.11) and (5.12) (if appropriate), and the procedure can be repeated for a new stress increment using a new current value of p'_o, if the yield locus was expanded on the previous increment.

The use of this model to predict the patterns of behaviour that can be expected in conventional drained and conventional undrained triaxial compression tests is described in the next sections. It will become clear that the procedure of the previous paragraph has to be modified under some circumstances. A quasi-graphical approach is used in which frequent reference is made to the progress of tests in the effective stress and compression planes. Such an approach is more enlightening than a direct onslaught using the compliance matrices (5.11) and (5.12), though of course the end result is the same.

5.3 Cam clay predictions: conventional drained triaxial compression

A conventional drained triaxial compression test is constrained by its stress path in the $p':q$ plane. Since the cell pressure is being kept constant,

$$\delta q = 3\,\delta p' \tag{5.13}$$

A typical increment of loading BC in a drained compression test is shown in Fig. 5.3. It is assumed that the increment starts from a stress state B lying on the current yield locus yl B (Fig. 5.3a). This yield locus has a tip pressure p'_{oB} and can be associated with an elastic unloading–reloading line url B ending at $p' = p'_{oB}$ on the isotropic normal compression line in the compression plane (Fig. 5.3b). Point B can thus be located by projection onto this unloading–reloading line.

The stress increment BC takes the stress state to C, which lies on a new, larger yield locus, yl C, with tip pressure p'_{oC} (Fig. 5.3a) and corresponding unloading–reloading line url C (Fig. 5.3b). Point C in the compression plane can be established by projection onto this unloading–reloading line. The change in volume from B to C is made up of a recoverable elastic part resulting from the change in p' between B and C and an irrecoverable plastic part resulting from the expansion of the yield locus and represented by the vertical separation δv^p_{BC} of the unloading–reloading lines url B and url C in the compression plane (Fig. 5.3b).

From the specific volume v_B of the soil at B, this irrecoverable change in volume can be converted to a plastic volumetric strain:

$$\delta\varepsilon^{p}_{pBC} = \frac{-\delta v^{p}_{BC}}{v_B} \tag{5.14}$$

The direction of the outward normal to the yield locus yl B at B (Fig. 5.3a) indicates the relative magnitudes of plastic shear strain and plastic volumetric strain. The actual magnitude of the plastic volumetric strain is known from (5.14), and hence the actual magnitude of the plastic shear strain can be calculated.

As further increments of drained compression are applied, CD, DE, and EF (Fig. 5.4a), the yield locus becomes progressively larger. As it does so, the position of the current stress state in relation to the tip of the current yield locus changes, so that the direction of the outward normal to the yield locus changes progressively (Fig. 5.4a). The ratio of plastic shear strain to plastic volumetric strain becomes steadily greater as the normal to the yield locus becomes more nearly parallel to the q axis, eventually

Fig. 5.3 Stress increment in conventional drained triaxial compression test ($\delta q = 3\,\delta p'$).

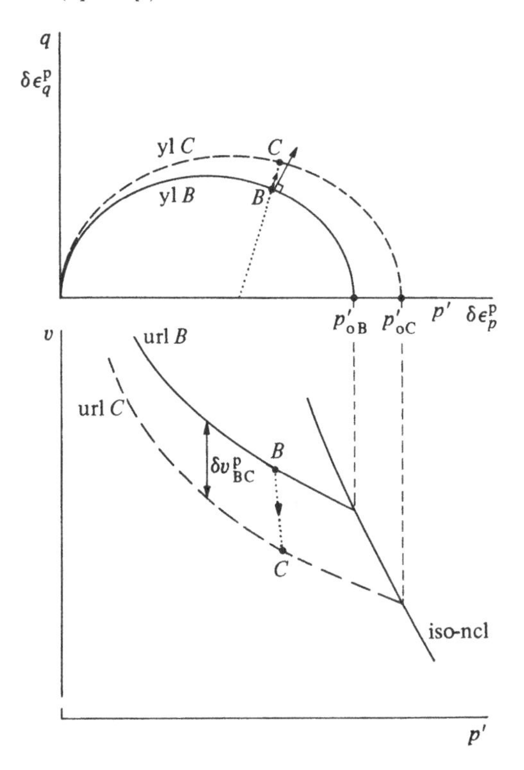

becoming infinite when the stress state reaches point F, at the crest of the yield locus yl F (Fig. 5.4a) [compare (5.7), for $\eta = M$]. At F, from Fig. 5.1 and expression (5.4),

$$\frac{q}{p'} = \eta = M$$

and

$$p' = \frac{p'_{oF}}{2}$$

At this point, unlimited plastic shear strains develop with no plastic volumetric strain. With no plastic volumetric strain, the yield locus yl F remains of constant size, $\delta p'_o = 0$ in (5.9). Since there is just one point where the effective stress path $BCDEF$ intersects this yield locus, plastic shearing continues at constant effective stresses, and the loading can

Fig. 5.4 Sequence of stress increments in conventional drained triaxial compression test: (a) p':q effective stress plane; (b) v:p' compression plane; (c) q:ε_q stress:strain plot; (d) v:ε_q volume:strain plot.

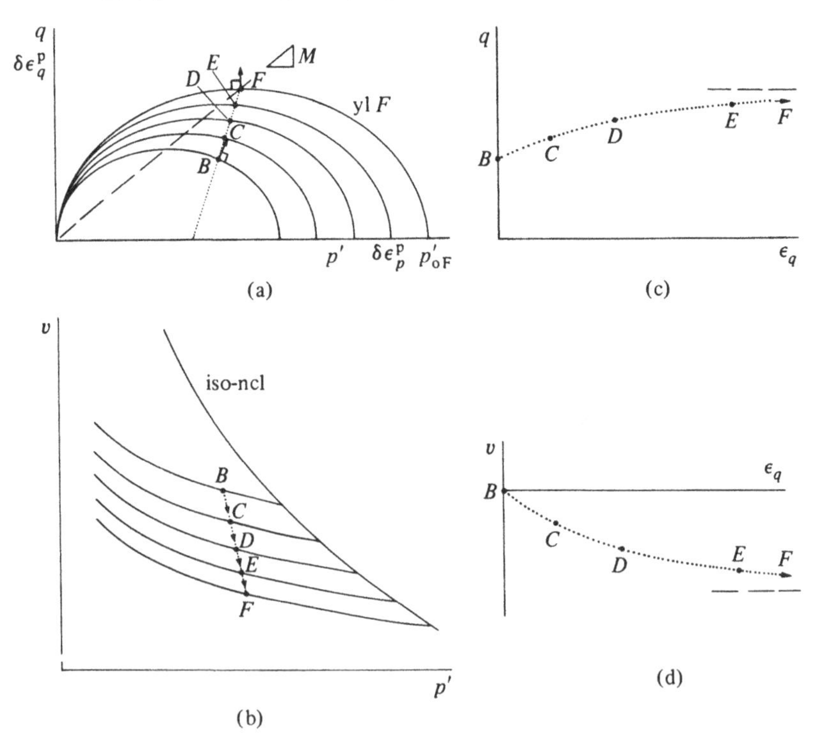

proceed no further. Looking at (5.12), for $\eta = M$, we have

$$\frac{\delta\varepsilon_p^{\mathrm{p}}}{\delta p'} = 0$$

and

$$\frac{\delta\varepsilon_q^{\mathrm{p}}}{\delta q} = \infty$$

The progress of the drained test has so far been described with reference only to the $p':q$ stress plane (Fig. 5.4a). At each point B, C, D, E, and F, a new yield locus can be drawn with its corresponding unloading–reloading line in the $p':v$ compression plane (Fig. 5.4b). The progress $BCDEF$ of the test in the compression plane can thus be deduced. The information about shear strains permits a stress:strain curve to be generated (Fig. 5.4c), showing the development of deviator stress q with triaxial shear strain ε_q.

Combining (5.13), which specifies the effective stress path, with the plastic stress:strain relation (5.12) produces the plastic shear strain:

$$\delta\varepsilon_q^{\mathrm{p}} = (\lambda - \kappa)\frac{2\eta(M^2 - \eta^2) + 12\eta^2}{3vp'(M^2 + \eta^2)(M^2 - \eta^2)}\delta q \tag{5.15}$$

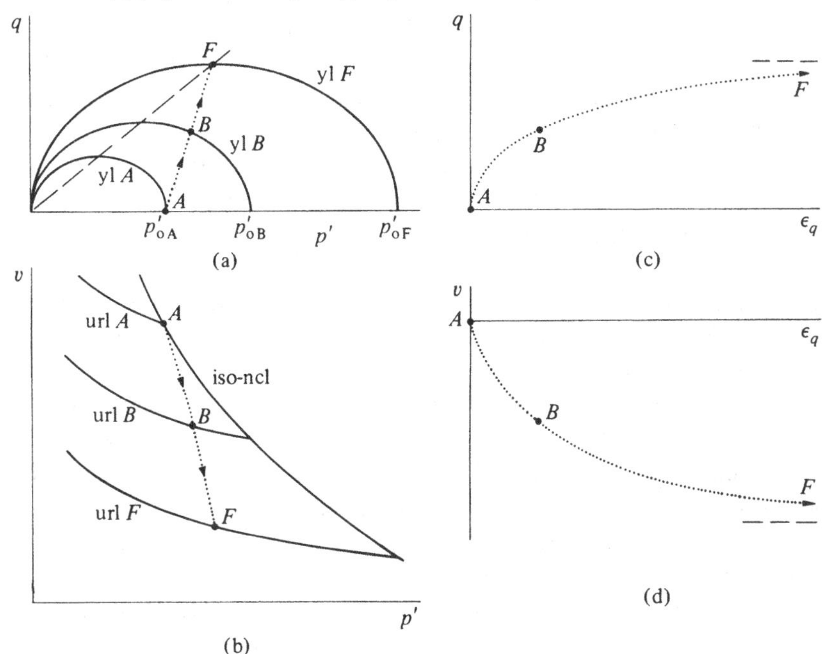

Fig. 5.5 Conventional drained triaxial compression test on normally compressed soil: (a) $p':q$ effective stress plane; (b) $v:p'$ compression plane; (c) $q:\varepsilon_q$ stress:strain plot; (d) $v:\varepsilon_q$ volume:strain plot.

As η increases towards M, the slope $\delta q/\delta\varepsilon_q^p$ decreases steadily towards zero (Fig. 5.4c) [the elastic shear strain contributes an unchanging stiffness $\delta q/\delta\varepsilon_q^e$ (5.11) as q increases].

Finally, the information about the test in Figs. 5.4b, c can be combined to show the changing specific volume of the soil as the shear strain increases (Fig. 5.4d). The change in volume is largely controlled by the irrecoverable plastic volumetric strain. From (5.7),

$$\frac{\delta\varepsilon_p^p}{\delta\varepsilon_q^p} = \frac{M^2 - \eta^2}{2\eta} \tag{5.7bis}$$

and as η increases towards M, the ratio $\delta\varepsilon_p^p/\delta\varepsilon_q^p$ and hence the slope $\delta v/\delta\varepsilon_q$ decrease steadily towards zero (Fig. 5.4d).

The route by which point B in Figs. 5.3 and 5.4 was reached has so far been left rather vague. Two simple histories can be proposed. Firstly, the soil could have been initially isotropically normally compressed to A in Fig. 5.5. Point A lies at the tip of yield locus yl A, and as soon as the drained compression begins, the yield locus has to expand to accommodate the new stress states. Plastic strains develop from the start of this drained compression, and both the stress:strain curve q:ε_q (Fig. 5.5c) and the volume:strain curve v:ε_q (Fig. 5.5d) show a continuous curve as the test proceeds.

Alternatively, the soil could have been isotropically normally compressed to $p' = p'_{oB}$ (Fig. 5.6a) and then unloaded isotropically to A, leaving the soil lightly overconsolidated. The yield locus yl B created by the normal compression is the yield locus through B: the isotropic unloading to A and the drained compression from A to B represent changes in stress lying inside this yield locus and are consequently purely elastic processes. Both A and B lie on the unloading–reloading line url B with its tip at $p' = p'_{oB}$ (Fig. 5.6b). The soil shows a stiff elastic response on drained compression from A to B; and when plastic strains start to develop as the soil yields at B, the stress:strain curve q:ε_q shows a sharp drop in stiffness (Fig. 5.6c). Similarly, the volume change from A to B is elastic, recoverable, and small; and the volume:strain relationship v:ε_q also shows a break point at B (Fig. 5.6d). (The sharpness of this break point depends on the values of the soil parameters λ, κ, G', and M and on the stress ratio η at which yielding occurs at B. It would be quite possible for the ratio of elastic volumetric and shear strain increments just before yielding at B to be the same as the ratio of elastic plus plastic volumetric and shear strain increments just after yielding at B.)

The soil sample in Fig. 5.6 was lightly isotropically overconsolidated. If the soil were isotropically compressed to K (Fig. 5.7a) and then unloaded

Fig. 5.6 Conventional drained triaxial compression test on lightly overconsolidated soil: (a) $p':q$ effective stress plane; (b) $v:p'$ compression plane; (c) $q:\varepsilon_q$ stress:strain plot; (d) $v:\varepsilon_q$ volume:strain plot.

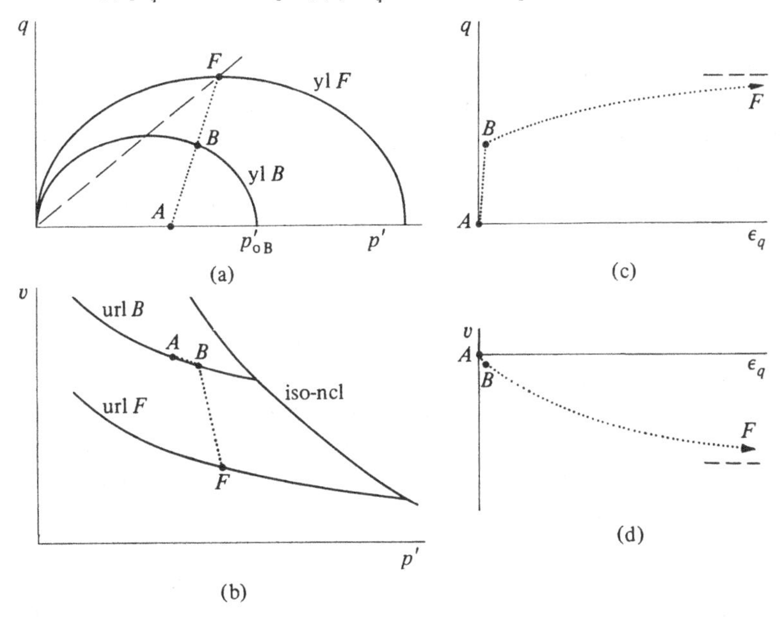

Fig. 5.7 Conventional drained triaxial compression test on heavily overconsolidated soil: (a) $p':q$ effective stress plane; (b) $v:p'$ compression plane; (c) $q:\varepsilon_q$ stress:strain plot; (d) $v:\varepsilon_q$ volume:strain plot.

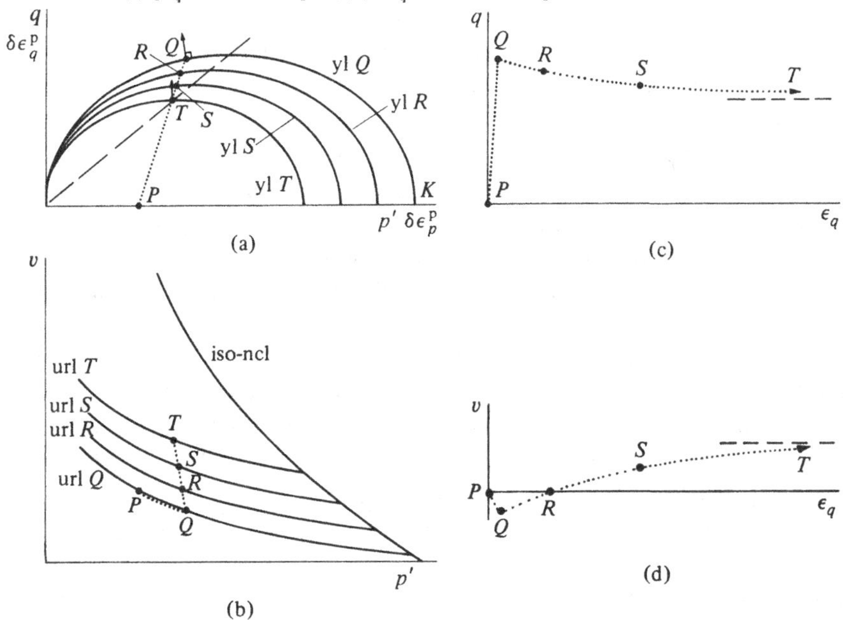

to a point such as P, on drained compression the response would be elastic until the yield locus yl Q was reached at a point Q lying to the left of the top of the yield locus. At Q the plastic strain increment vector points to the left, indicating that if a continued *increase* in shear strain is to occur, with plastic shear strains of the same sign as the preceding elastic shear strains, then this must be associated with *negative* plastic volumetric strain, in other words, plastic volumetric expansion. Expression (5.7) changes sign as η passes from $\eta < M$ to $\eta > M$.

Plastic volumetric expansion requires, from (5.9), that $\delta p'_o < 0$ and that the yield locus should contract. So, at Q the effective stress path has to retreat back towards P. Increments (or decrements) of p' and q are still linked through (5.13). The progress of the test in the compression plane (Fig. 5.7b) can be deduced by projecting the points R, S, and T in the stress plane (Fig. 5.7a) from their yield loci yl R, yl S, and yl T down to the corresponding unloading–reloading lines url R, url S, and url T.

As the yield loci contract, the position of the effective stress state Q, R, S, T relative to the top of the yield locus changes, and the direction of the plastic strain increment vector gradually approaches the vertical until at T, with $\eta = M$, plastic shear deformations can again continue without plastic change in volume. The stress:strain $q{:}\varepsilon_q$ and volume:strain $v{:}\varepsilon_q$ relationships can be constructed as before (Figs. 5.7c, d). After the initial elastic rise in q and decrease in volume (PQ), further plastic shearing ($QRST$) is associated with a drop in q and an increase in v towards the limiting values corresponding to point T.

The response of this heavily overconsolidated soil sample indicates that there is a problem in applying the general procedure for use of the model outlined at the end of Section 5.2 because, at points such as Q, R, and S in Fig. 5.7, with $\eta > M$, plastic strains are associated with contraction rather than expansion of the yield locus: plastic softening instead of plastic hardening. A little consideration shows that some uncertainty is associated with a change of stress QR (Fig. 5.8a). It could be associated with positive plastic shear strain QR in Fig. 5.8b, as in the previous example. However, stress state R could also be reached from Q by a purely *elastic* unloading, QR' in Fig. 5.8b, with no change in size of the yield locus. This aspect of the behaviour of the model is discussed further in Section 7.4.

The contrasting patterns of response predicted by the Cam clay model for conventional drained compression tests on normally compressed (Fig. 5.5) and heavily overconsolidated (Fig. 5.7) samples of soil can be compared with experimental data from tests on normally compressed and heavily overconsolidated samples of reconstituted Weald clay taken from Bishop and Henkel (1957) (Figs. 5.9 and 5.10). These experimental data are presented as plots of deviator stress against axial strain (q:ε_a) (Figs. 5.9a and 5.10a) and plots of volumetric strain against axial strain (ε_p:ε_a) (Figs. 5.9b and 5.10b). These plots are broadly comparable with the plots used for the Cam clay predictions in Figs. 5.5c, d and 5.7c, d.

The responses of the model and of the real soil are similar. The normally compressed sample shows a steady increase in deviator stress q with strain

Fig. 5.8 (a) Change in stresses QR associated with (b) irrecoverable plastic shear deformation QR or elastic unloading QR'.

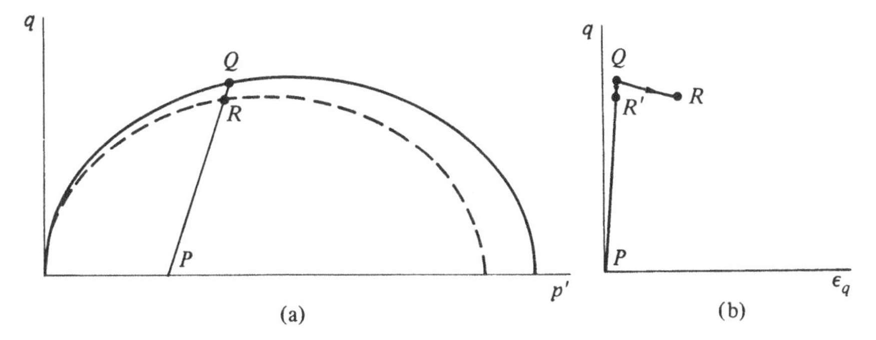

Fig. 5.9 Conventional drained triaxial compression test on normally compressed Weald clay [$\sigma'_r = 207$ kPa (30 lbf/in^2)]: (a) deviator stress q and axial strain ε_a; (b) volumetric strain ε_p and axial strain ε_a (after Bishop and Henkel, 1957).

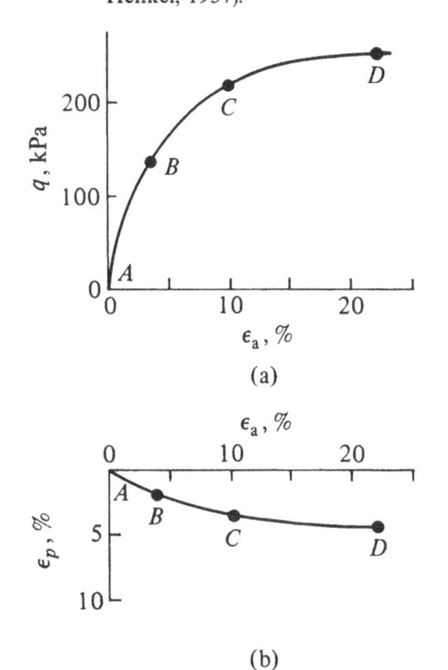

(Figs. 5.9a and 5.5c) and a steady decrease in volume (or increase in compressive volumetric strain) (Figs. 5.9b and 5.5d). The heavily overconsolidated sample shows a rise in deviator stress to a peak followed by a drop with continued shearing (Figs. 5.10a and 5.7c), and an initial decrease in volume (positive volumetric strain) followed by volumetric expansion (negative volumetric strain) (Figs. 5.10b and 5.7d). It is perhaps not surprising that the real soil has smoothed off the sharp peak proposed by the model in Fig. 5.7c; but apart from this detail, the features of the response predicted by the Cam clay model are those that are seen in the response of real soils and are features that one would wish to see incorporated in a successful model of soil behaviour.

5.4 Cam clay predictions: conventional undrained triaxial compression

The way in which counterbalancing, equal, and opposite elastic and plastic volumetric strains could combine to give zero nett change in volume was described in Section 4.3 for the general elastic–plastic model for soil. Since the elastic volumetric strain increment is

$$\delta\varepsilon_p^e = \kappa \frac{\delta p'}{vp'} \qquad (5.1\text{bis})$$

Fig. 5.10 Conventional drained triaxial compression test on heavily overconsolidated Weald clay [$\sigma'_r = 34$ kPa (5 lbf/in.^{2})]: (a) deviator stress q and axial strain ε_a; (b) volumetric strain ε_p and axial strain ε_a (after Bishop and Henkel, 1957).

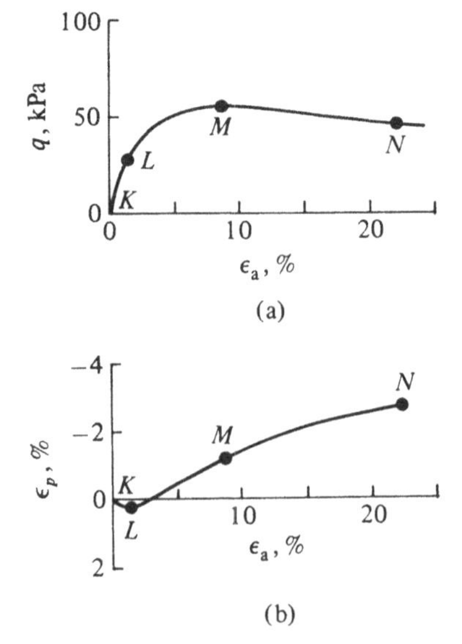

and the plastic volumetric strain increment is

$$\delta\varepsilon_p^{\mathrm{p}} = (\lambda - \kappa)\frac{\delta p_{\mathrm{o}}'}{vp_{\mathrm{o}}'} \tag{5.9bis}$$

the condition for the two to cancel out is

$$\delta\varepsilon_p^{\mathrm{e}} + \delta\varepsilon_p^{\mathrm{p}} = 0 \tag{5.16}$$

or

$$\kappa\frac{\delta p'}{p'} = -(\lambda - \kappa)\frac{\delta p_{\mathrm{o}}'}{p_{\mathrm{o}}'} \tag{5.17}$$

which forces a link between changes in mean effective stress p' and changes in the size of the yield locus, controlled by p_{o}'. The geometry of the yield locus provides a link between changes in p' and η (or q) which are causing yielding and the resulting changes in p_{o}', as indicated by (5.4). Thus, (5.17) provides a constraint on the changes of p' and η (or q) which lead to constant volume deformation of the soil. From (5.4a) and (5.17),

$$\frac{-\delta p'}{p'} = \frac{(\lambda - \kappa)}{\lambda}\frac{2\eta}{(M^2 + \eta^2)}\delta\eta \tag{5.18}$$

This expression can be integrated to give

$$\frac{p_{\mathrm{i}}'}{p'} = \left(\frac{M^2 + \eta^2}{M^2 + \eta_{\mathrm{i}}^2}\right)^{\Lambda} \tag{5.19}$$

where

$$\Lambda = \frac{\lambda - \kappa}{\lambda} \tag{5.20}$$

and p_{i}' and η_{i} define an initial effective stress state. Expression (5.19) specifies the shape of the undrained effective stress path in the $p'{:}q(=\eta p')$ plane (Fig. 5.11), provided yielding and plastic volumetric strains are occurring.

If the effective stress state lies inside the current yield locus, then there is no possibility of plastic volumetric (or shear) strains being generated, and the only way that (5.16) can be satisfied is for the elastic volumetric strain also to be zero, which implies, from (5.1), that for isotropic elastic soil the mean effective stress remains constant. Consequently, for any initially overconsolidated sample that is subjected to an undrained test, the effective stress path must rise at constant p' until the current yield locus is reached (*AB* in Fig. 5.12) and the elastic–plastic effective stress path of (5.19) can be joined (*BC* in Fig. 5.12).

It is clear from (5.17) that the changes in p' and p_{o}' must always be of opposite sign. The sign of $\delta p_{\mathrm{o}}'$ is linked with the sign of the plastic volumetric strain increment, which is linked with the slope of the yield curve (or, strictly, plastic potential) at the current stress state. If the soil is in a state where $\eta < M$ (AB in Fig. 5.13), then the yielding takes place with a plastic strain increment vector directed to the right, implying plastic volumetric *compression*. The soil wants to harden plastically, and the current yield locus must increase in size, $\delta p'_o > 0$. The mean effective stress must fall, $\delta p' < 0$, so that elastic expansion can balance the plastic compression (Fig. 5.13).

If the undrained increment of deformation is being applied with $\eta > M$ (QR in Fig. 5.14), then the yielding takes place with a plastic strain increment vector directed to the left, implying plastic volumetric *expansion*. The soil now wants to soften plastically, and the current yield locus must decrease in size, $\delta p'_o < 0$. The mean effective stress must now rise, $\delta p' > 0$, so that elastic compression can balance the plastic expansion (Fig. 5.13).

If the undrained increment of deformation is being applied at the top of the yield locus (plastic potential) with $\eta = M$ (G in Fig. 5.13), then yielding takes place with a plastic strain increment vector directed parallel

Fig. 5.11 Effective stress path for constant volume deformation, Equation (5.19).

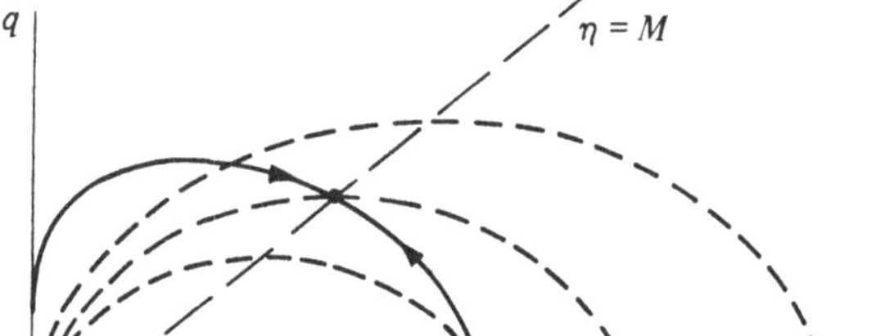

Fig. 5.12 Effective stress path for constant volume, undrained, deformation of lightly overconsolidated soil.

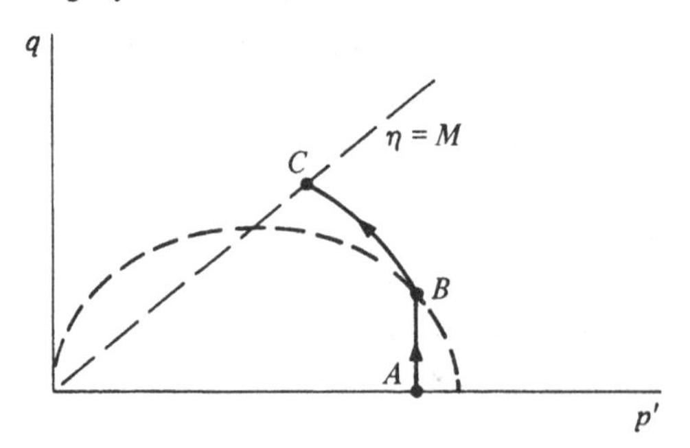

to the q axis, implying zero plastic volumetric strain, $\delta\varepsilon_p^p = 0$. Shearing now continues indefinitely without change in size of yield locus, $\delta p'_o = 0$ and hence, from (5.17), also without change in mean effective stress, $\delta p' = 0$. This particular stress ratio $\eta = M$ thus acts as a limit to the undrained effective stress path which cannot pass through this particular stress ratio (Fig. 5.11), though it can approach it from either direction.

The prediction of the *effective* stress path followed by a sample in an undrained test requires no knowledge of the *total* stress path that is being imposed. The same effective stress path is followed for *any* triaxial compression total stress path: the only difference to be seen in tests

Fig. 5.13 Increments of constant volume deformation: (a) $p':q$ effective stress plane; (b) $v:p'$ compression plane.

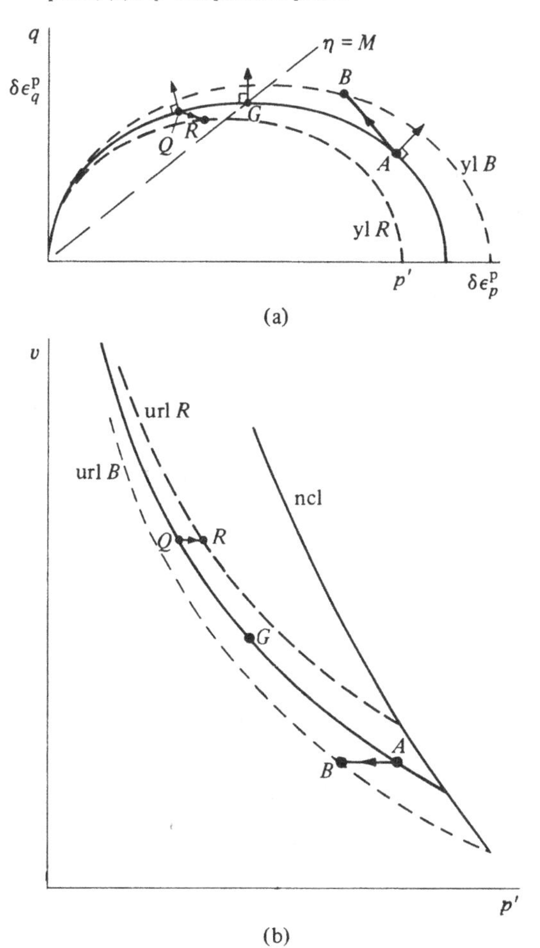

performed with different total stress paths is in the pore pressure that develops. If an increment of the effective stress path indicates a change in mean effective stress $\delta p'$, and the corresponding increment of the total stress path indicates a change in total mean stress δp, then the change in pore pressure δu is (Fig. 5.14)

$$\delta u = \delta p - \delta p' \tag{5.21}$$

In Section 1.6, a pore pressure parameter a was introduced to link pore pressure changes with changes in applied total stresses:

$$\delta u = \delta p + a\,\delta q \tag{1.65bis}$$

Comparison of (1.65) and (5.21) shows that a indicates the current slope of the undrained effective stress path:

$$a = \frac{-\delta p'}{\delta q} \qquad (5.22)(\text{cf. } 1.64)$$

In conventional undrained triaxial compression tests conducted with constant cell pressure, the total stress path is given by [cf. (5.13)]

$$\delta q = 3\,\delta p \tag{5.23}$$

If the soil is deforming purely elastically, with effective stress states lying inside the current yield locus, then

$$\delta p' = 0$$

and

$$\delta u = \delta p = \frac{\delta q}{3} \tag{5.24}$$

As previously noted in Section 2.2, this implies that the pore pressure parameter a is zero for such an elastic process.

If the soil is deforming plastically, then the value of a can be deduced from (5.18). Recall that

$$\delta q = \eta\,\delta p' + p'\,\delta\eta \tag{5.25}$$

Fig. 5.14 Change in pore pressure $\delta u = \delta p - \delta p'$.

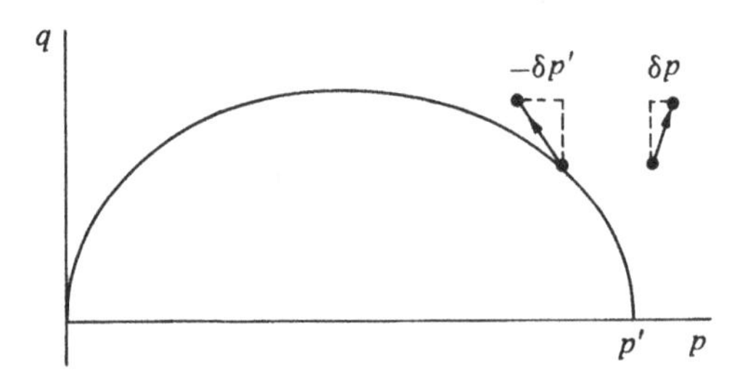

The resulting expression for a is then

$$a = \frac{2(\lambda - \kappa)\eta}{\lambda(M^2 + \eta^2) - 2(\lambda - \kappa)\eta^2} \tag{5.26}$$

An undrained test on an initially normally compressed sample is shown in Fig. 5.15. Because the path of the test is constrained in the $p':v$ compression plane to lie at constant volume, $\delta v = 0$, the progress of the test can be followed in graphical construction by choosing a sequence of points in the compression plane (Fig. 5.15b). Each point A, B, C, D, E, and F lies on a new unloading–reloading line and hence on a new yield locus.

Fig. 5.15 Conventional undrained triaxial compression test on normally compressed soil: (a) $p':q$ effective stress plane; (b) $v:p'$ compression plane; (c) $q:\varepsilon_q$ stress:strain plot; (d) $u:\varepsilon_q$ pore pressure:strain plot.

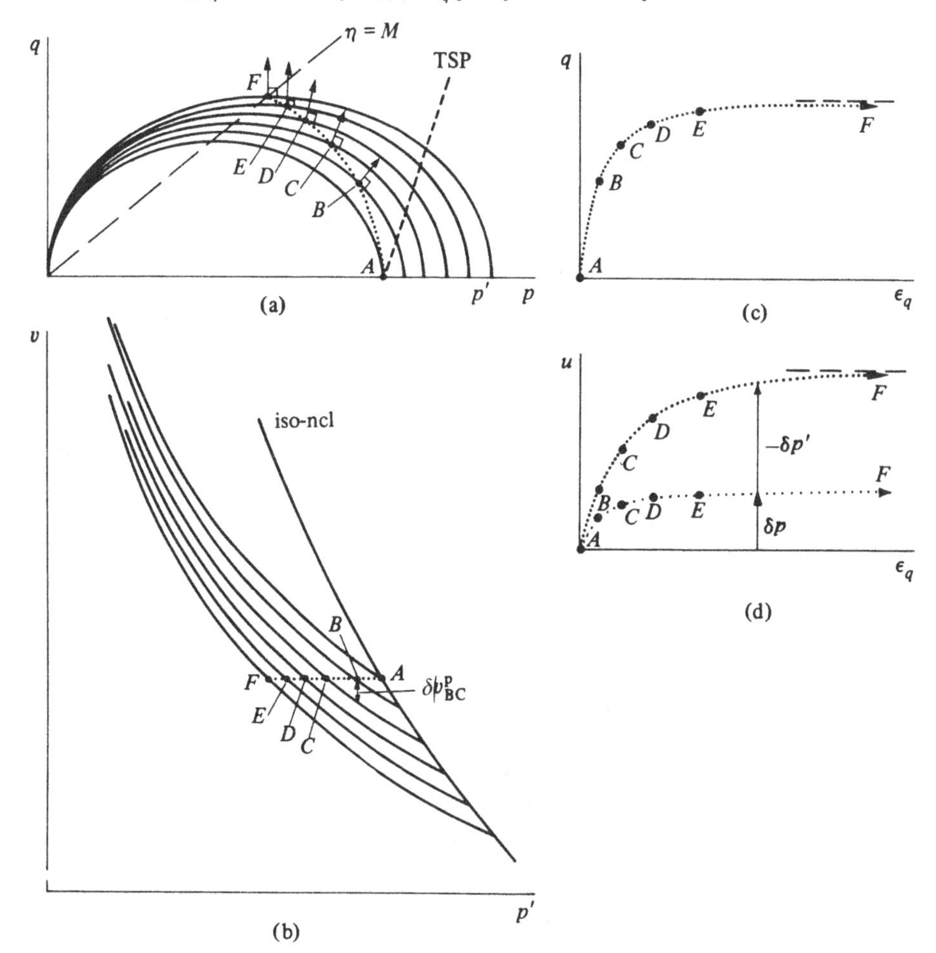

The path in the $p':q$ effective stress plane can then be generated graphically by projecting up from points A, B, C, D, E, and F to the corresponding yield loci (Fig. 5.15a). At each point the direction of the plastic strain increment vector can be plotted, and the ratio of components of plastic shear strain increment to plastic volumetric strain increment is thus specified. The magnitude of the plastic volumetric strain increment can be calculated from the vertical separation (δv^{p}) of successive unloading–reloading lines in the compression plane (Fig. 5.15b),

$$\delta\varepsilon_p^{\mathrm{p}} = \frac{-\delta v^{\mathrm{p}}}{v} \tag{5.27}$$

and hence the magnitude of the plastic shear strain increment can be determined. By adding the elastic shear strains, we can build up the deviator

Fig. 5.16 Conventional undrained triaxial compression test on lightly overconsolidated soil: (a) $p':q$ effective stress plane; (b) $v:p'$ compression plane; (c) $q:\varepsilon_q$ stress:strain plot; (d) $u:\varepsilon_q$ pore pressure:strain plot.

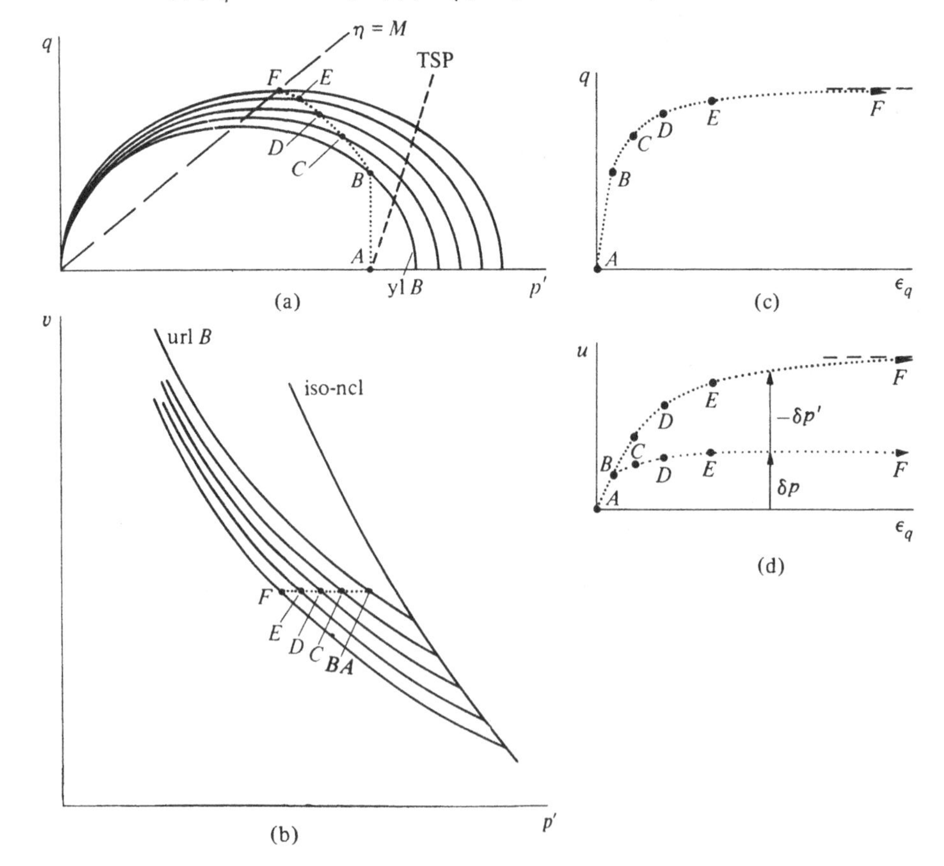

stress:shear strain relationship (Fig. 5.15c). The pore pressure at each stage can be deduced from the horizontal separation of the effective and total stress paths in the $p':q$ plane (Fig. 5.15a), and this can be plotted against the shear strain (Fig. 5.15d) to show the steady increase in pore pressure in this conventional undrained test on a normally compressed sample. The pore pressure has been separated into two parts, indicating the pore pressure due to change in total mean stress δp, over which the soil has no influence, and the pore pressure resulting from the suppressed volume change of the soil, $-\delta p'$. As discussed earlier, the test cannot proceed beyond the stress ratio $\eta = M$, and the implied values of deviator stress

Fig. 5.17 Conventional undrained triaxial compression test on heavily overconsolidated soil: (a) $p':q$ effective stress plane; (b) $v:p'$ compression plane; (c) $q:\varepsilon_q$ stress:strain plot; (d) $u:\varepsilon_q$ pore pressure:strain plot.

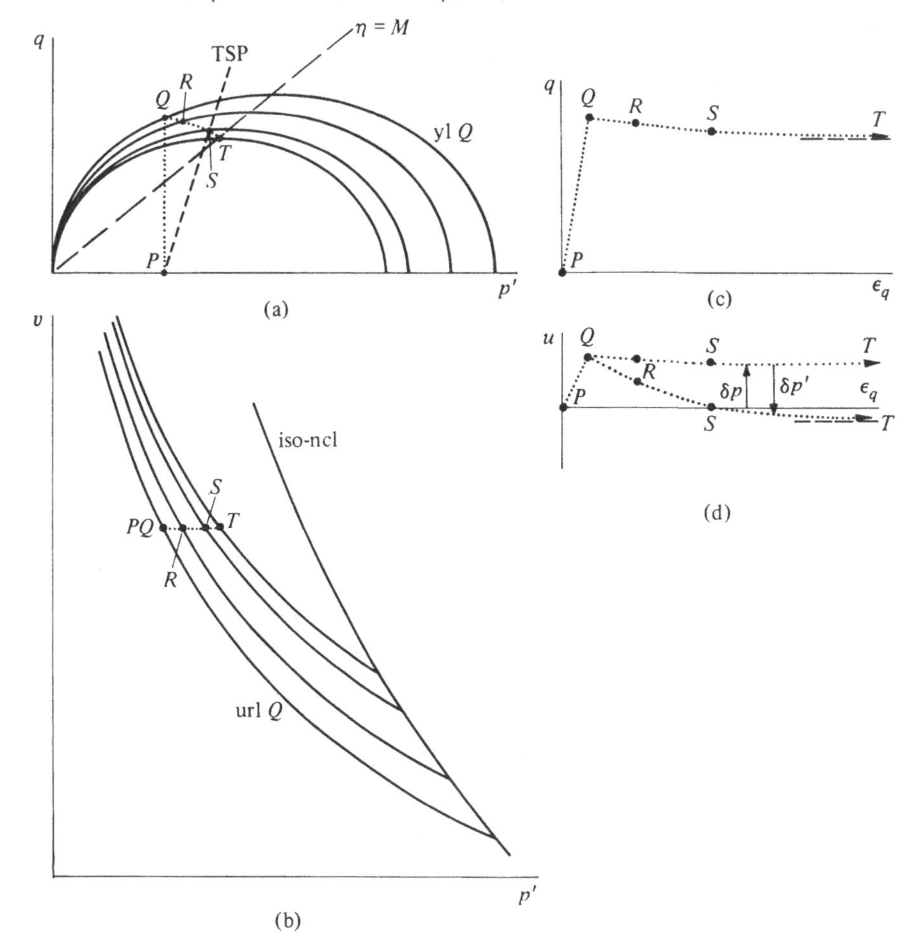

q and mean effective stress p' impose asymptotes towards which the stress: strain curve (Fig. 5.15c) and pore pressure:strain curve (Fig. 5.15d) tend.

An undrained test on an initially lightly overconsolidated sample is essentially similar (Fig. 5.16) except that it begins with an elastic phase AB at constant p' until the stress state reaches the initial yield locus yl B. A graphical deduction of this initial path is as follows. The test is constrained to lie along a constant v line in the compression plane; the test path cannot escape from its initial unloading–reloading line url B (Fig. 5.16b) unless plastic volumetric strains occur; plastic volumetric strains can occur only if the effective stress state lies on the current yield locus; hence the test path must lie at the intersection of the initial unloading–reloading line and the constant volume line; in other words, it must retain the initial value of p' until the value of deviator stress q has increased

Fig. 5.18 Conventional undrained triaxial compression test on normally compressed Weald clay [$\sigma'_r = 207$ kPa (30 lbf/in.^{2})]: (a) deviator stress q and triaxial shear strain ε_q; (b) pore pressure change Δu and triaxial shear strain ε_q (after Bishop and Henkel. 1957).

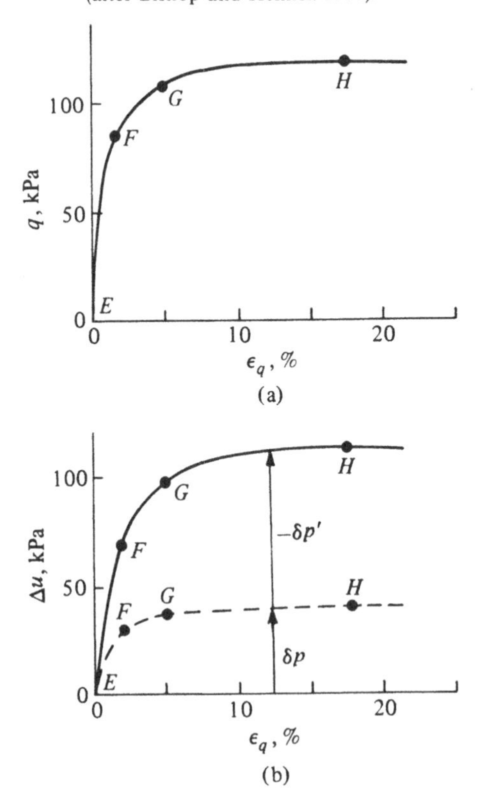

sufficiently to bring the effective stress state to the yield locus and allow plastic volumetric strains to occur.

The elastic stage AB (Fig. 5.16a) is associated only with elastic shear strains (5.2) (Fig. 5.16c) and with changes in pore pressure equal to the change in total mean stress (5.24) (Fig. 5.16d). When yielding begins at B, extra plastic shear strains occur and there is a sharp drop in stiffness (Fig. 5.16c). Also, when yielding begins, the effective stress path turns towards the left, $\delta p' < 0$, and there is in general a kink in the pore pressure: strain relationship (Fig. 5.16d).

In an undrained test on an initially heavily overconsolidated sample (Fig. 5.17), the first phase of loading is purely elastic PQ with no change in mean effective stress $\delta p' = 0$ until the initial yield locus yl Q is reached (Fig. 5.17a). The shear strains are purely elastic (Fig. 5.17c) and the pore pressures equal to the changes in total mean stress (Fig. 5.17d). Yielding, with $\eta > M$, is associated with shrinking of the yield locus (Fig. 5.17a) and with increase in mean effective stress. In the compression plane (Fig. 5.17b), the path moves across progressively higher unloading–reloading lines as plastic volumetric expansion occurs. The stress:strain curve (Fig. 5.17c)

Fig. 5.19 Conventional undrained triaxial compression test on heavily overconsolidated Weald clay [$\sigma_r' = 34$ kPa (5 lbf/in.^{2})]: (a) deviator stress q and triaxial shear strain ε_q; (b) pore pressure change Δu and triaxial shear strain ε_q (after Bishop and Henkel, 1957).

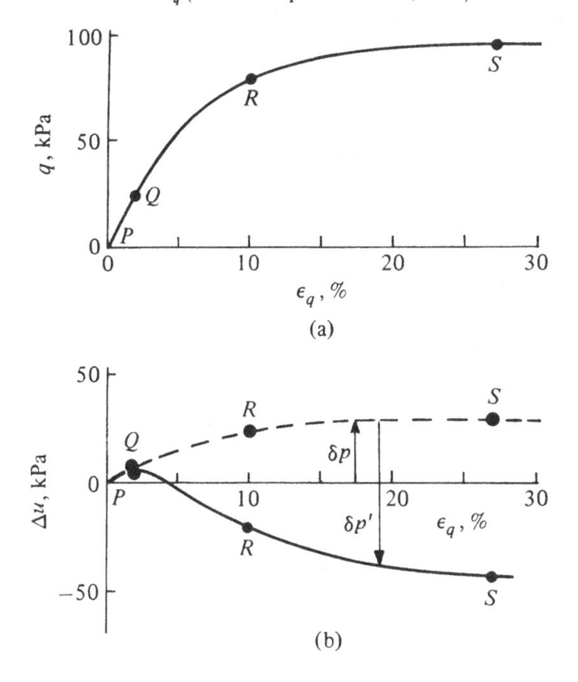

can be built up as before. The pore pressure (Fig. 5.17d) reduces from its maximum value, reached at the moment of yield (Q), and may actually have become negative by the time the end of the test is reached with $\eta = M$.

Experimental data of conventional undrained triaxial compression tests on normally compressed and heavily overconsolidated samples of reconstituted Weald clay are shown in Figs. 5.18 and 5.19, taken from Bishop and Henkel (1957). These can be compared with the predictions of the Cam clay model (Figs. 5.15c, d and 5.17c, d). Again, the responses of the model and of the real soil are similar. The principal feature of both the predicted and the observed behaviours is the contrast between the steady increase of pore pressure in the undrained tests on the normally compressed soil (Figs. 5.15d and 5.18b) and the initial increase followed by a steady decrease of pore pressure in the undrained tests on the heavily overconsolidated soil (Figs. 5.17d and 5.19b). It is not surprising that the real soil has smoothed out the peak proposed by the model in Fig. 5.17d.

The features of the response predicted by the model correspond to the features observed in the response of real soil; as in Section 5.3, these are features that one would wish to see incorporated in a successful model of soil behaviour.

5.5 Conclusion

Note that the distinguishing features of either drained or undrained tests on normally compressed and heavily overconsolidated soil arise from the same aspect of the model: the assumption about the plastic potential, which controls the relationship between the magnitudes of increments of plastic shear strain and plastic volumetric strain. If the effective stress state is in a position in which strain hardening with plastic volumetric compression occurs in a drained stress increment, which for Cam clay means $\eta < M$, then a decrease in mean effective stress (and increase in pore pressure) must necessarily occur in a corresponding undrained stress increment. Similarly, if the effective stress state is in a position in which strain softening with plastic volumetric expansion occurs in a drained stress increment, which for Cam clay means $\eta > M$, then an increase in mean effective stress (and decrease in pore pressure) must necessarily occur in a corresponding undrained stress increment.

The use of this simple model to predict response of soil samples in conventional triaxial drained and undrained compression tests has been discussed in the previous sections, and it has been shown that these predictions match several of the characteristics observed in conventional triaxial tests on Weald clay. Equations (5.11) and (5.12) (or their generalisations) can be readily applied to the prediction of the response of soil samples to any other changes of effective stress or of strain. The model is in fact complete; and although it is likely that the values of the parameters of the model κ [in (5.1)], G' [in (5.2)], M [in (5.3)], and N and λ [in (5.8)] will be deduced from experimental data obtained from triaxial or oedometer tests, the model is quite capable of being used for making statements about all possible increments of stress and strain even with this small number of parameters. The extent to which such predictions provide a satisfactory description of actual response is discussed in Chapters 11 and 12.

Exercises

E5.1. A sample of clay is set up in a triaxial apparatus and anisotropically compressed, without allowing radial strains, to a vertical effective stress $\sigma'_v = 156$ kPa. It is found that the value of the earth pressure coefficient (ratio of horizontal to vertical effective stresses) measured in this process is $K_0 = 0.65$. The sample is then unloaded to a vertical effective stress of 100 kPa (point A) and allowed to reach equilibrium with zero pore pressure. It is found that K_0 is now equal to 1.0.

The deviator stress is now increased to 60 kPa without allowing drainage (point B). Without changing the total stresses, drainage is now permitted, and the pore pressure is allowed to dissipate. It is observed that yield occurs during this process (point C) and that the final equilibrium state at zero pore pressure is point D.

Estimate the axial and radial strains occurring on each stage AB, BC, and CD of the test. Assume that the behaviour of the clay can be described by the Cam clay model with values of soil parameters $M = 0.9$, $\lambda = 0.19$, $\kappa = 0.06$, $N = 2.88$, and $G' = 2500$ kPa.

Compare the strengths that would be measured in conventional drained and undrained triaxial compression of the clay starting from state D.

E5.2. A saturated sample of clay in a triaxial cell is isotropically normally compressed under a cell pressure of 1000 kPa at 20°C with a specific volume $v = 3.181$. Investigate the effects of a sudden fall of temperature of 10°C on the subsequent undrained strength of the clay. Assume that the temperature change occurs so rapidly that no water is able to enter or leave the sample, but that pore pressures arising from the temperature change are allowed to dissipate before the sample is tested in undrained compression at the lower temperature.

Assume that the coefficients of thermal expansion for water

and for the soil particles are about 20×10^{-6} per °C and 200×10^{-6} per °C, respectively, and that the bulk modulus of water is 2.2 GPa. Assume that the behaviour of the clay can be described by the Cam clay model with $\lambda = 0.25$, $\kappa = 0.05$, $N = 3.64$, and $M = 0.9$.

E5.3. A sample of Cam clay is in equilibrium in a triaxial cell under effective stresses $\sigma'_r = 200$ kPa and $\sigma'_a = 300$ kPa. The sample was deforming plastically just before this effective stress state was reached.

The sample is now subjected to changes in effective stresses $\delta\sigma'_r = -1$ kPa, $\delta\sigma'_a = +4$ kPa. Estimate the increments in axial and radial strain that will result. Take values for the soil parameters $\lambda = 0.26$, $\kappa = 0.07$, $N = 3.52$, $M = 0.85$, and $G' = 1500$ kPa.

E5.4. A sample of Cam clay under an isotropic effective stress $p' = 200$ kPa is found to have a specific volume $v = 2.06$. Indicate possible histories of (i) isotropic compression and unloading, (ii) conventional drained compression and unloading, (iii) conventional undrained compression and unloading, and (iv) constant mean effective stress compression and unloading which would be compatible with the observed state of the sample. Take values of soil parameters $\lambda = 0.19$, $\kappa = 0.045$, $N = 3.12$, and $M = 0.93$.

6

Critical states

6.1 Introduction: critical state line

All of the tests in the previous chapter, for which the simple elastic–plastic model, Cam clay, was used to predict the response, eventually tended towards an ultimate condition in which plastic shearing could continue indefinitely without changes in volume or effective stresses. This condition of perfect plasticity has become known as a *critical state*, the attainment of which can be expressed by

$$\frac{\partial p'}{\partial \varepsilon_q} = \frac{\partial q}{\partial \varepsilon_q} = \frac{\partial v}{\partial \varepsilon_q} = 0 \tag{6.1}$$

These critical states were reached with an effective stress ratio

$$\frac{q_{cs}}{p'_{cs}} = \eta_{cs} = M \tag{6.2}$$

In drained or undrained tests on normally compressed (or lightly overconsolidated) soil (*AB* and *AC* in Fig. 6.1; cf. Figs. 5.5 and 5.15), yielding first occurs with stress ratio $\eta < M$. Continued loading, whether drained or undrained, is associated with plastic hardening, expansion of yield loci, and increase of stress ratio until ultimately the effective stress state is at the top of the current yield locus (yl *B* or yl *C*), the plastic strain increment vector is directed parallel to the q axis, $\delta\varepsilon^p_p/\delta\varepsilon^p_q = 0$, and a perfectly plastic critical state is reached with $\eta = M$.

In drained and undrained tests on heavily overconsolidated soil (*PQ* and *PR* in Fig. 6.1; cf. Figs. 5.7 and 5.17), yielding first occurs with $\eta > M$. Continued deformation is associated with plastic softening, contraction of yield loci, and decrease of stress ratio until ultimately the effective stress state is at the top of the current yield locus (yl *Q* or yl *R*), and a critical state is again reached with $\eta = M$.

A moment's consideration shows that this stress ratio $\eta = M$ produces an ultimate limiting condition, a critical state, in all tests, *provided plastic deformations are occurring*, because with $\eta = M$ the effective stress state is at the top of the current yield locus, and so indefinite plastic shearing can occur without further expansion or contraction of the yield locus. The caveat "provided plastic deformations are occurring" is important because, with the shape of yield locus assumed in the Cam clay model, it is quite possible for stress ratios equal to M to be reached elastically within the

Fig. 6.1 Conventional drained and undrained triaxial compression tests on normally compressed and heavily overconsolidated samples.

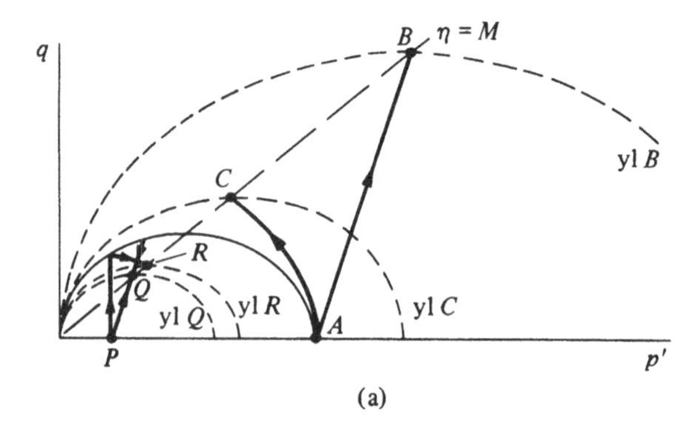

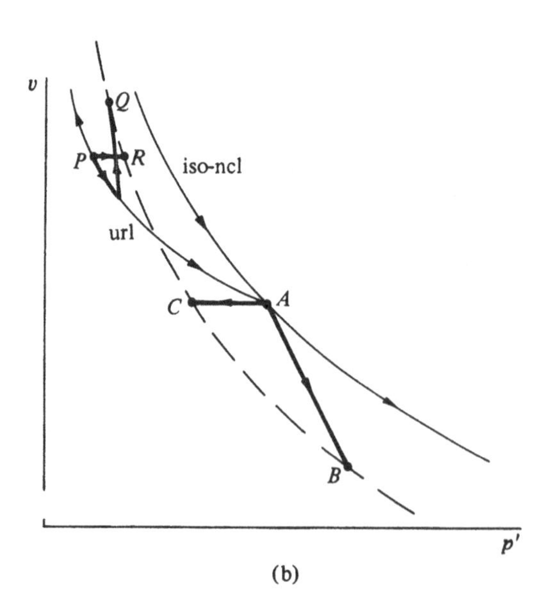

current yield locus (e.g. at X in Fig. 6.2) and without the serious consequences of a critical state developing.*

The locus of critical states in the p':q stress plane is the line joining the tops of yield loci (Fig. 6.3a), $\eta = M$ or

$$q_{cs} = Mp'_{cs} \tag{6.3}$$

this line is called the *critical state line* (csl). The proportions of the Cam clay yield loci are independent of their size. The general equation for the

Fig. 6.2 Points X and Y inside current yield locus and not at critical state.

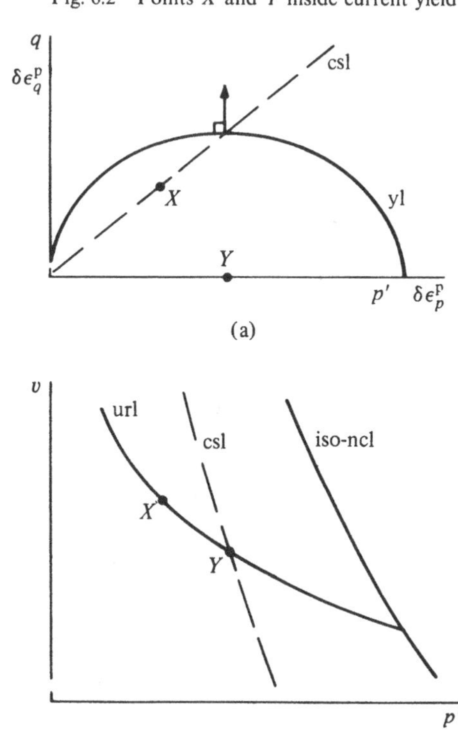

*Although it is usually convenient to contemplate changes in effective stress in relation to yield loci, these conclusions about the existence of critical states for $\eta = M$ result because the plastic potentials in the Cam clay model have normals parallel to the q axis, implying $\delta\varepsilon^p_p/\delta\varepsilon^p_q = 0$, for $\eta = M$. It so happens that these plastic potentials have the same shape as the yield loci, so the critical state is reached with the effective stress state at the top of the current yield locus. For other elastic–plastic models of the more general form discussed in Chapter 4, the plastic potentials and yield loci can be distinct, and there is then no necessity for the condition $\delta\varepsilon^p_p/\delta\varepsilon^p_q = 0$ to be associated with the top of the yield loci. Compare Exercise E4.4.

yield loci is

$$\frac{p'}{p'_o} = \frac{M^2}{(M^2 + \eta^2)} \qquad (6.4)(5.3\text{bis})$$

The size of a yield locus is controlled by p'_o, and the top of the yield locus, where $\eta = M$ (Fig. 6.3a), has

$$p'_{cs} = \frac{p'_o}{2} \qquad (6.5)$$

Fig. 6.3 Critical state line (csl) and intersection of yield loci with line $q/p' = \eta$.

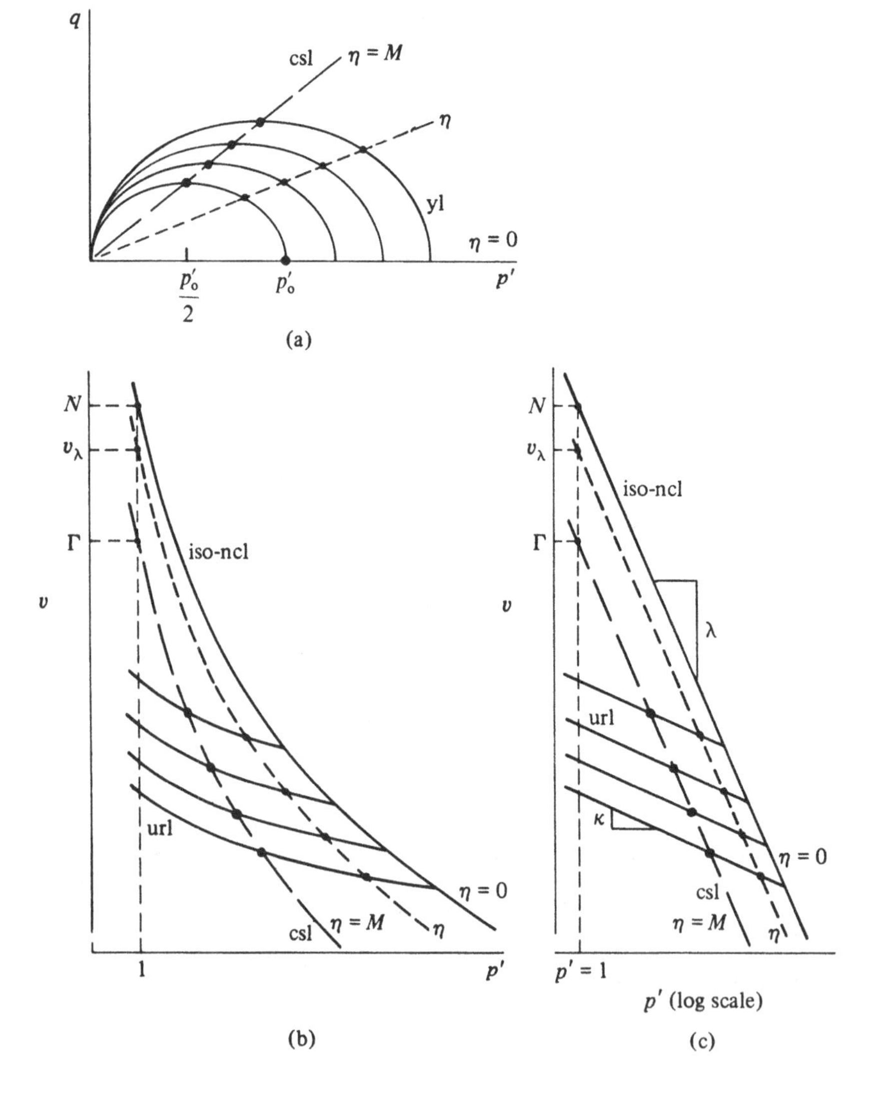

Each yield locus (yl) is associated with an unloading–reloading line (url) in the compression plane $p':v$ (Fig. 6.3b) which has its tip at $p' = p'_o$ on the isotropic normal compression line (iso-ncl). This normal compression line has the equation

$$v = N - \lambda \ln p' \qquad (6.6)(5.8\text{bis})$$

and is straight in the semi-logarithmic compression plane $v:\ln p'$ (Fig. 6.3c). The unloading–reloading lines are also straight in this form of the compression plane, with general equation

$$v = v_\kappa - \kappa \ln p' \qquad (6.7)(4.3\text{bis})$$

Thus, the particular unloading–reloading line which corresponds to the yield locus with size p'_o is

$$v = N - \lambda \ln p'_o + \kappa \ln \frac{p'_o}{p'} \qquad (6.8)$$

At a mean stress $p' = p'_{cs} = p'_o/2$, the specific volume is then

$$v_{cs} = N - \lambda \ln 2p'_{cs} + \kappa \ln 2$$

or

$$v_{cs} = N - (\lambda - \kappa)\ln 2 - \lambda \ln p'_{cs} \qquad (6.9)$$

Each critical state combination of p'_{cs} and q_{cs} in the effective stress plane is associated with a critical state combination of p'_{cs} and v_{cs} in the

Fig. 6.4 Three-dimensional view of normal compression line (ncl), critical state line (csl), and series of Cam clay yield loci.

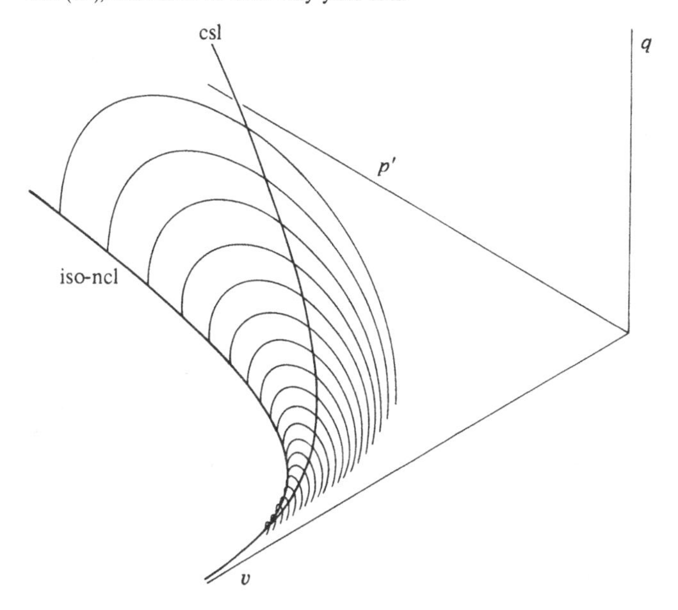

compression plane (Figs. 6.3a, b). The line joining these critical states has the expression (6.9), which can be rewritten as

$$v_{cs} = \Gamma - \lambda \ln p'_{cs} \tag{6.10}$$

where

$$\Gamma = N - (\lambda - \kappa) \ln 2 \tag{6.11}$$

This is a line in the compression plane at constant vertical v separation, $(\lambda - \kappa) \ln 2$, from the normal compression line (parallel to the normal compression line in the $v{:}\ln p'$ plane of Fig. 6.3c) which links the combinations of p'_{cs} and v_{cs} corresponding to the effective stress states in which plastic deformations are occurring with $\eta = M$. The constant Γ is the specific volume intercept at $p'_{cs} = 1$. Like N, its value, unfortunately, depends on the units of stress. Throughout this book the unit of stress is the kilopascal (kPa), so $v_{cs} = \Gamma$ for $p'_{cs} = 1$ kPa. The two expressions (6.3) and (6.10) provide a complete specification of the critical state line.

Combinations of p'_{cs}, q_{cs}, and v_{cs} which simultaneously satisfy (6.3) and (6.10) are critical states. Again, it is possible to discover combinations of p', q, and v which satisfy one but not both (6.3) and (6.10). Point X in Fig. 6.2a has $q = Mp'$; it appears to be on the critical state line in the stress plane but lies inside the current yield locus, so its position on the current unloading–reloading line in the compression plane (Fig. 6.2b) is not on the critical state line. Point Y in Fig. 6.2b, reached perhaps by isotropic compression and unloading, appears to lie on the critical state line in the compression plane but also lies inside the current yield locus, so its position in the stress plane (Fig. 6.2a) is not on the critical state line.

If p', q, and v are thought of as three orthogonal axes, the combinations of stresses and volume can be plotted in the three-dimensional space defined by these axes. The critical state line becomes a single curved line (Fig. 6.4) of which (6.3) and (6.10) are projections onto the $p'{:}q$ plane ($v = 0$) and $p'{:}v$ plane ($q = 0$), respectively. Only combinations of p', q, and v lying on this curved three-dimensional line are critical states.

6.2 Two-dimensional representations of $p'{:}q{:}v$ information

Display of information about (triaxial) states of stress $p'{:}q$ and about specific volumes v requires two two-dimensional plots – the stress plane and the compression plane – or one three-dimensional plot such as that shown in Fig. 6.4. The Cam clay model can point the way to two-dimensional devices for displaying this information.

The equation of the Cam clay yield locus is

$$\frac{p'}{p'_o} = \frac{M^2}{M^2 + \eta^2} \tag{6.4bis}$$

and involves only ratios of stresses. For a given ratio of deviator stress to mean stress ($q/p' = \eta$), there is a certain ratio of mean stress to tip stress (p'/p'_o). Constant ratios become constant separations in logarithmic plots. Therefore, just as the stress ratio $\eta = M$ which defines the tops of the yield loci produces the critical state line parallel to the normal compression line in the compression plane $v{:}\ln p'$ (Fig. 6.3c), so any other stress ratio which defines a series of geometrically similar points on a series of yield loci produces a line which is particular to that stress ratio and also parallel to the normal compression line (Fig. 6.3c).

Each of these lines has an equation of the form

$$v = v_\lambda - \lambda \ln p' \tag{6.12}$$

A way of converting specific volume and mean effective stress information to a single variable is to use

$$v_\lambda = v + \lambda \ln p' \tag{6.13}$$

or, in other words, to project the information of v and $\ln p'$ parallel to the normal compression line up to the line $p' = 1$. The value of v_λ for soil which is deforming plastically (yielding) depends only on the stress ratio $\eta = q/p'$; that is, the pair of variables $\eta{:}v_\lambda$ can be used to display information about the three quantities $p'{:}q{:}v$ defining the state of the soil. Evidently, by comparison with (6.6), for isotropic normal compression,

$$\eta = 0, \qquad v_\lambda = N$$

(N in Fig. 6.5); and by comparison with (6.10), at the critical state,

$$\eta = M, \qquad v_\lambda = \Gamma$$

(C in Fig. 6.5). The values of v_λ corresponding to other values of η can be found by combining the expression for the specific volume at a point on an unloading–reloading line,

$$v = N - \lambda \ln p'_o + \kappa \ln \frac{p'_o}{p'} \tag{6.8bis}$$

with the definition of v_λ (6.13) to give

$$v_\lambda = N - (\lambda - \kappa) \ln \frac{p'_o}{p'} \tag{6.14}$$

This, with the equation of the yield locus (6.4), gives

$$v_\lambda = N - (\lambda - \kappa) \ln \frac{M^2 + \eta^2}{M^2} \tag{6.15}$$

or

$$\frac{\eta^2}{M^2} = \exp\left(\frac{N - v_\lambda}{\lambda - \kappa}\right) - 1 \tag{6.16}$$

This relationship is plotted in Fig. 6.5 as curve NCX. Any stress state which is on a current yield locus lies on this curve.

The path of any conventional drained or undrained compression test on isotropically normally compressed soil starts at N and moves along curve NC to end at a critical state at C. The path of an undrained test on an isotropically overconsolidated soil starts at a point such as P (lightly overconsolidated) or R (heavily overconsolidated) in Fig. 6.5 with $\eta = 0$. Loading within the yield locus occurs at constant mean stress p'. Since v is also constant in an undrained test, from (6.13) v_λ remains constant until the yield locus is reached (PQ or RS in Fig. 6.5). The path then moves along the yield locus until a critical state is reached (QC or SC in Fig. 6.5).

The path of any conventional drained test on an isotropically overconsolidated soil also starts at a point such as P or R. Initial loading within the yield locus is associated with increase in p' and decrease in v down an unloading–reloading line. The slope κ of the unloading–reloading line is lower than the slope λ of the normal compression line, and, hence, this initial loading produces an increase in v_λ (PF and RG in Fig. 6.5) until the yield locus is reached. Then the yield locus is followed up or down to a critical state (FC and GC in Fig. 6.5).

The quantity v_λ converts data of p' and v to a constant mean stress section. An alternative which is equivalent to converting these data to a

Fig. 6.5 Paths of conventional drained and undrained triaxial compression tests plotted in terms of normalised volume v_λ and stress ratio η.

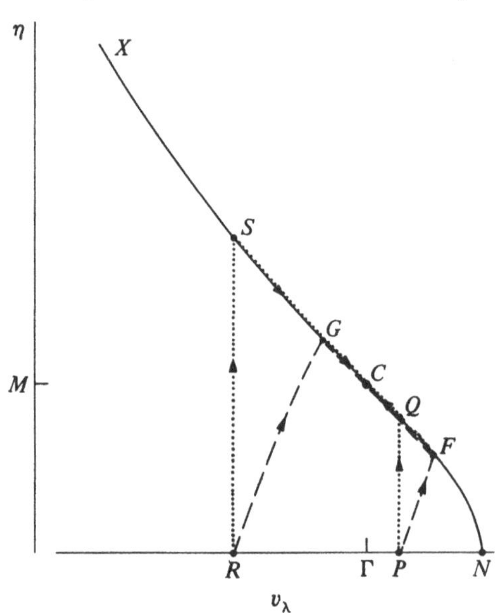

constant volume section is to normalise the effective stresses $p':q$ with respect to the so-called equivalent consolidation pressure p'_e. This equivalent consolidation pressure is the pressure which, in isotropic normal compression, would give the soil its current specific volume (Fig. 6.6). The isotropic normal compression line is

$$v = N - \lambda \ln p' \tag{6.6bis}$$

and hence p'_e is

$$p'_e = \exp \frac{N - v}{\lambda} \tag{6.17}$$

With a mean stress p' inside a current yield locus of size p'_o,

$$v = N - \lambda \ln p'_o + \kappa \ln \frac{p'_o}{p'} \tag{6.8bis}$$

and hence, after some manipulation,

$$\frac{p'}{p'_e} = \left(\frac{p'}{p'_o}\right)^{\Lambda} \tag{6.18}$$

where

$$\Lambda = \frac{\lambda - \kappa}{\lambda} \qquad (6.19)(5.20\text{bis})$$

Fig. 6.6 Equivalent consolidation pressure p'_e.

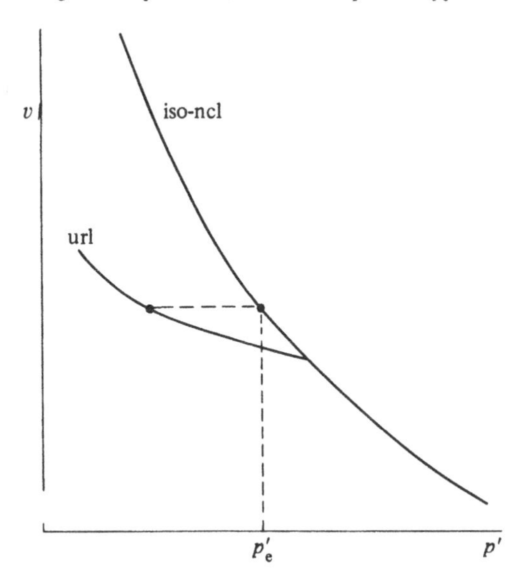

For stress states on the current yield locus,

$$\frac{p'}{p'_o} = \frac{M^2}{M^2 + \eta^2} \tag{6.4bis}$$

and

$$\frac{p'}{p'_e} = \left(\frac{M^2}{M^2 + \eta^2}\right)^{\Lambda} \tag{6.20}$$

which is essentially the same as (5.19), the equation of the undrained effective stress path. Evidently,

$$\frac{q}{p'_e} = \frac{\eta p'}{p'_e} \tag{6.21}$$

and the pair of expressions (6.20) and (6.21) can be used to generate the curve in the p'/p'_e:q/p'_e plane corresponding to the Cam clay yield locus (Fig. 6.7). There is a point N ($p'/p'_e = 1$:$q/p'_e = 0$), corresponding to isotropic normal compression, and a point C ($p'/p'_e = 2^{-\Lambda}$:$q/p'_e = M2^{-\Lambda}$), corresponding to the critical state.

Because p'_e remains unchanged during a constant volume undrained test, undrained effective stress paths scale directly into the p'/p'_e:q/p'_e diagram. So an undrained test on an overconsolidated sample which shows no change of mean effective stress p' until the current yield locus is reached rises at constant p'/p'_e until the yield locus is reached (PQ and RS in Fig. 6.7) and then moves round the yield curve in the p'/p'_e:q/p'_e plane to a critical state (QC and SC in Fig. 6.7).

A conventional drained test on a normally compressed sample follows

Fig. 6.7 Paths of conventional drained and undrained triaxial compression tests plotted in non-dimensional effective stress space p'/p'_e:q/p'_e.

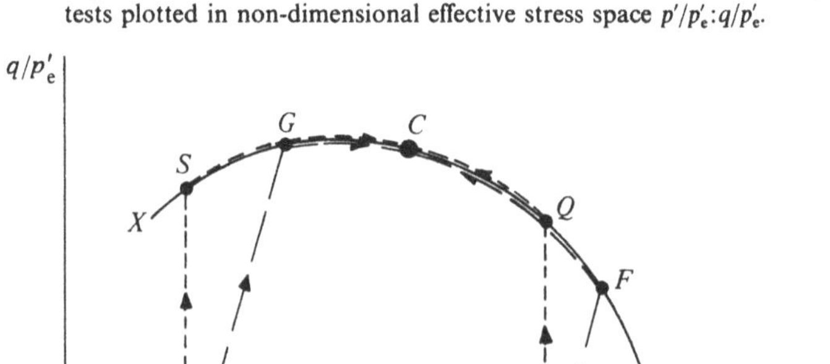

the path NC in the compression plane shown in Fig. 5.5b. The mean stress p' increases because of the imposed stress path, but the equivalent consolidation pressure increases more rapidly as the soil compresses and the ratio p'/p'_e falls. In fact, the test follows round the curve NC in the $p'/p'_e:q/p'_e$ plane (Fig. 6.7). (The point C is not at the top of the curve NCX for precisely the same reason that the effective stress path followed in an undrained test does not have zero slope, $\delta q/\delta p' \neq 0$, at the critical state, which is a consequence of the non-negligible elastic volumetric strains that are assumed to occur in the model. If there are no elastic volumetric strains, $\kappa = 0$ and $\Lambda = 1$, and the point C is then at the summit of the curve in Fig. 6.7.)

A conventional drained test on an overconsolidated sample begins with a small volumetric compression down the current unloading–reloading line as the stress path rises to the current yield locus (Figs. 5.6b, 5.7b). The equivalent consolidation pressure p'_e increases slightly with this volumetric compression, but not as rapidly as the mean stress; p'/p'_e increases, until the yield locus is reached (PF and RG in Fig. 6.7). Yielding of the lightly overconsolidated sample is associated with plastic volumetric compression to a critical state, p'/p'_e decreases (FC in Fig. 6.7; compare Fig. 5.6b). Yielding of the heavily overconsolidated sample is associated with plastic volumetric expansion to a critical state, p'/p'_e increases (GC in Fig. 6.7; compare Fig. 5.7b).

These two-dimensional representations bring together drained and undrained response into a single diagram and help to emphasise the fact that, considered in terms of effective stresses, different modes of testing are merely probing different parts of a single unified picture of soil behaviour.

6.3 Critical states for clays

Data from four conventional triaxial compression tests on samples of reconstituted Weald clay (taken from Bishop and Henkel, 1957) were used in Sections 5.3 and 5.4 to demonstrate that the simple Cam clay model was reproducing commonly observed features of soil response. These data were presented in the form of the standard plots that are used to display triaxial test results: plots of deviator stress q and volumetric strain ε_p or pore pressure change Δu against axial strain ε_a (Figs. 5.9, 5.10, 5.18, and 5.19). Axial strain is a quantity which can increase without limit, subject only to restrictions imposed by the apparatus. In the search for experimental evidence of critical states, it is more helpful to look at data in terms of quantities which are limited in their variation and to plot the paths of tests in the effective stress plane $p':q$ and the compression plane

$p':v$, or in the combined two-dimensional representations of these two planes that have been discussed in the previous section.

Enough information is usually available to achieve the conversion. The deviator stress q is given directly. The constant cell pressure at which a test is conducted is the initial mean effective stress p'_i (assuming zero pore pressure and zero deviator stress at the start of the test). At subsequent stages of the test,

$$p' = p'_i + \frac{q}{3} - \Delta u \tag{6.22}$$

where Δu is the measured change in pore pressure, which is zero for drained tests. Conversion of volumetric strains to specific volumes requires a value of a volumetric variable at some stage of the test, usually the initial water content w_i for saturated clays. The initial specific volume is then

$$v_i = 1 + G_s w_i \tag{6.23}$$

where G_s is the specific gravity of the soil particles; at subsequent stages of the test,

$$v = v_i(1 - \varepsilon_p) \tag{6.24}$$

For the reconstituted Weald clay, $G_s = 2.75$, and data of isotropic compression and unloading are reported by Henkel (1959) (Fig. 6.8). The two normally compressed samples (1 and 2) were compressed isotropically to a mean effective stress $p' = 207$ kPa (30 pounds force per square inch, lbf/in^2) before being sheared. The two overconsolidated samples (3 and

Fig. 6.8 Isotropic compression and unloading of Weald clay (after Henkel, 1959).

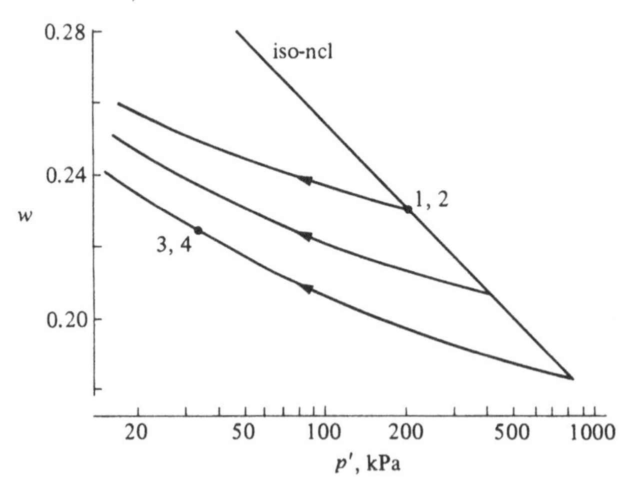

Table 6.1

Test	Point	w	ε_p (%)	v	q (kPa)	Δu (kPa)	p' (kPa)	Remarks
1	A	0.232	0	1.640	0		207	
	B		2.0	1.607	138		253	
	C		3.5	1.583	221		281	
	D		4.5	1.566	248		290	Failure
2	E	0.232		1.640	0	0	207	
	F			1.640	90	69	167	
	G			1.640	114	96	148	
	H			1.640	119	114	133	Failure
3	K	0.224	0	1.617	0		34	
	L		0.3	1.612	28		43	
	M		−1.3	1.638	55		53	Failure
	N		−2.7	1.661	45		50	
4	P	0.224		1.617	0	0	34	
	Q			1.617	24	7	36	
	R			1.617	83	−21	83	
	S			1.617	93	−45	110	Failure

4) were compressed isotropically to a mean effective stress $p' = 827\,\text{kPa}$ ($120\,\text{lbf/in}^2$) and then allowed to swell back isotropically to a mean effective stress $p' = 34\,\text{kPa}$ ($5\,\text{lbf/in}^2$), resulting in an overconsolidation ratio of 24.

To illustrate the conversion of the experimental data to the stress plane and the compression plane from the conventional plots of Figs. 5.9, 5.10, 5.18, and 5.19, four points have been marked for each test, and values of $p':q:v$ have been computed for each point. These values are tabulated in Table 6.1.

The paths of these tests are plotted in the $p':q$ effective stress plane and the $p':v$ compression plane in Fig. 6.9. The effective stress paths of the drained tests (1 and 3) both have to rise at gradient $\delta q/\delta p' = 3$ (Fig. 6.9a). However, whereas test 1 rises steadily, $ABCD$, test 3 rises to a peak KLM and then falls, MN. Test 1 shows a steady fall in specific volume (Fig. 6.9b) whereas test 3 shows an initial drop in volume KL, followed by expansion LMN.

The effective stress paths for drained tests 1 and 3 form the total stress paths for the undrained tests 2 and 4: the pore pressure at each stage is the horizontal (p') separation in the stress plane between the two corresponding paths. The effective stress path for test 2 lies to the left of its corresponding total stress path (test 1): positive pore pressures build up, and the mean effective stress falls. The effective stress path for test 4 lies initially to the left and then to the right of its corresponding total stress path (test 3): the pore pressure is initially positive but then becomes negative. The initial pore pressure variation in this test is slightly deceptive. Looking at the tabulated values for points P and Q, we can see that the mean effective stress p' is initially almost constant (as expected for undrained constant volume shearing of isotropic elastic soil, and Cam clay would predict that soil with this history of overconsolidation is initially elastic) in spite of the initial positive pore pressures. The pore

Fig. 6.9 Conventional drained and undrained triaxial compression tests on Weald clay: (a) $p':q$ effective stress plane; (b) $v:p'$ compression plane (data from Bishop and Henkel, 1957).

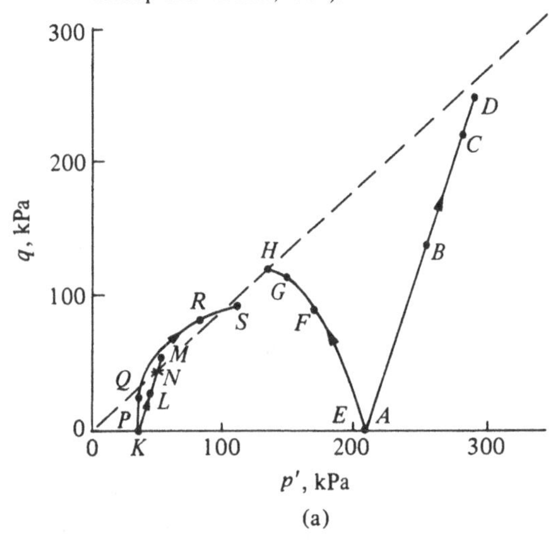

(a)

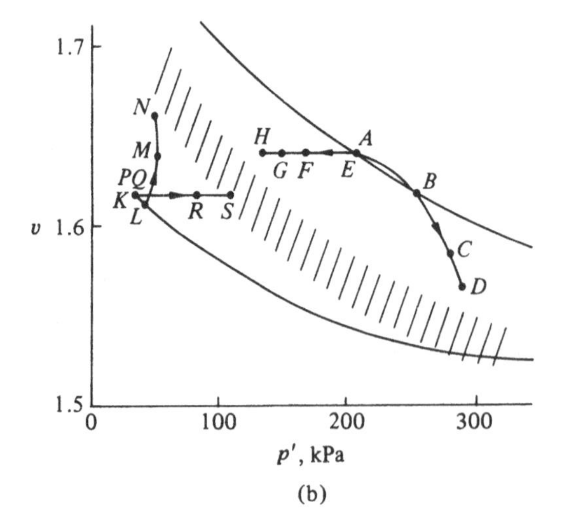

(b)

pressure more or less balances the increase in *total* mean stress imposed in this constant cell pressure test: $\delta u \approx \delta p = \delta q/3$, as shown in Fig. 5.19b.

It was noted in Chapter 5 that there is a parallelism between volume changes in drained tests and pore pressure changes in undrained tests on samples with the same consolidation history. Pore pressures develop in undrained tests only because the soil wants to change in volume as it is sheared but is prevented from doing so by the test configuration. The normally consolidated soil wants to collapse as it is sheared (test 1, Figs. 5.9 and 6.9b): positive pore pressures build up in the undrained test to reduce the mean effective stress carried by the soil particles. The resulting expansion just balances the collapse that could not occur (test 2, Figs. 5.18 and 6.9a). The heavily overconsolidated soil ultimately shows negative pore pressures in the undrained test (test 4, Figs. 5.19 and 6.9a): the soil skeleton is having to be held back by the pore pressure to prevent it from exploding (compare the drained test 3, Figs. 5.10 and 6.9b).

Do these test data support the existence of critical states for Weald clay? In the effective stress plane (Fig. 6.9a), a straight line can be drawn through the origin and through, or close to, the end points D, H, N, and S of the four tests. In the compression plane (Fig. 6.9b), these four tests do not define a line or curve of end points particularly closely. However, a curved zone can be indicated towards which the tests appear to be heading: test 1 by decrease in volume, test 2 by decrease in mean effective stress, test 3 by increase in volume, and test 4 by increase in mean effective stress.

To draw conclusions about general patterns of behaviour from the results of just four tests would be risky. Roscoe, Schofield, and Wroth (1958), however, gathered data of end points of a large number of drained and undrained triaxial compression tests on reconstituted Weald clay. Data from undrained tests are shown in the stress plane and the compression plane in Fig. 6.10, and data from drained tests in Fig. 6.11. In both cases, data from tests with overconsolidation ratios greater than 8 at the start of shearing have been excluded because there is a tendency for non-uniformities to develop in these tests, as is explained in Section 7.4. (Thus, the end points of tests 3 and 4 from Fig. 6.9 do not appear in Figs. 6.10 and 6.11.)

The end points from undrained tests on *normally compressed* samples have been used to define a locus of end points in both the stress plane (Fig. 6.10a), in which this locus is a straight line through the origin, and in the compression plane (Fig. 6.10b), in which the points lie on a smooth curve. It is clear, however, that the end points from *all* the undrained tests lie on or close to these lines.

These same lines have been drawn again in Figs. 6.11a, b for comparison with the data of end points from drained tests on Weald clay. It is clear that the same relationships match both sets of data. This does not, perhaps, now seem a surprising result. The response in drained and undrained tests can be predicted with a single model of soil behaviour based on effective stresses, such as Cam clay, and these tests are expected to fit into a single

Fig. 6.10 End points of conventional undrained triaxial compression tests on Weald clay (○ normally compressed samples; • overconsolidated samples): (a) $p':q$ effective stress plane; (b) $v:p'$ compression plane; (c) $v:p'$ compression plane (p' plotted on logarithmic scale) (after Roscoe, Schofield, and Wroth, 1958).

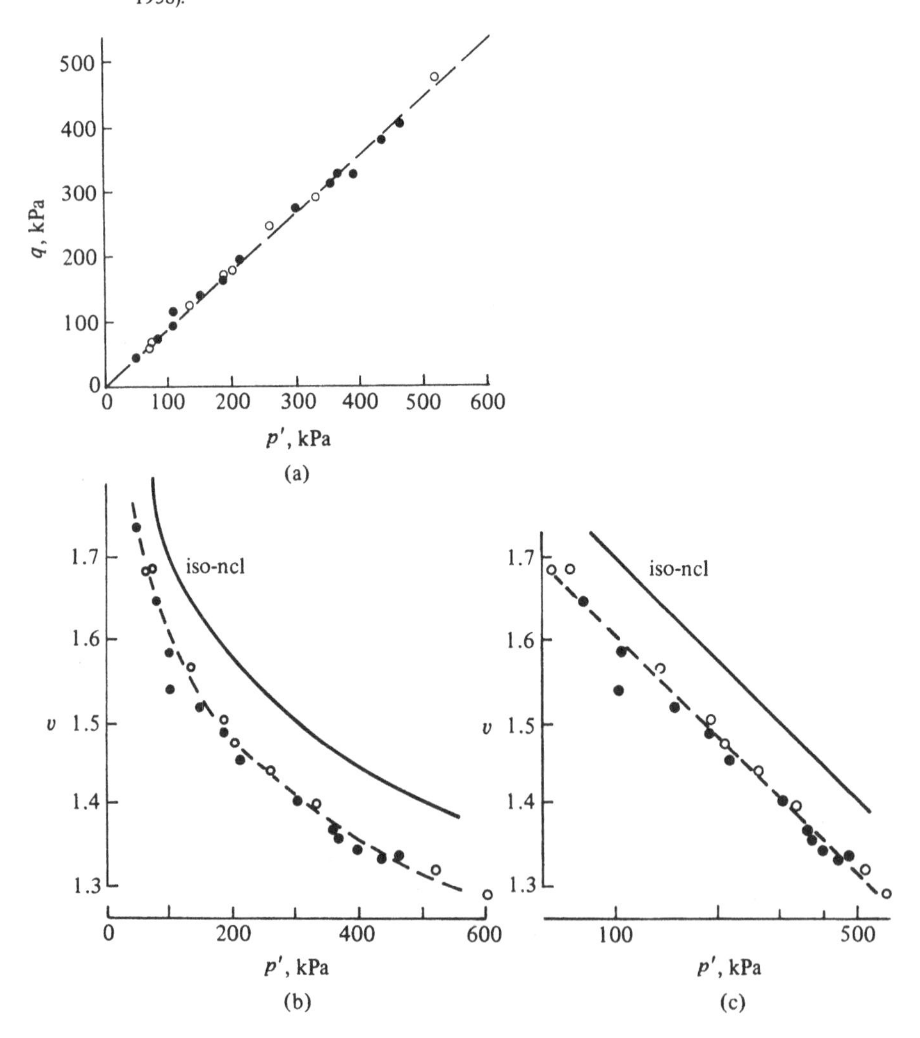

pattern of effective stress:strain response. Other drained tests could be conducted with other effective stress paths, so there is nothing particularly significant about the division between drained and undrained response. This finding should, however, be seen in the historical context of the state of soil mechanics thought in the 1950s when there was a tendency to regard the results of drained and undrained tests as quite unrelated and incompatible.

The collection of data in Figs. 6.10 and 6.11 presents more convincing evidence for the existence of a line of values of p', q, and v towards which all the tests have headed, irrespective of the type of test (which controls

Fig. 6.11 End points of conventional drained triaxial compression tests on Weald clay (○ normally compressed samples; • overconsolidated samples): (a) $p':q$ effective stress plane; (b) $v:p'$ compression plane (after Roscoe, Schofield, and Wroth, 1958).

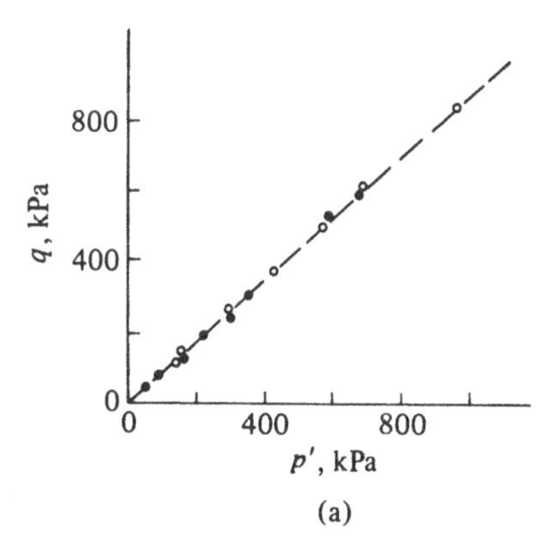

(a)

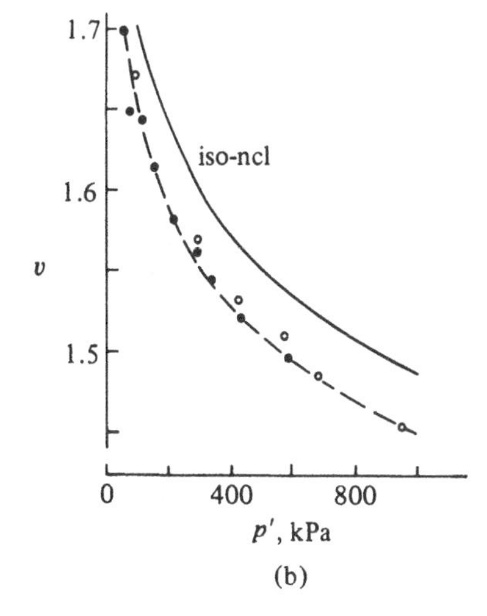

(b)

to some extent the path of the test) and irrespective of the consolidation history (which controls the position of the starting points of the tests relative to the line of end points). This line is described in the $p':q$ effective stress plane (Figs. 6.10a and 6.11a) by an equation,

$$q = 0.872p' \tag{6.25}$$

Drawing the line in a semi-logarithmic compression plane $v:\ln p'$ together with the isotropic normal compression line (Fig. 6.10b), we can see that the two lines are parallel and essentially straight and that the line of ultimate states can be described by an equation,

$$v = 2.072 - 0.091 \ln p' \tag{6.26}$$

The form of this line thus matches the form of the critical state line that emerged from the Cam clay model, (6.3) and (6.10); it is convenient to refer to it as the critical state line for Weald clay.

The critical state line operates as a limit on the changes of p', q, and v that occur in a test. The formal definition of a critical state requires that

$$\frac{\partial p'}{\partial \varepsilon_q} = \frac{\partial q}{\partial \varepsilon_q} = \frac{\partial v}{\partial \varepsilon_q} = 0 \tag{6.1bis}$$

Arrival at the condition described by (6.1) may require very large strains since, for states close to true critical states, the derivatives $\partial p'/\partial \varepsilon_q$, $\partial q/\partial \varepsilon_q$, and $\partial v/\partial \varepsilon_q$ (which include the tangent shear stiffness $\partial q/\partial \varepsilon_q$) are close to zero. Test conditions may not permit sample uniformity to be retained to

Fig. 6.12 (a) Conventional drained and (b) conventional undrained triaxial compression tests ending off the critical state line but still trying to head towards it.

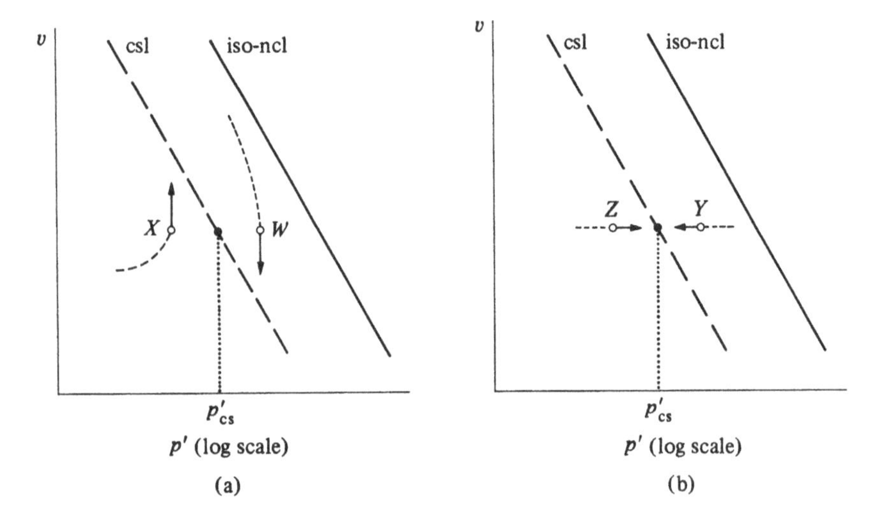

these large deformations and critical states may not actually be attained. Nevertheless the critical state line is still important as a line towards which tests are heading.

Parry (1958), for example, subjected the results of his triaxial tests on Weald clay to close scrutiny and concluded that when the quoted end points of tests did not seem to lie on the previously determined critical state lines, the states of the soils were in fact still moving towards critical states at the end of the tests. Schematically, certain drained tests ended to the right of the critical state line (at points such as W in Fig. 6.12a). The value of mean effective stress p'_f recorded at failure (or at the end of the test) was greater than the value of mean stress p'_{cs} on the critical state line at the same value of specific volume that the sample had at the end of the test. In such tests, the volume of the sample was observed to be decreasing at the end of the test and the state of the soil to be moving downwards towards the critical state line in the compression plane. That is, for $p'_{cs}/p'_f < 1$, $(\partial v/\partial \varepsilon_q)_f < 0$.

Conversely, when drained tests ended to the left of the critical state line (X in Fig. 6.12a), so that $p'_{cs}/p'_f > 1$, then the volume of the sample was observed to be increasing at the end of the test, $(\partial v/\partial \varepsilon_q)_f > 0$, and the state of the soil to be moving upwards towards the critical state line in the compression plane.

When undrained tests ended to the right of the critical state line (Y in Fig. 6.12b), so that $p'_{cs}/p'_f < 1$, then the pore pressure was observed to be increasing at the end of the test, $(\partial u/\partial \varepsilon_q)_f > 0$, and hence the mean effective stress to be falling, $(\partial p'/\partial \varepsilon_q)_f < 0$, and the state of the soil to be moving leftwards towards the critical state line in the compression plane. Conversely, when undrained tests ended to the left of the critical state line (Z in Fig. 6.12b), so that $p'_{cs}/p'_f > 0$, then the pore pressure was observed to be decreasing at the end of the test, and hence the mean effective stress

Fig. 6.13 End points of conventional triaxial tests on Weald clay: (a) rate of volume change in drained tests; (b) rate of change of pore pressure in undrained tests (after Parry, 1958).

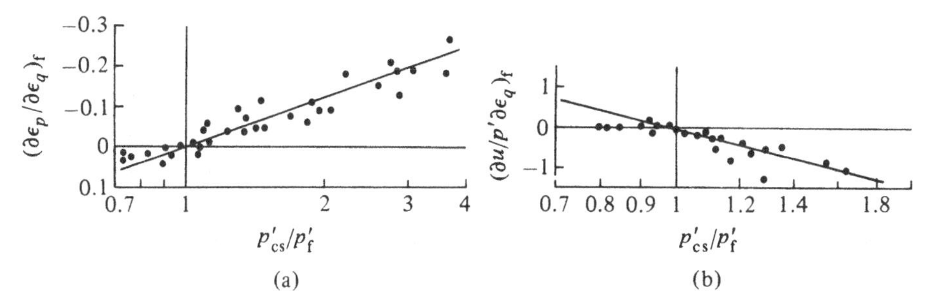

to be rising, $(\partial p'/\partial \varepsilon_q)_f > 0$, and the state of the soil to be moving rightwards towards the critical state line in the compression plane.

The data assembled by Parry (1958) are shown in Fig. 6.13 in terms of values of $(\partial \varepsilon_p/\partial \varepsilon_q)_f$ for drained tests (Fig. 6.13a) and of $(\partial u/p'\,\partial \varepsilon_q)_f$ for undrained tests (Fig. 6.13b). The way in which the critical state line separates these contrasting modes of behaviour at the ends of the tests is clear.

6.4 Critical state line and qualitative soil response

The critical state line emerged in Section 6.1 as a consequence of the elastic–plastic model of soil behaviour which was constructed in preceding chapters. Now experimental evidence has been produced for such a line of states towards which triaxial compression tests tend to head. The acceptance of a critical state line permits an assessment to be made of the expected *qualitative* response in any triaxial compression test on a soil with any consolidation history. Although this section may appear repetitive when read in conjunction with Sections 5.3 and 5.4, we make this assessment here without reference to (or acknowledgement of) any detailed elastic–plastic model, building only on an acceptance of the existence of critical state lines and not on an understanding of all the details of Cam clay. Demonstration of how this can be achieved illustrates the importance of considering soil response in the effective stress plane and in the compression plane concurrently.

Two pairs of tests are considered: firstly, a pair of tests on samples which have been isotropically normally compressed (A in Fig. 6.14), and secondly, a pair of tests on samples which have been isotropically overconsolidated but have the same initial mean stress as samples A (B in Fig. 6.14). All four tests thus start at the same point on the p' axis, in the $p':q$ effective stress plane (Figs. 6.14a, b), but their initial positions A and B in the $p':v$ compression plane are on opposite sides of the critical state line (Fig. 6.14c).

In an undrained test on a normally compressed sample, starting from point A, the end point U on the critical state line (Fig. 6.14c) is dictated by the constant volume condition in the $p':v$ plane. Hence, the end point U in the stress plane $p':q$ can be deduced on the critical state line at the corresponding value of mean stress p' (Fig. 6.14a). The route in the effective stress plane from A to U is not known, but a simple curve can be sketched. In a conventional triaxial compression test with constant cell pressure, the total stress path has gradient $\delta q/\delta p = 3$. This line AW lies to the right of the effective stress path, and hence positive pore pressures are expected in the undrained test.

The total stress path of the undrained test, from A towards W, becomes the effective (and total) stress path of a drained test on the other normally compressed sample. The end point of this test is governed by the intersection W of this stress path with the critical state line in the $p':q$ stress plane (Fig. 6.14a). The end point W in the $p':v$ compression plane

Fig. 6.14 Expected behaviour in conventional drained and undrained triaxial compression tests deduced from initial state of soil and location of critical state line: (a), (b) paths in $p':q$ effective stress plane; (c) paths in $v:p'$ compression plane.

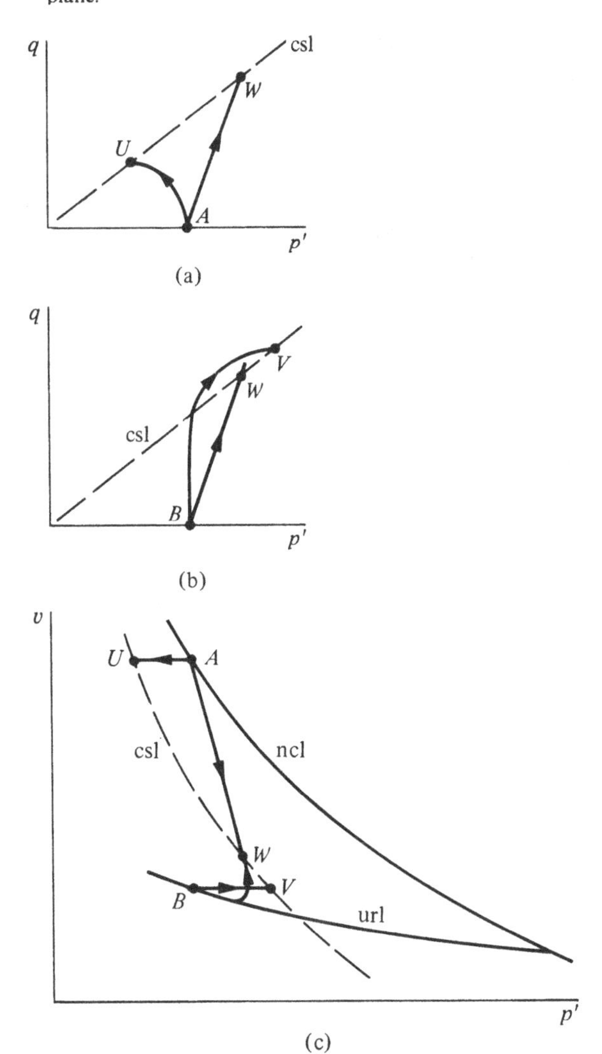

is then fixed on the critical state line at the same value of mean stress p' (Fig. 6.14c): volumetric compression is expected. Again, the shape of the path in the compression plane from A to W is not known, but a simple curve can be sketched.

The overconsolidated samples starting at point B (Figs. 6.14b, c) have the same initial effective stresses as the samples starting at point A. In a drained test, therefore, the effective (and total) stress path is the same, and consequently the critical state reached in this test is at W just as for the normally compressed sample. Point B lies well below point W in the compression plane (Fig. 6.14c), so this drained shearing must be associated with volumetric expansion. That much can be stated with knowledge only of the position of the critical state line. Other experience of soil behaviour could be incorporated to suggest that the stress path might overshoot W

Fig. 6.15 Triaxial compression tests on normally compressed soil: applied total stress path such that drained strength is lower than undrained strength; (a) paths in $p':q$ effective stress plane; (b) paths in $v:p'$ compression plane.

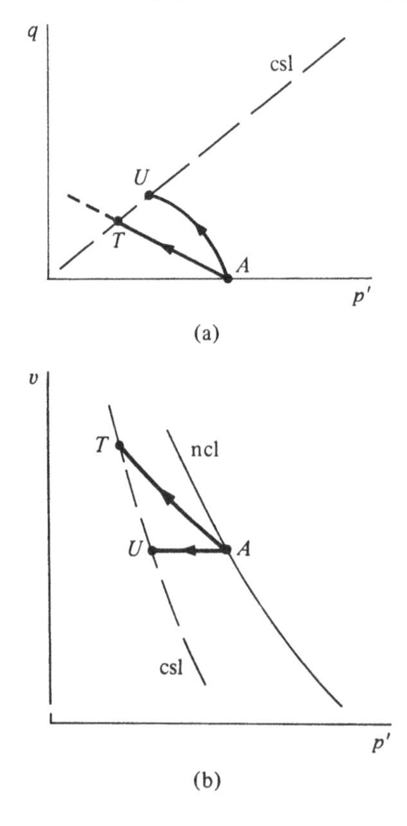

in the $p':q$ plane (Fig. 6.14b) (so that the stress:strain curve would show a peak), and that the sample might compress initially before expanding.

The end point reached on the critical state line in the undrained test on the overconsolidated sample is quite different from that reached in the undrained test on the normally compressed sample. The history of overconsolidation has left the soil at B with a much lower specific volume (Fig. 6.14c) and hence a much higher undrained strength. The position of the critical state point V relative to the starting point B in the compression plane (Fig. 6.14c) and the corresponding points in the effective stress plane (Fig. 6.14b) shows that an increase in mean effective stress p' has occurred during the test; in other words, a negative pore pressure is to be expected when the critical state is reached.

Fig. 6.16 X, tests on lightly overconsolidated samples with drained strength greater than undrained strength; Y, tests on heavily overconsolidated samples with drained strength lower than undrained strength: (a), (b) $p':q$ effective stress plane; (c) $v:p'$ compression plane.

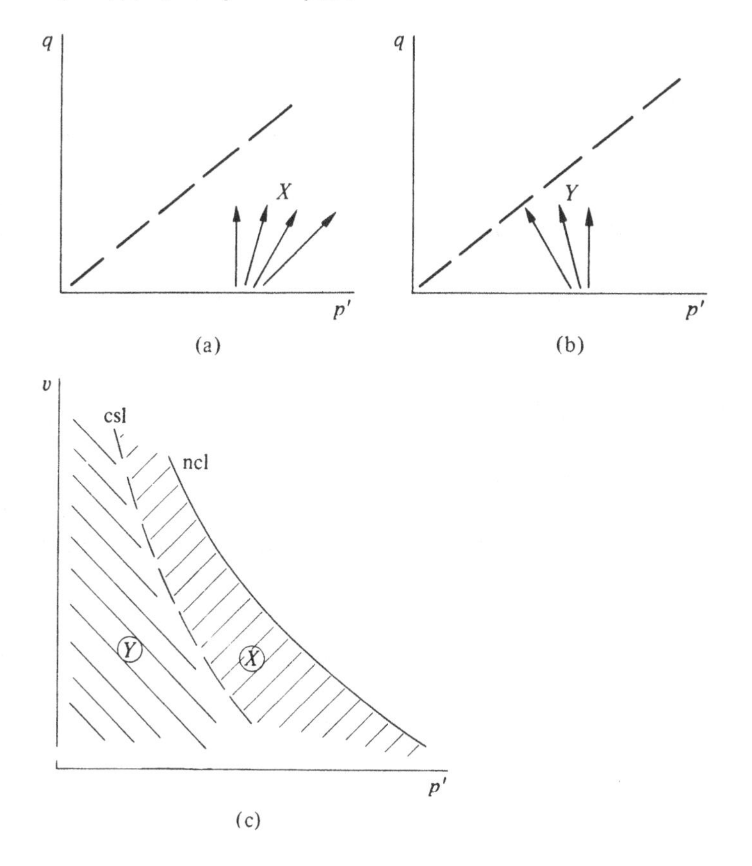

It is important to note that the critical state line must not be considered as a rigid dividing line in the compression plane between a region in which positive pore pressures and volumetric compression occur (for samples such as A in Fig. 6.14) and a region in which negative pore pressures and volumetric expansion occur (for samples such as B in Fig. 6.14). The precise response always depends on the total stress path which is applied. For example, an initially normally compressed sample subjected to a total stress path AT (Fig. 6.15), which has a considerable component of unloading of mean normal stress, shows *negative* pore pressures in an undrained test (AU) and volumetric *expansion* in a drained test (AT). The drained or long-term strength in this case is *lower* than the undrained strength.

However, it is correct to say that if normally compressed or lightly overconsolidated clays, which start above the critical state line in the compression plane, are subjected to total stress paths in which the mean stress p is kept constant (or increased) (X in Figs. 6.16a, c), then the undrained strength is lower than the drained strength. *Short-term* stability of geotechnical structures for which such stress paths are relevant would be a prime consideration since dissipation of the positive pore pressure set up during undrained shearing leads to volumetric compression and hardening of the soil. If heavily overconsolidated clays, which start below the critical state line in the compression plane, are subjected to total stress paths in which the mean stress p is kept constant (or reduced) (Y in Figs. 6.16b, c), then the undrained strength is higher than the drained strength. *Long-term* stability of geotechnical structures for which such stress paths are relevant would be a prime consideration since dissipation of the negative pore pressures set up during undrained shearing leads to volumetric expansion and softening of the soil. Actual stress paths associated with some typical geotechnical constructions are discussed in Chapter 10.

6.5 Critical states for sands and other granular materials

The qualitative picture that emerges from the discussion around Fig. 6.14 in the previous section, based only on experimental observation of the existence of a critical state line for clays, is that in drained tests

Fig. 6.17 (a) Loose packing of spherical particles compressing as it is sheared; (b) dense packing of spherical particles expanding as it is sheared.

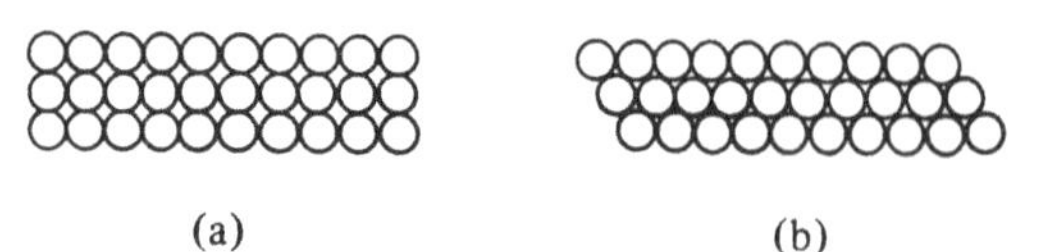

conducted at a given effective stress level, normally compressed clays, which have an initially high specific volume, contract when they are sheared whereas overconsolidated clays, which have an initially low specific volume, expand when they are sheared. It is a truth universally acknowledged that loose sands contract and dense sands expand when they are sheared. The similarity between these two statements suggests that the existence of critical states may be a rather general feature of soil behaviour.

The significance of the general phenomenon of *dilatancy* (the change in volume associated with distortion of granular materials) emerges in the writings of Osborne Reynolds (1885, 1886), but he himself notes (1886) that 'dilatancy has long been known to those who buy and sell corn'. The need for volume changes to occur when regular packings of spherical particles are deformed is clear. The loose packing of spheres in Fig. 6.17a is clearly unstable and will collapse as soon as any shear deformation is imposed; the dense packing in Fig. 6.17b can deform (neglecting the elastic stiffness of the spheres) only if spheres in each layer rise up over the spheres in the layer below. The arrangements of particles in a real granular material are much more irregular than the packings suggested in Fig. 6.17, but the modes of deformation are essentially the same.

Reynolds performed many experiments with granular materials and was able to ascribe the phenomena that occurred to the dilatancy of the materials, but he did not make any measurements of the strengths of materials with various densities of packing. Casagrande (1936), however, does describe the stress:strain curves which are expected when dense and loose samples of sand are sheared: for example, in a shear box (Fig. 6.18).

During the shearing test on the dense sand, 'the shearing stress reaches a maximum S_D (point B on the curve) and if the deformation is continued, the shearing stress drops again to a smaller value, at which value it remains constant for all further displacement. During this drop in shearing stress, the sand continues to expand (curve EG), finally reaching a critical {void ratio} at which continuous deformation is possible at the constant shearing stress S_L'.

When a loose sample of sand is subjected to a shearing test under constant normal pressure, however, 'the shearing stress simply increases until it reaches the shearing strength S_L, and if the displacement is continued beyond this point the resistance remains unchanged. Obviously, the volume of the sand in this state must correspond to the critical {void ratio} which we had finally reached when performing a test on the same material in the dense state. Therefore the curves representing the volume changes during shearing tests on material in the dense and the loose state must meet at the critical {void ratio} when the stationary condition is established'.

Casagrande concludes that 'every cohesionless soil has a certain critical {void ratio}, in which state it can undergo any amount of deformation or actual flow without volume change'. (Casagrande actually writes *critical density* but notes in an addendum that *critical void ratio* gives a more correct meaning.)

Some quantitative results in support of Casagrande's notion of critical void ratios are shown for the shearing of 1-mm diameter steel balls in Fig. 6.19. Samples of steel balls were prepared by Wroth (1958) at various initial void ratios and then sheared in an early simple shear apparatus. Just as for the Weald clay (Figs. 5.9 and 5.10), the results of shearing tests can be shown in terms of the change of volume or volumetric strain which occurs with deformation of the samples, indicated by relative displacement x. Such diagrams, Fig. 6.19a, show the pattern of changing volumetric

Fig. 6.18 Effect of shearing on the volume of dense and loose sands (after Casagrande, 1936).

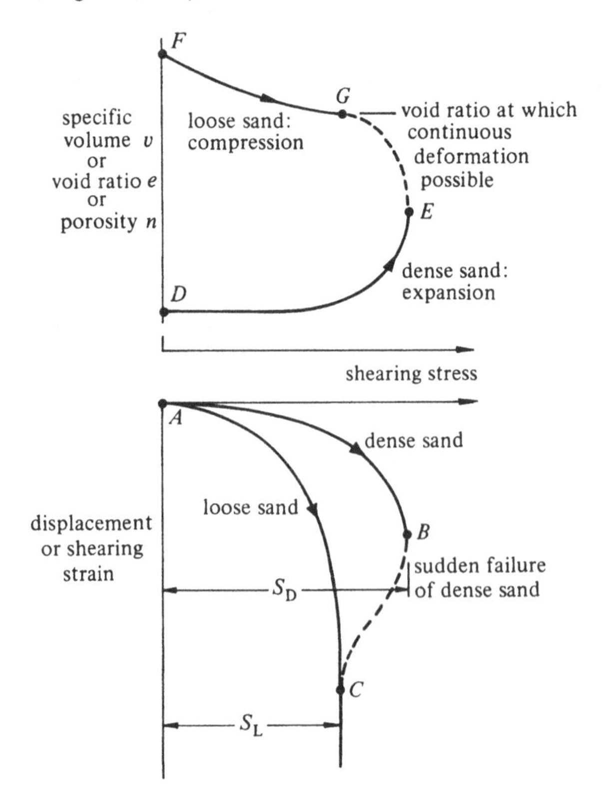

Fig. 6.19 Simple shear tests on 1-mm diameter steel balls with normal stress 138 kPa (20 lbf/in^2): (a) volume change Δv and shear displacement x; (b) specific volume v and shear displacement x; (c) end points of tests in $\sigma'_{yy}:\tau_{yx}$ effective stress plane; (d) end points of tests in $v:\sigma'_{yy}$ compression plane (after Wroth, 1958).

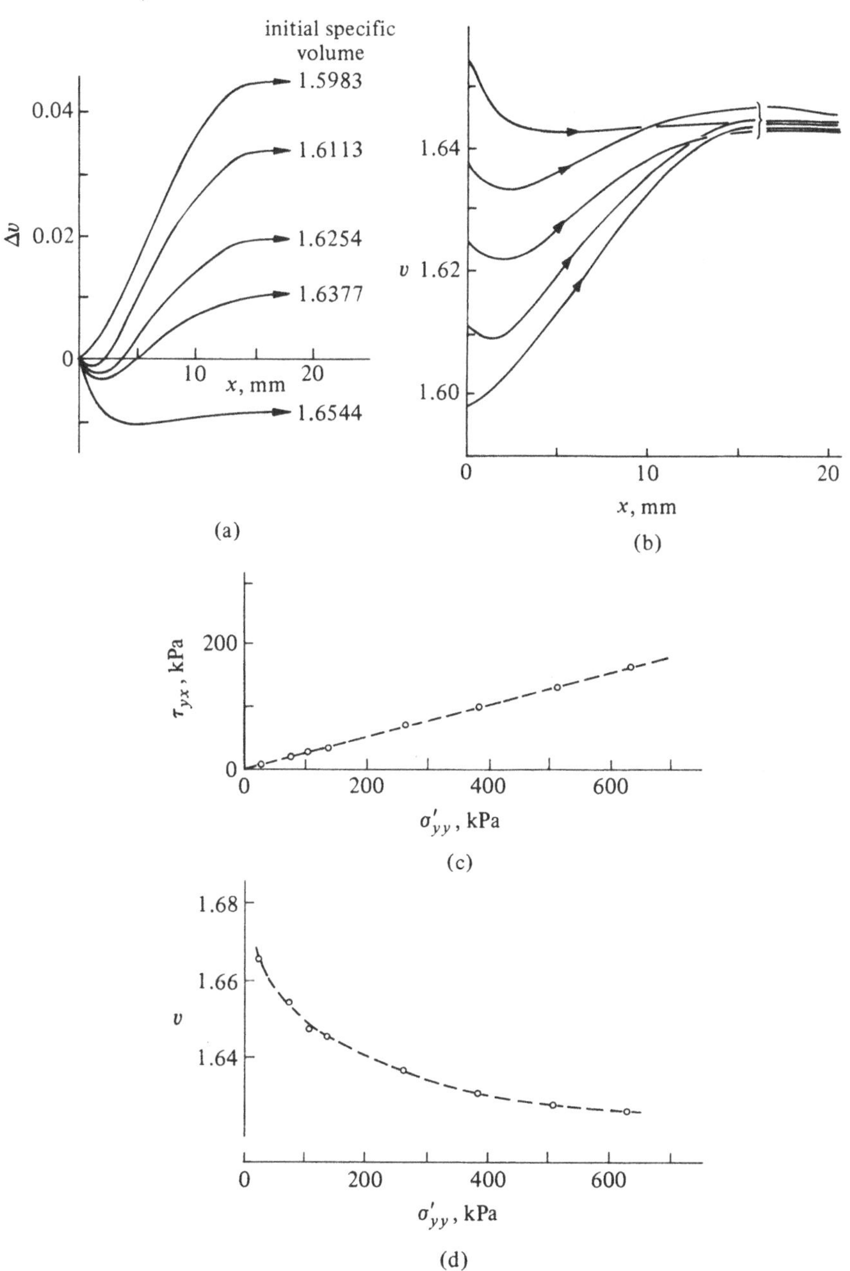

Fig. 6.20 Drained triaxial compression tests on Chattahoochee River sand with constant mean effective stress. Dense samples: A (○), $p' = 98$ kPa, $v_o = 1.69$; C (●), $p' = 2.07$ MPa, $v_o = 1.72$; D (△), $p' = 34.4$ MPa, $v_o = 1.69$. Loose sample: B (×), $p' = 98$ kPa, $v_o = 2.03$; (a) stress ratio q/p' and triaxial shear strain ε_q; (b) volumetric strain ε_p and triaxial shear strain ε_q; (c) specific volume v and triaxial shear strain ε_q (data from Vesić and Clough, 1968).

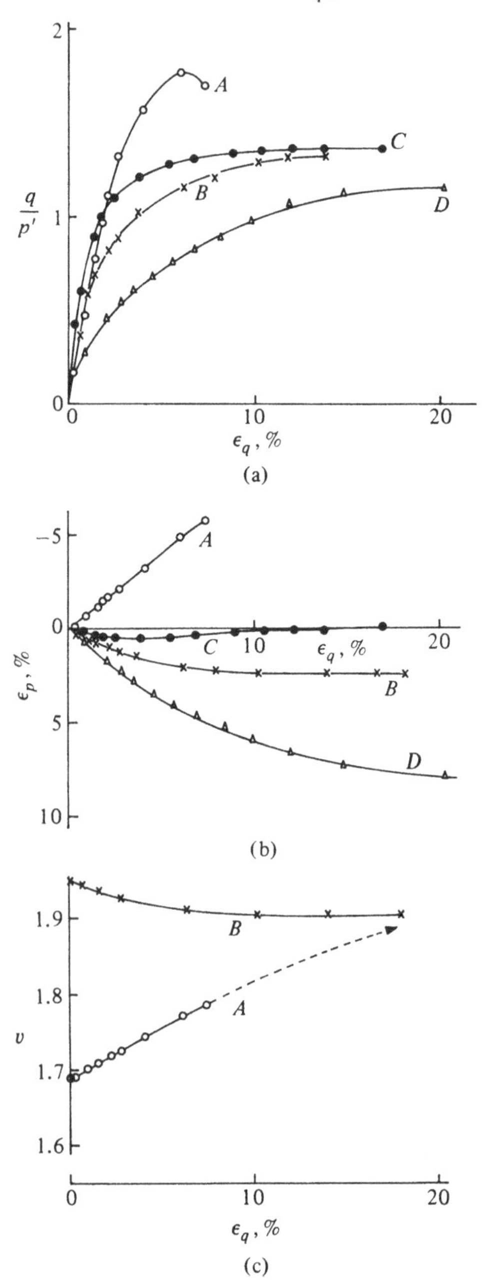

response as the initial void ratio of the sample changes: the initially dense samples with low void ratios expand, and the initially loose samples contract, as they are sheared. However, such diagrams do not help to demonstrate the existence of a critical void ratio. For this, absolute values of void ratio are required (Fig. 6.19b). In this figure, it is apparent that irrespective of the initial void ratio of the samples, at large shear displacement all samples are deforming at the same critical void ratio. All the tests shown in Figs. 6.19a, b were performed with a normal stress of 20 lbf/in^2 (138 kPa), and for this stress level a critical void ratio of about 0.64 would be deduced.

A similar pattern of response can be observed in the results of triaxial tests on samples of sand. The triaxial apparatus produces more complete information concerning the stress state in a soil sample than does the simple shear apparatus (or shear box); but as is seen in Section 7.4, it can be difficult to follow reliably the post-peak softening which is a feature of the response of dense samples. The results in Fig. 6.20 for Chattahoochee River sand (from Vesić and Clough, 1968) are typical. Note that these are not standard triaxial tests performed with constant cell pressure but are tests with constant mean effective stress p'. In Figs. 6.20a, b, c (curves *A* and *B*), results from tests on initially dense and initially loose samples with a mean effective stress of 98 kPa are shown. It is clear that testing difficulties have prevented the dense sample from attaining an ultimate condition with shearing continuing at constant volume (curve *A* in Fig. 6.20c); however, from the test on the loose sand, a critical void ratio of about 0.9 can be estimated. (The attainment of this critical void ratio would imply a volumetric expansion of about 17 per cent for the initially dense sample.)

The results of triaxial tests performed by Vesić and Clough at other constant mean effective stresses are also shown in Fig. 6.20. The effect of increasing the stress level on the volumetric strains which develop in samples prepared at essentially the same dense initial void ratio can be seen in Figs. 6.20a, b (curves *A*, *C*, and *D*). The effect of increasing the stress level is to eliminate the peak observed in the conventional response, seen in the test at the lowest stress level. The pattern which is observed is essentially the same as that in Fig. 6.19a. There, however, the response changed as the initial void ratio was varied at constant stress level; now, the response is changing as the stress level is varied at approximately constant void ratio.

Casagrande (1936) remarks that 'static pressure is relatively ineffective in reducing the volume of a sand; for example, it is not possible to change a loose sand into a dense sand by static pressure alone'. What is shown here is that it is possible to change a dense sand into a 'loose' sand by static pressure alone. Accepted notions of what constitutes a loose sand or a dense sand are rendered valueless unless accompanied by a statement about the stress level of interest. (The mean stress level at which the dense Chattahoochee River sand in Fig. 6.20 ceases to expand on shearing is about 2000 kPa. If a ratio of horizontal to vertical effective stress of about 0.4 is assumed, then this stress corresponds to a depth of about 200 m in dry sand or 330 m in submerged sand.)

A confirmation of this feature is provided by the test results shown in Fig. 6.20. The expected contrast appears between the responses of the initially dense and the initially loose samples when tested at the same low mean effective pressure ($p' = 98$ kPa) (curves A and B in Figs. 6.20a, b). However, the responses of the initially dense sample tested at a high mean effective stress ($p' = 34{,}433$ kPa) and of the initially loose sample tested at the low pressure are qualitatively the same (curves B and D in Figs. 6.20a, b).

Initially dense samples of sand sheared at low stress levels expand. Initially dense samples of sand sheared at high stress levels contract. The act of increasing the stress level reduces the volume or void ratio of the sand; consequently, the critical void ratio attained when the sand is sheared

Fig. 6.21 (a) Isotropic compression of dense and loose Chattahoochee River sand; paths of triaxial compression tests with constant mean effective stress (end points: • dense, ○ loose); and possible location of critical state line; (b) end points and critical state line in $p':q$ effective stress plane; (c) end points and critical state line in $v:p'$ compression plane (data from Vesić and Clough, 1968).

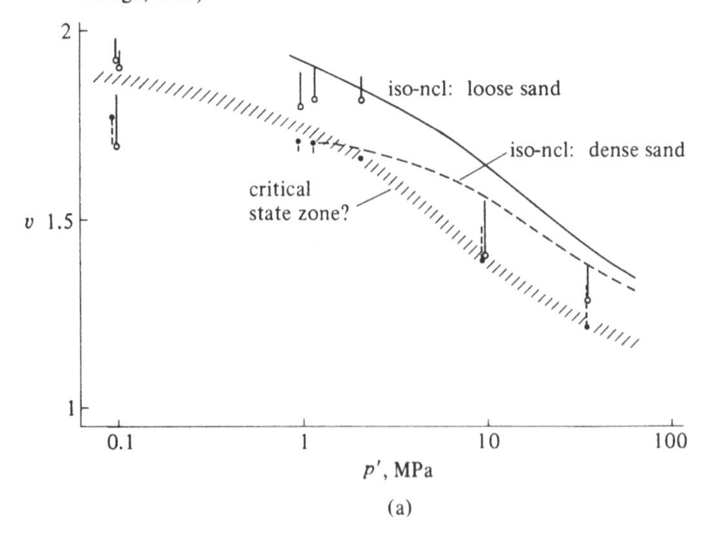

at a high stress level must be significantly lower than the critical void ratio attained when the sand is sheared at a low stress level. Casagrande's (1936) conclusion that 'Every cohesionless soil has a certain critical {void ratio}, in which state it can undergo any amount of deformation or actual flow without volume change' is seen to be too simple. Just as for clays, critical void ratios for sands are stress level dependent.

The paths followed in the tests on Chattahoochee River sand conducted by Vesić and Clough (1968) can be plotted in the compression plane v:ln p' (Fig. 6.21a), with the mean effective stress plotted on a logarithmic scale

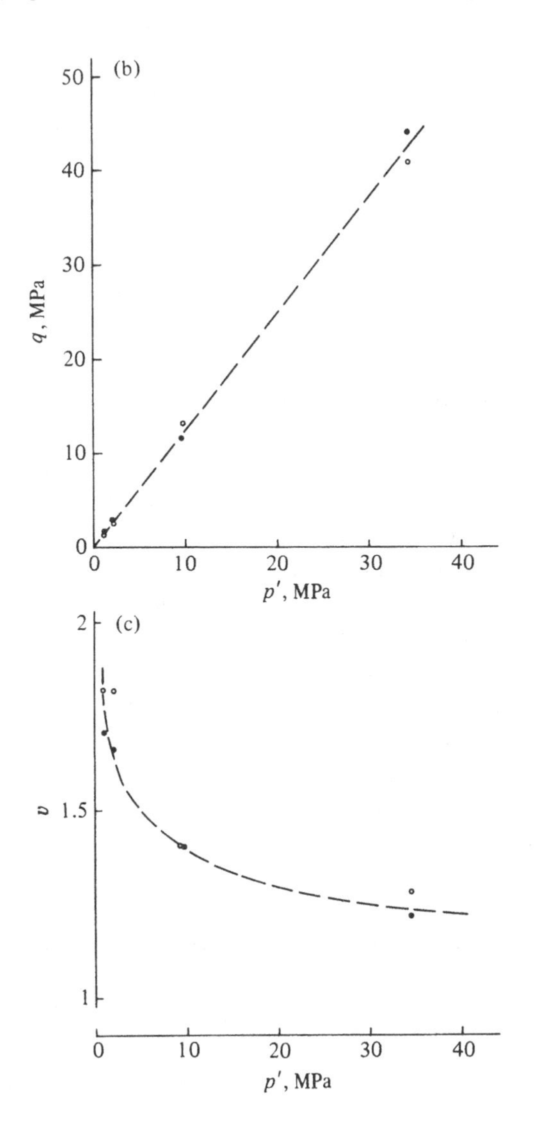

to accommodate large ranges of pressure. The paths of tests on loose and dense samples head towards a somewhat diffuse, but clearly pressure dependent, zone of critical void ratios.

Mathematical description of such a diffuse band is not particularly helpful. Upper and lower constraints on critical void ratios may be noted. The absolute minimum value of void ratio must be zero, implying a minimum specific volume of unity. For sand to shear at constant volume at a near-zero void ratio, a mechanism other than the rolling and sliding of grains, which is associated with low-pressure shearing, has to be available. Grading curves are shown by Vesić and Clough for Chattahoochee River sand (Fig. 6.22) in its initial condition, after isotropic compression to 62.1 megapascals (MPa), and after shearing in a conventional triaxial compression test at that constant cell pressure. A considerable increase in the proportion of material finer than 0.1 mm occurs both during isotropic compression and during shearing, indicating that particle crushing is a primary mechanism of deformation at high stress levels. Thus, the material that emerges from the test at a high stress level is very different from the material that went into the tests at low stress levels.

At the other extreme, at extremely low pressures, reliable data are difficult to obtain because of the inevitable gravitational variation of stress in a sample, though this could be avoided by performing experiments on board an orbiting space laboratory. Indirect measurements made as part of hopper flow experiments by Crewdson, Ormond, and Nedderman (1977) appear to indicate a levelling off of the zone of critical void ratios to a roughly constant void ratio at very low pressures. This region of the compression plane may be of importance in studies of liquefaction of soils, in which flow of soil under effective stresses which are near zero, because of the presence of high pore pressures, is of importance.

Also plotted in the compression plane (Fig. 6.21a) are the average isotropic compression curves for dense and loose Chattahoochee River sand. These curves illustrate the difficulty of exploring the compression plane using only isotropic stresses: the loose structure of loose sand is not disturbed by isotropic compression. Considerable particle rearrangement is required for the initial structures of loose and dense sands to become the same, and this rearrangement can occur readily only in the presence of shear stresses. It is only at high isotropic pressures that the compression curves for loose and dense samples start to converge. Isotropic compression is, at these high stresses, associated with particle crushing, but still the original structure of the sand is not completely eliminated.

Rockfill can be considered as an extreme granular material, but it is a material which shows the same pattern of response that has been demonstrated for finer granular materials. Problems of obtaining good quality test data from tests on rockfill arise because of the large specimen which is needed in order that the complete grading of particle sizes will be included and that the particles will not be unduly restrained by the proximity of the boundaries. Some triaxial tests on rockfill materials are reported by Marachi, Chan, and Seed (1972), and paths in the compression

Fig. 6.22 Grading curves for Chattahoochee River sand (1) before testing, (2) after isotropic compression to 62.1 MPa, and (3) after conventional drained triaxial compression to failure with cell pressure $\sigma_r = 62.1$ MPa (after Vesić and Clough, 1968).

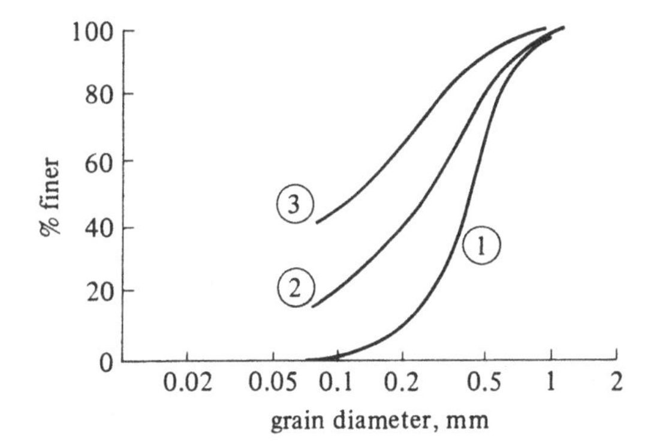

Fig. 6.23 Paths in compression space of conventional drained triaxial compression tests on Pyramid Dam rockfill material (data from Marachi, Chan, and Seed, 1972).

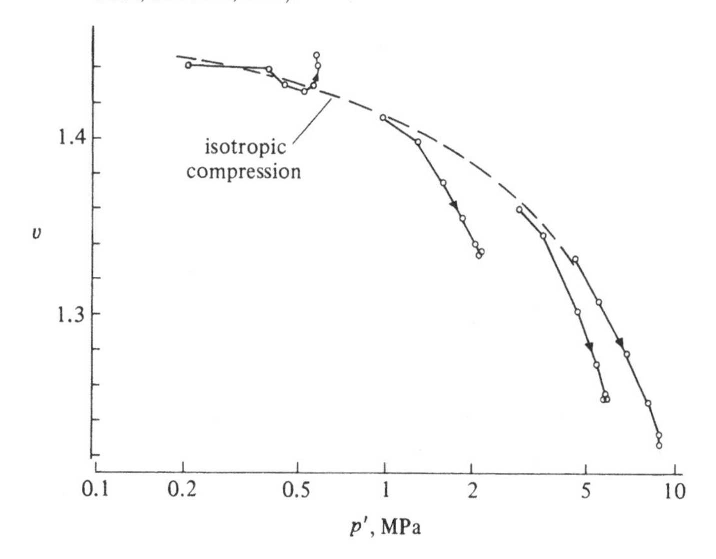

plane are shown in Fig. 6.23 for the tests on one of these materials: rockfill from the Pyramid Dam quarry. For the tests shown, the samples were 0.91 m (3 ft) in diameter and 2.29 m (7.5 ft) high. The change in the character of the response that occurs as the confining pressure of the triaxial tests is increased is apparent. Dilation at low pressure becomes marked compression at higher pressures. For this material, the transition from 'dense' to 'loose' response occurs at a stress level of about 1000 kPa corresponding to a depth of the order of 100 m in a dry mass of the rockfill.

In presenting data of critical states for clays, parallel use was made of the $p':q$ stress plane and the $p':v$ compression plane. The critical state lines predicted by the Cam clay model (Fig. 6.3) and deduced for the Weald clay (Fig. 6.10) show an ultimate critical state deviator stress which is proportional to the ultimate critical state mean effective stress. The parameter M introduced into the Cam clay model implies an ultimate purely frictional strength.

The ultimate failure and flow of sands and other granular materials are governed almost entirely by frictional factors. If a critical void ratio is reached and shearing continues at constant volume and constant stress level, then the shearing resistance is linked to the stress level by a coefficient of friction. This frictional relationship can be illustrated by plotting results in a stress plane: shear stress:normal stress for simple shear or shear box tests and deviator stress q:mean normal stress p' for triaxial tests. The conditions associated with critical void ratios observed in simple shear tests on steel balls (Wroth, 1958) are shown in a stress plane (Fig. 6.19c) and a compression plane (Fig. 6.19d). For this ideal material, a fairly well-defined line of critical void ratios is obtained in both the stress plane and the compression plane. The data concerning the critical void ratio for a single normal stress level (Figs. 6.19a, b) provide one point on this line of critical void ratios.

Critical void ratios are not so well defined for the Chattahoochee River sand tested by Vesić and Clough (1968), but some of the relevant data are collected in similar plots in Figs. 6.21b, c.

The term *critical state* was used in Section 6.1 to describe a combination of effective stresses and specific volume at which shearing of soil could continue indefinitely, and the idea of a critical state line linking critical states was introduced. The lines of end points (Figs. 6.19 and 6.21) corresponding to attainment of critical void ratios for various granular materials can be thought of, similarly, as critical state lines. It is apparent that the pattern of critical states which is observed is equivalent to that which emerged from the Cam clay model in Section 6.1 and which was discovered for Weald clay in Section 6.3.

6.6 Conclusion

This chapter has shown that there emerges from the Cam clay model a set of combinations of effective stresses and specific volume $p':q:v$ at which indefinite shearing (perfect plasticity) occurs. These critical states emerge merely because the plastic potentials assumed in the model have a slope in the stress plane $\delta q/\delta p' = 0$, at a particular stress ratio $q/p' = \eta = M$. Critical states emerge automatically from this elastic–plastic soil model and do not require any further assumptions to be made.

When data from shear tests on geotechnical materials are examined in terms of 'limited' quantities such as effective stresses and specific volume (as opposed to the 'unlimited' quantity shear strain), it is found that all tests tend towards such critical states. Critical states are a major feature of observed response.

In some cases, rather simple relationships exist between the combinations of effective stresses and specific volume at which critical states are attained: it is then possible to make powerful quantitative statements about expected patterns of soil behaviour without recourse to the complexities of Cam clay or other elastic–plastic models.

Exercises

E6.1. For a particular soil, the critical state line is defined by the following soil parameters: $M = 0.9$, $\lambda = 0.25$, $\Gamma = 3.0$.

Find the end states in terms of $p':q:v$ for conventional drained and undrained triaxial compression tests ($\Delta\sigma_r = 0$) on soils with the initial states tabulated. For the undrained tests, find also the pore pressure at failure.

	p'_i (kPa)	q_i (kPa)	v_i
a	100	0	2.049
b	100	0	1.700
c	100	0	1.849
d	200	0	1.875
e	100	50	1.875
f	50	30	2.000

E6.2. A stress-controlled compression test is carried out on a sample of normally compressed clay with a total confining stress of 490 kPa, which is held constant throughout the test. Each load increment is left on for sufficient time to allow the sample to reach a new state of equilibrium, as recorded in the table below. For the first three increments AB, BC, and CD, the sample is undrained. The drainage connection is then opened, and the sample reaches equilibrium again at E. Further load increments are applied, with drainage allowed, until the sample reaches failure at F, and the sample is then in a critical state.

Plot the progress of the test in the p':q effective stress plane and in the p':v compression plane and estimate the position of the critical state line for this clay.

Two additional samples of this clay (X and Y) are normally compressed under an isotropic pressure $p' = 350$ kPa. If a conventional drained triaxial compression test is performed on X, predict the values of p', q, and v at failure. If a conventional undrained triaxial compression test is performed on Y, predict the values of p', u, and q at failure.

	q (kPa)	u (kPa)	v
A	0	0	2.270
B	39.2	28.0	2.270
C	78.4	70.0	2.270
D	154.0	200.9	2.270
E	154.0	0	2.205
F	623.0	0	1.990

E6.3. Study the models described in Exercises E4.3, E4.4, E4.5, and E4.6 to discover whether or not each model predicts the existence of a unique critical state line.

E6.4. Obtain data for a natural clay from oedometer and triaxial tests performed as part of a site investigation. Plot the oedometer data in the compression plane, and estimate values for λ and κ for the soil. From the ends of the triaxial tests, estimate the slope M of the critical state line in the effective stress plane and its position Γ in the compression plane. From the initial stages of the triaxial tests, estimate a value for the shear modulus G'.

Use the Cam clay model to estimate the complete stress:strain response for one of the triaxial tests, and compare the estimated response with the observed response.

7

Strength of soils

7.1 Introduction: Mohr–Coulomb failure

To most geotechnical engineers the phrase 'strength of soils' conjures up images of Mohr–Coulomb failure criteria. This chapter presents strength of soils in a rather more general way to show how classical notions of Mohr–Coulomb failure can be reconciled with the patterns of response that have been developed in earlier chapters. For convenience, discussion of the link between dilatancy and strength is deferred to Chapter 8; however, it is one of the messages of critical state soil mechanics that these two aspects of soil behaviour are inextricably entwined.

Mohr–Coulomb failure is concerned with stress conditions on potential rupture planes within the soil. The Mohr–Coulomb failure criterion says that failure of a soil mass will occur if the resolved shear stress τ on any plane in that soil mass reaches a critical value. It can be written as

$$\tau = \pm(c' + \sigma' \tan \phi') \tag{7.1}$$

This defines a pair of straight lines in the $\sigma':\tau$ stress plane (Fig. 7.1). If Mohr's circle of effective stress touches these lines, then failure of the soil will occur. It is assumed that, for sliding to occur on any plane, the shear stress has to overcome a frictional resistance $\sigma' \tan \phi'$, which is dependent on the effective normal stress σ' acting on the plane and on a friction angle ϕ', together with a component c', which is independent of the normal stress. This component c' is often called *cohesion* but is more usefully regarded merely as an intercept on the shear stress axis which defines the position of the Mohr–Coulomb strength line. [Wroth and Houlsby (1985) prefer to use the symbol s rather than c in order to escape from the physical associations of the 'cohesion' intercept.]

Mohr–Coulomb failure can also be defined in terms of principal stresses.

From Fig. 7.1 the limiting relationship between the major and minor principal effective stresses σ'_{I} and σ'_{III} is

$$\frac{\sigma'_{\mathrm{I}} + c' \cot \phi'}{\sigma'_{\mathrm{III}} + c' \cot \phi'} = \frac{1 + \sin \phi'}{1 - \sin \phi'} \tag{7.2}$$

(Previously, principal stresses have been written with subscripts 1, 2, and 3 without any declaration about the relative magnitudes of the three principal stresses. Where it is important to specify the major, intermediate, or minor principal stress, then the subscripts I, II, and III, respectively, will be used.)

Mohr–Coulomb failure, because it deals with conditions on a plane, is opaque to the value of the intermediate principal effective stress σ'_{II}, which plays no part in (7.2). Thus, in Fig. 7.1, this intermediate principal stress can be equal to the minor principal stress $\sigma'_{\mathrm{II}} = \sigma'_{\mathrm{III}}$ (Fig. 7.1a), equal to

Fig. 7.1 Mohr–Coulomb failure. Intermediate principal stress (a) equal to minor principal stress, (b) truly intermediate, and (c) equal to major principal stress.

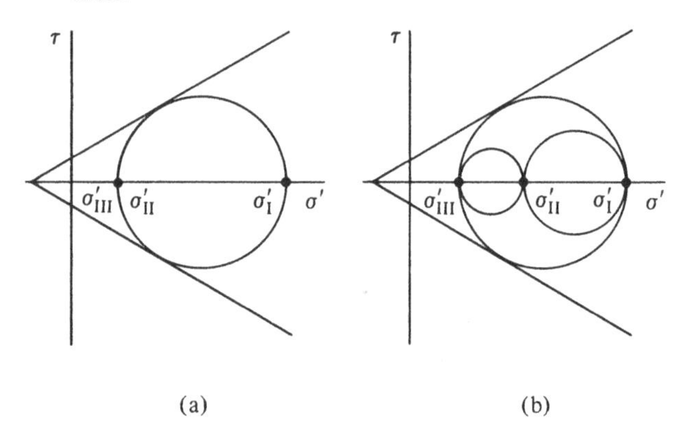

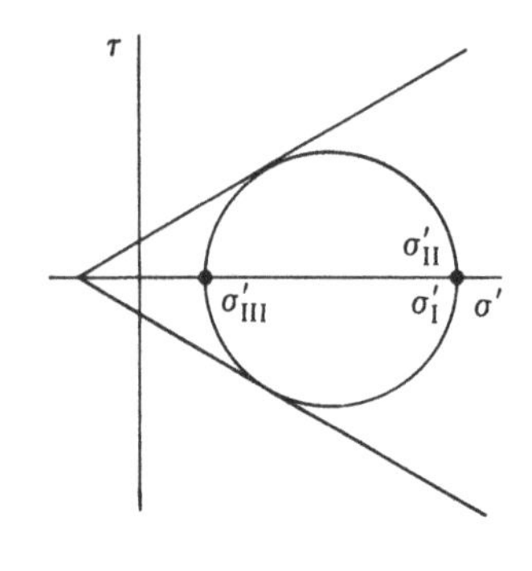

the major principal stress $\sigma'_{II} = \sigma'_I$ (Fig. 7.1c), or take some truly intermediate value $\sigma'_I > \sigma'_{II} > \sigma'_{III}$ (Fig. 7.1b) without affecting the position of the largest Mohr circle, containing the major and minor principal stresses, which controls the attainment of failure.

The stress conditions illustrated in Fig. 7.1a with $\sigma'_{II} = \sigma'_{III}$ correspond to triaxial compression in which the cell pressure provides the minor (and equal intermediate) principal stress. Expression (7.2) can be rewritten in terms of triaxial stress variables $p':q$, where

$$p' = \frac{\sigma'_I + 2\sigma'_{III}}{3} \tag{7.3}$$

$$q = \sigma'_I - \sigma'_{III} \tag{7.4}$$

and becomes (Fig. 7.2)

$$\frac{q}{p' + c' \cot \phi'} = \frac{6 \sin \phi'}{3 - \sin \phi'} \tag{7.5}$$

The stress conditions illustrated in Fig. 7.1c with $\sigma'_{II} = \sigma'_I$ correspond to triaxial extension in which the cell pressure provides the major (and equal intermediate) principal stress. The triaxial stress variable p' then becomes

$$p' = \frac{2\sigma'_I + \sigma'_{III}}{3} \tag{7.6}$$

and (7.2) can be written as

$$\frac{q}{p' + c' \cot \phi'} = \frac{-6 \sin \phi'}{3 + \sin \phi'} \tag{7.7}$$

This is also plotted in Fig. 7.2, where, for convenience, a negative sign has been assigned to values of q in triaxial extension.

The critical states described in Chapter 6 and produced in triaxial

Fig. 7.2 Mohr–Coulomb failure criterion.

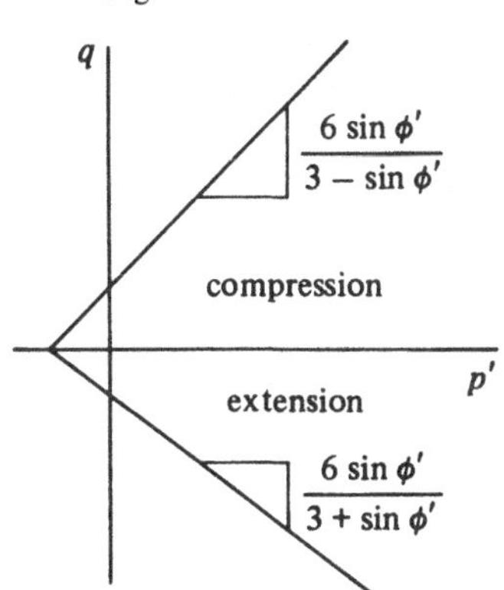

compression tests defined lines given for many soils by

$$\eta = \frac{q}{p'} = M \tag{7.8}$$

Comparison of (7.8) and (7.5) suggests that soils are failing in a *purely frictional* manner at the critical state, that is, with $c' = 0$. The deformations have been so large that the soil has been thoroughly churned up, and any bonding between particles which might have led to some cohesive strength has broken down. With $c' = 0$, comparison of (7.8) and (7.5) shows that for triaxial compression,

$$M = \frac{6 \sin \phi'}{3 - \sin \phi'} \tag{7.9}$$

or

$$\sin \phi' = \frac{3M}{6 + M} \tag{7.10}$$

Comparison of (7.7) and (7.5), however, shows that if the soil knows about a particular angle of friction ϕ', and if this soil reaches the critical states given by (7.8) in both triaxial compression and triaxial extension, then it is not possible for the value of M to be the same in triaxial compression and extension. In triaxial extension,

$$M^* = \frac{6 \sin \phi'}{3 + \sin \phi'} \tag{7.11}$$

or

$$\sin \phi' = \frac{3M^*}{6 - M^*} \tag{7.12}$$

Conversely, if the soil knows about a particular value of M at the critical state, by whatever mode of deformation that critical state may have been reached, then with $M = M^*$, the corresponding angles of friction must, from (7.10) and (7.12), be different. However, experimental evidence (e.g. from Gens, 1982) suggests that the critical state angle of friction is the same under conditions of triaxial compression and triaxial extension (and plane strain), and it would therefore be inappropriate to force $M = M^*$ in (7.10) and (7.12) to describe these ultimate failure conditions.

Expressions (7.9) and (7.11) have been plotted in Fig. 7.3. Although the concern here is with ultimate failure conditions, it will be useful subsequently to interpret them also as relationships between values of stress ratio $\eta = q/p'$ and mobilised angles of friction ϕ' at any stage of triaxial compression or extension. Expression (7.9) for triaxial compression can be written approximately as

$$M \approx \frac{\phi'}{25} \tag{7.13}$$

and (7.11) can be written rather less accurately as

$$M^* \approx \frac{\phi'}{35} \tag{7.14}$$

where, in each expression, ϕ' is measured in degrees.

7.2 Critical state line and undrained shear strength

So far, discussion of the Mohr–Coulomb failure criterion has been in terms of effective stresses. If a soil sample is not allowed to drain, then the Mohr circle of effective stress at failure (E in Fig. 7.4) can be associated with an infinite number of possible total stress circles ($T_1, T_2, \ldots$ in Fig. 7.4) displaced along the normal stress axis by an amount equal to

Fig. 7.3 Relationship between stress ratio $\eta = q/p'$ and mobilised angle of shearing resistance ϕ' in triaxial compression and extension.

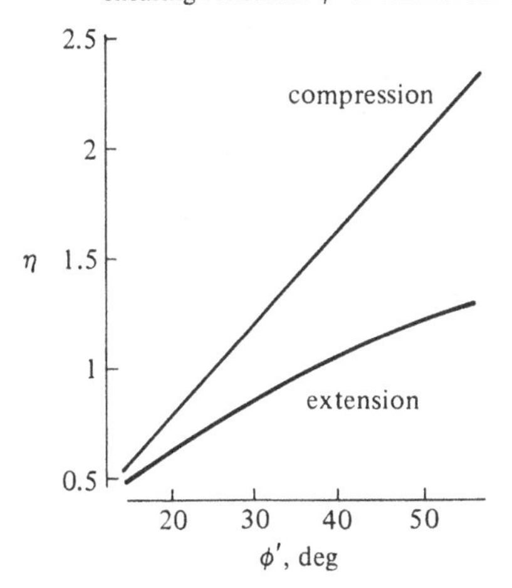

Fig. 7.4 Mohr's circles of total stress and effective stress.

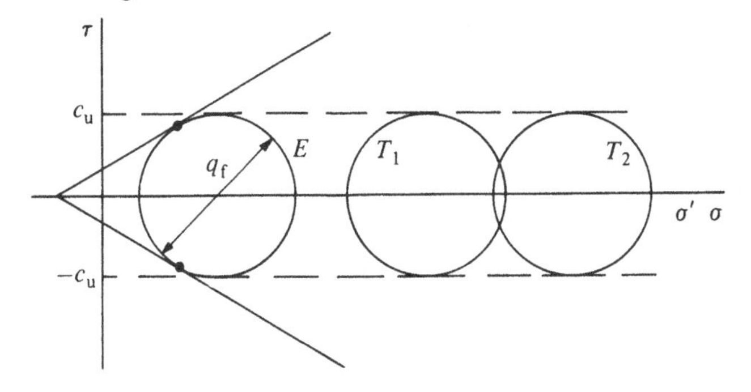

the pore pressure. The pore pressure does not affect the differences of stresses or shear stresses, so all these total stress circles must have the same size. Clay soils are often loaded sufficiently fast for drainage of shear-induced pore pressures to be impossible. It is then convenient to perform analysis of the stability of clay masses in terms of an undrained shear strength c_u, which is the radius of all the Mohr circles in Fig. 7.4, whether effective stress or total stress circles. This undrained strength is the maximum shear stress that the soil can withstand, and the failure criterion for undrained conditions becomes [compare (7.1)]:

$$\tau = \pm c_u \tag{7.15}$$

A unique undrained shear strength for a given specific volume has here been presented as a consequence of the principle of effective stress. A critical state line provides a unique relation between specific volume (or void ratio or water content, for a saturated soil) and the ultimate value of deviator stress q_f for soil sheared in a particular mode, for example, triaxial compression. This ultimate value of deviator stress is the difference between the axial and radial total or effective stresses at the end of the test, which is the diameter of the Mohr circle of effective or total stress (Fig. 7.4) and is hence twice the undrained shear strength c_u. The simple form of experimentally observed critical state lines implies a similarly simple mathematical description of the relationship between undrained

Fig. 7.5 Conventional undrained triaxial compression test on overconsolidated soil: (a) $p':q$ effective stress plane (ESP, effective stress path; TSP, total stress path); (b) $v:p'$ compression plane.

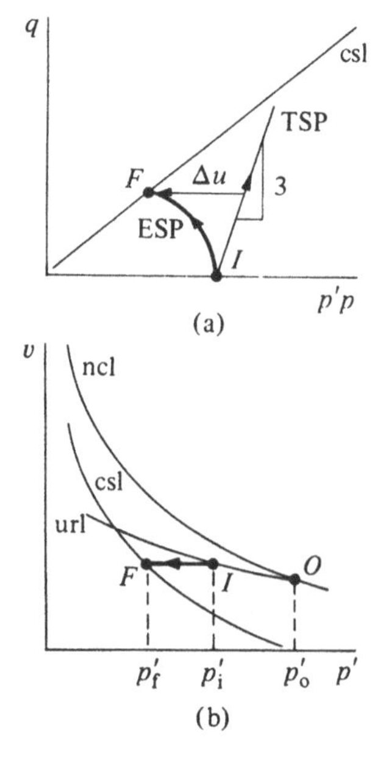

strength and specific volume and also permits the undrained strength of soil subjected to any specified consolidation history to be estimated.

A soil with a specific volume v (Fig. 7.5), no matter how that volume was arrived at, will end on the critical state line at a mean effective stress p'_f when tested in undrained triaxial compression. From (6.10),

$$p'_f = \exp\left(\frac{\Gamma - v}{\lambda}\right) \tag{7.16}$$

This implies an ultimate value of deviator stress

$$q_f = Mp'_f \tag{7.17}$$

and hence an undrained shear strength

$$c_u = \frac{Mp'_f}{2} = \frac{M}{2}\exp\left(\frac{\Gamma - v}{\lambda}\right) \tag{7.18}$$

To link undrained shear strength with consolidation history, some idealisations concerning compression and unloading of clays must be made. It was suggested in Chapter 6 that both the critical state line and normal compression lines could be assumed, from experimental evidence, to be reasonably straight and parallel in a semi-logarithmic compression plane v:ln p' (Fig. 7.6) over a reasonable range of mean effective stress. A normal compression line, not necessarily for isotropic normal compression, can be written as

$$v = v_\lambda - \lambda \ln p' \qquad (7.19)(4.2\text{bis})$$

[For isotropic normal compression, v_λ would be equal to N; compare (5.8).] The critical state line can be written as

$$v = \Gamma - \lambda \ln p' \qquad (7.20)(6.10\text{bis})$$

It is convenient to assume that the unloading and reloading of soil are described by a similar expression,

$$v = v_\kappa - \kappa \ln p' \qquad (7.21)(4.3\text{bis})$$

Fig. 7.6 Normal compression line (ncl), unloading–reloading line (url), and critical state line (csl).

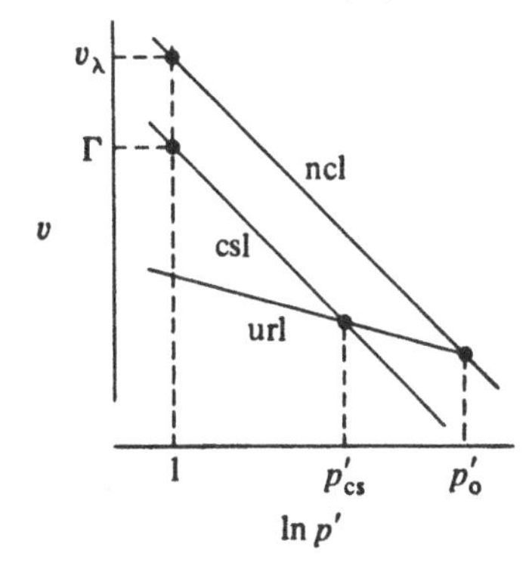

giving straight unloading–reloading lines in the v:ln p' compression plane (Fig. 7.6).

The volume separation of the normal compression and critical state lines is $v_\lambda - \Gamma$ (Fig. 7.6). It will be useful to describe the separation of these lines also in terms of the ratio r of the pressures on any particular unloading–reloading line between the normal compression and the critical state lines. Referring to Fig. 7.6, one has

$$r = \frac{p'_o}{p'_{cs}} \tag{7.22}$$

or

$$r = \exp\left(\frac{v_\lambda - \Gamma}{\lambda - \kappa}\right) \tag{7.23}$$

This pressure ratio r can be regarded as an extra soil parameter. In the description of elastic–plastic models for soil in Chapter 4, the value of r is implicitly fixed by the relative geometry of yield loci and plastic potentials. In particular, for the Cam clay model described in Chapter 5, if isotropic normal compression is being considered, so that $v_\lambda = N$, then r is the ratio of tip pressure to critical state pressure for any yield locus and is fixed at $r = 2$.

Now consider an undrained test on a sample which has been normally compressed to O (Fig. 7.5) with $p' = p'_o$ and then unloaded and allowed to swell to I with $p' = p'_i$. Even if this loading and unloading history is not isotropic, an isotropic overconsolidation ratio n_p can be defined as the ratio of these two mean effective stresses:

$$n_p = \frac{p'_o}{p'_i} \tag{7.24}$$

The specific volume of the sample at I is

$$v_i = v_\lambda - \lambda \ln p'_o + \kappa \ln n_p \tag{7.25}$$

Hence, from (7.18), the undrained strength reached at point F on the critical state line (Fig. 7.5) is

$$c_u = \frac{M}{2} \exp\left[\frac{(\Gamma - v_\lambda)}{\lambda} + \ln p'_o - \left(\frac{\kappa}{\lambda}\right) \ln n_p\right] \tag{7.26}$$

which, from (7.23) and (7.24), can be written as

$$\frac{c_u}{p'_i} = \frac{M}{2}\left(\frac{n_p}{r}\right)^\Lambda \tag{7.27}$$

where

$$\Lambda = \frac{\lambda - \kappa}{\lambda} \qquad (7.28)(6.19\text{bis})$$

In (7.27), the undrained strength is made non-dimensional by dividing it by the mean effective stress which the soil experienced at the end of compression (or unloading) just before the undrained test was begun. This strength ratio is linked with the isotropic overconsolidation ratio n_p. the other parameters in (7.27), M, Λ, and r, are all soil constants describing different aspects of the effective stress response of the soil. Expression (7.27) links a total stress characteristic of soils, the undrained shear strength, with effective stress parameters and the consolidation history of the soil.

The parameter Λ describes the relative slopes of the normal compression and unloading–reloading lines for the soil. For a typical clay with $\Lambda = 0.85$, $r = 2$, and $M = 1$ [implying $\phi' \approx 25°$ from (7.13)], the variation of c_u/p'_i with n_p is shown in Fig. 7.7. When both c_u/p'_i and n_p are plotted on logarithmic scales, a straight-line relationship emerges with slope Λ.

For normally compressed samples, $n_p = 1$ and

$$\left(\frac{c_u}{p'_i}\right)_{nc} = \left(\frac{M}{2}\right) r^{-\Lambda} \tag{7.29}$$

The ratio of the values of c_u/p'_i for overconsolidated and normally compressed samples is independent of M and r:

$$\frac{c_u/p'_i}{(c_u/p'_i)_{nc}} = n_p^{\Lambda} \tag{7.30}$$

This analysis has been presented only in terms of the isotropic overconsolidation ratio $n_p = p'_o/p'_i$. A conventional overconsolidation ratio $n = \sigma'_{vo}/\sigma'_{vi}$, defined in terms of the ratio of past and present vertical effective stresses, would be more familiar to practising engineers. The relationship between n_p and n, of course, depends on the type of past loading. If the soil has known only isotropic stresses through laboratory testing, then $n = n_p$. If the soil has a history of one-dimensional loading and unloading,

Fig. 7.7 Ratio of undrained strength to initial mean effective stress (c_u/p'_i) varying with isotropic overconsolidation ratio.

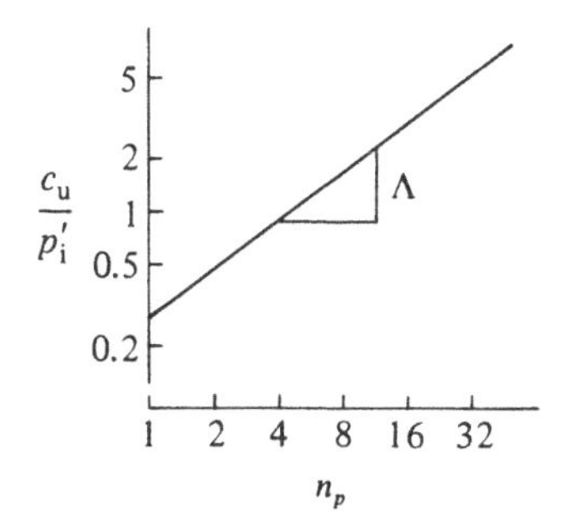

Fig. 7.8 Ratio of undrained strength to initial vertical effective stress (c_u/σ'_{vi}) varying with overconsolidation ratio n: (1) Drammen clay ($I_P = 0.30$) (after Andresen, Berre, Kleven, and Lunne, 1979); (2) Maine organic clay ($I_P = 0.34$); (3) Bangkok clay ($I_P = 0.41$); (4) Atchafalaya clay ($I_P = 0.75$); (5) AGS CH clay ($I_P = 0.41$); (6) Boston blue clay ($I_P = 0.21$); (7) Connecticut Valley varved clay ($I_P = 0.39/0.12$) (after Ladd, 1981).

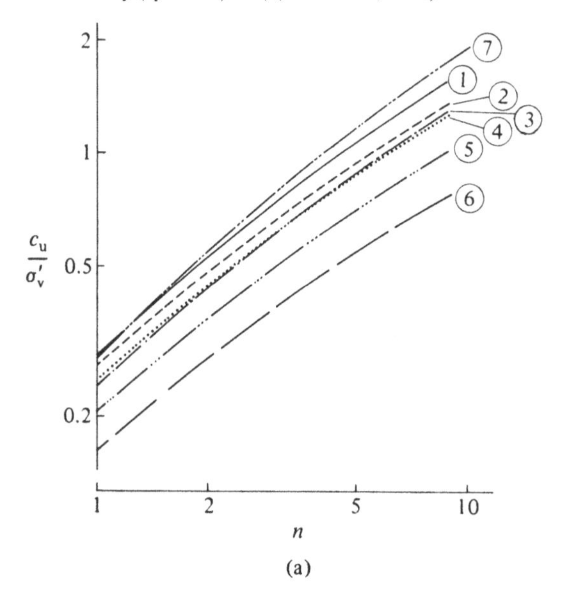

(a)

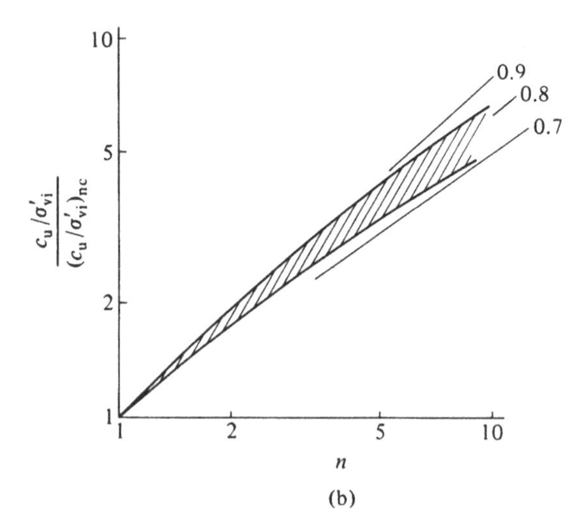

(b)

then the link between n and n_p is more complex (it is discussed in Section 10.3). Written in terms of vertical effective stresses instead of mean effective stresses, (7.30) becomes

$$\frac{c_u/\sigma'_{vi}}{(c_u/\sigma'_{vi})_{nc}} = \left(\frac{n}{n_p}\right) n_p^{\Lambda} \tag{7.31}$$

Data of the strength ratio c_u/σ'_{vi} from undrained simple shear tests on six clays, reported by Ladd and Edgers (1972) and Ladd, Foott, Ishihara, Schlosser, and Poulos (1977), are shown in Fig. 7.8a. The samples in these tests were initially one-dimensionally compressed, and the strength ratio is plotted against the overconsolidation ratio n. The same data are presented in Fig. 7.8b in the form of the ratio $(c_u/\sigma'_{vi})/(c_u/\sigma'_{vi})_{nc}$ plotted against n. Ladd et al. remark that all the data then fit into a narrow band which is reasonably well defined by the expression

$$\frac{c_u/\sigma'_{vi}}{(c_u/\sigma'_{vi})_{nc}} = n^{\mu} \tag{7.32}$$

where $\mu \approx 0.8$, though a better fit is obtained if μ reduces from 0.85 to 0.75 with increasing overconsolidation.

Although the expressions (7.30) and (7.31) were deduced with reference to undrained strengths in triaxial compression tests, the *ratio* of strengths of overconsolidated and normally compressed soils contained in this expression should not depend on the type of test used to measure the strengths (the same test being used for all strengths). It will emerge in Section 10.3.2 that the values of n^{Λ} and $(n/n_p)n_p^{\Lambda}$ are typically fairly close for values of n_p less than 16. That the empirical expression (7.32) has the same *form* as (7.30) provides support for the analysis presented in this section. The value of the exponent Λ depends on the soil being investigated; it may be coincidence that the seven clays studied by Ladd et al. provide such a narrow band of data in Fig. 7.8b.

It will be seen in Section 9.4.5 that there is a problem in determining reliable values of κ, the slope in the v:ln p' compression plane of unloading–reloading lines. Consequently, Mayne and Swanson (1981) have made use of (7.30) [or, strictly, expression (7.32) with μ set equal to Λ, since they have analysed data from 95 clays subjected to a variety of tests on isotropically and one-dimensionally consolidated samples] to deduce the average value of Λ of which each soil is aware. They report a large range of values of Λ (see Section 9.4.5). The value of Λ thus deduced can in principle be used in other analyses using models based within the framework of critical state soil mechanics.

7.3 Critical state line and pore pressures at failure

In Section 1.6, a pore pressure parameter a was introduced to link pore pressure changes with changes in applied total stresses:

$$\delta u = \delta p + a\,\delta q \tag{7.33)(1.65bis}$$

It was shown in Sections 1.6 and 5.4 that a is an indication of the current slope of the undrained effective stress path. Since undrained effective stress paths will not usually be straight, the value of a will certainly not be a soil constant but will depend on the current effective stress state in the soil, as well as on the history of consolidation of the soil. The link between effective stress and total stress descriptions of soil strength requires some knowledge of the pore pressure at failure. Even though the pore pressure parameter a may not be a constant, an average value a_{f} can be defined as the average slope of the effective stress path during the undrained loading:

$$a_{\mathrm{f}} = \frac{-\Delta p'}{\Delta q} \tag{7.34}$$

It has already been shown (Section 7.2) that the mere proposal of the existence of a critical state line permits an estimate to be made of the undrained triaxial compression strength of overconsolidated samples. The pore pressure that would be observed when a critical state was reached in a conventional undrained test, and the corresponding value of a_{f}, can be calculated by an extension of the same analysis.

Consider a sample with initial isotropic overconsolidation ratio n_p at I (Fig. 7.5):

$$n_p = \frac{p'_{\mathrm{o}}}{p'_{\mathrm{i}}} \tag{7.24bis}$$

When subjected to an undrained compression test, the path in the compression plane moves from I to F, on the critical state line (Fig. 7.5b). The end point of the test is thus fixed in the stress plane too (Fig. 7.5a), though the shape of the effective stress path between I and F is not known. The value of deviator stress q_{f} at F is given from (7.27) as

$$q_{\mathrm{f}} = Mp'_{\mathrm{i}}\left(\frac{n_p}{r}\right)^{\Lambda} \tag{7.35}$$

For a conventional triaxial compression test, the total stress path (TSP in Fig. 7.5a) has a gradient $\Delta q/\Delta p = 3$. Thus, when the undrained test reaches failure, the total stress changes will be

$$\Delta q = q_{\mathrm{f}} \tag{7.36a}$$

$$\Delta p = \frac{q_{\mathrm{f}}}{3} \tag{7.36b}$$

The effective mean stress at the end of the test, p'_f is

$$p'_f = \frac{q_f}{M} = p'_i\left(\frac{n_p}{r}\right)^{\Lambda} \tag{7.37}$$

and the pore pressure change during the test will be (Fig. 7.5a)

$$\Delta u = p'_i + \Delta p - p'_f$$

$$= p'_i + q_f\left(\frac{1}{3} - \frac{1}{M}\right)$$

$$\Delta u = p'_i\left[1 + M\left(\frac{n_p}{r}\right)^{\Lambda}\left(\frac{1}{3} - \frac{1}{M}\right)\right] \tag{7.38}$$

Then, from (7.34),

$$a_f = -\frac{(p'_f - p'_i)}{q_u}$$

$$a_f = \frac{(n_p/r)^{-\Lambda} - 1}{M} \tag{7.39}$$

Data of pore pressures at failure in triaxial compression tests on samples of Weald clay tested at different isotropic overconsolidation ratios are given by Bishop and Henkel (1957) and shown in Fig. 7.9 in terms of the pore pressure parameter a_f. Values of a_f have been calculated and plotted for comparison, using (7.39), with the soil parameters $\lambda = 0.091$ and

Fig. 7.9 Dependence of pore pressure parameter a_f at failure on overconsolidation ratio for isotropically overconsolidated Weald clay (after Bishop and Henkel, 1957).

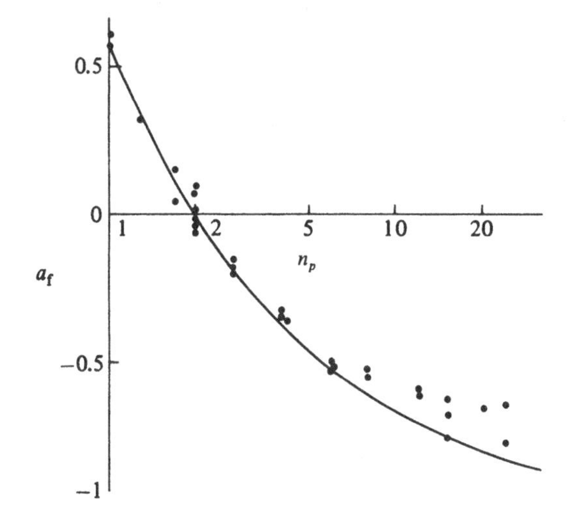

$M = 0.95$ (as found in Section 6.3), $\kappa = 0.034$ (from an average slope of the unloading lines in Fig. 6.8), and $r = 2.0$ (according to the Cam clay model). Then, from (7.28), $\Lambda = 0.63$. Agreement is good, particularly at lower overconsolidation ratios.

For isotropically normally consolidated Weald clay ($n_p = 1$), $a_f \approx \frac{2}{3}$. This value is typical of many clays and indicates, from comparison of (7.34) and (7.36), that for the conventional triaxial compression test in which the deviator stress is increased at constant cell pressure, the pore pressure at failure is approximately equal to the deviator stress at failure (Fig. 7.10). It follows that the average slope of the effective stress path from I to F is about $-\frac{3}{2}$. For other total stress paths, the pore pressure at failure would be quite different even if the values of deviator stress and pore pressure parameter a_f were the same, because the contribution to the pore pressure of the change of total mean stress Δp would be quite different.

For $n_p = 2$, the value of a_f in Fig. 7.9 is calculated and observed to be zero. For this overconsolidation ratio, the Cam clay model predicts a purely elastic response all the way from the initial state to failure at a critical state, and there is no tendency for dilatancy of the soil to contribute to the observed pore pressure.

7.4 Peak strengths

In the previous two sections, the strength of soil has been associated with conditions at the critical state. At a critical state, shear deformations can continue at constant effective stresses and constant volume: the soil is being continuously churned up or remoulded. This is in principle an ultimate state, and a critical state strength should be an ultimate strength. (Practical exceptions will be discussed in Section 7.7.)

According to the Cam clay model described in Chapter 5, all tests

Fig. 7.10 Particular case of conventional undrained triaxial compression test with pore pressure u equal to deviator stress q at failure.

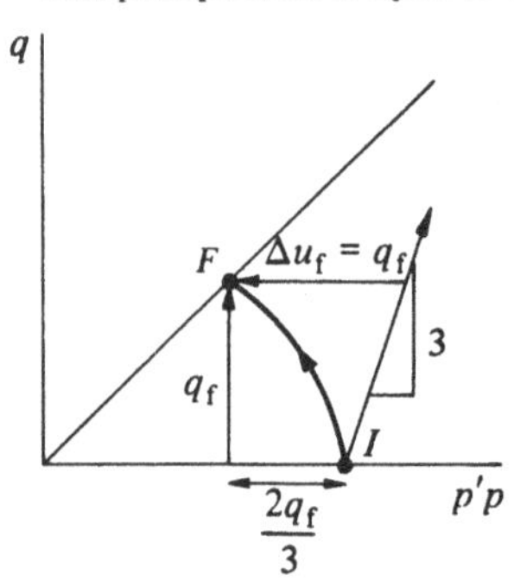

eventually reach the critical state line. However, it was seen in Section 5.3 and Fig. 5.7 that drained tests on heavily overconsolidated samples pass through a peak value of deviator stress followed by a subsequent drop in deviator stress to a critical state. This occurs because the yield locus in the $p':q$ stress plane lies above the critical state line at high values of overconsolidation ratio. If conventional drained triaxial compression tests were performed on a large number of samples, each subjected to the same maximum past isotropic consolidation pressure but unloaded to different overconsolidation ratios, then the locus of the maximum values of deviator stress reached in these tests would be as shown in Fig. 7.11a. This locus of peak points is made up of two sections:

1. Normally compressed and lightly overconsolidated samples (up to an overconsolidation ratio of about 2 according to the Cam

Fig. 7.11 Points of peak deviator stress q in conventional drained triaxial compression tests on isotropically overconsolidated samples having same past maximum preconsolidation pressure: (a) $p':q$ effective stress plane; (b) $v:p'$ compression plane.

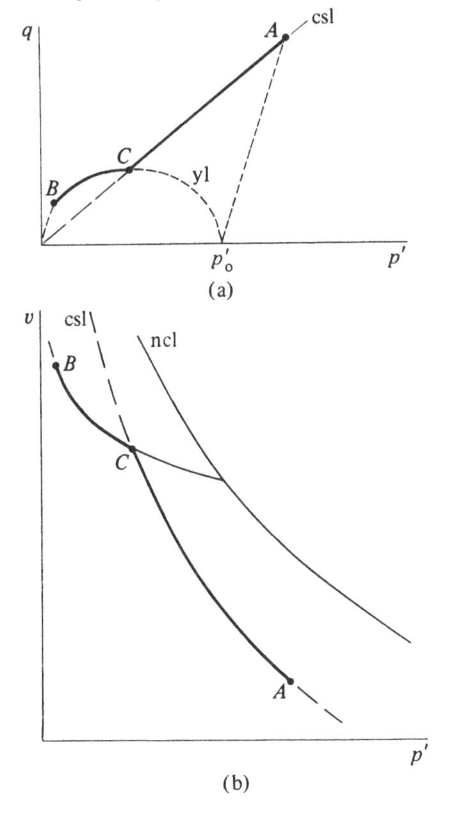

clay model) end up on the critical state line between C and A, and the critical state condition represents the maximum value of deviator stress that they have experienced.

2. For more heavily overconsolidated samples, the peak deviator stress comes at the intersection of the applied stress path ($\delta q/\delta p = 3$ for conventional drained compression) with the initial yield locus between C and B. (The point B would be reached for a sample tested at zero cell pressure.)

Clearly, it would be inappropriate to try to fit a single Mohr–Coulomb strength equation such as (7.5) to this set of data.

These data represent soil samples with a single past maximum consolidation pressure. Similar sets of tests on soil samples with different past maximum consolidation pressures would produce similar loci of peak deviator stresses, with normally compressed and lightly overconsolidated samples still ending on the critical state line, but with more heavily overconsolidated samples reaching a peak on distinct segments of different critical yield loci (Fig. 7.12). Conventional drained tests on samples with *all* possible histories would produce peak points lying in the fan-shaped region TOC (Fig. 7.12), and it would not be rational to try to pick a single Mohr–Coulomb strength line to represent an average of all these data.

One way in which the peak strength locus BCA in Fig. 7.11 could be made to apply to all soils would be to make the axes non-dimensional by dividing by the maximum past consolidation pressure p'_o. All the results of Fig. 7.12 would then become a single locus BCA in the $p'/p'_o:q/p'_o$ plane (Fig. 7.13). Given a particular sample of soil, it is not easy to deduce what the past maximum consolidation pressure might have been. However, it is feasible to measure the water content (for saturated soil) and thus, with

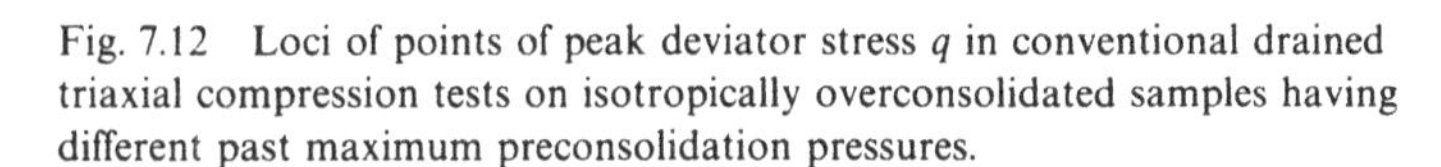
Fig. 7.12 Loci of points of peak deviator stress q in conventional drained triaxial compression tests on isotropically overconsolidated samples having different past maximum preconsolidation pressures.

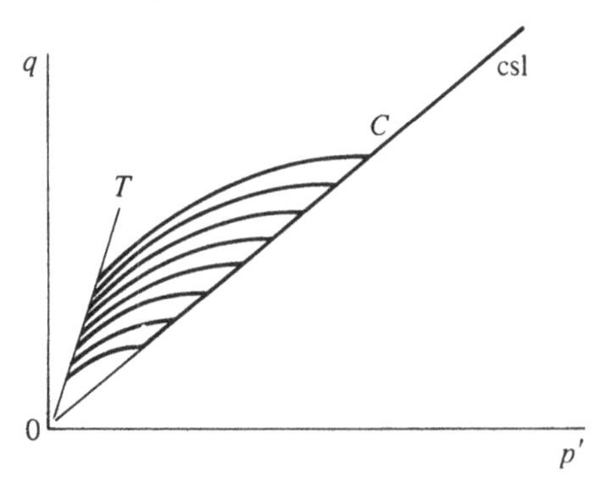

knowledge of the normal compression characteristics of the soil, to deduce the equivalent consolidation pressure p'_e (Fig. 6.6 and Section 6.2).

The data of peak deviator stress were discussed in Fig. 7.11a without reference to paths in the compression plane. The tests that reach their peak on the critical state line between C and A in the stress plane will, at that peak deviator stress, be on the critical state line between C and A in the compression plane (Fig. 7.11a, b). The tests that reach their peak on the initial yield locus between B and C in the stress plane will, at that peak, be on the initial unloading–reloading line between B and C in the compression plane (Fig. 7.11a, b). The two-part nature of this locus of peak points is again clear. The specific volume (or water content) of each sample changes during these drained tests. When the stresses $p'{:}q$ at the point of peak deviator stress are divided by the value of the equivalent consolidation pressure p'_e appropriate to the volume of the sample at that point, the data of $p'/p'_e{:}q/p'_e$ again lie on a single, curved segment (CB in Fig. 7.14). As described in Section 6.2, the critical state line becomes a single point C in this plot. All the tests that reach their peak on the critical state line, on OC in Fig. 7.12, plot at point C in Fig. 7.14. All the tests that reach their peak within the fan TOC in Fig. 7.12, plot on the curve CB in Fig. 7.14. The use of the non-dimensional stress quantities $p'/p'_e{:}q/p'_e$ takes account of differences in water content at failure in different samples and makes apparently incompatible information compatible.

In the predictions of the response of heavily overconsolidated soil in drained triaxial compression made using the Cam clay model, the peak deviator stress was followed by a drop of deviator stress (towards a critical state) as deformation was continued (see Figs. 5.7 and 7.15). Such a falling stress:strain curve can only be followed to the critical state (FC) in a test

Fig. 7.13 Data of peak deviator stress normalised with respect to past maximum preconsolidation pressure p'_o.

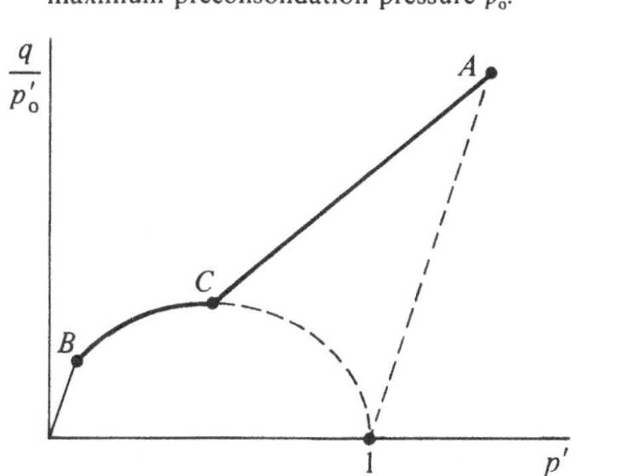

in which the axial compression ε_a of the sample is steadily increased, a strain-controlled test. In a test in which the deviator stress q is steadily increased (a stress-controlled test), catastrophic failure occurs as soon as the peak deviator stress is reached (*FG* in Fig. 7.15). Even in a strain-controlled test, uniform deformation of the sample is unlikely beyond the peak. Theoretical treatment of the development of non-uniform deformations in test specimens is complex (e.g. see Vardoulakis, 1978; Vermeer, 1982), but a qualitative discussion is possible with reference to conditions on a potentially critical plane within the specimen.

Just as the critical state line in the $p':q$ plane can be equated with a purely frictional strength criterion *OC* in the normal stress:shear stress $\sigma':\tau$ plane (Fig. 7.16a), so the peak strength envelope for heavily overconsolidated samples, *BC* in Fig. 7.11, can be converted into a strength criterion *BC* in the $\sigma':\tau$ plane (Fig. 7.16a). If behaviour in terms of σ' and τ is thought of as being broadly equivalent to behaviour in terms of p' and q, then the response expected on this potentially critical plane is as follows. The initial stresses are isotropic, so the initial shear stress at *P* is zero. As the drained

Fig. 7.14 Data of peak deviator stress normalised with respect to equivalent consolidation pressure p'_e.

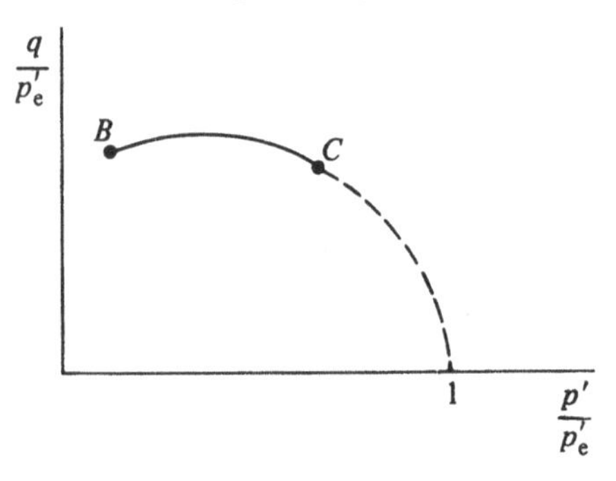

Fig. 7.15 Strain softening response after peak.

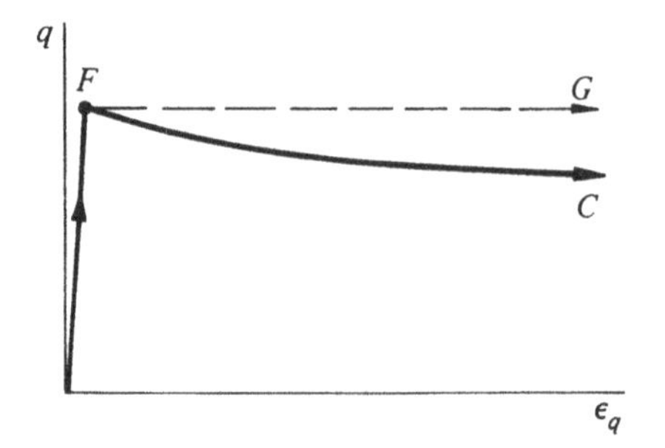

Fig. 7.16 Bifurcation of response as deformation becomes concentrated in thin region: (a) $\tau{:}\sigma'$ effective stress plane; (b) $v{:}\sigma'$ compression plane; (c) shear stress τ and specific volume v; (d) shear stress τ and shear displacement x; (e) distribution of specific volume before formation of failure plane; (f) distribution of specific volume after formation of failure plane.

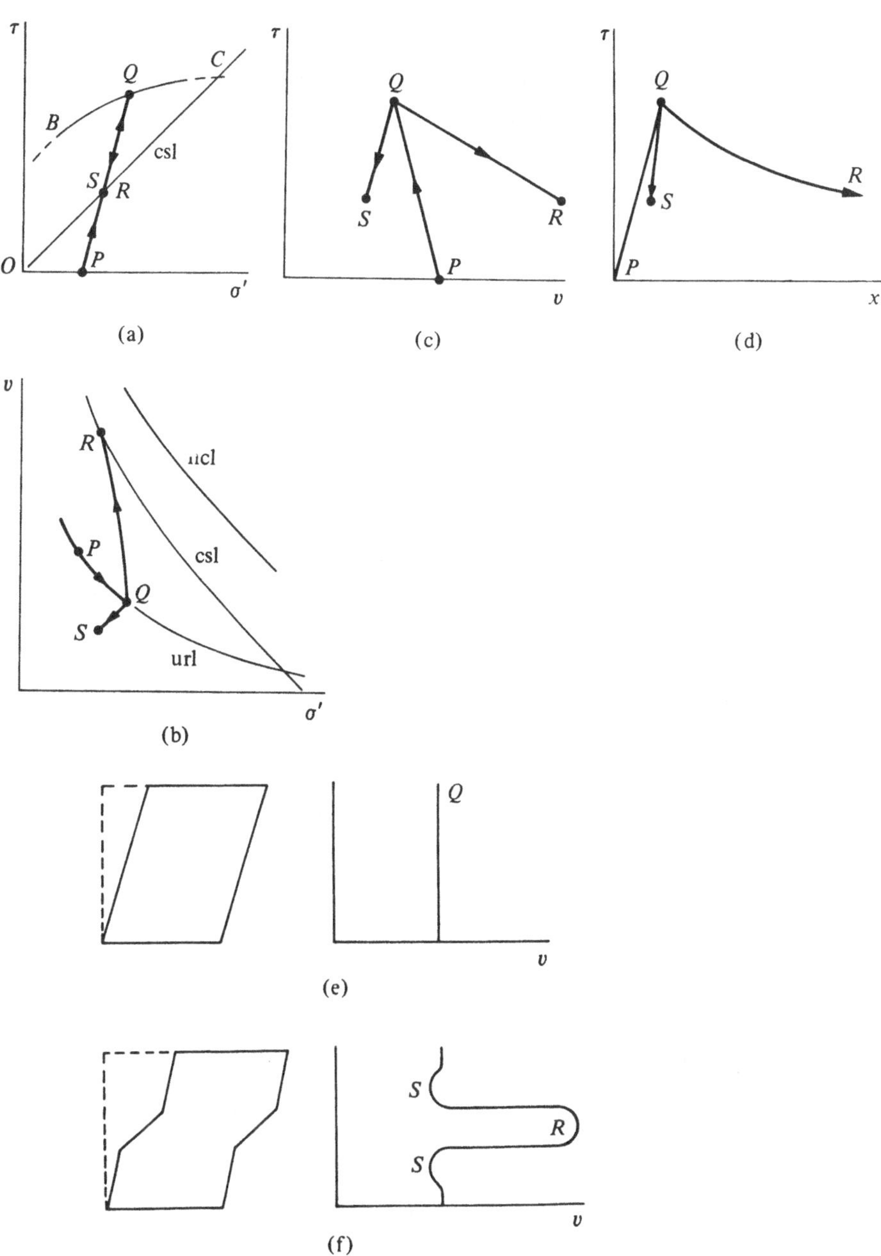

compression test proceeds, the shear stress and the normal stress on this plane increase, *PQ* (Fig. 7.16a). This is associated with essentially elastic changes in specific volume (Fig. 7.16b, c) and elastic deformations between points on opposite sides of the plane (Fig. 7.16d). After the peak strength envelope *BC* has been reached at *Q*, then softening to the critical state at *R* requires increase in specific volume or water content of the soil. As the water content or specific volume of the soil increases, the corresponding strength curve changes (compare Fig. 7.12), and the soil becomes weaker. This will be an unstable process because it is extremely unlikely that the soil sample will be absolutely homogeneous; it only requires that some parts of the sample should start this post-peak softening fractionally ahead of other parts for further deformation to become concentrated in the weaker parts of the sample. Shearing then occurs almost solely on critical planes in the soil. In Fig. 7.16, the soil in the critical plane (which is likely to be a thin zone of failure) may continue to deform towards the critical state *QR*. If the soil is saturated, then the expansion in volume in this failure region may occur at the expense of a decrease in volume of the adjacent soil, which will unload stiffly, *QS*. There is a bifurcation of the response similar to that which occurs when a strut buckles and similar to the uncertainty of response illustrated in Fig. 5.8. The expected spatial distribution of water content in a saturated soil before and after this bifurcation of response and localisation of deformation is shown in Figs. 7.16e, f. A field observation of just such a distribution of water content will be discussed in Section 7.7.

A nice example of the contrast between soil which deforms steadily to a critical state and soil which develops failure planes after a peak strength is shown in Fig. 7.17, taken from Taylor (1948). The loose sample of sand

Fig. 7.17 Failed samples of (a) dense and (b) loose Fort Peck sand tested in conventional drained triaxial compression (from Taylor, 1948).

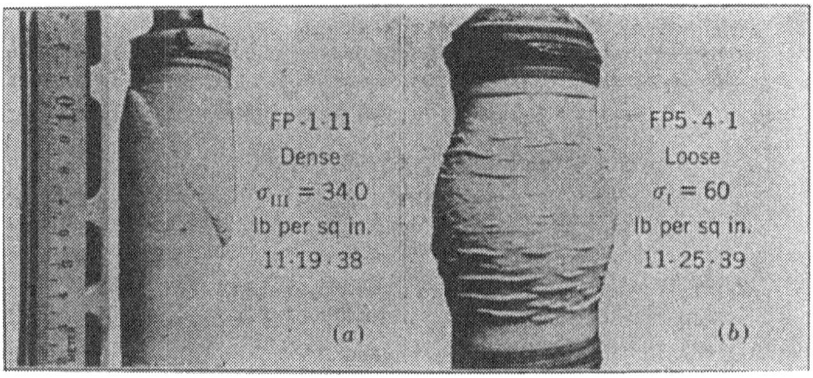

(Fig. 7.17b) hardens as it is sheared and bulges symmetrically at large deformations. No critical failure planes are being mobilised in this sample; and although the restraint provided by the end platens is producing some axial non-uniformities, variables based on principal stresses and strains such as $p':q$ and $\varepsilon_p:\varepsilon_q$ will be appropriate for describing the behaviour to large deformations. The dense sample of the same sand tested under similar conditions (Fig. 7.17a) develops a failure plane. Although variables such as $p':q$ and $\varepsilon_p:\varepsilon_q$ may be appropriate for describing the behaviour before the failure plane develops, once it has developed, it is only logical to describe subsequent response in terms of the normal and shear stresses acting on the plane and the components of strain within the thin failure zone, or the deformations across the failure plane.

Such a study of stress and deformation conditions obtaining in a failure zone formed in a dense sand sample has been carried out by Vardoulakis (1978); some typical results are shown in Fig. 7.18. In Fig. 7.18a, the friction mobilised on the failure plane is plotted against the relative sliding

Fig. 7.18 (a) Mobilised friction and (b) volume change of thin failure zone shearing to critical state (medium grained Karlsruhe sand, $v_o = 1.57$) (after Vardoulakis, 1978).

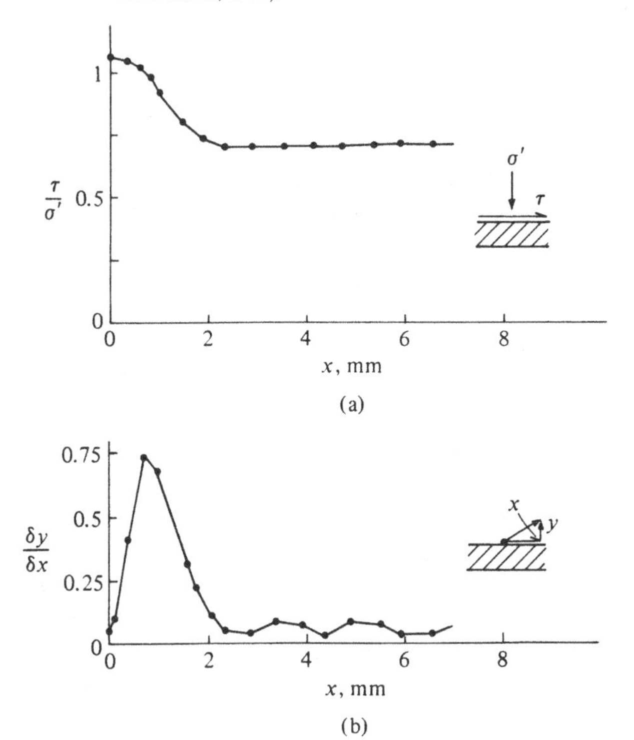

movement x that occurs across the plane. The friction is initially high but drops to a steady value as deformation continues. In Fig. 7.18b, $\delta y/\delta x$, the rate of change of separation y across the failure plane with sliding movement x, is plotted against x. This ratio is a measure of the volume change occurring in the failure zone and is high as sliding begins but falls to zero as deformation continues. At larger deformations, the dense sand has actually reached a critical state at which it deforms at constant volume and constant mobilised friction.

7.4.1 *Peak strengths for clay*

Once localisation of deformation occurs in a test sample (Fig. 7.16), then statements about deformations and specific volumes made on the basis of boundary measurements will no longer be reliable. Hvorslev (1937) studied the strength of clays using a direct shear box. He was aware of the localisation of deformation that occurred and took care to dismantle his apparatus rapidly after each test so that samples of soil could be obtained from the region of failure, for water-content determination. He was then able to plot his data of failure in both a stress plane and a compression plane.

In the shear box (Fig. 1.24), the stress components that are measured are the average normal effective stress, $\sigma'_v = P/A$, and the average shear stress, $\tau_h = Q/A$, acting on the horizontal plane of failure. Typically, shear box tests are performed with the normal stress σ'_v held constant. The stress plane in which data from shear box tests can be plotted has axes of normal stress σ'_v and shear stress τ_h, and the compression plane also has abscissa σ'_v. One set of results obtained by Hvorslev (1937), for Vienna clay, is shown in Fig. 7.19. The peak shear stresses that were observed are shown in the $\sigma'_v:\tau_h$ stress plane in Fig. 7.19a. The data were obtained from specimens sheared from various one-dimensionally normally compressed states or at various overconsolidation ratios on a single unloading–reloading loop. The pattern found is broadly similar to that shown for the Cam clay model in Fig. 7.11a.

The water contents measured in the thin failure zone of the shear box are plotted in the $\sigma'_v:w$ compression plane in Fig. 7.19b. Again, the pattern can be related to that shown for the Cam clay model. Normally compressed samples, 1 and 2, show equal drops in water content from the one-dimensional normal compression line to the critical state line; such equal drops are implied by the parallelism of these lines assumed in the Cam clay model. Overconsolidated samples reach failure at points in the compression plane below the critical state line; compare Figs. 7.19b and 7.11b.

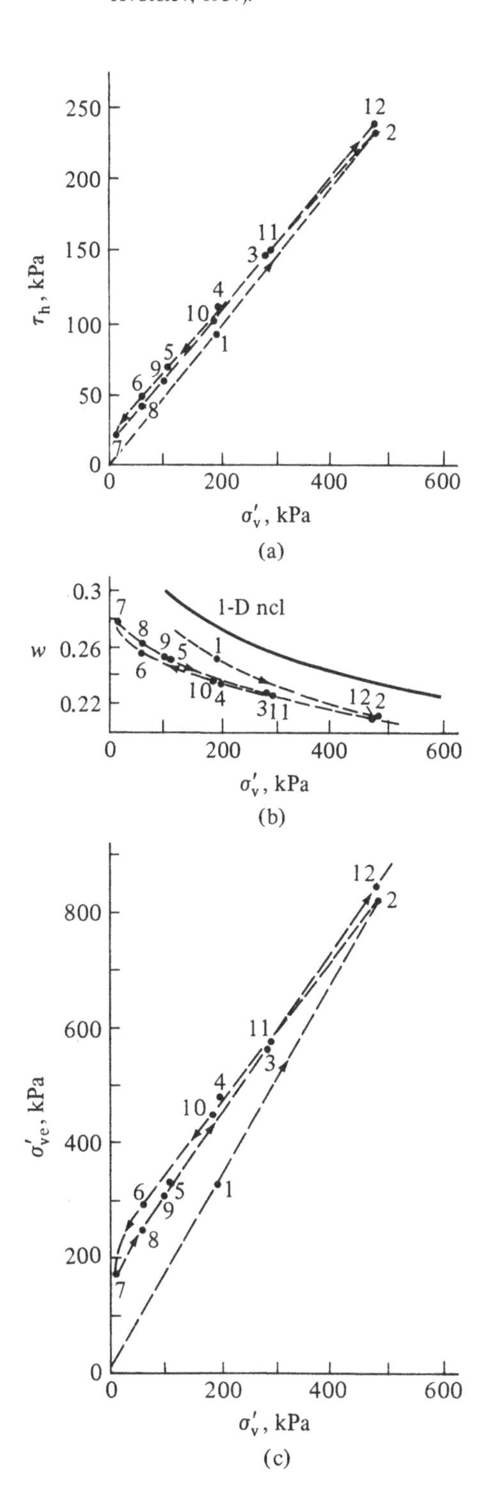

Fig. 7.19 Shear box tests on normally compressed and overconsolidated Vienna clay: (a) $\tau_h:\sigma'_v$ effective stress plane; (b) $w:\sigma'_v$ compression plane; (c) equivalent consolidation pressure σ'_{ve} and vertical effective stress σ'_v (after Hvorslev, 1937).

Since complete information about stress states is not available from this apparatus, it is not possible to calculate values of mean effective stress at any stage of a test, and hence it is not possible to calculate values of the equivalent consolidation pressure p'_e. However, an equivalent one-dimensional consolidation pressure σ'_{ve} can be defined as shown in Fig. 7.20 (compare Fig. 6.6). Values of the equivalent consolidation pressure σ'_{ve} relevant at failure in these shear box tests are shown in Fig. 7.19c. It is immediately apparent that the pattern of variation of σ'_{ve} at failure is essentially the same as the pattern of variation of τ_h at failure.

By plotting the ratios σ'_v/σ'_{ve} and τ_h/σ'_{ve} at failure, the data are brought together into a single picture which is equivalent to that shown in Fig. 7.14 for Cam clay. The data for Vienna clay are shown in Fig. 7.21. The data lie on a straight line of the form

$$\frac{\tau_h}{\sigma'_{ve}} = c'_{ve} + \frac{\sigma'_v}{\sigma'_{ve}} \tan \phi'_e \tag{7.40}$$

where c'_{ve} and ϕ'_e are soil parameters. Comparing this with the conventional

Fig. 7.20 Equivalent consolidation pressure σ'_{ve} on one-dimensional normal compression line (1-D ncl).

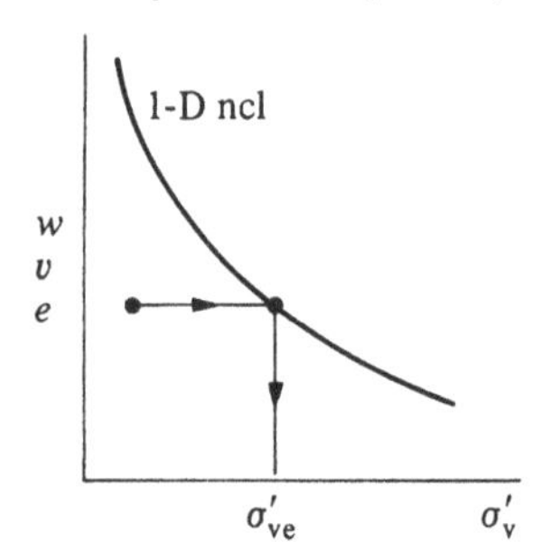

Fig. 7.21 Failure data from shear box tests on Vienna clay (after Hvorslev, 1937).

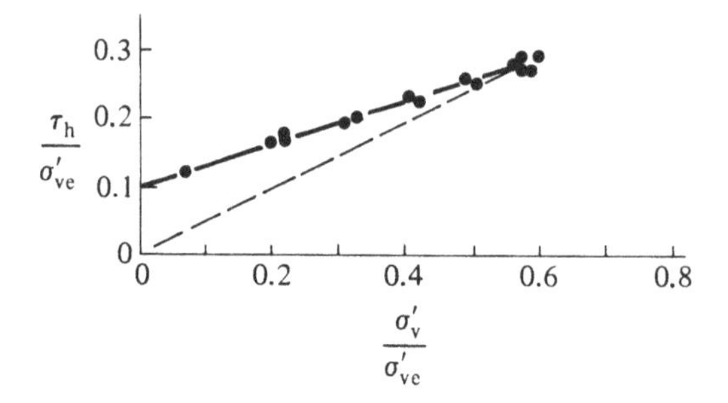

Mohr–Coulomb failure criterion,

$$\tau = c' + \sigma' \tan \phi' \tag{7.1bis}$$

we see that

$$c' = c'_{ve}\sigma'_{ve} \tag{7.41}$$

and

$$\phi' = \phi'_{e} \tag{7.42}$$

In other words, the apparent cohesion to be used in the Mohr–Coulomb strength equation depends linearly on the equivalent consolidation pressure; hence, if a linear relation is assumed between water content and logarithm of pressure during normal compression, the cohesion increases exponentially as the water content decreases.

This provides a good illustration of the need to consider a volumetric quantity as well as effective stress variables when trying to assemble strength data. If strength data follow the relationship found by Hvorslev (7.40), then it is only samples which fail at the same water content or specific volume – and hence at the same equivalent consolidation pressure – that will lie on a single, simple Mohr–Coulomb line (7.1). A series of samples which have been unloaded from a single maximum past pressure, or a series of samples taken from a single profile in the ground, can have quite different water contents and hence can produce failure points which lie each on a quite different strength line.

The normally compressed and lightly overconsolidated samples tested by Hvorslev did actually fail and reach their greatest strength on what, in the light of discussions in Chapter 6, can be called a critical state line. At the critical state, the strength seen in the shear box should be purely frictional and of the form

$$\tau_h = \sigma'_v \tan \phi'_{cs} \tag{7.43}$$

This critical state condition imposes a limit on the range of validity of (7.40), and Fig. 7.21 shows that there is a cluster of points at the right-hand end of the line. The values of the strength quantities c'_{ve}, ϕ'_e, and ϕ'_{cs} in (7.40) and (7.43) for Vienna clay are given in Table 7.1, together with values for the Little Belt clay, also tested by Hvorslev.

The significance of the limited extent of the Hvorslev failure line (7.40) may become clearer when peak strength points from triaxial tests are considered and a diagram which is directly comparable with Fig. 7.14 can be obtained. For triaxial tests on isotropically normally compressed and overconsolidated samples, the equivalent consolidation pressure to be used to normalise the stress results is p'_e, as defined in Fig. 6.6 and Section 6.2. The peak strength data obtained by Parry (1956) in a large number of

Table 7.1. *Hvorslev strength parameters*

Soil	c'_{ve}	c'_{pe}	ϕ'_e(deg)	ϕ'_{cs}(deg)
Vienna clay	0.100	0.141[a]	17.5	26.0
Little Belt clay	0.145	0.187[a]	9.9	19.6
Weald clay	0.034[a]	0.046	18.8	22.4

[a]These values have been calculated assuming that $p'_e/\sigma'_{ve} = (1 + 2K_{0nc})/3 = (3 - 2\sin\phi'_{cs})/3$, where K_{0nc} is the ratio of horizontal to vertical effective stress during one-dimensional normal compression.

triaxial tests on Weald clay are shown in Fig. 7.22 in the dimensionless stress plane p'/p'_e:q/p'_e. These tests include drained and undrained, compression and extension tests with various total stress paths, but the failure points lie closely around two straight lines: one for compression tests and one for extension tests. (Extension tests, having axial stress less than radial stress, are plotted with negative values of $q = \sigma_a - \sigma_r$.) Although these lines have slopes of different magnitude in the p'/p'_e:q/p'_e plane, they are in fact matched by a single set of Hvorslev strength parameters [refer to Section 7.1 and expressions (7.5) and (7.7)], as shown by Schofield and Wroth (1968). Expression (7.40) is now slightly modified to

$$\frac{\tau}{p'_e} = c'_{pe} + \frac{\sigma'}{p'_e}\tan\phi'_e \tag{7.44}$$

Fig. 7.22 Failure data from triaxial compression and extension tests on Weald clay (data from Parry, 1956).

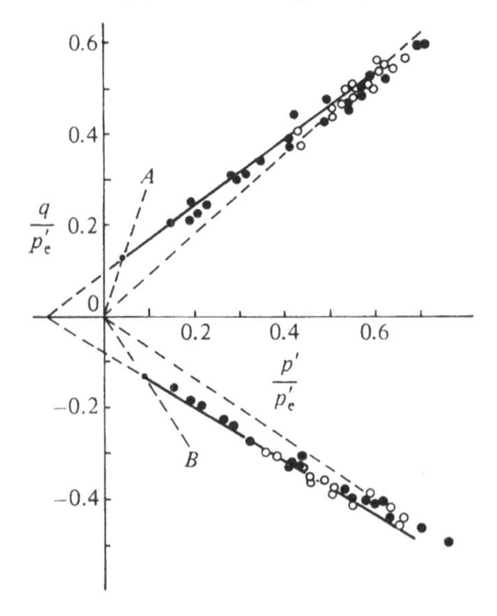

so that (7.41) becomes

$$c' = c'_{pe} p'_{e} \tag{7.45}$$

In triaxial compression, the failure data lie on the line

$$\frac{q}{p'_e} = h_c\left(g + \frac{p'}{p'_e}\right) \tag{7.46}$$

and in triaxial extension the failure data lie on the line

$$\frac{q}{p'_e} = -h_e\left(g + \frac{p'}{p'_e}\right) \tag{7.47}$$

with

$$g = c'_{pe} \cot \phi'_e \tag{7.48}$$

and, by comparison with (7.5) and (7.7),

$$h_c = \frac{6 \sin \phi'_e}{3 - \sin \phi'_e} \tag{7.49}$$

and

$$h_e = \frac{6 \sin \phi'_e}{3 + \sin \phi'_e} \tag{7.50}$$

The values of the strength parameters c'_{pe}, ϕ'_e, and ϕ'_{cs} for Weald clay are also given in Table 7.1. The final figure, for ϕ'_{cs}, comes from (6.25) via (7.9). The critical state condition again provides a limit to the line, just as the critical state point C provided a limit to the peak strength line predicted by the Cam clay model in Fig. 7.14. Normally compressed and lightly overconsolidated samples are expected to proceed stably to a critical state without a premature peak.

There will be a limit to the extent of the Hvorslev failure line also at low values of p'/p'_e. If, for example, it is supposed that the soil can withstand no tensile effective stresses, then the condition of zero effective radial stress defines a limiting line in triaxial compression OA in Fig. 7.22,

$$q = 3p' \tag{7.51}$$

and the condition of zero effective axial stress defines a limiting line in triaxial extension OB in Fig. 7.22,

$$q = -\frac{3p'}{2} \tag{7.52}$$

The Hvorslev lines then span between the critical state points and the no-tension lines.

Comparison of Fig. 7.14 with Fig. 7.22 (and indirectly with Fig. 7.21) suggests that although the Cam clay model as formulated in Chapter 5 is correctly indicating that peak strengths are to be expected for heavily overconsolidated samples, the peak strengths that are being predicted by the model look too high.

A complete picture of the triaxial compression response of the Weald clay can be produced in the dimensionless $p'/p'_e:q/p'_e$ plane, as shown in Fig. 7.23a. The constant volume locus of the Cam clay model (Fig. 6.7) has been drawn between the isotropic normal compression point N and the critical state point C, using the same values of soil parameters as in Section 7.3 ($M = 0.872$, $\lambda = 0.091$, and $\kappa = 0.034$). The Hvorslev line from Fig. 7.22 has been drawn through C to meet the no-tension line OT at T.

Fig. 7.23 Triaxial test paths and Hvorslev failure for Weald clay (+, drained tests; ×, undrained tests): (a) $p'/p'_e:q/p'_e$; (b) $\eta:v_\lambda$ (data from Bishop and Henkel, 1957).

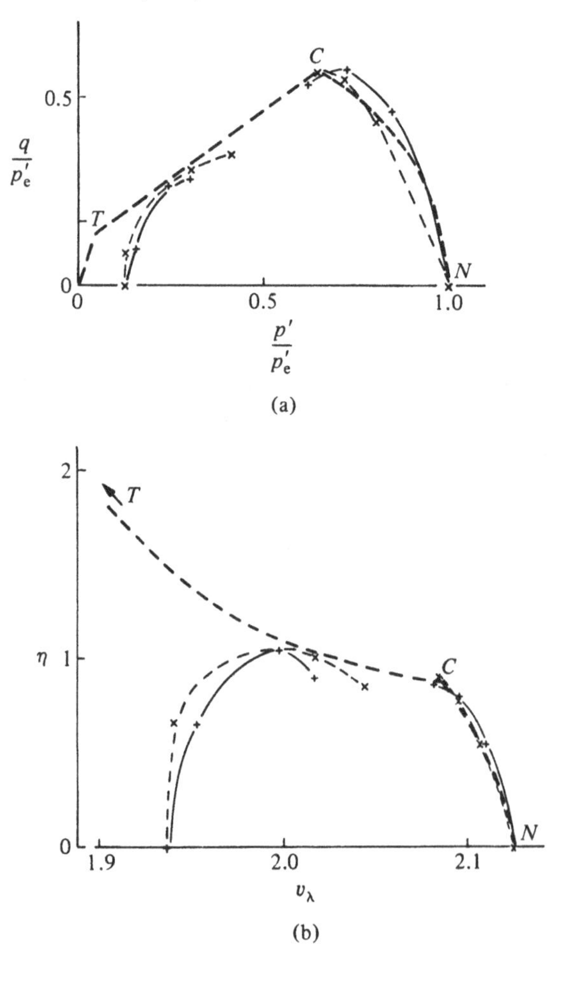

The data of the two drained and two undrained tests on Weald clay that were replotted in the stress plane and compression plane in Fig. 6.9 have also been included in Fig. 7.23a.

It is clear that in this two-dimensional plot the data of drained and undrained tests are brought together. The two tests on the normally compressed samples lie close to the Cam clay curve *NC*. The two tests on the heavily overconsolidated samples rise towards the Hvorslev line *TC* and curl over towards the critical state point, which they fail to reach before the tests are terminated.

This two-dimensional diagram can be used, following Schofield (1980), to illustrate the regions of different expected characters of response. At high values of p'/p'_e, the soil yields stably according to an elastic–plastic soil model such as Cam clay and eventually reaches a critical state C. At values of p'/p'_e less than about $\frac{2}{3}$ for Weald clay (the values for other clays depend on the relevant values of M, λ, and κ), rupture of the soil can occur on critical planes according to a Hvorslev strength criterion (7.44), CT in Fig. 7.23 – or Mohr–Coulomb strength criterion (7.1) with strength parameters given by (7.45) and (7.42). At very low values of p'/p'_e (perhaps less than about $\frac{1}{20}$), fracture of the soil may occur as the soil fails in tension, TO in Fig. 7.23.

Writing the Hvorslev expression for the peak strength using the form (7.46) or (7.44) suggests an apparent cohesion dependent on current water content, through an equivalent consolidation pressure p'_e. Equation (7.46) can be rearranged as

$$\eta = \frac{q}{p'} = h_c\left(1 + g\frac{p'_e}{p'}\right) \tag{7.53}$$

which suggests a peak stress ratio, or peak mobilised angle of friction, which is dependent on mean effective stress. The alternative two-dimensional plot of Section 6.2 uses stress ratio η and the volumetric variable v_λ:

$$v_\lambda = v + \lambda \ln p' \tag{6.13bis}$$

If the equivalent consolidation pressure p'_e is determined on the isotropic normal compression line, then from (6.13) and (6.17),

$$v_\lambda = N + \lambda \ln \frac{p'}{p'_e} \tag{7.54}$$

and (7.53) becomes

$$\eta = h_c\left(1 + g \exp \frac{N - v_\lambda}{\lambda}\right) \tag{7.55}$$

Then, since at the critical state $v_\lambda = \Gamma$ and $\eta = M$, the four soil parameters

M, Γ, g, and h_c are related through

$$M = h_c\left(1 + g\exp\frac{N - \Gamma}{\lambda}\right) \tag{7.56}$$

and only three of the four parameters are in fact independent.

The Weald clay data of Fig. 7.23a have been replotted in terms of η and v_λ in Fig. 7.23b, and the coalescence of data of drained and undrained tests is again apparent. (The value of Γ implied in Fig. 7.23b is different from that deduced in Section 6.3. The separation $N - \Gamma$ of the critical state line and isotropic normal compression line is found experimentally to be greater than that deduced from Cam clay with $\lambda = 0.091$ and $\kappa = 0.034$. This value of κ was deduced from the average slope of the unloading curves in Fig. 6.8. The observed value of $N - \Gamma$ is consistent with a much lower value of κ, perhaps corresponding to the initial slope of the unloading curves where they leave the normal compression line.)

The two-dimensional plots $p'/p'_e{:}q/p'_e$ or $\eta{:}v_\lambda$ were presented in Section 6.2 as convenient ways of displaying information about three variables: p', q, and specific volume v. The two-dimensional diagrams of Fig. 7.23 can be transformed back into a three-dimensional surface in $p'{:}q{:}v$ space (Fig. 7.24). The segments NC, CT, and TO now become surfaces corresponding to Cam clay yielding, Hvorslev rupture, and tensile fracture; these essentially limit the combinations of stress and specific volume that can be reached in triaxial compression tests. This composite surface can be called a state boundary surface.

Fig. 7.24 Limiting states for soil plotted in three-dimensional space $p'{:}q{:}v$.

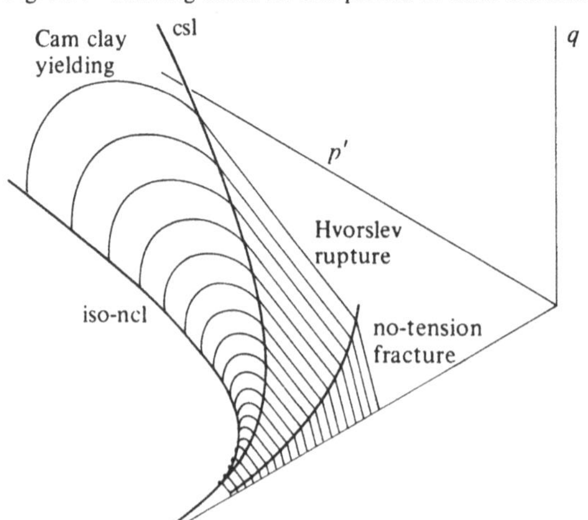

7.4.2 Interpretation of peak strength data

This section discusses some strength data from triaxial tests on 42 samples (3 samples from each of 14 locations and depths) of London clay to illustrate how the Hvorslev strength equation (7.44) can help to focus attention on the factors controlling the strength of the soil. These data have been discussed also by Wroth and Houlsby (1985).

All the tests were consolidated drained triaxial compression tests. A standard procedure was adopted of taking sets of three samples from a single depth and testing them with three cell pressures. A traditional way of interpreting the failure data is to draw a straight line touching as nearly as possible the three Mohr circles of effective stress at failure (Fig. 7.25a). This line is then treated as a Mohr–Coulomb strength criterion of the form (7.1), giving values of apparent cohesion c' and friction angle ϕ' for

Fig. 7.25 Selection of strength parameters from drained triaxial tests on London clay: (a) Mohr-Coulomb failure line fitted to Mohr's circles of effective stress; (b) range of values of cohesion c' and angles of friction ϕ'; (c) failure points in $p':q$ effective stress plane.

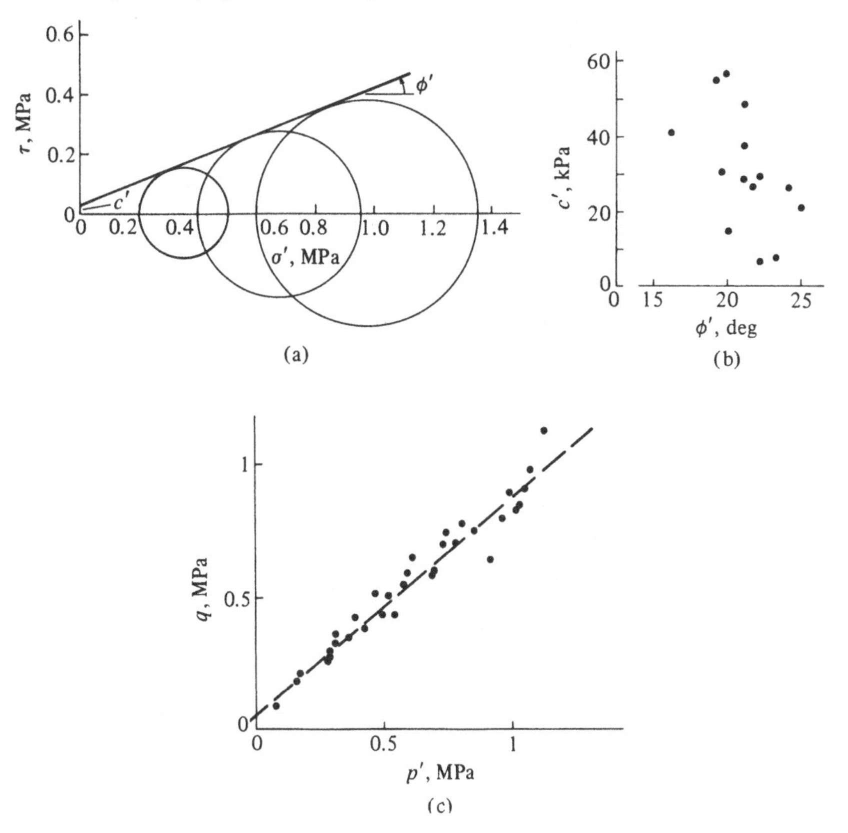

this trio of samples. The range of Mohr–Coulomb strength values deduced from the 14 sets of samples is shown in Fig. 7.25b; the range of apparent cohesions is from 6.9 to 56.2 kPa with an average of 30.8 kPa, and the range of angles of friction is from 16.2° to 25.0° with an average of 21.2°.

An alternative, fairly traditional way of interpreting the failure data is to plot the failure values of mean effective stress p' and deviator stress q for all 42 samples (Fig. 7.25c) and then to fit the best straight line through all these points. The slope and intercept of this line can be converted, from (7.5), to values of apparent cohesion $c' = 27.4$ kPa and angle of friction $\phi' = 21.3°$.

Neither of these methods of interpretation makes allowance for the fact that the various sets of samples may be different even though they are samples from one geological stratum and one locality, so that the different strengths seen in Fig. 7.25 may be related to differences in the past histories of the soil. One manifestation of differences in the past history of the soil is the different values of water content of the various samples. If information about the normal compression of the clay is available, then water content differences can be accommodated through the equivalent consolidation pressure p'_e. The 42 failure points are plotted in Fig. 7.26a as values of $p'/p'_e:q/p'_e$, and again the best line can be fitted through these data. Using (7.48) and (7.49) produces $c'_{pe} = 0.031$ and $\phi'_e = 19.7°$.

The angle of friction has dropped a little from the previous averages, but the value of c'_{pe} can be converted to a relationship between apparent

Fig. 7.26 Failure data for London clay (a) plotted in normalised effective stress plane $p'/p'_e:q/p'_e$; (b) implied dependence of cohesion on water content.

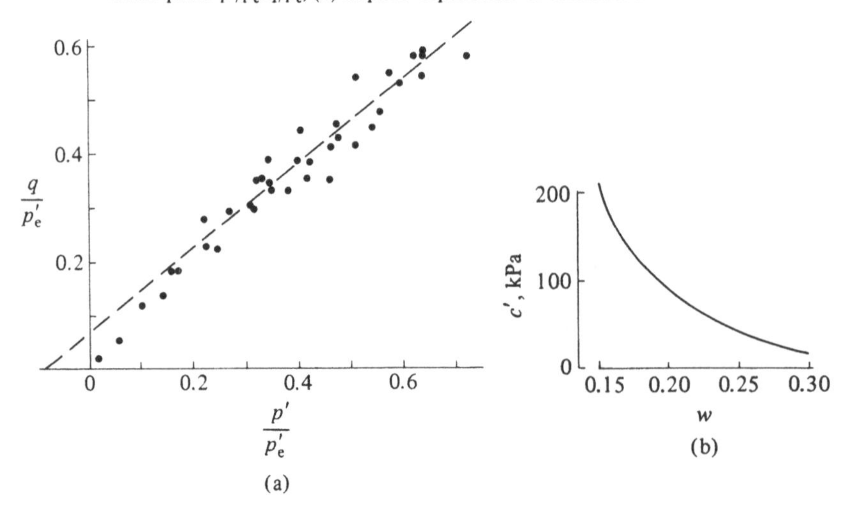

cohesion c' in (7.1) and water content (Fig. 7.26b). On the one hand, this suggests that a little of the scatter of the strength data can be ascribed to variations in water content, though the variations of water content between the various samples were not great, and the amount of scatter is perhaps not significant; a case could probably be made for supporting either interpretation. On the other hand, whether or not this final interpretation is the preferred interpretation, the discussion of Section 7.4.1 should serve to emphasise the importance of controlling water content if peak cohesive strengths are to be replied upon.

7.4.3 *Peak strengths for sand*

The message that has emerged from Section 7.4.1 is that peak strengths of clay can be properly understood only if account is taken of both the effective mean stress p'_f and the specific volume v_f at failure, since both of these will influence the failure value of deviator stress q_f. The interdependence of these three quantities was explored in two ways. The first exploration was in a plot of p'/p'_e:q/p'_e, using the Hvorslev analysis and interpreting the data as an indication that clays have an apparent cohesion that is dependent on water content. The second exploration was in a plot of η:v_λ. In this plot, since $\eta = q/p'$ is equivalent to mobilised friction, the data can be interpreted to indicate an angle of friction dependent on the failure state, as summarised by the variable $v_{\lambda f} = v_f + \lambda \ln p'_f$, which combines the information of mean effective stress and of specific volume. The two interpretations are entirely equivalent, and it may ultimately be personal preference which will guide the choice of variable apparent cohesion or of variable friction as a means of describing the failure conditions. Sands are usually thought of as frictional materials, and it is an adaptation of this second approach that appears to lead to the most convenient description of the strength of sands. The picture will be built up in stages.

'Loose sands and gravels are known to have less resistance to shear than the same soils in a dense state' (Winterkorn and Fang, 1975). A first estimate of the peak angle of friction of a sand might be obtained from charts such as those produced by Winterkorn and Fang (1975) or the U.S. Department of the Navy (1971), which require knowledge only of the packing of the sand (and some basic information about particle shape and size). The packing of the sand is indicated by its relative density I_D:

$$I_D = \frac{v_{max} - v}{v_{max} - v_{min}} \tag{7.57}$$

where v_{max} and v_{min} are so-called maximum and minimum values of specific volume, determined by standard procedures (e.g. see Kolbuszewski, 1948). This dependence on relative density is illustrated for Chattahoochee River sand in Fig. 6.20, where the dense samples were prepared with $I_D = 0.84$ and the loose sample with $I_D = 0.14$.

It is evident from Fig. 6.20, however, that relative density on its own is not sufficient since the dense sample tested at a high stress level shows a much lower strength, close to that of the loose sample. The data shown in Fig. 6.20 were obtained from tests in which the mean stress was kept constant. Most test data for soils have come from conventional triaxial compression tests in which the cell pressure is held constant, with the consequence that in a drained test the mean stress level increases from the start of the test until failure occurs. If only the strength of sands in conventional triaxial compression tests is of concern, then it may be acceptable to seek correlation of strength with initial densities and confining stresses.

The steady decrease in the peak angle of friction of dense Chattahoochee River sand as the mean effective stress at failure increases is shown in Fig. 7.27 (from Vesić and Clough, 1968). At high stresses, the dense sand shows no peak strength and proceeds to a critical state angle of friction ϕ'_{cs}, as the loose sand does at all stress levels. The character of this response has been examined for many sands by Bolton (1986), and he has produced the expression

$$\phi' - \phi'_{cs} = 3I_D(10 - \ln p'_f) - 3 \qquad (7.58)$$

Fig. 7.27 Variation of peak angles of shearing resistance ϕ' with mean effective stress p' for initially dense (•) and initially loose (○) samples of Chattahoochee River sand (after Vesić and Clough, 1968) with expression (7.58) superimposed.

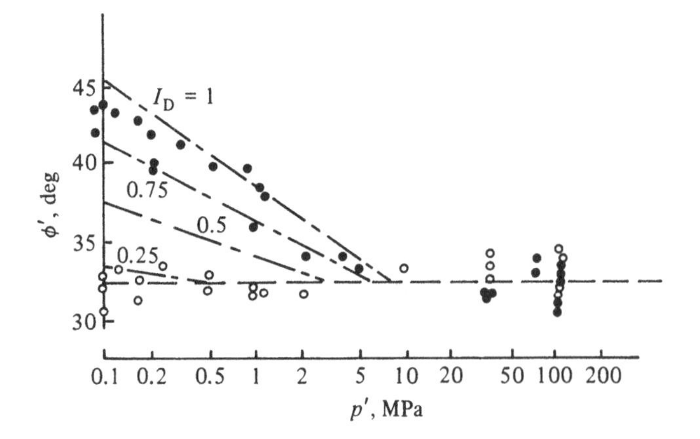

as a best fit to a wide range of data. In this expression, mean stress has to be measured in kilopascals and ϕ' in degrees. Bolton suggests that it should be used only where it leads to values of ϕ' in the range $12^\circ > (\phi' - \phi'_{cs}) > 0$. The resulting chart is superimposed on the data in Fig. 7.27.

The problem with the use of relative density as an index of sand behaviour is that it is conventionally computed using the specific volume of the sample as it has been prepared, with no confining pressure. Consequently, it does not reflect the changes in volume that may occur either as an initial stress state is applied or as the sand is sheared. For clays, the composite volumetric variable v_λ was used to reflect the current volumetric and stress state. The data in Fig. 7.23b can be described by the expression

$$\phi' - \phi'_{cs} = f(v_\lambda - \Gamma) \tag{7.59}$$

If a critical state line for a sand of the same form

$$v = \Gamma - \lambda \ln p' \tag{7.20bis}$$

can be located in the $p':v$ compression plane, then the difference

$$v_\lambda - \Gamma = v + \lambda \ln p' - \Gamma \tag{7.60}$$

can be calculated at any stage of a test.

An extensive study of the use of the quantity $(v_\lambda - \Gamma)$ to characterise the strength (and dilatancy) of sands has been made by Been and Jefferies (1985, 1986). They have managed to locate straight critical state lines in the $v:\ln p'$ compression plane for many sands and sandy silts and have calculated values of $(v_{\lambda i} - \Gamma)$, which they call the 'state parameter', from the volume and mean stress obtaining when the sand is about to be sheared in a triaxial test. Thus, they include the effect of the volume change that has occurred as the sample is compressed but not the effect of dilatancy during shear. The strength data for sand that they have accumulated reveal a fairly narrow spread (Fig. 7.28). As a result, if the initial value of the state parameter $(v_{\lambda i} - \Gamma)$ is known, then the peak strength to be expected in triaxial compression tests can be estimated to an accuracy of $\pm 2.5^\circ$. Subsequent work (Been, Crooks, Becker, and Jefferies, 1986; Been, Jefferies, Crooks, and Rothenburg, 1987) has indicated that this state parameter $(v_{\lambda i} - \Gamma)$ is useful also in understanding results of cone penetration tests in sands and sandy silts.

However, the use of initial values of $(v_\lambda - \Gamma)$ is not satisfactory if a rational picture of sand response is to be built up because, in general, volume changes and stress changes on relevant field stress paths (which may bear little resemblance to triaxial compression stress paths) will lead to continuous and major variations in v_λ.

Consideration of the data for the Chattahoochee River sand in Fig. 6.21a shows that the critical state line that has been estimated for this sand can be considered only locally straight in the $v:\ln p'$ compression plane. If the specific volume v_{cs} on the critical state line, of whatever actual shape, at the current mean effective stress can be determined, then the quantity $(v - v_{cs})$ becomes a more general state variable which has wider application than $(v_\lambda - \Gamma)$. If the critical state line is straight in the $v:\ln p'$ compression plane, then

$$v_{cs} = \Gamma - \lambda \ln p' \tag{7.61}$$

Fig. 7.28 Variation of peak angles of shearing resistance of sands with state parameter (after Been and Jefferies, 1986).

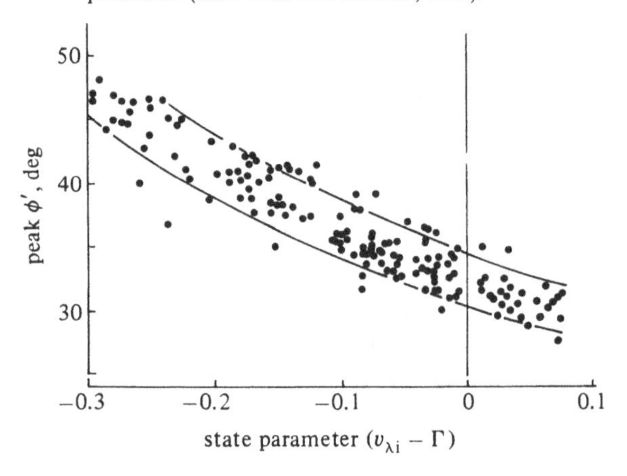

Fig. 7.29 Isotropic compression and conventional triaxial compression tests (●→) on Sacramento River sand (after Lee and Seed, 1967).

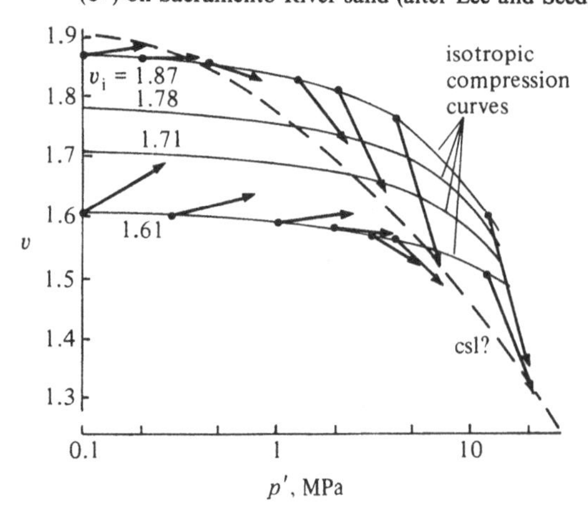

and, from (7.60),

$$v - v_{cs} = v_{\lambda} - \Gamma \tag{7.62}$$

Examples of the use of v_{λ} to show the progress of triaxial and simple shear tests on sand have been given by Stroud (1971) and Atkinson and Bransby (1978). The use of the state variable $(v - v_{cs})$ to produce a diagram for a sand similar to Fig. 7.23b for Weald clay will now be described, using data for Sacramento River sand reported by Lee and Seed (1967).

Lee and Seed report results of conventional triaxial compression tests performed at constant cell pressures between 98 kPa and 12 MPa. They report triaxial tests on samples prepared at two initial densities and isotropic compression tests on samples prepared at four initial densities. All the available compression plane information is shown in Fig. 7.29, where the arrows indicate the progress from initial state to failure of the triaxial tests. The steady reduction in dilatancy of the dense samples is apparent. The loose samples also show some dilatancy when tested at the lowest stress level.

An approximate location for a curved critical state line in the compression plane is suggested in Fig. 7.29; a straight line would not fit well with these data. The isotropic compression tests have a smaller slope than the critical state line at the same stress level. Major volume changes only occur in isotropic compression at very high mean stress levels, when particle crushing becomes a major feature of the response (compare Fig. 6.22). At very high stress levels, greater than about 10 MPa for this Sacramento River sand, the isotropic compression and critical state lines may be becoming approximately parallel, and the behaviour starts to resemble the compression behaviour of clays. At lower stress levels, the changing slope of the critical state line is significant, and in this detail the behaviours of sand and clay diverge, though the general pattern is broadly similar.

The curved critical state line in Fig. 7.29 has been used to calculate values of v_{cs} for particular values of mean effective stress. The failure data reported by Lee and Seed are presented in Fig. 7.30 in terms of peak angle of friction as a function of $v_f - v_{cs}$, calculated from the failure values of specific volume. An approximate description of these failure data is

$$\phi' - \phi'_{cs} \approx -55(v_f - v_{cs}) \qquad \text{for } v_{cs} > v_f \tag{7.63a}$$

$$\phi' - \phi'_{cs} = 0 \qquad \text{for } v_{cs} < v_f \tag{7.63b}$$

where ϕ' is measured in degrees and ϕ'_{cs} is the critical state value. The expectation from the work of Been and Jefferies (1985, 1986) is that these relationships should hold for samples of this sand prepared at any initial density.

Complete test paths for two tests on initially dense and two tests on initially loose samples are shown in a $(v - v_{cs})$:η plot in Fig. 7.31, which is equivalent to Fig. 7.23b for Weald clay. The loose sample tested at low pressure and the dense sample tested at a moderately high pressure show essentially no change in state variable $(v - v_{cs})$ as they are sheared; they start and remain very close to the critical state line in the v:ln p' plane. The dense sample tested at low pressure rises to a peak and then heads down towards the critical state. The loose sample tested at high pressure

Fig. 7.30 Dependence of peak angles of shearing resistance on state variable at failure for Sacramento River sand: •, dense, $v_o = 1.61$; ○, loose, $v_o = 1.87$ (data from Lee and Seed, 1967).

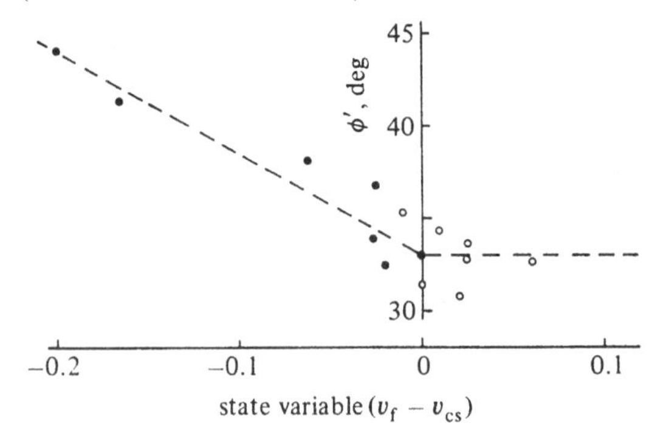

Fig. 7.31 Triaxial test paths for Sacramento River sand: (1) $v_o = 1.609$, $\sigma'_r = 98.1$ kPa; (2) $v_o = 1.576$, $\sigma'_r = 293.3$ kPa; (3) $v_o = 1.870$, $\sigma'_r = 98.1$ kPa; (4) $v_o = 1.769$, $\sigma'_r = 393.4$ kPa (data from Lee and Seed, 1967).

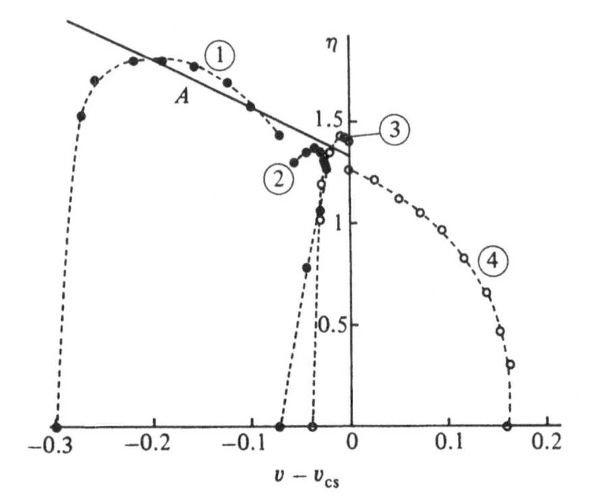

rises steadily towards the critical state. In effect, (7.63b) is redundant because samples with $v > v_{cs}$ are not expected to show a peak before the critical state is reached.

Plotting information in terms of η and $(v - v_{cs})$, where v_{cs} is deduced from a curved critical state line, amounts to a geometric distortion of the $\eta : v_\lambda$ plot used as a two-dimensional representation of effective stress and specific volume information for clays but serves the same purpose in bringing together data from samples with widely differing densities (corresponding to samples of clay with widely differing histories of overconsolidation). The state variable $(v - v_{cs})$ introduces mean effective stress in a rational way, and a limiting relationship such as (7.63a) (shown as curve A in Fig. 7.31) could be converted back into a limiting surface in $p':q:v$ space for the sand, very similar to Fig. 7.24 for a clay.

Relative density is not a sufficient quantity for characterising sand behaviour. It might be suggested that $(v - v_{cs})$ should be normalised by dividing it by $(v_{max} - v_{min})$ to produce a composite state variable which can bring together data for sands of differing mineralogy, angularity, and particle size (compare Hird and Hassona, 1986; Been and Jefferies, 1986). The specific volume range $(v_{max} - v_{min})$ gives an indication of the range of packings available at low stress levels but does not appear to relate directly either to the slope of the critical state line at low stress levels or to the slope of isotropic compression curves at higher stress levels, where particle crushing starts to become important; these are both factors that, through the state variable $(v - v_{cs})$, appear to have a controlling influence on sand behaviour in general and on the strength of sands in particular.

7.5 Status of stability and collapse calculations

Values of strength parameters for soils are required for analysis of the stability and collapse of geotechnical structures. There are essentially three principal approaches to the estimation of the loads that cause collapse of geotechnical structures; they have been well described and applied by Atkinson (1981) and are only briefly summarised here. The first two, stress fields and collapse mechanisms, are well founded in the theory of plasticity; the third, limit equilibrium, has no such theoretical basis but has been found to provide plausible results in many applications and is widely used.

1. *Stress fields.* If a distribution of stress within the soil can be found which is in equilibrium with the applied loads and which does not violate the failure criterion for the soil, then the applied loads will not cause collapse. It may be supposed that the soil will always be cleverer in distributing its stresses and hence that a

human estimate of safe applied loads will always be a *lower bound* to the actual collapse load of the soil mass.

2. *Collapse mechanisms.* If a mechanism of collapse for the soil can be postulated, then the corresponding applied loads can be calculated as the loads necessary to drive the collapse mechanism, that is, to provide a work input which just balances the work absorbed by the collapse mechanism in the soil. However, the soil will always be cleverer in finding other, more efficient modes of collapse than those that we postulate, and hence a human estimate of collapse loads will always be an *upper bound* to the actual collapse load of the soil mass.
3. *Limit equilibrium.* In the limit-equilibrium method, the soil is typically divided into a number of blocks separated by failure planes. It is assumed that the stresses on these failure planes cannot violate the failure criterion for the soil. Calculations of equilibrium of the blocks are linked together to estimate, from equilibrium considerations alone, the collapse loads. The mechanism of collapse – the arrangement of the failure planes – can be varied and optimised to obtain the most pessimistic estimate of the collapse load. There is no attempt to generate distributions of stress within the sliding blocks and no attempt to satisfy any kinematic constraints in choosing the mechanisms.

That the first two methods do indeed provide lower and upper bounds to collapse loads can be readily proved (e.g. see Calladine, 1985; Davis, 1968), but the proof is subject to the condition that the plastic deformations of the soil that occur at collapse should be *associated* with the failure criterion. The concept of normality or associated flow was discussed in Section 4.4.3. It was noted that to postulate associated flow for soils is often convenient but is not always justifiable; an example of a sliding frictional block which did not follow a rule of associated flow was introduced in Section 4.4.1. Normality was incorporated into the elastic–plastic models for soil in Chapter 4 in terms of normality of plastic strain increment vectors to yield loci.

The critical state line provides a locus of failure points, states of stress at which indefinite shearing can occur, but the plastic deformations that occur at failure are associated not with this critical state line, which provides the failure criterion, but with the yield loci. Normality to the critical state line would imply very high rates of dilation, volumetric expansion, at failure (Fig. 7.32a). (The rates of dilation that might be associated with peak strengths rather than ultimate critical state strengths will be discussed in Chapter 8.) In general, therefore, there is a difficulty in applying the theorems of plastic collapse to justify the designation of estimates of collapse loads for soil structures as upper or lower bounds.

One exception exists, however, in the analysis of the undrained collapse of clay masses. Analyses of undrained failure are usually conducted in terms of total stresses. It was shown in Section 7.2 that a consequence of the principle of effective stress is that the undrained strength of soil is independent of the applied total stresses. In the p:q *total* stress plane, the undrained failure criterion becomes a straight line parallel to the p axis (Fig. 7.32b). When the critical state is reached in undrained shearing, plastic shear deformation continues at constant effective stresses and constant volume. Thus vectors of plastic strain increment plot normal to the failure criterion ($\delta\varepsilon_p^p = 0$, Fig. 7.32b), and the condition of associated flow is satisfied. For this case only, techniques of collapse analysis based on study of stress fields and collapse mechanisms do indeed lead to demonstrable lower or upper bounds to the collapse loads.

7.6 Total and effective stress analyses

Although it has been emphasised throughout previous chapters that the response of soils should be studied in terms of effective stresses, it is often easier to analyse equilibrium of geotechnical structures in terms of total stresses. For example, at depth z in a soil deposit (Fig. 7.33), the total vertical stress will be

$$\sigma_v = \gamma z \tag{7.64}$$

where γ is the unit weight of the overlying soil. However, the effective vertical stress cannot be known without some information about the pore

Fig. 7.32 (a) Normality of plastic strain increment vectors to critical state line (csl) implies continuing large volumetric expansion at failure; (b) total stress strength criterion for undrained failure and corresponding plastic strain increment vectors.

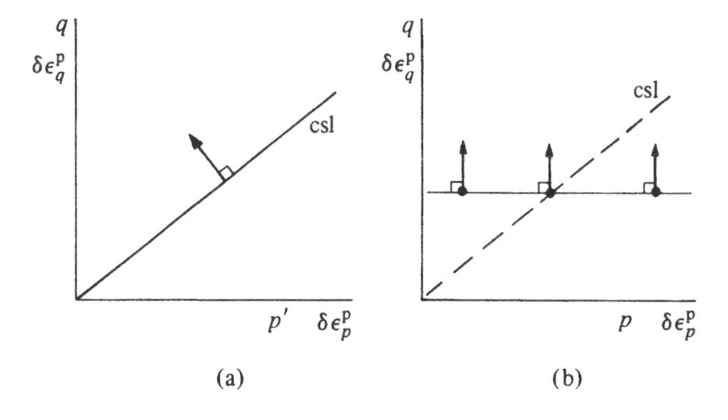

pressure, and the pore pressure may consist of separable parts due to the presence of a water table or a steady seepage flow regime, and due to the shearing of the soil.

Stability analyses can be carried out in terms of either total stresses or effective stresses. Clearly, a geotechnical structure that is on the point of collapse should appear critical whether the stability analysis is carried out in terms of total stresses or effective stresses. For the effective stress analysis to be correct, it is necessary that the distribution of pore pressures should be correctly known; for this reason, the idea of pore pressure parameters linking changes in pore pressure with changes in applied total stresses was introduced in Section 1.6. For the total stress analysis to be correct, it is necessary that the variation of soil strength should be correctly known.

Total stress analyses are usually performed for situations in which undrained loading or response is of interest and a single undrained shear strength can be assigned to the soil. Then, at different points within the soil mass, different total stress Mohr circles may be operating at failure, but they will all have the same radius and will be displaced by appropriate pore pressures from a single effective stress Mohr circle, as shown in Fig. 7.4. Once some drainage has been permitted, then changes of water content or specific volume will have occurred, and it may no longer be reasonable to argue that the total stress Mohr circles should all have the same size: where the water content has fallen, the strength will have increased and vice versa.

An example of parallel total and effective stress stability calculations is provided by back-analysis of a slope failure which occurred at Jackfield, Shropshire after a period of heavy rain, during the winter 1952–3 (Henkel and Skempton, 1955). A plan and section of the landslide are shown in Fig. 7.34. About 300,000 tons of overconsolidated clay soil slid between 10 and 20 m down a plane slope inclined at 10.5° to the horizontal. Investigation revealed the existence of a thin but extensive plane of failure parallel to the plane of the slope and at a uniform depth of about 5 m.

An element of clay of depth z from the surface of an infinite slope of angle β, width b, unit thickness, and weight $W = \gamma bz$, where γ is the total unit weight of the clay, is shown in Fig. 7.35a. Consideration of

Fig. 7.33 Element at depth z below ground surface.

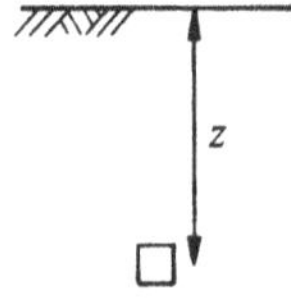

equilibrium of the forces acting on the element shows that the total normal stress σ and shear stress τ acting on a plane at depth z parallel to the slope (Fig. 7.35a) are

$$\sigma = \gamma z \cos^2 \beta \tag{7.65}$$

$$\tau = \gamma z \cos \beta \sin \beta \tag{7.66}$$

A total stress approach to the analysis of the stability of the slope would say that failure would be expected if the shear stress (7.66) reached the undrained strength of the clay.

The clay at Jackfield had liquid limit $w_L = 0.45$, plastic limit $w_P = 0.2$, and natural water content, away from the failure zone, $w = 0.2$. The undrained strength of this clay was 76.6 kPa. The saturated unit weight of the clay was 20.4 kN/m^3. At a depth of 5 m the shear stress would have been $\tau = 18.3$ kPa, giving an apprarent factor of safety of 4.2 by comparison with the strength of the bulk of the clay.

Fig. 7.34 Landslide at Jackfield, Shropshire (after Henkel and Skempton, 1955).

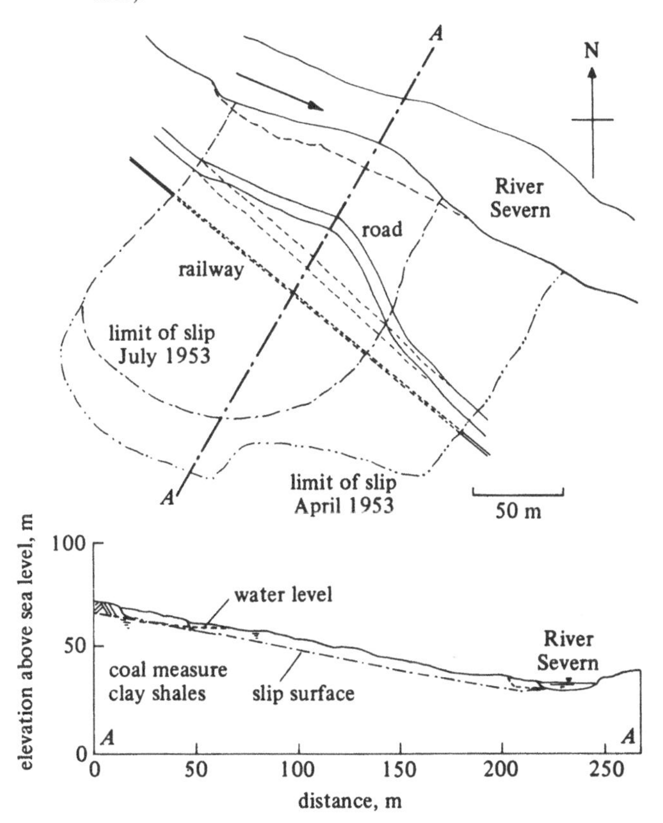

Tests on specimens of clay from the failure zone showed that the soil there had a water content $w = 0.30$ and an undrained strength 21.5 kPa. Using this strength, we obtain an apparent factor of safety of 1.17, which is sufficiently close to unity to confirm concern for the stability of the slope.

Drained triaxial tests gave effective stress data from which Henkel and Skempton deduced that the clay failed according to a Mohr–Coulomb failure criterion

$$\tau = c' + \sigma' \tan \phi' \qquad (7.1\text{bis})$$

with $c' = 7.2$ kPa and $\phi' = 21°$. The shear strength on any plane depends on the effective normal stress σ', where

$$\sigma' = \sigma - u \qquad (7.67)$$

and u is the pore pressure. The total stress on the failure plane can be calculated from (7.65) as $\sigma = 98.6$ kPa, and the necessary pore pressure to precipitate failure according to the failure criterion with these strength parameters is then $u = 69.7$ kPa, corresponding to a head of 7.1 m of water on the failure surface.

High positive pore pressures are unlikely to result from the shearing of an overconsolidated clay (see Section 7.3). The maximum steady in situ pore pressures occur when there is a state of steady seepage parallel to the slope, down the hillside (Fig. 7.35b). The flowlines for this seepage

Fig. 7.35 (a) Element of soil in infinite slope; (b) seepage parallel to slope.

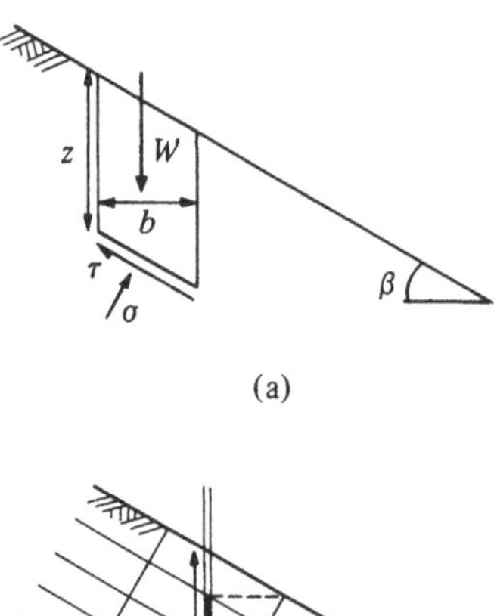

(a)

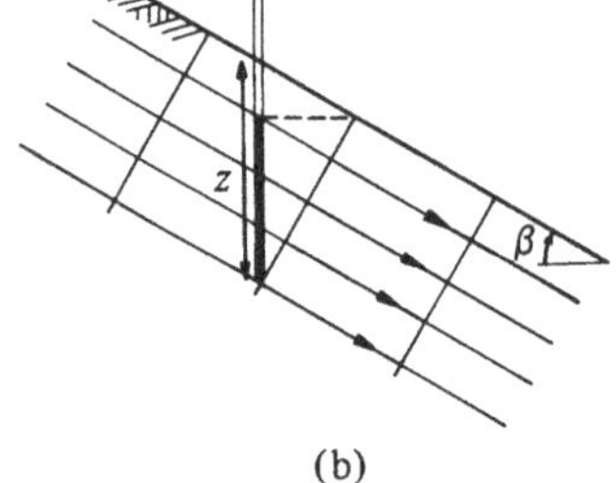

(b)

flow are then parallel to the slope, the equipotentials orthogonal to the slope, and the total pore pressure at depth z is (from Fig. 7.35b)

$$u = \gamma_w z \cos^2 \beta \quad (7.68)$$

At a depth of 5 m, then, a maximum pore pressure of $u = 47.4$ kPa could be anticipated.

The values of the strength parameters c' and ϕ' quoted above relate to peak strengths seen in drained tests. It seems from the water content evidence that in the failure zone the clay had an opportunity to soften towards an ultimate critical state condition, for which it would be expected that a failure criterion with $c' = 0$ would be more appropriate. Accepting the angle of friction $\phi' = 21°$ as a critical state angle of friction and assuming that a pore pressure $u = 47.4$ kPa was present on the failure surface, we can calculate a shear strength of 19.6 kPa, giving a factor of safety of 1.07 by comparison with the shear stress $\tau = 18.3$ kPa and also confirming an expectation of instability. (The critical state angles of friction ϕ'_{cs} quoted in Section 7.4.1 are in general somewhat higher than Hvorslev angles of friction ϕ'_e used to describe peak strength conditions. However, the strength parameters quoted by Henkel and Skempton have not been deduced from a plot in which water content differences have been taken into account, and the angle of friction quoted is probably intermediate between ϕ'_e and ϕ'_{cs}.)

Thus total and effective stress analyses of this slope failure both give sensible results, provided the appropriate undrained strength is used in the former and the appropriate strength parameters and appropriate pore pressures are used in the latter. A moral from the effective stress analysis is that peak strengths may not be reliable for design purposes: their existence relies on softening of the soil not occurring or being prevented.

7.7 Critical state strength and residual strength

When a cutting is excavated in a stiff, heavily overconsolidated clay and a retaining wall is constructed to support the remaining soil, negative pore pressures are left in the clay behind the wall as a result of the reduction of the lateral stress and the associated shearing (and repressed dilation, assuming the construction process to be rapid) of the soil. A negative pore pressure makes effective stresses higher than total stresses and contributes beneficially to the strength of the clay. However, with time, these negative pore pressures tend to cause water to be drawn in from the nearby soil at a rate dependent on the swelling and permeability characteristics of the clay and on the structure of the soil. The water content or specific volume of the clay increases and its strength decreases.

If the design of the wall has not taken this potential softening into account, then failure is likely to occur some time after the construction of the wall.

An example is provided by the failure in 1954 of a retaining wall built in 1912 on a London underground railway line, described by Watson (1956) and quoted by Schofield and Wroth (1968). A section through the wall is shown in Fig. 7.36a. Investigations indicated the presence of discontinuous slip planes behind the wall, and water content studies reported by Henkel (1956) showed significant increases of water content

Fig. 7.36 Failure of retaining wall near Uxbridge: (a) section (after Watson, 1956); (b) water content variations around slip zone (after Henkel, 1956).

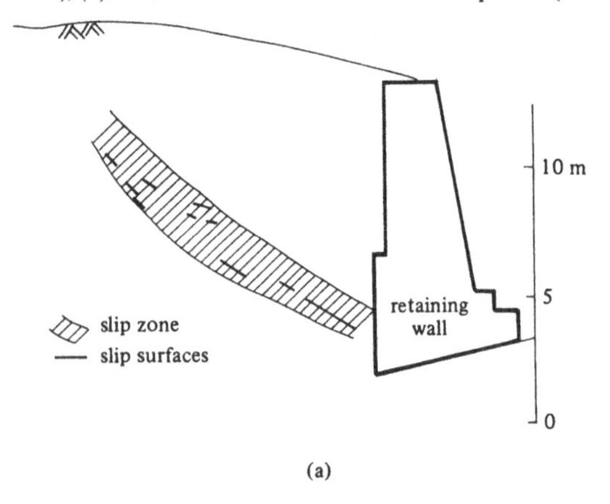

(a)

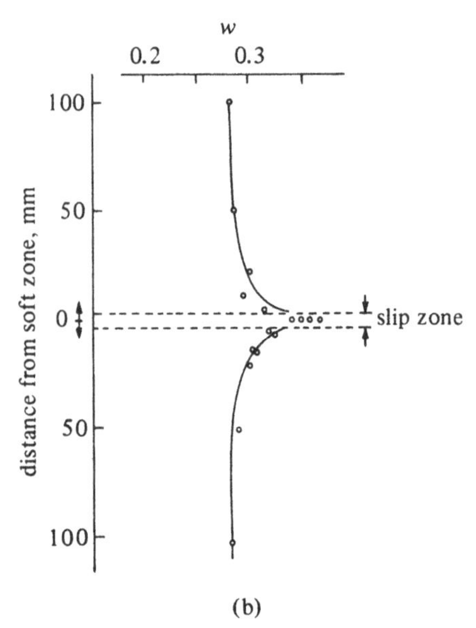

(b)

in the slip zone (Fig. 7.36b), of a form similar to those proposed on theoretical grounds in Fig. 7.16 and Section 7.4. Just as for the slope at Jackfield, described in the previous section, calculations of the stability of the wall that do not allow for the softening of the soil to a critical state in the failure region do not lead to sensible conclusions.

Data gathered by Skempton (1970b) for a number of slope failures in London clay in general confirm this picture (Fig. 7.37): failures are consistent with a critical state angle of friction $\phi'_{cs} = 20°$ and zero cohesion

Fig. 7.37 Peak strength envelope, critical state line, and residual strength envelope for London clay (adapted from Skempton, 1970b).

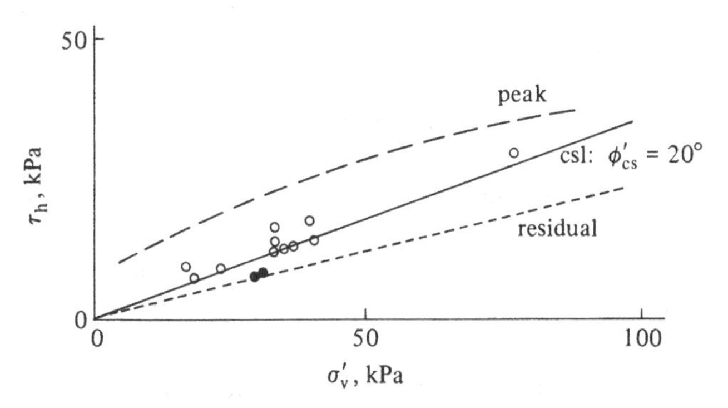

Fig. 7.38 (a) Diagram of ring shear test; (b) drop of shearing resistance to residual value in ring shear test on Kalabagh clay ($I_P = 0.36$) (after Skempton, 1985); (c) orientation of clay particles on residual sliding surface.

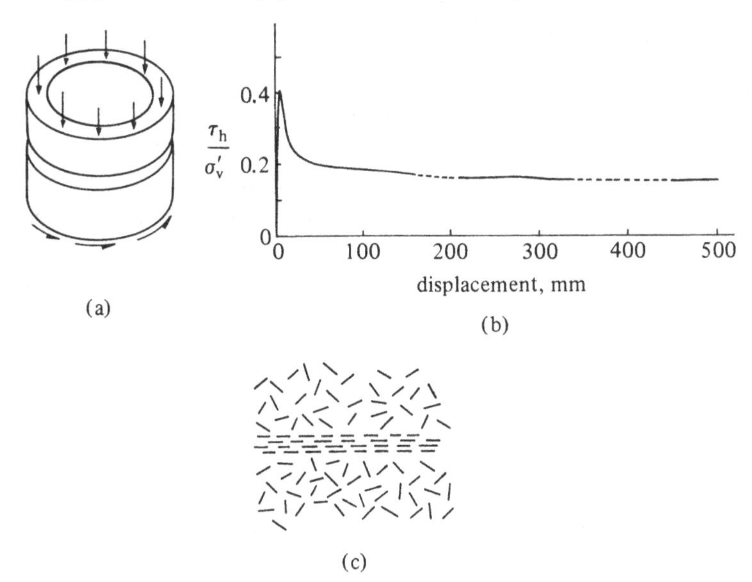

$c' = 0$. Exceptions are provided, however, by slope failures which turn out to be reactivations of previous slope instabilities within historic or geologic past. For these, the angle of friction mobilised is considerably lower than the critical state value.

Such low angles of friction can be reproduced in the laboratory if ring shear tests are carried out in which very large displacements between two parts of a clay sample are applied. The ring shear apparatus is rather like a long shear box in which the two ends of the box have been bent round and joined together (Fig. 7.38a) so that, like the shear box, failure is forced to occur in a thin central region but, unlike the shear box, the failure surface through the soil specimen has no ends to it. Typical results of a ring shear test on Kalabagh clay taken to large displacement are shown in Fig. 7.38b from Skempton (1985): after a displacement of 100–200 mm, the angle of friction has dropped from a peak value of about 22° to a residual value $\phi'_r \sim 9°$. Skempton notes that water content changes in the shear zone seem generally to have ceased after displacements of 5–10 mm. Skempton (1970b) notes that although sections of failure planes were found at the retaining wall failure in Fig. 7.36, there was no continuous failure surface; movements prior to failure had been small and insufficient to produce a deterioration of friction to the residual value.

When the mechanisms of residual shearing and shearing to a critical state are compared, it becomes clear that they are very different. When soil is at a critical state, it is being continuously remoulded and churned up, and its structure remains random. When a clay is sheared to a residual state on a failure surface, the deformations have been so large that the clay particles on both sides of the failure surface have become oriented parallel to the failure (Fig. 7.38c), so it is not surprising that the friction generated on such polished, slickensided surfaces is much lower than the friction mobilised when soil particles are still being stirred up.

The critical state strength is thus not always a lower bound to the strength of the soil; but whether residual strength is actually important depends on whether it is possible for a smooth, continuous failure surface to form through the soil. This possibility depends on the composition of the soil. Most soils are made up a range of particle sizes and shapes. The formation of a smooth failure surface requires the presence of an adequate proportion of clay particles: platelike particles which are capable of realignment. A soil with only a small proportion of clay particles is not able to form a continuous failure surface because the dominant rotund particles get in the way.

Lupini, Skinner, and Vaughan (1981) suggest that the most satisfactory correlation is with a volumetric variable, granular specific volume v_g, where, for saturated soil [compare (1.6)],

$$v_g = 1 + \frac{\text{volume of water} + \text{volume of platelike particles}}{\text{volume of rotund particles}} \tag{7.69}$$

This variable is broadly equivalent to plasticity or clay content; but, unlike these quantities, granular specific volume can make some allowance for the change in volume occupied by clay particles that may occur when the stress level is changed. A plot of residual friction against granular specific volume for a number of soils – some natural soils and some artificial sand and clay or glacial till and clay mixtures – is shown in Fig. 7.39a. Low residual angles of friction are expected for granular specific volumes in excess of about 3. To confirm this observation, Lupini et al. performed microscopic examinations of thin sections of the soil samples which had been sheared to large displacements. They were able to distinguish the different modes of failure that are illustrated in Figs. 7.39b, c. With a low clay content and low granular specific volume, no preferred orientation of clay particles developed, and the structure showed evidence of 'turbulent' flow to a critical state (Fig. 7.39b). With a high clay content and high granular specific volume, slickensided failure surfaces were found with strong orientation of clay particles (Fig. 7.39c). The observations of turbulent, transitional, or sliding behaviour are noted on Fig. 7.39a.

Residual strength is of particular importance where a new geotechnical structure is likely to load soil which has been previously sheared to very

Fig. 7.39 (a) Angles of shearing resistance and modes of failure related to granular specific volume v_g (○, turbulent; ×, transitional; •, sliding); (b) and (c) mixtures of London clay and Happisburgh till; (b) clay fraction 0.2; (c) clay fraction 0.4 (after Lupini, Skinner, and Vaughan, 1981).

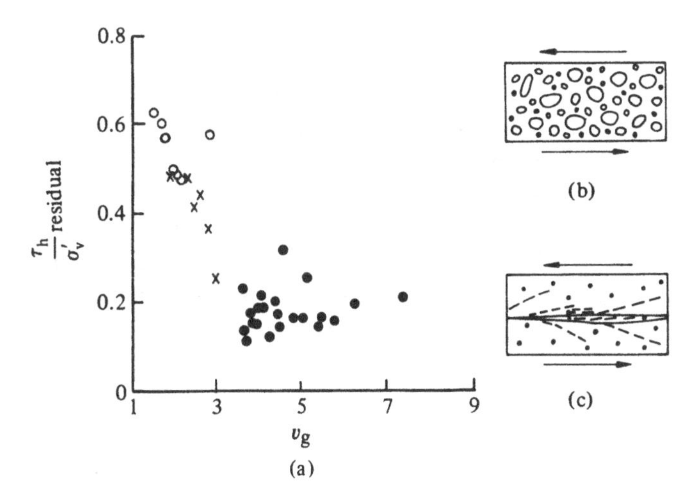

large deformations; frequently this will be the result of landslides which have occurred in the geological history of a deposit, but it may also result from solifluction, erosion, or other more recent processes. An engineer will need to be careful to look out for possible existing shear surfaces during the site investigation and need to study the geological record in detail.

Residual strength may not control the occurrence of first-time failures in previously intact clay, but the movements that occur when a slip takes place may be all the larger because the strength that can be mobilised on a failure plane drops dramatically as deformations develop.

7.8 Conclusion

Various aspects of the strength of soils have been discussed. The existence of a critical state line and its relationship to a normal compression line allow statements to be made about the variation of undrained strength with overconsolidation ratio. It has been shown that the form of the elastic–plastic models of soil behaviour developed in Chapters 4 and 5 leads to the existence of peak strengths. The need for strain softening to occur as the soil weakens from a peak strength to a critical state makes it very likely that localisation of deformation in thin rupture zones will occur, and this will tend to obscure overall observation of attainment of critical states. The critical state represents in many ways a lower bound to the strength of soils, but a lower, residual strength may be seen if it is possible for orientation of particles parallel to a failure plane to develop. It is, however, very important to distinguish between the drop of strength which arises because of particle reorientation and the drop of strength which occurs in overconsolidated soils as they suck in water and soften on shearing. That softening emerges naturally in an elastic–plastic model of soil behaviour such as Cam clay. Once a localised plane or thin zone of failure has developed, then such continuum models of soil behaviour lose their attraction; subsequent response must be described in terms of stresses on and displacements across that thin zone. All of this discussion indicates the importance of a volumetric variable as well as stresses: it makes little sense to obtain a single set of strength parameters from test data which are not comparable because the specific volumes or water contents of various samples are widely different. Apparent scatter of strength data may be understood when variations of volume are recognised.

Exercises

E7.1. A set of samples of Weald clay, for which the compression parameter $\lambda = 0.091$ and the specific gravity $G_s = 2.75$, have been isotropically normally compressed under a pressure $p' = 827\,\mathrm{kPa}$ ($120\,\mathrm{lbf/in^2}$), and the samples have been allowed to swell to a number of pressures along the lowest swelling line in Fig. 6.8. The samples have been subjected to conventional drained or conventional undrained triaxial compression tests, and their conditions at failure have been examined. Failure has been defined as the condition when the deviator stress reaches its maximum value q_f; this is found experimentally for Weald clay to be related to the mean effective stress at failure p'_f by the expression

$$q_f = 0.72(p'_f + 0.13p'_e)$$

where p'_e is the equivalent pressure on the normal compression line corresponding to the water content of the sample at failure.

Plot a diagram showing the variation, with overconsolidation ratio, of the ratio of strengths in conventional drained and undrained triaxial compression tests. What implications does this diagram have for the design engineer? Does the point at which the ratio of strengths is equal to 1 have any special significance?

E7.2. A soil fails according to a Hvorslev surface described by parameters c'_{pe} and ϕ'_e. For soil which has been isotropically normally compressed to a maximum pressure p'_c and then unloaded isotropically to a given overconsolidation ratio n, derive an expression for the deviator stress q_f at failure in a conventional undrained triaxial compression test, in terms of the current mean effective stress p'_i (assumed constant during the undrained test), n, $\Lambda = 1 - \kappa/\lambda$, and the Hvorslev strength parameters.

For $c'_{pe} = 0.046$, $\phi'_e = 18.8°$, and $\Lambda = 0.63$, calculate values of q_f for $p'_c = 400\,\mathrm{kPa}$ and values of $n = 2, 4, 8, 16$, and 32. Compare these values with peak values obtained using the Cam clay model, taking $M = 0.87$. Plot the two strength envelopes in the shear stress:effective normal stress plane ($\tau:\sigma'$).

E7.3. A sample of saturated clay is normally compressed in a triaxial cell by increasing the cell pressure and holding the axial length constant. The axial effective stress is K_1 times the radial effective stress, and the specific volume v is given by

$$v = v_\lambda - \lambda \ln p'$$

where v_λ and λ are soil constants, and p' is the mean normal stress. The samples are then subjected to undrained compression tests with the cell pressure held constant. Derive an expression for the normalised pore pressure at failure (u_f/q_f), in terms of K_1, v_λ, λ, and the soil constants M and Γ which define the positions of the critical state line for the clay.

8

Stress–dilatancy

8.1 Introduction

A general framework on which to create elastic–plastic models of soil behaviour was set up in Chapter 4. The basic features of this framework included yield surfaces, which bound elastically attainable states of stress, and plastic potentials, which control the mode or mechanism of plastic deformation that occurs when the soil yields. Although examples of yield loci deduced from triaxial tests on soils were presented in Chapter 3, few data were produced to guide the choice of plastic potentials because, it was noted, it is often convenient to make plastic potentials and yield surfaces coincident. Indeed in the Cam clay model described in detail in Chapter 5, we assumed that the plastic potentials and yield surfaces were identical, so that plastic deformation of this model soil obeys an associated flow rule.

In this chapter, the forms of the plastic potentials which emerge from various models of material response are presented, and some of the factors controlling the modes of plastic deformation are discussed with reference to various sets of experimental data.

8.2 Plastic potentials, flow rules, and stress–dilatancy diagrams

Plastic potentials were introduced in Section 4.4.2 as curves in the p':q effective stress plane to which, by definition, the vectors of plastic strain increment $\delta\varepsilon_p^p$:$\delta\varepsilon_q^p$ are orthogonal (Fig. 8.1). Many of the models of soil behaviour that have been proposed make the mechanism of plastic deformation, the value of the ratio $\delta\varepsilon_q^p/\delta\varepsilon_p^p$, dependent only on stress ratio $\eta = q/p'$ and not on the individual values of q or p'. The ratio $\delta\varepsilon_q^p/\delta\varepsilon_p^p$ is by definition in the direction of the normal to the plastic potential, and if this is a function only of stress ratio η, all plastic potential curves for a soil, drawn in the p':q effective stress plane, can be collapsed onto a single curve. Any particular plastic potential curve can be obtained from any other one by radial scaling from the origin of the $p':q$ plane.

An alternative way of presenting information about plastic potentials is to plot stress ratio η against the plastic strain increment ratio $\delta\varepsilon_p^p/\delta\varepsilon_q^p$ or $\delta\varepsilon_q^p/\delta\varepsilon_p^p$. However, use of the ratios $\delta\varepsilon_p^p/\delta\varepsilon_q^p$ or $\delta\varepsilon_q^p/\delta\varepsilon_p^p$ themselves is inconvenient if situations are to be included which involve yielding with no plastic shear strain ($\delta\varepsilon_q^p = 0$) or with no plastic volumetric strain ($\delta\varepsilon_p^p = 0$), respectively. A quantity which expresses the ratio between increments of plastic volumetric strain and shear strain but which always remains finite is the angle β between the strain increment vector and the p' axis (Fig. 8.1), where

$$\tan\beta = \frac{\delta\varepsilon_q^p}{\delta\varepsilon_p^p} \tag{8.1}$$

The relationship between plastic strain increment ratio and stress ratio is known as a flow rule governing the mode or mechanism of plastic deformation or flow of the soil. The ratio $\delta\varepsilon_p^p/\delta\varepsilon_q^p$ is the plastic dilatancy

Fig. 8.1 Stress ratio $\eta = q/p'$ and dilatancy angle $\beta = \tan^{-1}\delta\varepsilon_q^p/\delta\varepsilon_p^p$.

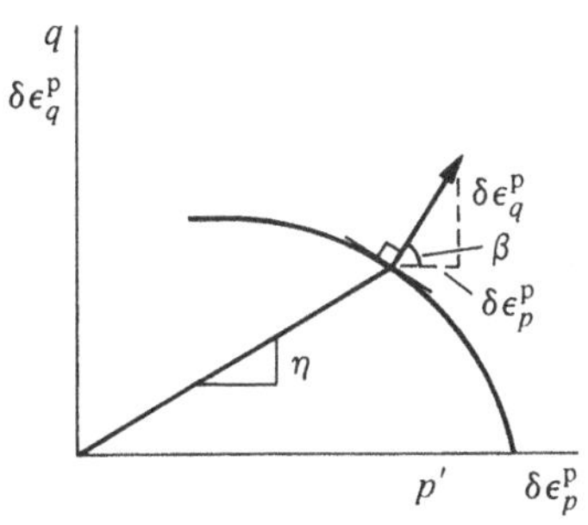

Fig. 8.2 (a) Plastic strain increments for isotropic compression and critical states plotted in $p':q$ effective stress plane; (b) stress–dilatancy diagram $\eta:\beta$; isotropic compression ($I:\eta = 0$) and critical state ($C:\eta = M$).

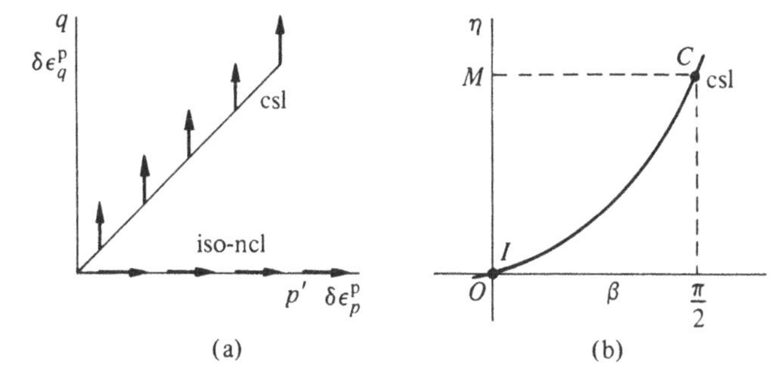

of the soil, and the resulting plots in terms of β and stress ratio η (Fig. 8.2b) are called *stress–dilatancy diagrams.*

This method of presentation links strain increments with stresses, a familiar feature of plastic behaviour, but something which is alien to usual notions of elastic behaviour. It would not be appropriate to plot an elastic strain increment ratio on such a diagram because at any given stress ratio η, the ratio of elastic strain increments depends on the applied ratio of stress increments, which can take any value. It is not appropriate, either, to plot a ratio of total strain increments since this includes both plastic and elastic elements. Frequently, however, separation of elastic and plastic components of strain is not straightforward, and total strain increments

Fig. 8.3 (a) Plastic potential and (b) stress–dilatancy relationship for Cam clay model (drawn for $M = 1$).

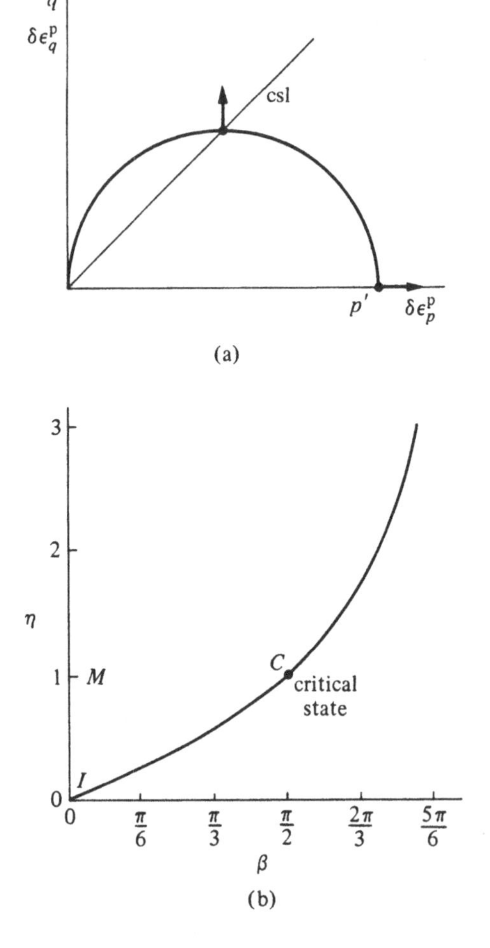

are plotted in a stress–dilatancy diagram. For many situations the contribution of elastic strains to total strains may be negligible when yielding is occurring, and the difference between a plastic strain increment ratio and a total strain increment ratio may be small. However, though this may seem attractive, it is necessary to view total strain increment plots with caution: an undrained or constant volume test has $\delta\varepsilon_p/\delta\varepsilon_q = 0$ throughout; but if the soil is yielding, elastic and plastic volumetric strains are of opposite sign and, in general, are certainly not zero in magnitude, so that $\delta\varepsilon_p^p/\delta\varepsilon_q^p \neq 0$.

For soil which is being sheared without plastic volumetric strain, corresponding to the attainment of a critical state according to the Cam clay model, $\eta = M$ and $\delta\varepsilon_p^p/\delta\varepsilon_q^p = 0$. The plastic strain increment vectors are directed parallel to the q axis (Fig. 8.2a) and, from (8.1), $\beta = \pi/2$ (point C in Fig. 8.2b). A soil which has experienced only isotropic stresses in its past and is now being compressed isotropically ($q = 0$) changes in volume without changing in shape, $\delta\varepsilon_q^p = 0$. The plastic strain increment vectors are directed parallel to the p' axis (Fig. 8.2a), and since $\delta\varepsilon_q^p = 0$, $\beta = 0$ (point I in Fig. 8.2b). A plastic potential for isotropic soil can be converted into a flow rule which links points C and I in the $\beta{:}\eta$ diagram, but the precise form of the curve between C and I is a matter for assumption in any particular soil model.

The Cam clay model described in Chapter 5 is a model of particular relevance for isotropically compressed soil. The shape of the plastic potentials assumed in that model (identical to the elliptical yield loci) (Fig. 8.3a) leads to a flow rule:

$$\frac{\delta\varepsilon_p^p}{\delta\varepsilon_q^p} = \frac{M^2 - \eta^2}{2\eta} \qquad (8.2)(5.7\text{bis})$$

or

$$\tan\beta = \frac{2\eta}{M^2 - \eta^2} \qquad (8.3)$$

This flow rule is plotted in Fig. 8.3b.

8.3 Stress–dilatancy in plane strain

The Cam clay model has been applied so far only to the description of the behaviour of soils in triaxial tests. Models which illustrate the interconnection between stress ratios and rates of dilation can be described more simply for conditions of plane strain. Parallels can then be drawn to devise related flow rules for triaxial conditions.

A simple analogy of dilating soil is provided by the interlocking saw blades shown in Fig. 8.4. Because the soil is expanding as it is sheared, it is supposed that sliding within the soil takes place, not on horizontal planes, but on planes inclined at an angle of dilation ψ to the horizontal (Fig. 8.4a). Sliding between adjacent soil particles occurs on these planes. Looking at the forces involved in this sliding process (Fig. 8.4b), if the angle of friction resisting sliding on the inclined planes is ϕ'_{cs}, then the apparent externally mobilised angle of friction on horizontal planes, ϕ'_m, is

$$\phi'_m = \phi'_{cs} + \psi \tag{8.4}$$

If it is supposed that the angle of friction resisting motion between layers of soil particles is always ϕ'_{cs}, then ϕ'_{cs} can be seen as a soil constant, and (8.4) becomes a stress–dilatancy relation linking the mobilised friction ϕ'_m with an angle of dilation ψ. [Equation (8.4) is not a rigorous statement about the equilibrium of the saw tooth shown in Fig. 8.4 because it results from an argument based on forces when an argument based on stresses would be more appropriate. However, it does immediately suggest that some relationship between mobilised friction, or stress ratio, and dilatancy is to be expected.]

The saw-tooth failure or sliding surface of Fig. 8.4 could be imagined forming across a direct shear box test on a soil sample (Fig. 8.5). The quantities measured in such a test are the normal load P, the shear load Q, and the corresponding displacements y and x of the boundaries of the shear box. The work done by the applied loads P and Q on the soil sample during incremental displacements δy and δx is

$$\delta W_T = P\,\delta y + Q\,\delta x \tag{8.5}$$

The term $Q\,\delta x$ represents the work done in shearing the sample. The term $P\,\delta y$ represents the work that is done because the soil sample is changing in volume as it is sheared. For a soil such as dense sand that is dilating as it is sheared, $\delta y < 0$ and the normal load P is being lifted up as the

Fig. 8.4 (a) Sliding of interlocking saw blades on inclined rough surfaces; (b) resultant force on inclined surface.

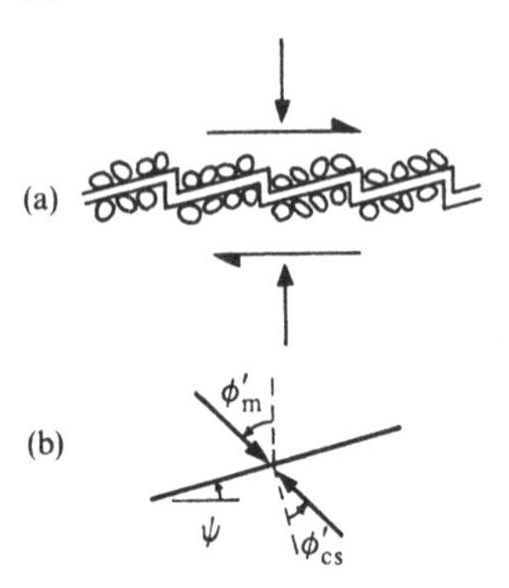

soil is sheared. The soil expands because of the interlocking of the particles: deformation can proceed only if some particles are able to ride up over other particles. Because of this interlocking, not all of the shearing work $Q\,\delta x$ is absorbed by the soil; some of it is required to lift the normal load and overcome the interlocking of the soil particles. The remainder, the nett work input δW_T goes into the sample; some may be stored in elastic deformations of the soil particles, but most is dissipated in frictional resistance between the grains as they roll and slide on each other.

Following Taylor (1948), a simple stress–dilatancy equation can be obtained if it is assumed that *all* of this nett work input is dissipated in friction (no energy is stored in elastic deformations) and if it is also assumed that this frictional dissipation is controlled by the normal load P and a frictional constant μ, so that

$$\delta W_T = \mu P\,\delta x \tag{8.6}$$

Putting (8.5) and (8.6) together and rearranging gives

$$\frac{Q}{P} + \frac{\delta y}{\delta x} = \mu \tag{8.7}$$

The first term is the externally mobilised friction on horizontal planes

$$\frac{Q}{P} = \tan\phi'_m \tag{8.8}$$

The second term describes the dilatancy of the sample; by comparison with Fig. 8.4:

$$\frac{\delta y}{\delta x} = -\tan\psi \tag{8.9}$$

When the soil particles are sliding and rotating in such a way that the volume of the soil remains constant, so that the soil has reached a critical state, then $\delta y/\delta x = 0$ and $Q/P = \mu$. Expression (8.7) can be written as

$$\tan\phi'_m = \mu + \tan\psi \tag{8.10}$$

or, strength equals friction plus dilatancy. (The relationship between μ

Fig. 8.5 Inclined rough surfaces in shear box.

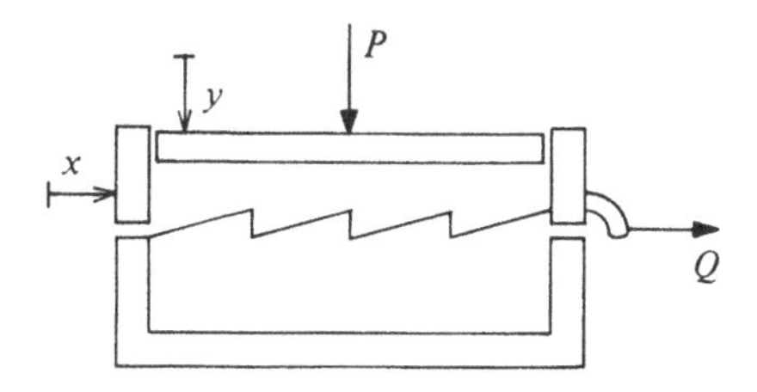

and a critical state angle of friction ϕ'_{cs} has been left deliberately vague: the details of the link between the two require a number of assumptions to be made; see, for example, Airey, Budhu, and Wood, 1985.)

Relationship (8.10) can be tested on data from direct shear box tests on dense and loose sands reported by Taylor (1948) (Fig. 8.6). The slope of the graph of displacement data $x:y$ (Fig. 8.6b) is used to generate values of $\delta y/\delta x$, which are plotted with the ratio of shear load to normal load Q/P in Fig. 8.7a. The sum $Q/P + \delta y/\delta x$ is plotted against horizontal displacement x in Fig. 8.7b. It is evident that the initial points in Figs. 8.7a, b do not match the simple expression (8.7) or (8.10). But a flow rule should

Fig. 8.6 Direct shear tests on Ottawa sand with normal stress 287 kPa (×, dense, $v_o = 1.562$; •, loose, $v_o = 1.652$): (a) moblised friction Q/P on horizontal plane and shear displacement x; (b) vertical displacement y and shear displacement x (after Taylor, 1948).

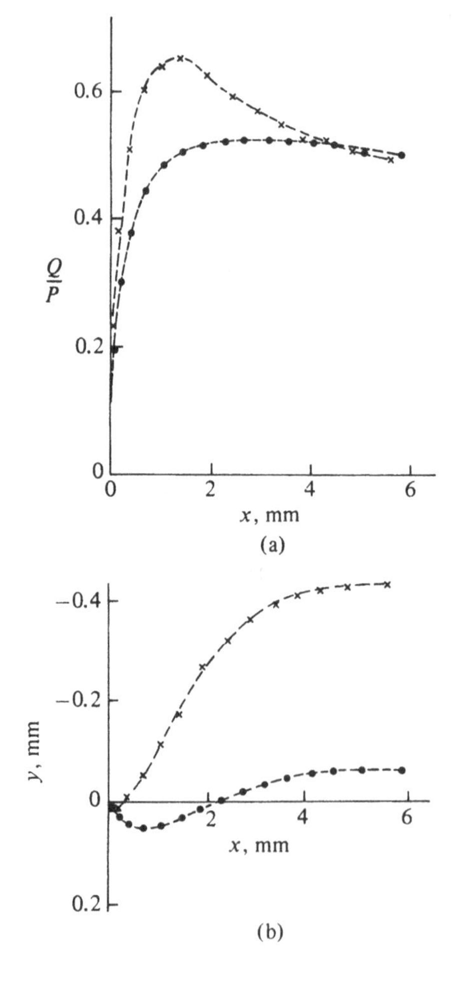

contain information about plastic deformations rather than total deformations; and in the initial stages of a shear test, some work is probably being done by the shear load in causing elastic deformation of the soil particles, so some deviation from (8.7) or (8.10) is to be expected. Beyond the points of peak load ratio (or peak mobilised friction) the data are

Fig. 8.7 Stress ratio and dilatancy in direct shear tests on Ottawa sand with expression (8.7) superimposed (×, dense; •, loose): (a) friction Q/P and dilatancy $\delta y/\delta x$; (b) $Q/P + \delta y/\delta x$ and shear displacement x (data from Taylor, 1948).

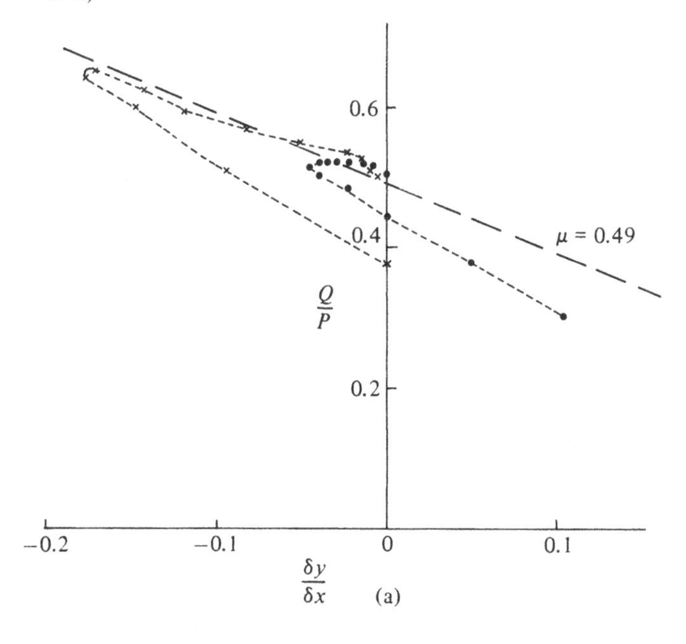

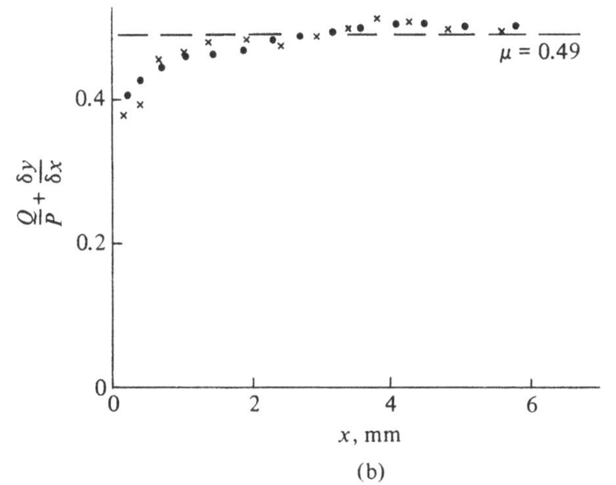

more consistent, and a value of $\mu \approx 0.49$ might be proposed. Expression (8.7), with $\mu = 0.49$, is plotted in Fig. 8.7a and describes fairly well the softening of the sand as the stress ratio drops from its peak value.

Although external measurements of forces P and Q and displacements x and y can be made in tests on soils in a direct shear box (Fig. 8.5), interpretation of these quantities in terms of stresses and strains is not feasible because the soil in the shear box is clearly not deforming homogeneously. The simple shear apparatus is another plane strain apparatus which attempts to impose a more uniform deformation on soil samples. This mode of deformation is shown in Fig. 8.8a: the length of the sample remains constant, but its height may change as the sample is sheared. Stress conditions in the simple shear apparatus are not particularly uniform (see Section 1.4.2), but with suitable instrumentation, measurements of the normal and shear stresses σ_{yy} and τ_{yx} acting on the soil can be made. Measurements of horizontal and vertical displacements x and y can be converted to strains:

$$\delta\varepsilon_{yy} = \frac{\delta y}{h} \tag{8.11}$$

and

$$\delta\gamma_{yx} = \frac{\delta x}{h} \tag{8.12}$$

Fig. 8.8 (a) Simple shear deformation and (b) definition of angle of dilation ψ; (c) mobilised angle of friction ϕ'_m.

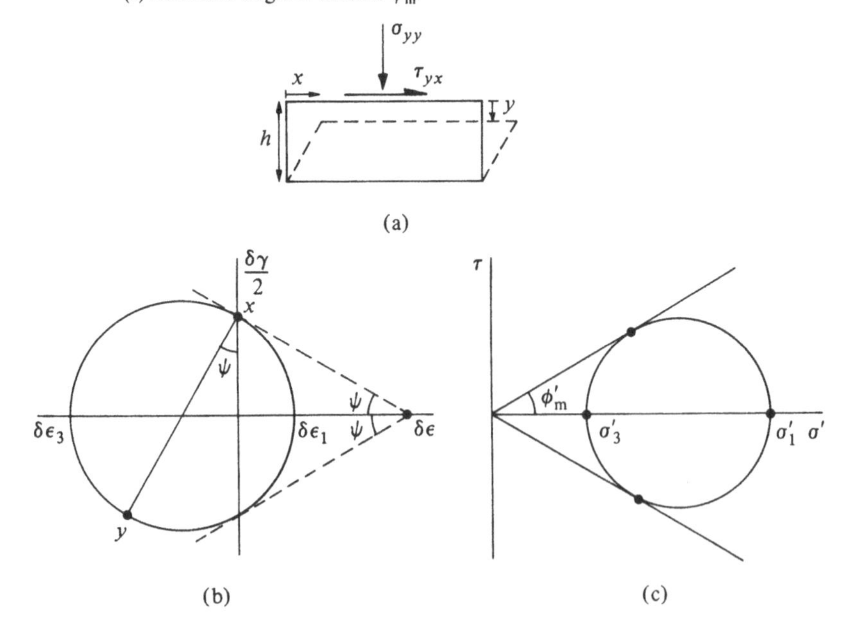

where h is the height of the sample. The constant length of the sample fixes $\delta\varepsilon_{xx} = 0$. The angle ψ, which for the shear box was the direction of movement of the top platen of the shear box relative to the bottom platen, now has a particular geometrical significance in the Mohr circle of strain increment (Fig. 8.8b),

$$\tan\psi = \frac{-\delta\varepsilon_{yy}}{\delta\gamma_{yx}} \tag{8.13}$$

or

$$\sin\psi = \frac{-\delta\varepsilon_s}{\delta\varepsilon_t} \tag{8.14}$$

where $\delta\varepsilon_s$ is the increment of volumetric strain and $\delta\varepsilon_t$ the increment of shear strain (that is, the diameter of the Mohr circle in Fig. 8.8b; these strain variables for plane strain situations were introduced in Section 1.5). The angle ψ is called the angle of dilation; it is also, as seen in Fig. 8.8b, the slope of the tangent to this Mohr circle at the points where the circle cuts the line of zero direct strain, $\delta\varepsilon = 0$. It is often helpful to think of ψ as a strain increment equivalent of angle of friction ϕ' (Fig. 8.8c). It is, however, only useful under conditions of plane strain; and although, following (8.14), an angle θ might be defined for triaxial conditions as

$$\sin\theta = \frac{-\delta\varepsilon_p}{\delta\varepsilon_q} \tag{8.15}$$

Fig. 8.9 Sum of stress ratio and dilatancy in simple shear tests on Leighton Buzzard sand (•, dense, $v_o = 1.53$; ×, loose, $v_o = 1.78$) (after Stroud, 1971).

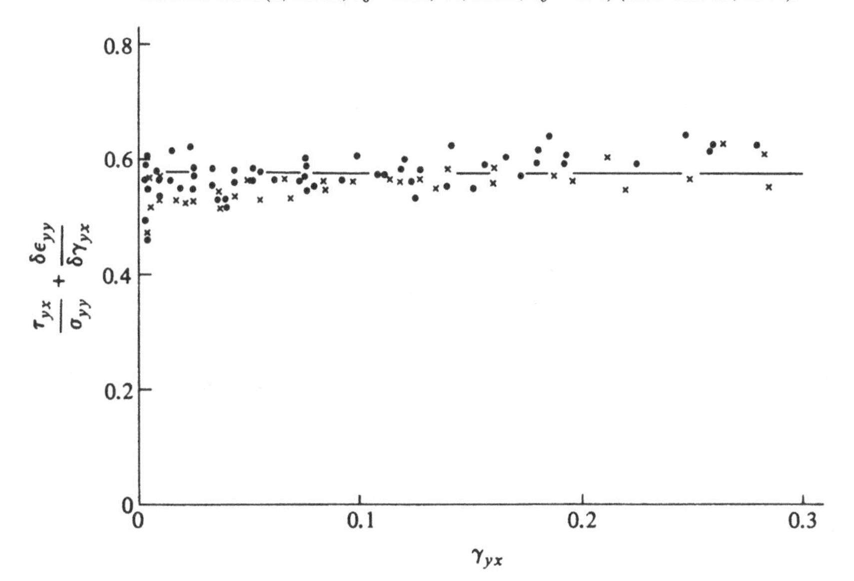

neither this nor the angle β that was used in Section 8.2 has any geometrical interpretation in Mohr circles of strain increment for conditions of axial symmetry.

In terms of the stress and strain quantities that can be determined in the simple shear apparatus, (8.7) becomes

$$\frac{\tau_{yx}}{\sigma_{yy}} + \frac{\delta\varepsilon_{yy}}{\delta\gamma_{yx}} = \mu \tag{8.16}$$

Data from simple shear tests on dense and loose Leighton Buzzard sand performed by Stroud (1971) have been plotted in Fig. 8.9 as values of the sum $\tau_{yx}/\sigma_{yy} + \delta\varepsilon_{yy}/\delta\gamma_{yx}$ against measured values of shear strain γ_{yx}. This plot is equivalent to Fig. 8.7b for the direct shear box data. The simple shear data seem to support more closely a constant value for $\mu \approx 0.575$, though there is a slight tendency for the sum to increase with increasing shear strain. The quantity $\delta\varepsilon_{yy}/\delta\gamma_{yx}$ has been calculated in terms of total strain increments with no allowance for elastic recoverable deformations. Nevertheless, both these examples show that when volume change and stress:strain data are brought together, the response of initially dense and initially loose samples of sand again falls into a single clear picture.

8.4 Work equations: 'original' Cam clay

Equations (8.5) and (8.6) describe the way in which the work done by the forces acting on the boundaries of the shear box is dissipated as friction in the soil. The resulting stress–dilatancy equation (8.7) is specific to the shear box, but it can be used to suggest how a stress–dilatancy equation might be generated which is appropriate for the axisymmetric conditions of the triaxial test.

The obvious parallels to be drawn are between

normal load P and mean effective stress p',

shear load Q and deviator stress q,

shear deformation x and triaxial shear strain ε_q, and

volumetric deformation y and volumetric strain ε_p.

The total work input per unit volume to a triaxial sample supporting stresses p':q as it undergoes strains $\delta\varepsilon_p$:$\delta\varepsilon_q$ is

$$\delta W = p'\delta\varepsilon_p + q\,\delta\varepsilon_q \tag{8.17)(1.32bis}$$

but of this work input, part is stored in elastic deformations of the soil. The energy available for dissipation is

$$\delta E = p'\,\delta\varepsilon_p^{\mathrm{p}} + q\,\delta\varepsilon_q^{\mathrm{p}} \tag{8.18}$$

Following Taylor's (1948) analysis of the shear box, we might assume that this energy is dissipated entirely in friction according to a simple expression

$$\delta E = Mp'\,\delta\varepsilon_q^{\mathrm{p}} \tag{8.19}$$

Then the combination of (8.18) and (8.19) can be rearranged in a form equivalent to (8.7) or (8.16):

$$p'\,\delta\varepsilon_p^{\mathrm{p}} + q\,\delta\varepsilon_q^{\mathrm{p}} = Mp'\,\delta\varepsilon_q^{\mathrm{p}} \tag{8.20}$$

or

$$\frac{q}{p'} + \frac{\delta\varepsilon_p^{\mathrm{p}}}{\delta\varepsilon_q^{\mathrm{p}}} = M \tag{8.21}$$

The critical state parameter M is appropriate because at the critical state $\delta\varepsilon_p^{\mathrm{p}}/\delta\varepsilon_q^{\mathrm{p}} = 0$ and $q/p' = \eta = M$. Expression (8.21) can again be broadly interpreted as proposing that friction plus dilatancy equals a constant.

Expression (8.21) can be plotted as a stress–dilatancy relationship in a $\eta:\beta(=\tan^{-1}\delta\varepsilon_q^{\mathrm{p}}/\delta\varepsilon_p^{\mathrm{p}})$ diagram: curve 2 in Fig. 8.10. The corresponding shape of plastic potentials in the $p':q$ plane can be obtained by integration since the direction of the strain increment vector, controlled by the ratio $\delta\varepsilon_q^{\mathrm{p}}/\delta\varepsilon_p^{\mathrm{p}}$, is by definition the same as the direction of the normal to the plastic potential. The equation to be integrated then becomes

$$\frac{q}{p'} - \frac{dq}{dp'} = M \tag{8.22}$$

and the equation of the plastic potential is then

$$\frac{\eta}{M} = \ln\frac{p'_{\mathrm{o}}}{p'} \tag{8.23}$$

where p'_{o} merely indicates the size of a particular plastic potential curve and is the value of p' for $\eta = 0$. This curve is plotted in Fig. 8.11.

Fig. 8.10 Stress–dilatancy relationships: (1) Cam clay, (2) original Cam clay, and (3) Rowe's stress–dilatancy (drawn for M = 1).

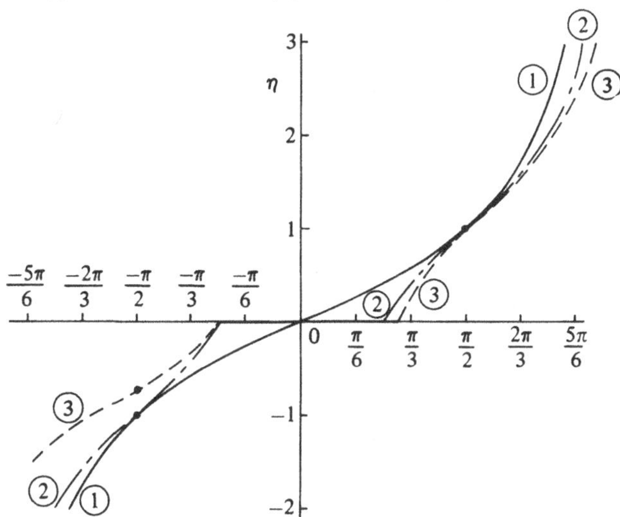

If it is assumed that a soil whose plastic flow is governed by (8.21) also obeys the principle of normality, then yield loci for the soil also have the shape given by (8.23). If these yield loci and identical plastic potentials are then placed in the framework of volumetric hardening elastic–plastic models of soil behaviour described in Chapter 4, then the 'original' Cam clay model, which was described by Roscoe and Schofield (1963) and mentioned in passing in Section 5.1, is recreated.

The flow rule of the Cam clay model described in Chapter 5,

$$\frac{\delta\varepsilon_p^{\mathrm{p}}}{\delta\varepsilon_q^{\mathrm{p}}} = \frac{M^2 - \eta^2}{2\eta} \tag{8.2bis}$$

can also be recast as a plastic dissipation equation,

$$p'\,\delta\varepsilon_p^{\mathrm{p}} + q\,\delta\varepsilon_q^{\mathrm{p}} = p'\sqrt{(\delta\varepsilon_p^{\mathrm{p}})^2 + (M\,\delta\varepsilon_q^{\mathrm{p}})^2} \tag{8.24}$$

though this lacks the simplicity of (8.20). The flow rule and elliptical plastic potential for the Cam clay model are shown in Fig. 8.10 (curve 1) and Fig. 8.11 for comparison with original Cam clay. The most obvious difference appears at low values of stress ratio η: the original Cam clay flow rule does not pass through the isotropic point ($\eta = 0$, $\beta = 0$) in Fig. 8.10. This implies that significant plastic shear strains will develop even for isotropic compression at zero stress ratio.

Schofield and Wroth (1968) cover this problem by suggesting that the plastic potentials in triaxial extension are mirror images of the curves in triaxial compression (Fig. 8.11) so that the flow rule has a discontinuity at $\eta = 0$ (Fig. 8.10). The soil then makes the theoretically acceptable decision to adopt the mid-value at the discontinuity, i.e. $\beta = 0$. Roscoe,

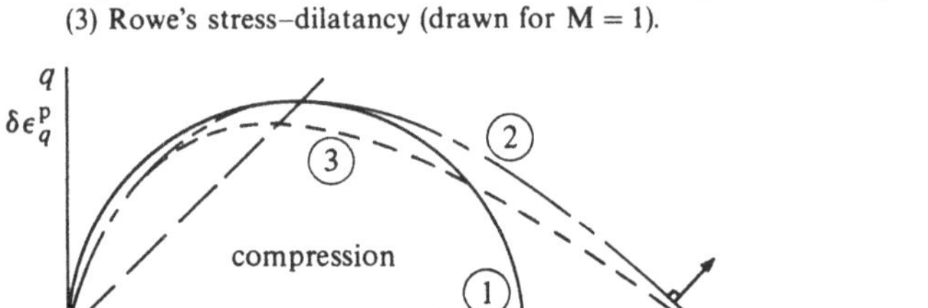

Fig. 8.11 Plastic potentials: (1) Cam clay, (2) original Cam clay, and (3) Rowe's stress–dilatancy (drawn for M = 1).

Schofield, and Thurairajah (1963) propose heterogeneous soil deformations and suggest that 'A sample in this state may contain local regions in which plastic distortion occurs in different directions in such a manner that the whole sample does not exhibit any overall distortion'. Neither of these proposals is particularly satisfactory from a homogeneous computational point of view because the model still implies that for very small but non-zero stress ratios, significant plastic shear strains will develop. The discontinuous plastic potential function can also cause computational difficulties in numerical analyses; careful study of the relevant subroutines in the finite element program CRISP described by Britto and Gunn (1987) shows that, for practical use in numerical calculations, the point of the original Cam clay plastic potential has to be rounded off and the discontinuity eliminated.

8.5 Rowe's stress–dilatancy relation

An alternative justification of a flow rule is provided by the stress–dilatancy relation proposed by Rowe (1962). Rowe produces an expression which states that, for a soil sample that is being sheared, the ratio of the work done by the driving stress to the work done by the driven stress in any strain increment should be a constant. This constant K is supposed to be the same for triaxial and plane strain conditions:

$$\frac{\text{work put in by driving stress}}{\text{work taken out by driven stress}} = -K \tag{8.25}$$

The constant K is related to an angle of soil friction ϕ'_f by the expression

$$K = \tan^2\left(\frac{\pi}{4} + \frac{\phi'_f}{2}\right) = \frac{1 + \sin\phi'_f}{1 - \sin\phi'_f} \tag{8.26}$$

Rowe suggests that the angle ϕ'_f lies in the range

$$\phi_\mu \leqslant \phi'_f \leqslant \phi'_{cs} \tag{8.27}$$

where ϕ'_{cs} is the critical state angle of friction for constant volume shearing, and ϕ_μ is the angle of interparticle sliding friction.

In triaxial compression, the axial stress σ'_a is the driving stress (with an associated compressive strain increment $\delta\varepsilon_a$), and the radial stress σ'_r is the driven stress (with an associated tensile strain increment $-\delta\varepsilon_r$). Rowe's stress–dilatancy relation then states that for triaxial compression,

$$\frac{\sigma'_a\ \delta\varepsilon_a}{-2\sigma'_r\ \delta\varepsilon_r} = K \tag{8.28}$$

Equation (8.28) can be rewritten in terms of the stress variables p' and q (which appear only as stress ratio $\eta = q/p'$) and the strain increment variables $\delta\varepsilon_p$ and $\delta\varepsilon_q$ which are preferred in this book:

$$\frac{\delta\varepsilon_p}{\delta\varepsilon_q}=\frac{3\eta(2+K)-9(K-1)}{2\eta(K-1)-3(2K+1)} \tag{8.29}$$

Neglecting elastic strains, we can integrate (8.29) to give a plastic potential

$$\frac{p'}{p'_o}=3\left[\frac{3-\eta}{(2\eta+3)^K}\right]^{1/(K-1)} \tag{8.30}$$

It is convenient to fix the value of ϕ'_f at the ultimate critical state value ϕ'_{cs}. Then, since ϕ'_{cs} and M are related for triaxial compression by

$$\sin\phi'_{cs}=\frac{3M}{6+M} \qquad (8.31)(\text{cf. } 7.10)$$

expression (8.26) becomes

$$K=\frac{3+2M}{3-M} \tag{8.32}$$

and expression (8.29) can be written as

$$\frac{\delta\varepsilon_p}{\delta\varepsilon_q}=\frac{9(M-\eta)}{9+3M-2M\eta} \tag{8.33}$$

which has similarities to the original Cam clay flow rule obtained by rearranging (8.21):

$$\frac{\delta\varepsilon_p^p}{\delta\varepsilon_q^p}=M-\eta \tag{8.34}$$

The flow rule (8.33) is plotted in terms of η and β (ignoring the difference between total and plastic strains) in Fig. 8.10 (curve 3). The plastic potential curve (8.30) is plotted in the p':q effective stress plane in Fig. 8.11.

For conditions of triaxial extension, the radial stress is now the driving stress and the axial stress is the driven stress, and (8.25) becomes

$$\frac{2\sigma'_r\,\delta\varepsilon_r}{-\sigma'_a\,\delta\varepsilon_a}=K \tag{8.35}$$

The equivalent of (8.29) for triaxial extension is then

$$\frac{\delta\varepsilon_p}{\delta\varepsilon_q}=\frac{3\eta(2K+1)+9(K-1)}{2\eta(1-K)-3(K+2)} \tag{8.36}$$

and the plastic potential, equivalent of (8.30), is

$$\frac{p'}{p'_o}=3\left[\frac{3+2\eta}{(3-\eta)^K}\right]^{1/(K-1)} \tag{8.37}$$

Evidently, the plastic potential and flow rule do not have the same shapes in triaxial extension and triaxial compression. The corresponding curves, with $\eta < 0$ and $q < 0$, have been plotted in Figs. 8.11 and 8.10 (curve 3 in each figure). Like original Cam clay, the plastic potential has a vertex and the flow rule has a discontinuity for $\eta = 0$.

For conditions of plane strain, provided principal axes of strain increment and of stress are coaxial, (8.25) becomes

$$\frac{\sigma_1' \ \delta\varepsilon_1}{-\sigma_3' \ \delta\varepsilon_3} = K \tag{8.38}$$

since the intermediate principal stress does no work. Now, from the Mohr

Fig. 8.12 Conventional drained triaxial compression test on dense Fontainebleau sand ($v_0 = 1.61$, $\sigma_r = 100$ kPa): (a) stress ratio η and triaxial shear strain ε_q; (b) volumetric strain ε_p and triaxial shear strain ε_q (data from Luong, 1979).

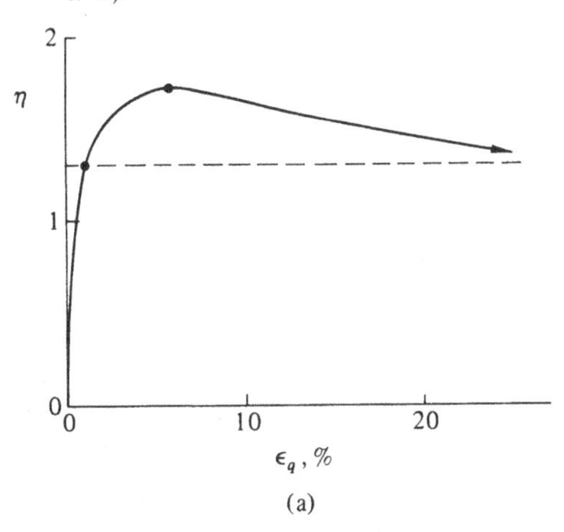

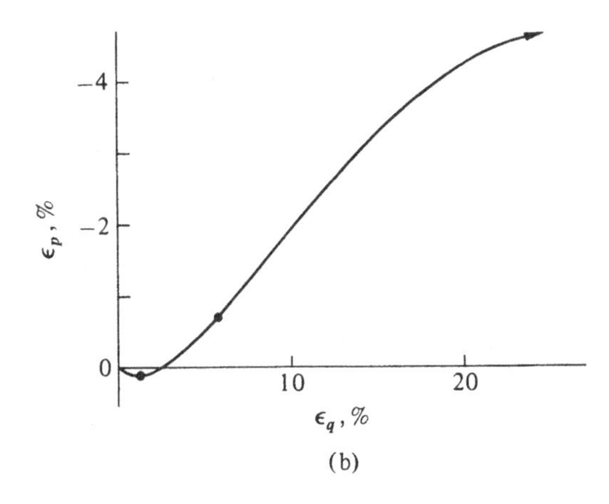

circle of effective stresses (Fig. 8.8c),

$$\frac{\sigma'_1}{\sigma'_3} = \frac{1 + \sin\phi'_m}{1 - \sin\phi'_m} \tag{8.39}$$

where ϕ'_m is the mobilised friction in the soil. From the Mohr circle of strain increments (Fig. 8.8b),

$$\frac{-\delta\varepsilon_1}{\delta\varepsilon_3} = \frac{1 - \sin\psi}{1 + \sin\psi} \tag{8.40}$$

where ψ is the angle of dilation for the soil. From (8.26) with $\phi'_f = \phi'_{cs}$, the critical state angle of friction, we obtain

$$K = \frac{1 + \sin\phi'_{cs}}{1 - \sin\phi'_{cs}} \tag{8.41}$$

Equation (8.38) can then be rearranged to give

$$\sin\phi'_m = \frac{\sin\phi'_{cs} + \sin\psi}{1 + \sin\phi'_{cs}\sin\psi} \tag{8.42}$$

Considerations of dilatancy, volume change on shearing, through the variables β or ψ, rather than ratios of principal strain increments such as $\delta\varepsilon_a/\delta\varepsilon_r$ in a triaxial test, give a better feel for the way in which the soil is responding to a particular state of stress: that is, whether it is choosing to expand or contract as it is sheared. The stress–dilatancy relation in the form (8.33) makes it clear that the volumetric strain increments are zero when the stress ratio $\eta = M$. Dense sands usually show a peak strength $\eta > M$ before deforming to an ultimate critical state with $\eta = M$ (Fig. 8.12a and see Sections 6.5 and 7.4.3). The stress ratio has passed through the value $\eta = M$ at an early stage of the test before the peak is reached. If the soil follows the stress–dilatancy relation (8.33) throughout the test (and the assumption of a certain form of flow rule or plastic potential places no restrictions on the form of the yield locus), then at this stage too the sand is deforming instantaneously at constant volume, $\delta\varepsilon_p/\delta\varepsilon_q = 0$ (Fig. 8.12b).

Studies of the behaviour of sand under cyclic loading such as those of Luong (1979) and Tatsuoka and Ishihara (1974b) show that this stress ratio, at which the sand deforms instantaneously at constant volume, plays an important role in governing the behaviour of the sand, controlling whether cyclic loading tends to stabilise the soil or leads to catastrophe. Tatsuoka and Ishihara call this stress ratio the 'phase transformation stress ratio', and Luong calls it the 'characteristic stress ratio'; but according to stress–dilatancy ideas, it should be the same as the critical state stress ratio. (Luong notes that his characteristic stress ratio is easier to determine because it is reached at an early stage of tests, whereas the critical state is usually reached only by extrapolation at a late stage.)

The role played by this stress ratio in controlling response of sand to small cycles of loading is illustrated in Fig. 8.13. In drained cyclic loading (Fig. 8.13a), cycling below the critical stress ratio tends to produce cumulative volumetric compression and densification of the sand; cycling above the critical stress ratio tends to produce cumulative volumetric expansion. In undrained cyclic loading (Fig. 8.13b) cycling below the critical stress ratio tends to produce positive pore pressures, because volumetric compression is prevented. These positive pore pressures reduce the effective mean stress and lead to liquefaction or related phenomena of large deformations. Cycling above the critical stress ratio tends to

Fig. 8.13 Phenomena observed in (a) drained and (b) undrained cyclic loading of sand (S, shakedown; I, incremental collapse; C, cyclic mobility; L, liquefaction of loose sand) (after Luong, 1979).

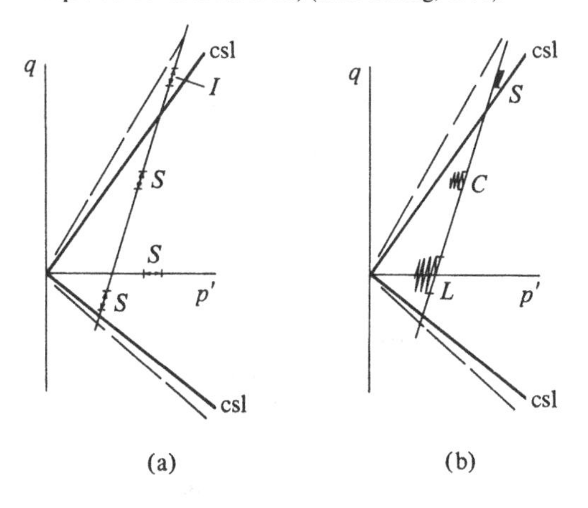

Fig. 8.14 Inclined toothed plane of separation in triaxial sample (after de Josselin de Jong, 1976).

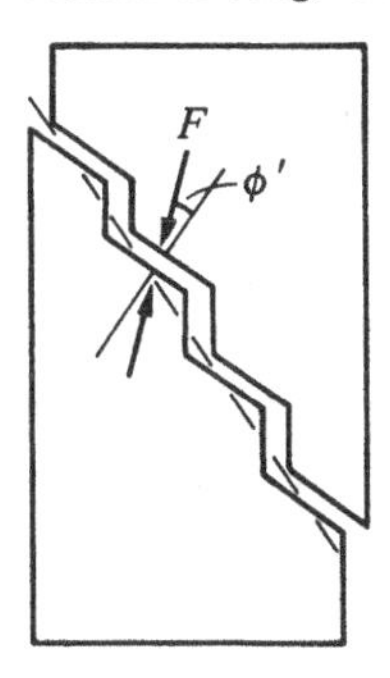

produce negative pore pressures because volumetric expansion is prevented. These negative pore pressures increase the effective mean stress and strengthen the soil.

Rowe originally deduced his stress–dilatancy relation from minimum energy considerations of particle sliding. However, de Josselin de Jong (1976) shows that the relation emerges also from combining considerations of equilibrium and kinematics of particles sliding on inclined saw-tooth surfaces (Fig. 8.14). The important feature is that the relation arises from analysis of sliding deformations in granular materials and hence that it is likely to be of most relevance in describing deformations of soils in which *sliding* between particles provides the principal contribution to the deformation. It is thus not surprising that this flow rule gives a paradoxical prediction of significant shear strains for purely isotropic stresses ($\eta = 0$), where particle sliding does *not* dominate; the flow rule is being stretched beyond its region of legitimate application. Given the similarity between the flow rules of Rowe's stress–dilatancy and original Cam clay, which is evident in Fig. 8.10, one might suggest that original Cam clay is also out of its region of legitimate application at low stress ratios, $\eta \approx 0$. The difference in concept between the two flow rules is that original Cam clay is based on continuum considerations, whereas Rowe's stress–dilatancy is based on considerations of equilibrium and kinematics of particular planes of sliding.

8.6 Experimental findings

So far, the presentation of flow rules has proceeded almost entirely along theoretical lines, with a brief interlude for the re-analysis of some direct shear and simple shear data following Taylor's (1948) simple flow rule. In this section, several sets of published data are produced for comparison with these theoretical relationships. It must be emphasised again that in presentation of data for exploration of flow rules, most authors either explicitly or implicitly ignore the presence of elastic deformations and base calculations of dilatancies on total strain increments. This approach tends to be most in error when the effective stress path followed in a test is moving most nearly tangentially to the current yield surface.

Rowe's stress–dilatancy relation has met with greatest success in describing the deformation of sands and other granular media. Data from simple shear tests on Leighton Buzzard sand performed by Stroud (1971) have already been introduced. The apparatus used by Stroud was heavily instrumented so that he was able to compute values of principal stresses in his tests. For these tests, an interpretation in terms of sin ϕ'_m and

$\sin\psi$ is possible, where ϕ'_m is the current angle of friction mobilised in the soil, which is not usually the same as the angle of friction mobilised on horizontal planes parallel to the top platen. The narrow spread of all his data on initially loose and dense samples is shown in Fig. 8.15. It is apparent that the agreement with the stress–dilatancy relation (8.42) is

Fig. 8.15 Range of data from simple shear tests on Leighton Buzzard sand compared with Rowe's stress–dilatancy relationship (*R*) (after Stroud, 1971).

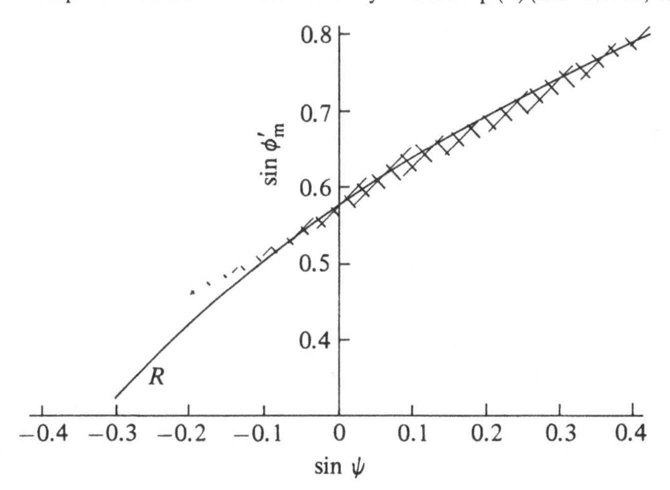

Fig. 8.16 Data from drained plane strain compression tests on Mersey River quartz sand (•, loose, $v_o = 1.66$; +, dense, $v_o = 1.54$) and feldspar (○, loose, $v_o = 1.79$; ×, dense, $v_o = 1.64$) (data from Rowe, 1971).

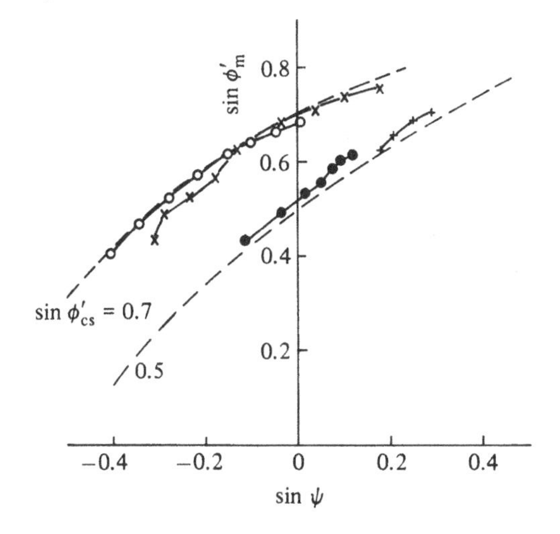

good, even though the principal axes of strain increment and of stress are rotating and not coincident during the tests.

Data from plane strain tests in which no rotations of principal axes occur are reported by Rowe (1971). These have been replotted in terms of $\sin\phi'_m$ and $\sin\psi$ in Fig. 8.16. Equation (8.42) has been plotted in this figure with two values of $\sin\phi'_{cs}$ (0.5 and 0.7) to show the shape of Rowe's stress–dilatancy relationship. The data refer to two sands, each sheared from two initial densities. The data for each sand lie within a narrow band.

Data from triaxial tests on sands are also reported by Rowe (1971). These have been replotted in terms of η and β in Fig. 8.17. For comparison, the Cam clay flow rule (8.3) and Rowe's stress–dilatancy relationship, deduced from (8.33), have been plotted for $M = 1.0$ and 1.5; the data seem to be following a trend which is closer to that deduced from (8.33). Although these values of β have been calculated from total strain increments, it is widely accepted that Rowe's stress–dilatancy relationship, suitably interpreted, provides a reasonable description of the plastic flow of sands, particularly when particle sliding is the dominant mechanism of irrecoverable deformation.

The Cam clay flow rule has not been included in Fig. 8.16. Rowe's stress–dilatancy relationship specifically ignores all elastic deformations. However, to apply the Cam clay flow rule to plane strain situations, an assumption about the elastic behaviour of the soil is required. Although the total strain increments $\delta\varepsilon_2$ are zero, these are made up of elastic and

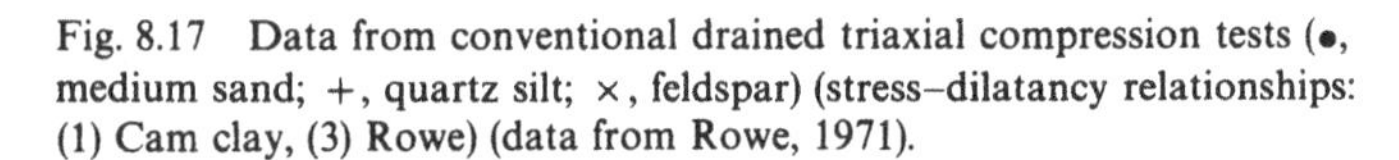
Fig. 8.17 Data from conventional drained triaxial compression tests (•, medium sand; +, quartz silt; ×, feldspar) (stress–dilatancy relationships: (1) Cam clay, (3) Rowe) (data from Rowe, 1971).

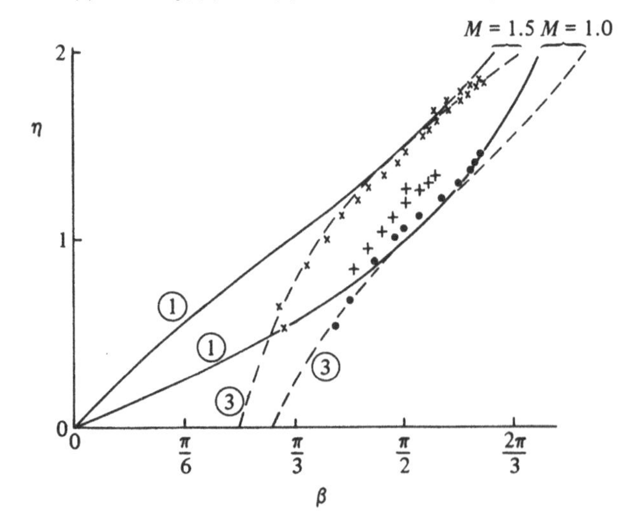

plastic components $\delta\varepsilon_2^e$ and $\delta\varepsilon_2^p$, and there is no reason why these individual components should also be zero. Looking at the relative shapes of the Cam clay and Rowe flow rules for triaxial compression in Fig. 8.17, we cannot expect the plane strain version of Cam clay (even if elastic strains are ignored) to be more successful than Rowe's stress–dilatancy relationship in matching the plane strain data in Fig. 8.16.

For clays, the picture is less clear. Roscoe, Schofield, and Thurairajah (1963) have presented data from triaxial tests on isotropically normally compressed kaolin ($w_L = 0.7$, $w_P = 0.4$) in support of the original Cam clay flow rule (8.34). Their data have been replotted in terms of η and β in Fig. 8.18. These data come from conventional drained and conventional undrained compression tests; elastic volumetric strains but not elastic shear strains have been taken into account. The elastic shear strains are likely to be most significant at low stress ratios, where the effective stress path progresses most nearly tangentially to the initial yield locus, and can be expected to be more significant in the undrained test than the drained test. Overestimation of $\delta\varepsilon_q^p$ leads to overestimation of β (8.1).

A more direct study of flow rules for clay is reported by Lewin (1973). Lewin prepared samples of silty clay ($w_L = 0.31$, $w_P = 0.18$) in the laboratory by consolidation from a slurry of slate dust and water. One group of samples was compressed isotropically (OA in Fig. 8.19a), another group was compressed one-dimensionally ($\delta\varepsilon_r = 0$) (OP in Fig. 8.19b). For

Fig. 8.18 Data from conventional drained (○) and undrained (•) triaxial compression tests on isotropically normally compressed spestone kaolin [stress–dilatancy relationships: (1) Cam clay, (2) original Cam clay] (data from Roscoe, Schofield, and Thurairajah, 1963).

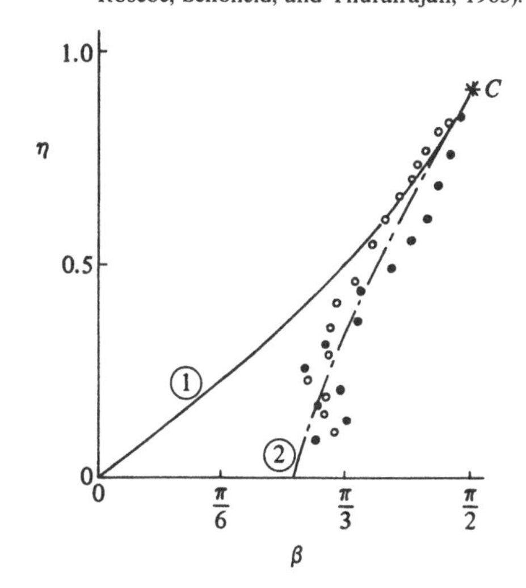

each sample the stress ratio $\eta = q/p'$ was then changed at constant mean effective stress p' (AB, AC, etc., in Fig. 8.19a; PQ, PR, etc., in Fig. 8.19b), and the strains were measured on a subsequent stress probe at constant stress ratio (BB', CC', etc., in Fig. 8.19a; QQ', RR', etc., in Fig. 8.19b). Lewin states that elastic strains were small so that total strain increments could be assumed to be essentially equal to plastic strain increments. Since the stress increments with which he was probing the yield surfaces in Fig. 8.19 are not far from being orthogonal to probable yield loci, plastic effects are likely to dominate.

Fig. 8.19 (a), (b) Stress paths used to investigate plastic potential and (c) resulting stress–dilatancy relationships for Llyn Brianne slate dust; (a) and (c, ○) isotropically compressed; (b) and (c, •) one-dimensionally compressed (data from Lewin, 1973).

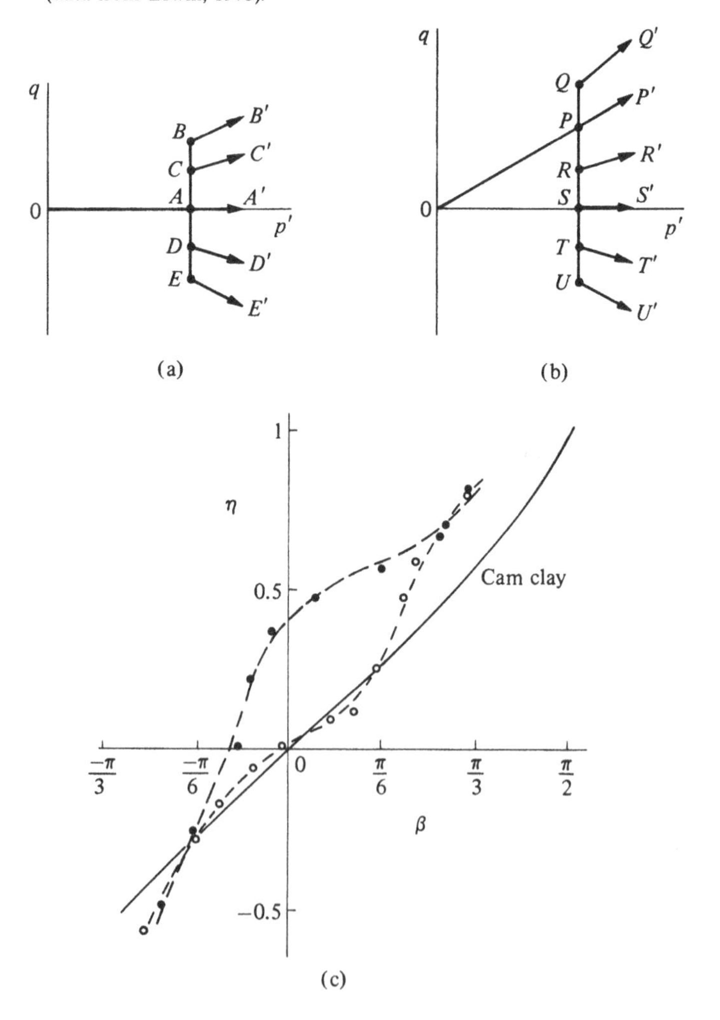

The results are shown in terms of η and β for the two sets of samples in Fig. 8.19c. For this clay there is a large difference between the results for the isotropically compressed and the one-dimensionally compressed clays, and the degree of anisotropy that the one-dimensional history produces in the flow rule is very significant. For example, from Fig. 8.19c, non-distortional compression ($\delta\varepsilon_q = 0$, $\beta = 0$) corresponds to a stress ratio $\eta \approx 0.4$, and deformation under isotropic stresses ($q = 0$, $\eta = 0$) corresponds to a value of $\beta \approx -20^\circ$ and a strain increment ratio $\delta\varepsilon_q/\delta\varepsilon_p \approx -0.36$.

Natural soils have certainly experienced non-isotropic stresses in their past, and although they may show a stress ratio, $\eta = M$, at which yielding occurs without plastic volumetric strain, there is no necessity for increases of isotropic stress in the absence of deviator stress to cause only compression without distortion.* Flow rules for natural soils can thus be expected to pass through point $C\,(\eta = M, \beta = \pi/2)$ in Fig. 8.2b but not necessarily to pass through the origin $I\,(\eta = 0, \beta = 0)$.

Strain increment vectors were shown in Fig. 4.14a for natural clay ($w_L \approx 0.77$, $w_P \approx 0.26$) from Winnipeg, Canada. For this clay Graham, Noonan, and Lew (1983) separated plastic and elastic components of strain increment and plotted plastic strain increment vectors. These data have been transferred to a $\eta{:}\beta$ stress–dilatancy diagram in Fig. 8.20.

The flow rule for this Winnipeg clay is well defined from a large number

Fig. 8.20 Stress–dilatancy relationship $\eta{:}\beta$ observed for undisturbed Winnipeg clay (data from Graham, Noonan, and Lew, 1983).

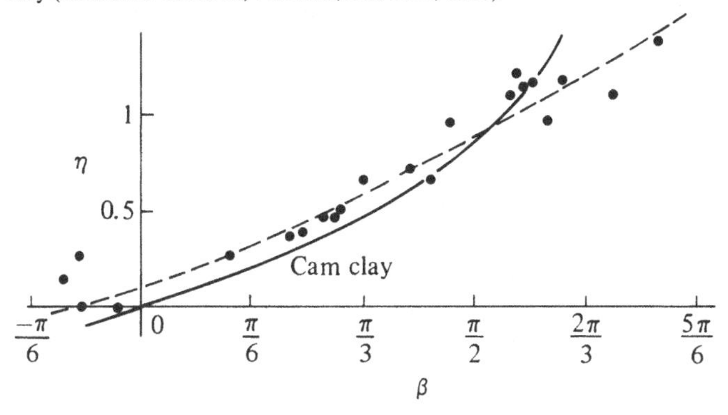

*The stress ratio at which yielding occurs without plastic volumetric strain can be loosely termed a critical state in the context of the present models: it is a feature of the initial yielding of the undisturbed soil. If major changes in the structure of the soil occur as the soil yields, which is the case for sensitive natural soils, then the basic assumption that yielding occurs without change of shape of the yield loci or plastic potentials may no longer be reasonable; and an ultimate critical state is a feature of a subsequent plastic potential rather than the initial plastic potential.

of triaxial compression tests. The data extend well beyond the point of zero plastic volumetric strain increment $\eta = M$, $\beta = \pi/2$, and $\delta\varepsilon_q^p/\delta\varepsilon_p^p = \infty$. At higher stress ratios, values of β greater than $\pi/2$ indicate that shearing is accompanied by plastic volumetric expansion. Although the data show a certain amount of scatter, they do suggest that the flow rule does not pass through the origin. Purely compressive, non-distortional plastic deformation $\beta = 0$ occurs for a small positive stress ratio, and yielding under isotropic stresses $\eta = 0$ corresponds to a small negative value of β, indicating plastic volumetric compression with negative plastic shear strain ($\delta\varepsilon_r^p > \delta\varepsilon_a^p$). On the other hand, the Cam clay flow rule (8.3) with $M = 0.88$ is also plotted in Fig. 8.20 and provides a moderately good fit to these data; in fact, the margin by which the flow rule defined by the experimental data misses the origin I ($\eta = 0$, $\beta = 0$) in Fig. 8.20 is really very small. The error in assuming that plastic flow could be described by the Cam clay flow rule would be small. The anisotropy that this implies in the plastic potentials is smaller than that evident in the yield loci (Fig. 3.22).

These data were used in Section 4.4.3 in support of a framework for elastic–plastic soil models based on the hypothesis of associated flow, identity of yield loci, and plastic potentials. Perhaps for clays too an improved description would be obtained by allowing non-associated flow – but the departure from normality does not seem to be as great as for sands.

No particular conclusion can be drawn about the appropriateness of any one theoretical flow rule for matching experimental data for clays. Choice of flow rule must be guided by some experimental observation. It is usually important to make some allowance for the anisotropy that has been introduced during the depositional compression of the clay, though the extent of this anisotropy may be related to the plasticity of the clay.

8.7 Strength and dilatancy

It was shown in Section 7.4 that heavily overconsolidated clays and dense sands can show peak strengths before ultimate critical states are reached. At the critical state, by hypothesis, shearing continues at constant volume. The flow rules or stress–dilatancy relations that govern the deformation behaviour of soils indicate that shearing at a stress ratio not equal to the critical state stress ratio ($\eta \neq M$) is accompanied by volume change. All the flow rules that have been presented here are in broad agreement that the amount of that volume change depends on how far the stress ratio is from the critical state value. For soils showing a peak strength, deforming with $\eta > M$, the shearing is accompanied by volumetric expansion.

Data of peak strength of sands were presented in Section 7.4.3. Rowe's stress–dilatancy relation (Section 8.5) was found to describe the dilatancy of many granular materials quite satisfactorily, and, in particular, it can be used to estimate the volumetric expansion expected for any observed peak stress ratio. The stress–dilatancy behaviour of sands is not much influenced by stress level, but, as discussed in Section 7.4.3, stress level has a strong influence on the peak strength of sands. The peak dilatancy $(\delta\varepsilon_p/\delta\varepsilon_q)$ is dependent on stress level only because the peak strength is dependent on stress level.

More comment can now be given about the desirability of relying on peak strengths in calculating the stability of geotechnical structures. Peak strengths can exist as a soil is deformed only if volumetric expansion is prevented. The examples of back analysis of slope and retaining-wall failures in Sections 7.6 and 7.7 showed that if it is possible for expansion (and consequent softening) to occur in a thin failure region, then a critical state strength rather than a peak strength is the only strength that can be relied upon in the long term.

Another comment can also be made about the status of analyses of plastic collapse of drained soils showing frictional strength characteristics. It was noted in Section 7.5 that the validity of plasticity analyses, which could potentially provide upper or lower bounds to actual collapse loads of geotechnical structures, depends on the deformations occurring during plastic failure being linked with the stress states under which plastic failure can occur. Most plasticity analyses are performed for situations of plane strain. For plane strain this requirement for associated flow implies that the angle of dilation ψ (Fig. 8.8b) and angle of friction ϕ' (Fig. 8.8c) should be equal. Clearly, when the soil is deforming at a critical state $\phi' = \phi'_{cs}$ and no volume change is occurring so that $\psi = 0$, and the condition of associated flow is not satisfied.

When the soil is deforming at mobilised angles of friction greater than the critical state angle $\phi'_m > \phi'_{cs}$, a stress–dilatancy relation such as (8.42) could be used to estimate expected angles of dilation. A typical example of this stress–dilatancy relation for $\phi'_{cs} = 35°$ was shown in Fig. 8.16. A peak angle of dilation of, say, 20° might be measured; the corresponding mobilised angle of friction would be, from (8.42), $\phi'_m = 49.9°$. Angles of dilation are always considerably lower than angles of friction.

8.8 Conclusion

This chapter has been concerned with stress–dilatancy relationships or flow rules. The primary intention has not been to champion any particular flow rule or stress–dilatancy relation but to illustrate the principle, which is generally accepted, that the volume-change characteristics and the stress:strain characteristics of soils are linked. Whether this link is thought of as a description of the way in which energy is dissipated as a soil is sheared, or as a shape of a curve (plastic potential) in an effective stress plane which indicates the ratio of components of (plastic) deformation, is secondary.

Experimental data have been presented in support of some of the theoretical flow rules or stress–dilatancy relationships, but Fig. 8.10 shows that since a critical state point is common to Cam clay, original Cam clay, and Rowe's stress–dilatancy, the likelihood of significant differences between these theories that can be detected experimentally is small for stress ratios in the range, say, of $0.6M < \eta < 1.8M$.

The importance of past history on the nature of plastic flow has been noted, and for clay a contrast has been drawn between soil which has a history of isotropic compression and soil which has a history of one-dimensional compression. The structure of a sand sample is determined by the sample preparation procedure, which in nature means the manner of deposition of the sand. Sample preparation usually takes place under a gravitational stress field which is directional and not isotropic. Deformations of sands are very much governed by the ways in which their structure can change, hence the success of flow rules such as Rowe's stress–dilatancy relation. It is not easy to eradicate evidence of the initial structure until the sand has been made to flow at a critical state. It may not be feasible to prepare truly isotropic samples of sands (that is, samples which have an unbroken history of isotropic compression) except in an orbiting space laboratory.

Finally, it is important to remember that a flow rule only describes the mode of plastic deformation that occurs when yielding takes place at any particular state of stress; it does not say anything about whether yielding will actually occur. This is the distinction between yield loci and plastic potentials that was drawn in Chapter 4. Consequently, experimental data in support of or defining any particular flow rule cannot provide evidence either for or against the possibility that the soil may obey the postulate of normality: the identity of yield loci and plastic potentials.

Exercises

E8.1. Separate versions of Rowe's stress–dilatancy relationship were produced for triaxial compression and for plane strain conditions in Section 8.5. For conditions of plane strain, an expression relating mobilised angles of dilation and friction has been quoted (8.42). Show that, for this expression to be valid also for conditions of triaxial compression, the angle ψ would have to be defined as (Tatsuoka, 1987)

$$\sin\psi = \frac{-3\,\delta\varepsilon_p}{6\,\delta\varepsilon_q - \delta\varepsilon_p}$$

E8.2. Translation between plane strain and triaxial compression conditions requires some assumed model of soil response. An alternative hypothesis to Rowe's stress–dilatancy, used in exercise E8.1, would be that, *at any particular mobilised angle of friction*, the Cam clay flow rule (8.2) was appropriate, irrespective of the test conditions, with generalised strain increments $\delta\varepsilon_q$ and $\delta\varepsilon_p$ being defined as in Section 1.4.1.

Neglecting elastic strains, deduce the plane strain form of the Cam clay flow rule and compare this with Rowe's stress–dilatancy relationship.

(Note that if we assume dependence of dilatancy on angle of friction, the value of the intermediate principal stress under plane-strain conditions is not of concern.)

E8.3. Shibata (1963) presents experimental data for a normally compressed Japanese clay tested in drained triaxial compression with constant mean effective stress p'. He found that the volume decrease was proportional to the effective stress ratio

$$\frac{-\delta v}{v_i} = k\,\delta\eta$$

where v_i is the initial specific volume, and k is observed to be a constant. Show that this finding implies that this clay has yield loci of the form given by the original Cam clay model (8.23), provided that

$$k = \frac{\lambda - \kappa}{M v_i}$$

Assume that the clay is to be modelled within the elastic–plastic framework of Chapter 4, with yield loci of size p'_o which varies only with irrecoverable volumetric strain:

$$\delta p'_o = \frac{v p'_o}{\lambda - \kappa}\,\delta\varepsilon_p^p$$

Assume also that the clay follows the principle of normality or associated flow.

E8.4. In discussion of the existence of critical states for clays in Section 6.3, some data obtained by Parry (1958) were presented (Fig. 6.13) which indicated that soil samples – which in drained triaxial compression tests failed at values of mean effective stress p'_f different from the critical state value p'_{cs} appropriate to the failure value of specific volume of the sample – were still changing in volume and still heading towards the critical state at failure, according to an expression

$$\frac{\delta\varepsilon_p}{\delta\varepsilon_q} = k \ln \frac{p'_f}{p'_{cs}}$$

Ignoring elastic deformations, deduce the implied flow rule or stress–dilatancy relation

$$\frac{\delta\varepsilon_p}{\delta\varepsilon_q} = k \ln \frac{M - h_c}{\eta - h_c} \qquad \text{for } \eta > M$$

It will be necessary to refer to the failure data for Weald clay which were presented in Section 7.4.1. These data, in terms of effective stresses at failure, supported an expression

$$\frac{q}{p'_e} = h_c\left(g + \frac{p'}{p'_e}\right)$$

Use the requirement that the critical state point ($\eta = M$, $p' = p'_{cs}$) must lie on this failure line to obtain an expression for p'_{cs}/p'_e.

Taking values of k, M, and h_c from the data presented in Sections 6.3 and 7.4.1, sketch this flow rule in a $\eta{:}\beta$ diagram and compare it with the Cam clay flow rule.

E8.5. Investigate the following flow rule for describing the plastic behaviour of clay:

$$\frac{\delta\varepsilon^p_q}{\delta\varepsilon^p_p} = \frac{\eta}{M^2 - \eta^2}$$

Plot this as a stress–dilatancy relationship and compare it with other flow rules discussed in this chapter. Derive an expression for the corresponding plastic potential, sketch the curve, and compare it with that for Cam clay.

For one-dimensional normal compression, derive an expression for the constant stress ratio η_K, assuming (i) that the elastic shear strains are negligible and (ii) that the changes in volume are given by the usual linear relationships between specific volume and the logarithm of pressure. Adopting values of $\lambda/\kappa = 5$ and $M = 1$, calculate the relevant value of $K_{0nc} = \sigma'_h/\sigma'_v$ during one-dimensional normal compression and comment on whether this is a reasonable value for a clay with these soil parameters.

E8.6. The results of constant stress ratio tests on Newfield clay performed by Namy (1970) fit the following flow rule:

$$\frac{\delta\varepsilon_q^{\mathrm{p}}}{\delta\varepsilon_p^{\mathrm{p}}} = A\eta$$

where A is a constant and $\delta\varepsilon_q^{\mathrm{p}}$, $\delta\varepsilon_p^{\mathrm{p}}$, and η have their usual meanings. Plot this flow rule in a $\eta:\beta$ stress–dilatancy diagram and compare it with the other flow rules introduced in this chapter.

Comment on the compatibility of this flow rule with the possible existence of critical states for this clay. Discuss the errors in the observations of strain increments that could have resulted from the fact that Namy actually performed his constant stress ratio tests by applying alternate large increments of cell pressure and axial stress.

9

Index properties

9.1 Introduction

Determining reliable values for soil parameters, either for use in an elastic–plastic soil model such as Cam clay or to obtain values of strength and stiffness for some less elaborate analysis, usually requires that laboratory tests such as triaxial or oedometer tests be performed on undisturbed samples of soil. To obtain good quality undisturbed samples is usually expensive and frequently difficult. Performance of good-quality triaxial and oedometer tests requires time and skill and is also expensive. It is possible to characterise or classify soils with quicker, less-sophisticated tests which do not require undisturbed samples of the soils. This characterisation for cohesive, clayey soils is achieved by using index tests which determine the natural water content and the so-called liquid limit and plastic limit of the soil.*

Although the procedures which have been adopted for performing these index tests may appear quaint, it is possible, using the ideas of critical state strengths and models of soil behaviour such as Cam clay, to relate values of index properties to other properties of engineering importance. Empirical correlations between index properties and strengths and compressibilities have been used for many decades: critical state soil mechanics points the way to a rational basis for many of these correlations.

It would be extreme to suggest that the availability of these correlations makes it unnecessary to perform any tests more sophisticated than the index tests. Their value lies in two areas. Where it is difficult to obtain good soil samples or where results of other tests are not yet available,

*The term *plastic* is traditionally used to describe soils for which it is possible to determine liquid and plastic limits. This is a confusing term since it is a theme of this book that elastic–plastic soil models are useful for describing the behaviour of soils such as sands, which according to this definition would be classified as non-plastic.

these correlations can be used to give preliminary values of soil properties for use in feasibility studies. At a later stage, when more data are available from a wide range of tests, such correlations can be used to check the internal consistency of the data that have been obtained and to draw attention to apparent anomalies.

The fall-cone test is described in Section 9.2. An analysis of this test in terms of critical state strengths shows it to be an extremely powerful index test. The range of correlations which emerge is presented in Section 9.3, together with examples of their applications. Subsequent sections then provide the theoretical background to and comment on these correlations. Readers who are concerned only to know the correlations themselves need proceed no further than Section 9.3.

9.2 Fall-cone test as index test

Static indentation tests have been used since 1900 to provide quick estimates of the hardness of metals (Tabor, 1951). Creation of an

Fig. 9.1 Fall-cone apparatus (Statens Järnvägars Geotekniska Kommission, 1922).

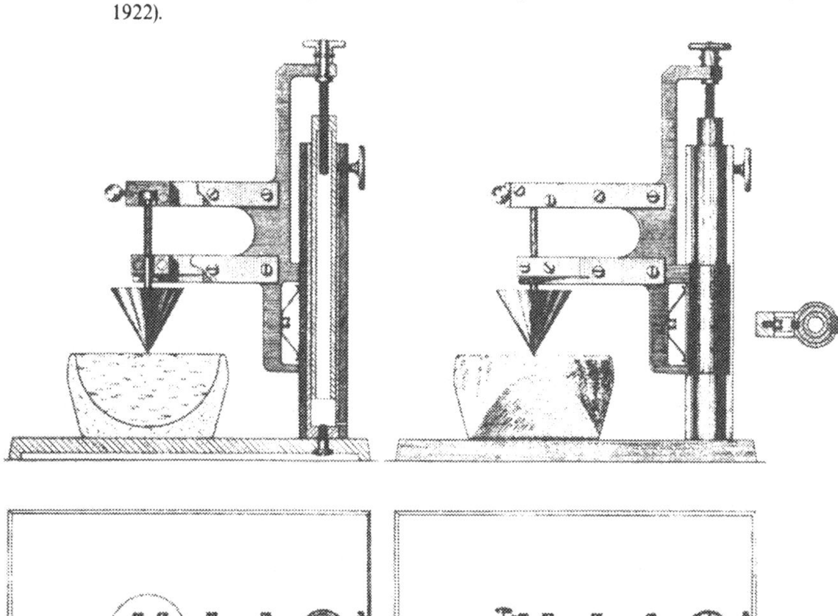

indentation is an almost entirely plastic process, and the 'hardness' of a metal, determined from the indentation force and the dimensions of the indentation (spherical, conical, or pyramidal), is directly related to its yield strength. Indentation tests have been used in Scandinavia since 1915 to provide a measure of the 'consistency' of clays in routine geotechnical investigations (Bjerrum and Flodin, 1960). Olsson (1921) describes the fall-cone apparatus which he developed for this purpose (Fig. 9.1): a cone is allowed to fall freely under its own weight from a position at rest with the point of the cone just touching the surface of the clay. This is no longer a static indentation test; the cone accelerates initially and then decelerates to rest and penetrates a distance which is greater than the penetration at which the soil resistance is equal to the weight of the cone (where the acceleration of the fall-cone is zero).

A detailed study of the relationship between cone penetration and soil strength is reported by Hansbo (1957). The variables governing the problem are (Fig. 9.2) the mass m and tip angle α of the cone, the penetration d, and the undrained shear strength of the soil c_u. Dimensional analysis (Wood and Wroth, 1978) then shows that

$$\frac{c_u d^2}{mg} = f(\alpha, \chi) \tag{9.1}$$

where the parameter χ allows for surface effects between the soil and cone (for example, friction or adhesion), and g is the acceleration due to gravity ($9.81\,\mathrm{m^2/s}$). For a given material of cone and for related soils (perhaps soils of similar activity, see Section 9.4.3), the cone angle is the dominant factor in $f(\alpha, \chi)$, so (9.1) can be rewritten as

$$\frac{c_u d^2}{mg} = k_\alpha \tag{9.2}$$

where k_α, the cone factor, certainly depends on the angle of the cone.

In Chapter 6, critical state lines for clay were proposed, supported by

Fig. 9.2 Fall-cone test.

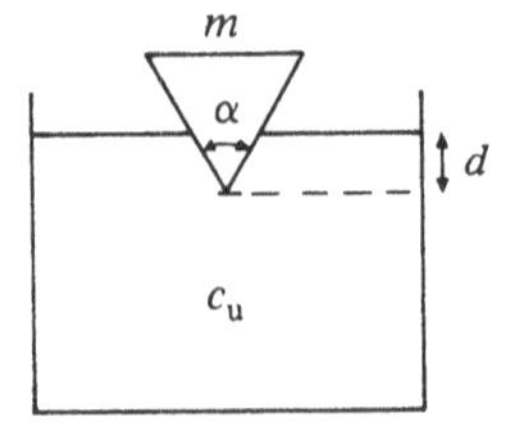

experimental evidence, of the form

$$v_{cs} = \Gamma - \lambda \ln p'_{cs} \tag{9.3)(6.10bis}$$

$$q_{cs} = M p'_{cs} \tag{9.4)(6.3bis}$$

Since undrained strength c_u is just half the deviator stress at failure,

$$c_u = \frac{q_{cs}}{2} \tag{9.5}$$

and since in undrained deformation the specific volume does not change, expressions (9.3)–(9.5) can be combined to give a general relationship between undrained strength c_u and specific volume v:*

$$v = \left[\Gamma + \lambda \ln\left(\frac{M}{2}\right)\right] - \lambda \ln c_u \tag{9.6}$$

or

$$c_u = \frac{M}{2}\exp\left(\frac{\Gamma - v}{\lambda}\right) \tag{9.7}$$

For saturated soils, a direct relationship exists between water content w and specific volume v:

$$v = 1 + G_s w \tag{9.8}$$

where G_s is the specific gravity of the soil particles. Combination of equations (9.2)–(9.8) then shows that if a series of fall-cone tests is performed using the same cone (with m and α chosen) on samples of the same clay prepared at various water contents, then the measured penetrations d are related to water content w by

$$w = \frac{2\lambda}{G_s}\ln d + \frac{\Gamma - 1}{G_s} + \frac{\lambda}{G_s}\ln\left[\frac{M}{2k_\alpha mg}\right] \tag{9.9}$$

A plot of water content w against logarithm of cone penetration d should give a straight line with slope $2\lambda/G_s$ (Fig. 9.3), and the compressibility λ of the soil can be determined from these cone tests.

If cone tests are performed with two geometrically similar cones of different masses m_1 and m_2, then plots of water content against cone penetration should produce two parallel lines with a water content

*The relationship between strengths measured with various devices is discussed in Section 10.6; in this chapter it is assumed that this strength represents the maximum shear stress which the soil can support, independent of the mode of shearing. The justification for this is that we are trying to establish approximate correlations rather than exact analyses in this chapter. However, this assumption is in essence the Tresca criterion discussed in Section 3.2 and is basic to many of the plasticity analyses of undrained loading of soils (Atkinson, 1981; Houlsby, 1982.)

separation Δw (Fig. 9.3) where, from (9.9),

$$\Delta w = \frac{\lambda}{G_s} \ln \frac{m_2}{m_1} \tag{9.10}$$

and this gives a second route by which λ can be deduced from the results of fall-cone tests. The slight advantage of (9.10) over (9.9) is that, because the fall-cone test is a dynamic rather than a static test, there may be effects associated with the rate at which the clay is being deformed which should cancel out if the geometry of the penetration process is kept constant by comparing results obtained at identical penetrations. Satisfactory results have been obtained for Cambridge gault clay (Fig. 9.4) using a ratio of cone masses $m_2/m_1 = 3$. Working from the gradients of the parallel lines (9.9) and taking $G_s = 2.75$ gives a value of $\lambda = 0.30$. Working from the spacing between the lines (9.10) gives an independent value of $\lambda = 0.31$.

These equations lead to two ways in which fall-cone tests can be used

Fig. 9.3 Relationship between water content and cone penetration.

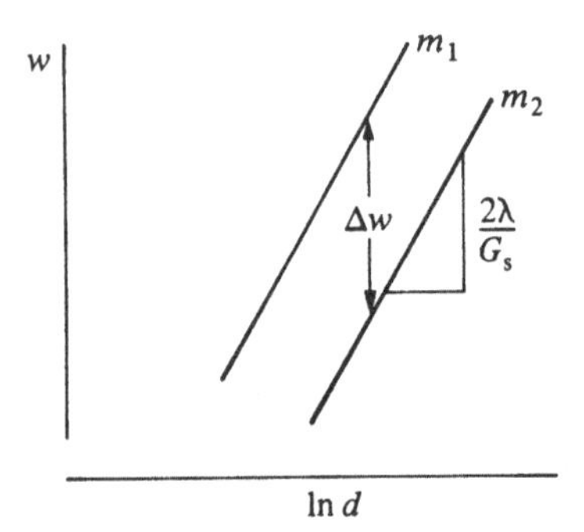

Fig. 9.4 Fall-cone tests on Cambridge gault clay.

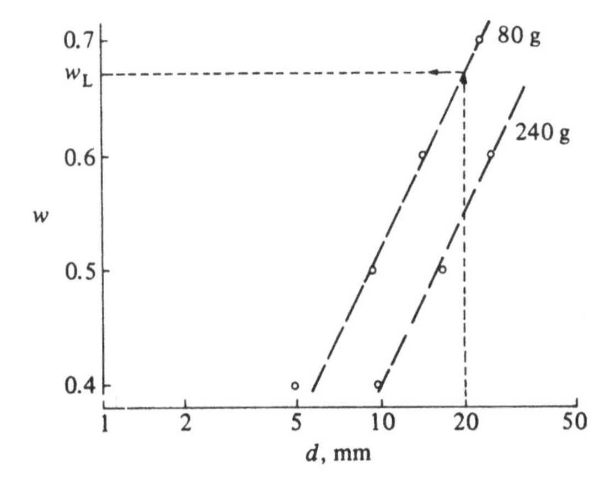

to provide strength indices for soils. Olsson (1921) devised a strength number (*hållfasthetstalet*) proportional to the mass of a cone of tip angle $\alpha = 60°$ which would produce a penetration $d = 10$ mm in the soil. Of course, only a certain number of combinations of cone mass and cone angle were available, and charts and tables were produced from which the penetration of any given cone could be converted to a strength number. A strength number of 10 was assigned to a cone mass of 60 g. From (9.2), the strength number is directly proportional to undrained strength c_u.

Olsson went further and defined an index property, fineness number (*finlekstalet*), as the water content of soil having a strength number 10, in other words, the fineness number is the water content at which a 60° cone of mass 60 g would penetrate 10 mm when allowed to fall under its own weight. This fineness number is now accepted in Sweden and other Scandinavian countries as equivalent to the liquid limit (Karlsson, 1977). Other cone-penetration tests have been used elsewhere to estimate liquid limits of soils (Sherwood and Ryley, 1970), and the test which is now the preferred British Standard test 2(A) for determining the liquid limit (BS 1377, 1975) is a fall-cone test using a 30° cone of mass 80 g. The liquid limit w_L has thus become the water content at which soil has a standard strength. (The background to the measurements of liquid limit is discussed in Section 9.4.1.)

Karlsson (1977) states that a 10-mm penetration of a 60° cone of mass 60 g corresponds to a soil strength of 1.7 kPa. Fall-cone and miniature vane tests performed by Wood (1985a) produce average values of cone factors,

$$k_\alpha = 0.85 \qquad \text{for } \alpha = 30°$$

and

$$k_\alpha = 0.29 \qquad \text{for } \alpha = 60°$$

which lead from (9.2) to estimates of strength at the liquid limit of 1.71 kPa according to the Swedish definition and 1.67 kPa according to the British definition. For the purposes of the approximate correlations which are set out in Section 9.3, however, we assume that the standard undrained shear strength of soils at their liquid limit c_L is approximately 2 kPa.

Another use of fall-cone tests proceeds from (9.10) to obtain an index of the way in which strength changes with water content. Specifically, if the same penetration d is obtained when cones of two different masses penetrate two different samples of the same soil, then, from (9.2), the ratio of the strengths of the soil samples is equal to the ratio of the masses of the cones. The change in water content Δw in (9.10) is that necessary to change the strength of the soil by the factor m_2/m_1. A useful index might then be the water content change Δw_{100} required to produce a 100-fold change in strength. Direct determination of this would require cone tests to be performed with a ratio of cone masses $m_2/m_1 = 100$. The standard cones used in Scandinavia and in the United Kingdom have masses $m_1 = 60$ g and 80 g, and so the parallel series of tests would have to be performed with $m_2 = 6$ kg or 8 kg. Such cones would be cumbersome to use, and it is consequently more appropriate to extrapolate to Δw_{100} using a smaller ratio of cone masses. Then from (9.10),

$$\Delta w_{100} = \Delta w \frac{\ln 100}{\ln (m_2/m_1)} \tag{9.11}$$

The results for Cambridge gault clay (Fig. 9.4) were obtained using a ratio of cone masses $m_2/m_1 = 3$. Since $\ln 3 \simeq (\ln 100)/4$, this implies an approximate 4-fold extrapolation, $\Delta w_{100} \simeq 4\Delta w$.

The determination of this strength change index by extrapolation relies on the value of λ in (9.6) being a constant. The value of λ that is being determined with (9.9) or (9.10) is valid for water contents close to the liquid limit. It typically drops with increasing pressure and decreasing water content (the water content after all cannot fall below zero, so the line $w = 0$ must provide an asymptote to the volume change characteristics), and so the extrapolation implied in (9.11) could be expected to yield values of Δw_{100} which are too high.

The fall-cone test is being used as an index test to give direct information about strengths and about change in strength with water content, which from critical state lines can be interpreted as information about compressibility such as the value of the parameter λ.

9.3 Properties of insensitive soils

Many of the values of liquid limit quoted in published work have been obtained using the Casagrande apparatus described in Section 9.4.1. It is assumed that these values also correspond to a strength of 2 kPa, so that as far as the relations between soil properties to be described here are concerned, liquid limit values obtained with the various devices are assumed to be interchangeable.

The other index test that is widely used for classifying clayey soils is the so-called plastic limit test, described in Section 9.4.2. This plastic limit test is probably a less consistent strength test than the fall-cone liquid-limit test, but study of the data of undrained shear strength and water content for remoulded soils collected by Mitchell (1976) (Fig. 9.5) suggests that at the plastic limit (with water content w_P), also, the range of strengths is reasonably small, having an approximate average value of around 200 kPa, that is about 100 times the strength at the liquid limit. The difference in water content between the liquid limit and plastic limit is known as the *plasticity index* I_P; it appears consequently that this is approximately equivalent to the quantity Δw_{100} introduced in the previous section.

The data of water contents in Fig. 9.5 have been plotted in terms of liquidity index I_L which scales water contents of soils in a standard way between their liquid and plastic limits:

$$I_L = \frac{w - w_P}{w_L - w_P} = \frac{w - w_P}{I_P} \tag{9.12}$$

Liquidity index thus takes values of 1 and 0 when the water content of a soil is at the liquid or plastic limit, respectively.

Two simple relationships between index properties and other properties of insensitive or remoulded soils can now be deduced. Setting $I_P = \Delta w_{100}$, we find that from (9.10) and (9.11) compressibility λ can be linked with

Fig. 9.5 Variation of remoulded undrained strength c_u with liquidity index I_L for (1) Horten clay ($w_L = 0.30$, $w_P = 0.16$); (2) London clay ($w_L = 0.73$, $w_P = 0.25$); (3) Shellhaven clay ($w_L = 0.97$, $w_P = 0.32$); (4) Gosport clay ($w_L = 0.80$, $w_P = 0.30$) (after Skempton and Northey, 1953); range of strength data collected by Mitchell (1976) shown shaded.

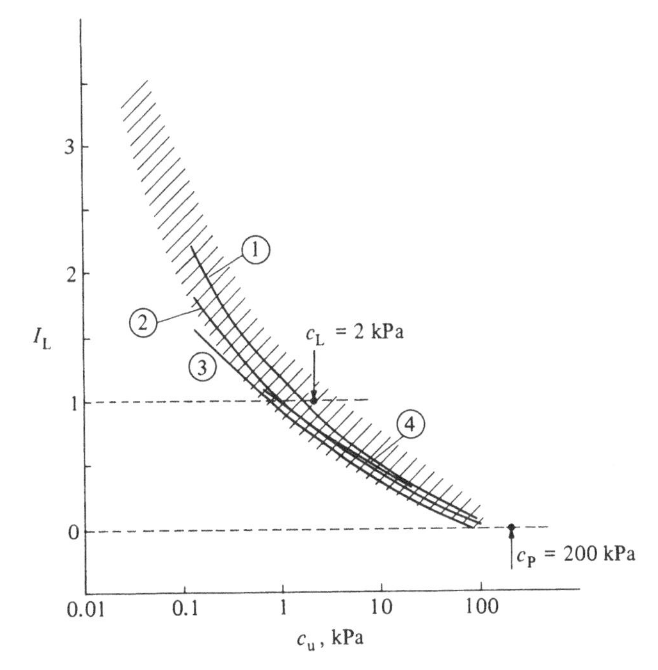

plasticity I_P:

$$\lambda = \frac{I_P G_s}{\ln 100} \tag{9.13}$$

$$\simeq 0.6 I_P. \tag{9.14}$$

using a typical value for $G_s \simeq 2.7$.

The parameter λ is the compressibility of the soil in the expression

$$v = v_\lambda - \lambda \ln p' \qquad (9.15)(4.2\text{bis})$$

for a normal compression line. The compressibility C'_c, which is the slope of the normal compression line when written as

$$v = v_\lambda - C'_c \log_{10} p' \tag{9.16}$$

is perhaps more often quoted. The two compressibilities are simply related,

$$C'_c = \lambda \ln 10 = 2.303\lambda \tag{9.17a}$$

Fig. 9.6 Values of compressibility (C'_c and λ) and plasticity index (I_P) for soils from the Mississippi delta and Gulf of Mexico (+) (data from McClelland, 1967); Egyptian soils (○) (data from Youssef, el Ramli, and el Demery, 1965); seabed soils off Tel Aviv (△) (data from Almagor, 1967); soils from Britain and elsewhere (●) (data from Skempton, 1944).

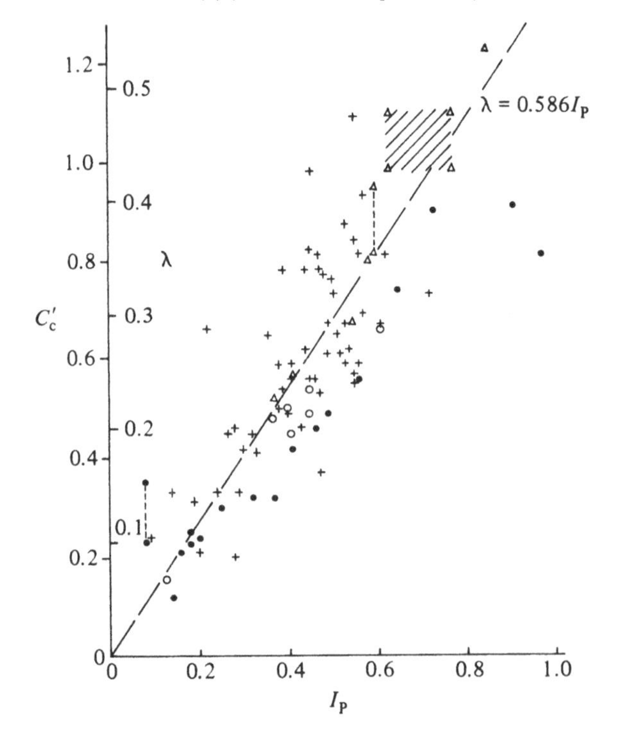

or

$$\lambda = 0.4343C'_c \tag{9.17b}$$

and (9.14) becomes

$$C'_c = \frac{I_P G_s}{2} \tag{9.18}$$

$$\simeq 1.35 I_P \tag{9.19}$$

Data of compressibility λ or C'_c and plasticity I_P in Fig. 9.6 have been culled from a number of published sources: soils from the Mississippi delta and Gulf of Mexico (McClelland, 1967), Egyptian soils (Youssef, el Ramli, and el Demery, 1965), seabed soils of Tel Aviv (Almagor, 1967), and various soils from Britain and elsewhere (Skempton, 1944). Relationship (9.14) is included and provides a reasonable fit to the data though there is, of course, a lot of scatter.

It is important to note that the data of undrained strengths in Fig. 9.5 were obtained from remoulded soils, and the mention of a critical state line at the beginning of this section also implies that remoulded strengths are being considered. The compressibility which is calculated using (9.14) or (9.19) is a compressibility for a remoulded, structureless soil. Undisturbed soils usually have a higher value of compression index than remoulded soils because of the extra compression associated with destruction of their structure. Schmertmann (1955), for example, shows

Fig. 9.7 Effect of disturbance on slope of one-dimensional normal compression line for marine organic silty clay (after Schmertmann, 1955).

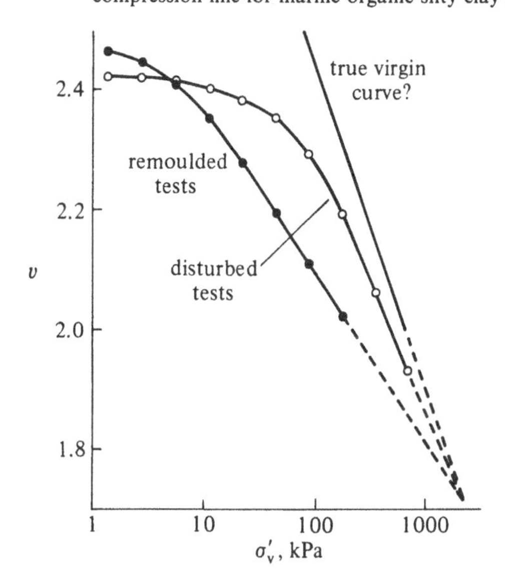

that disturbance of soil samples has an effect on compression index equivalent to partial remoulding (Fig. 9.7): a steeper normal compression line implies a higher value of λ or C'_c, and data of measured compressibilities are expected to congregate in the upper part of Fig. 9.6.

Having assigned strengths of 2 kPa and 200 kPa to soils at their liquid and plastic limits, if the spread and curvature of the liquidity: logarithm-of-strength relationship can be neglected, we can estimate the remoulded strength of a soil with knowledge only of its liquidity index, in other words, with knowledge only of its liquid and plastic limits and natural water content, by using the expression

$$c_u = 2 \times 100^{(1 - I_L)} \quad \text{kPa} \tag{9.20}$$

An example of the application of (9.20) is shown in Fig. 9.8 for a site on the Thistle field in the North Sea. Some undrained strength data from unconfined triaxial compression tests ($\sigma_r = 0$) are included for comparison. The agreement between these measured strengths and estimates made on the basis of values of liquidity index is reasonable, but even more encouraging is the match of the pattern of observed and estimated variations of strength with depth. The measured strength profile (which is typical of some North Sea sites) shows an initial rapid rise in strength to a depth of about 5 m, followed by a drop to a minimum at a depth of about 15 m, and followed by a rapid increase in strength at a depth of about 20 m; all these variations are reflected in the profile of liquidity

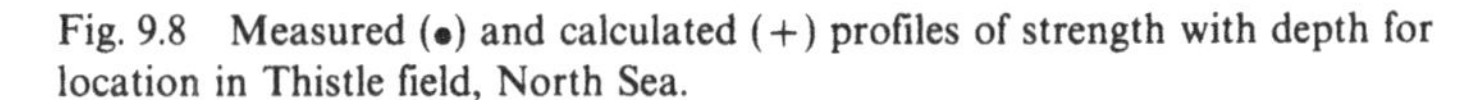

Fig. 9.8 Measured (•) and calculated (+) profiles of strength with depth for location in Thistle field, North Sea.

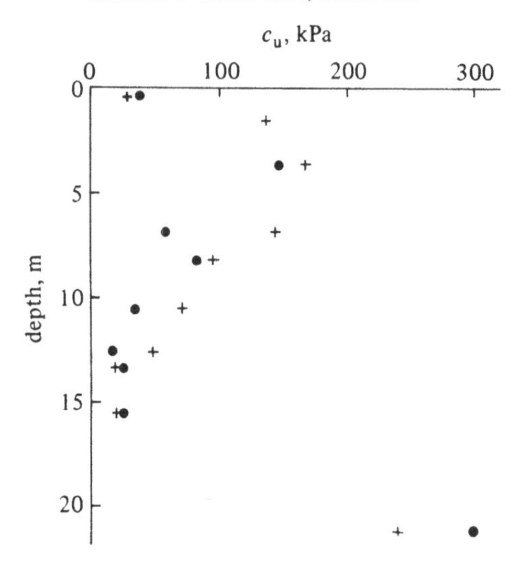

index with depth. It is precisely at such offshore locations that it may be difficult to retrieve good soil samples for direct determination of in situ strength. An independent estimate of strength using (9.20) thus provides an invaluable confirmation of those measured trends that are observed.

Expression (9.20) is sensitive to errors in liquidity index. Water contents and index properties are typically quoted to 0.01 (1%), so for soils of low plasticity, estimates of liquidity index, and thence of strength can very easily be in error. Expression (9.20) is plotted as line *A* in Fig. 9.9.

Fig. 9.9 Approximate unique relationship expected between liquidity index I_L and remoulded undrained shear strength c_u (line *A*).

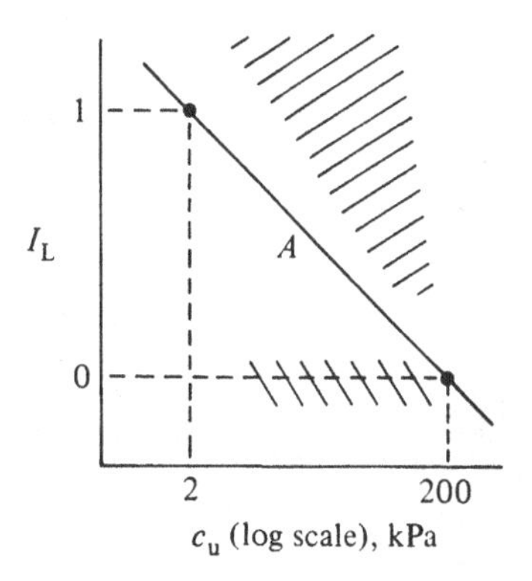

Fig. 9.10 Profile of water content w, index properties w_L and w_P, and undrained shear strength c_u with depth for boring at Drammen, Norway (after Bjerrum, 1954).

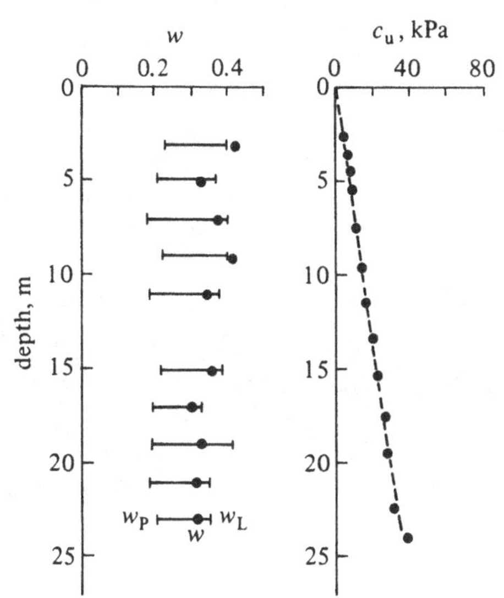

Fig. 9.11 (a) Sample of undisturbed Norwegian quick clay supporting load of 4 kgf (39.2 N); (b) same sample of Norwegian quick clay after remoulding at natural water content (photographs courtesy of Norwegian Geotechnical Institute, Oslo).

Terzaghi (1936), introducing the parameter liquidity index (9.12), notes that the value of liquidity index 'indicates the consistency of the clay after remoulding it without changing the water content'. Critical state strength too is a remoulded soil strength. Actual field or laboratory determinations of strength may, in a strength:liquidity diagram (Fig. 9.9), appear inconsistent with the simple expression (line A) if the strength that is determined is not properly a remoulded strength. Two possibilities have been indicated on Fig. 9.9.

A typical profile of water content and strength from a boring in soft clay at Drammen, Norway is shown in Fig 9.10 (from Bjerrum, 1954). Here the water content is near the liquid limit ($w \sim w_L$, $I_L \sim 1$) near the surface, dropping below the liquid limit with depth. However, the strengths measured with in situ vane tests would plot well to the right of line A in Fig. 9.9. These are undisturbed strengths, and this is a sensitive clay, that is, the natural clay has some structure which is destroyed by remoulding so the ratio of undisturbed to remoulded strength is significantly greater than unity (for this clay, the sensitivity is typically around 10 when the liquidity is close to 1). A pedagogic example of the character of a sensitive clay is shown in Fig. 9.11: an undisturbed sample can support a sizeable load, but once the structure is disturbed and the soil remoulded, the sample flows like a liquid with extremely low strength. We should regard the

Fig. 9.12 Profile of water content w, index properties w_L and w_P, and undrained shear strength c_u with depth for boring at Paddington, London (after Skempton and Henkel, 1957).

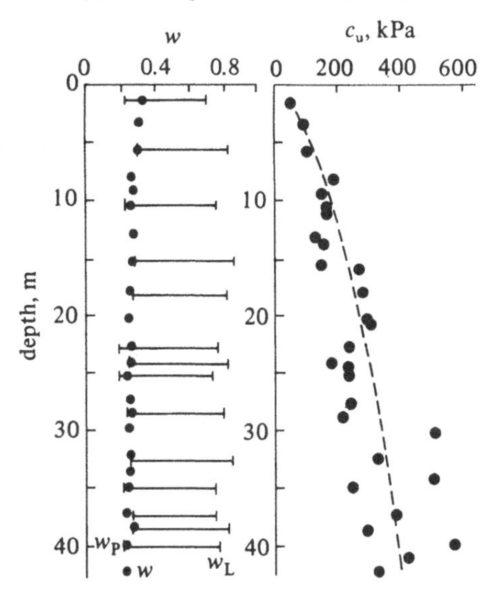

undisturbed strength data plotting to the right of line *A* in Fig. 9.9 with some caution. The strength that is appropriate to the in situ liquidity of the soil is the remoulded strength; it may be dangerous to rely on the peak, undisturbed strength for design purposes. Some properties of sensitive soils are discussed in Section 9.5.

A typical water content and strength profile from a boring in heavily overconsolidated London clay is shown in Fig. 9.12 (from Skempton and Henkel, 1957). The water contents below the top few metres are around the plastic limit ($w \sim w_P$, $I_L \sim 0$) and the strengths, measured in undrained triaxial tests, around and above 200 kPa. These strengths and liquidities plot close to line *A* in Fig. 9.9. However, it was noted in Section 7.7 that when old failure surfaces are present in heavily overconsolidated clays, the clay structure adjacent to the failure surfaces may have been modified by the sliding so that particles are oriented parallel to the failure surface. The residual strength which could be mobilised on such surfaces can then be lower than the strength of the remoulded soil, in which the clay structure is completely random. The *operational* strength of such soils which can be relied upon for design purposes plots to the left of line *A* in Fig. 9.9. Such heavily overconsolidated soils are likely to have water contents close to their plastic limits, $I_L \sim 0$, so these strengths plot in the lower shaded region in Fig. 9.9.

So far, liquidity index has been linked with undrained strength, that is, strength defined in terms of total stresses. If a soil is experiencing a combination of total stresses such that a particular undrained strength is being mobilised, it must also be experiencing a combination of effective stresses such that a criterion of effective stress failure is being satisfied. In

Fig. 9.13 (a) Mohr's circles of total stress (T) and effective stress (E) for soil failing with mobilised undrained strength c_L; (b) effective stress state on critical state line (csl).

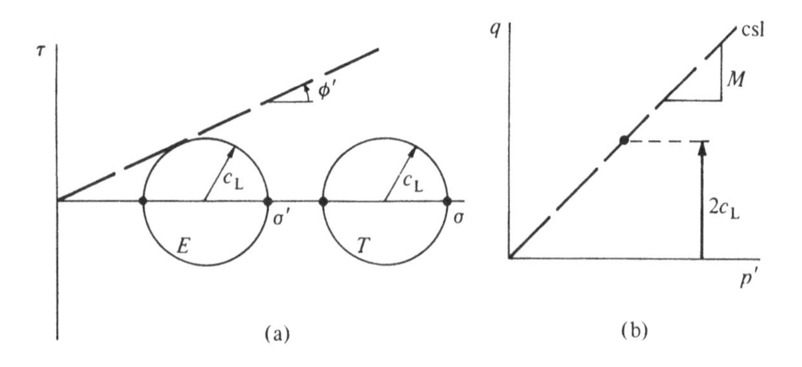

other words, a pore pressure (or pore suction) must exist in order that total and effective stress failure conditions can occur simultaneously.

Section 9.2 showed that soils with a water content equal to their liquid limit ($w = w_L$) had an undrained shear strength $c_L \sim 2\,\text{kPa}$. If a soil with a water content equal to its liquid limit is being sheared in undrained triaxial compression, then the undrained strength c_L defines the radii of the Mohr circles of total and effective stress (Fig. 9.13a) and specifies the deviator stress $q_L = 2c_L$ mobilised on the critical state line for the soil (Fig. 9.13b). Calculation of the effective mean stress

$$p'_L = \frac{q_L}{M} = \frac{2c_L}{M} \tag{9.21}$$

requires a value for the slope of the critical state line M which is related to the effective angle of friction ϕ':

$$M = \frac{6 \sin \phi'}{3 - \sin \phi'} \qquad (9.22)(7.9\text{bis})$$

Data of angle of friction (plotted as $\sin \phi'$) and plasticity index I_P have been assembled by Mitchell (1976) and are shown in Fig. 9.14; a corresponding scale of values of M for triaxial compression, calculated from (9.22), is included. Mitchell observes that 'the peak value of ϕ' decreases with increasing plasticity index and activity'. The spread is large but can be approximately described by the equation

$$\sin \phi' = 0.35 - 0.1 \ln I_P \tag{9.23}$$

However, for present purposes an average value $M \sim 1.0$ (corresponding to $\phi' = 25.4°$) can be taken. Then with $c_L = 2\,\text{kPa}$, the mean effective stress

Fig. 9.14 Relationship between M or $\sin \phi'$ and plasticity index I_P for normally compressed soils [after Mitchell, 1976, with additional data (•) from Brooker and Ireland, 1965].

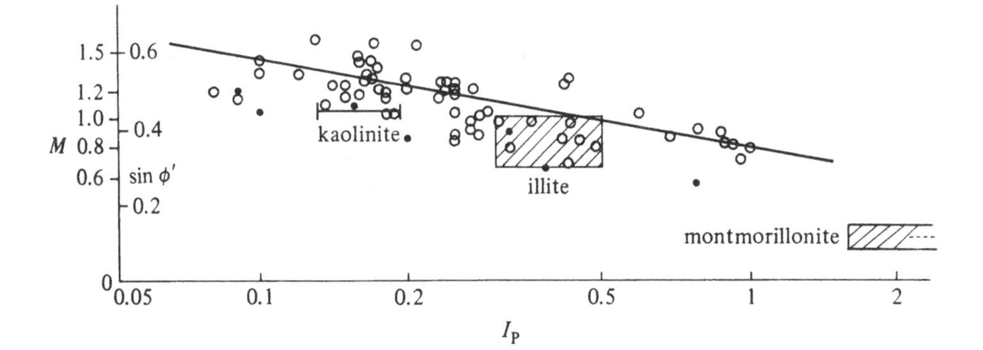

on the critical state line at a water content equal to the liquid limit is, from (9.21), $p'_L \sim 4\,\text{kPa}$.

This statement about a mean effective stress at the critical state for soil having a particular liquidity can be extended to a proposition about liquidity index and effective stress in normal compression if a model of soil behaviour such as one of the elastic–plastic models discussed in Chapters 4 and 5 is available. The relative positions in the compression plane of the critical state line and a normal compression line are controlled by the geometries of the yield loci and plastic potential curves in the $p':q$ plane. As an example, details of the calculation which can be made using the Cam clay model of Chapter 5 are presented in Section 9.4.5. It emerges there that an approximate estimate of the vertical effective stress in one-dimensional normal compression at a given specific volume, water content, or liquidity index can be obtained by doubling the mean effective stress on the critical state line at the same value of the volumetric variable.

If the chosen value of water content is the liquid limit w_L, then the corresponding mean effective stress on the critical state line is $p'_L \sim 4\,\text{kPa}$, and the deduced corresponding vertical effective stress in one-dimensional normal compression is $\sigma'_{vL} \sim 8\,\text{kPa}$. All these calculations are independent of the chosen value of water content. At the plastic limit, the remoulded undrained strength is assumed to be 100 times higher than at the liquid limit, and the mean effective stress on the critical state line and the vertical effective stress on the one-dimensional normal compression line are similarly 100 times higher. Assuming a constant compressibility between the liquid limit and plastic limit implies a relationship similar to (9.20) linking liquidity and vertical effective stress:

$$\sigma'_v = 8 \times 100^{(1-I_L)}\,\text{kPa} \qquad (9.24)$$

'Sedimentation compression curves' (i.e. data of vertical effective stress and specific volume) for one-dimensionally normally compressed 'argillaceous deposits' at 21 sites gathered by Skempton (1970a) are shown in Fig. 9.15a. The slope of each group of data points is an indication of the compressibility of that particular deposit, and the wide range of compressibility apparent in Fig. 9.15a is an indication of a similarly wide range of values of plasticity index. The data in Fig. 9.15a are brought together into a narrower band when the volumetric parameter is converted to liquidity index for each sample (Fig. 9.15b). These data confirm the proposition of an approximately unique relationship between vertical effective stress and liquidity index. Expression (9.24), plotted in Fig. 9.15b, provides a reasonable match to the assembled data.

Most soil deposits are not normally compressed, but are overconsolidated because of erosion of overburden or other effects. A typical soil element such as A in Fig. 9.16 may have experienced the history shown in the compression plane: normal compression to A_1 followed by swelling to A_2 with removal of overburden. An average one-dimensional normal compression line relating liquidity index and vertical effective stress for

Fig. 9.15 (a) Data of specific volume v or water content w and vertical effective stress σ'_v for normally compressed argillaceous sediments; (b) spread of data from (a) with water contents normalised to liquidity index I_L (after Skempton, 1970a).

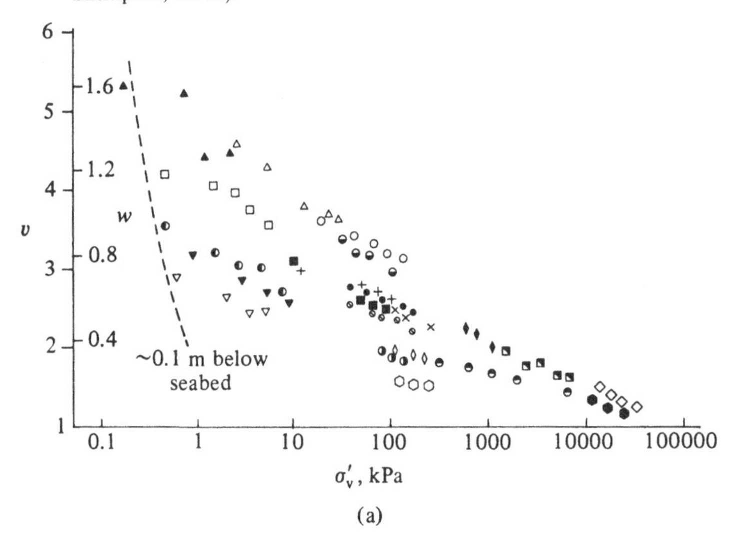

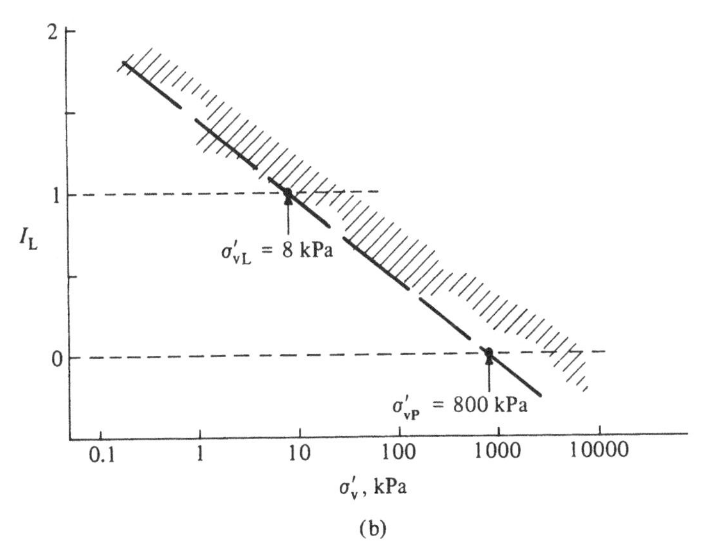

insensitive soils has been proposed. A soil which is overconsolidated has a combination of liquidity index and vertical effective stress which does not plot on this line. The compression and swelling processes shown in the $\sigma'_v:v$ compression plane in Fig. 9.16 can just as well be plotted in a $\sigma'_v:I_L$ compression plane (Fig. 9.17), and the position of point A_2 relative to the normal compression line LA_1P can be used to estimate the in situ overconsolidation ratio n of the soil.

It is convenient to assume that volume changes occur on unloading according to an expression of the form

$$v = v_s - \kappa^* \ln \sigma'_v \tag{9.25}$$

In other words, the line linking A_2 and A_1 in Figs. 9.16 and 9.17 is straight. (The details of this one-dimensional unloading process and the relationship of κ^* to other soil properties are considered in Section 10.3.2.) If the ratio of slopes of unloading and normal compression lines κ^*/λ is known, the point A_1 can be re-established by projection at the appropriate slope from

Fig. 9.16 (a) Normal compression of soil at A; (b) overconsolidation of soil at A caused by erosion of overburden.

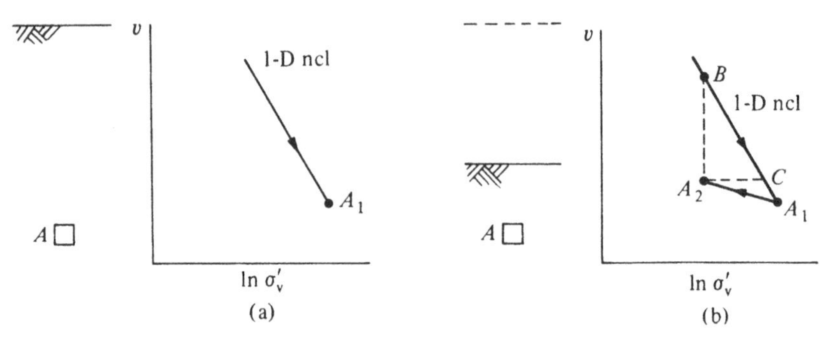

Fig. 9.17 Estimation of past maximum vertical effective stress from in situ combination of liquidity index and vertical effective stress.

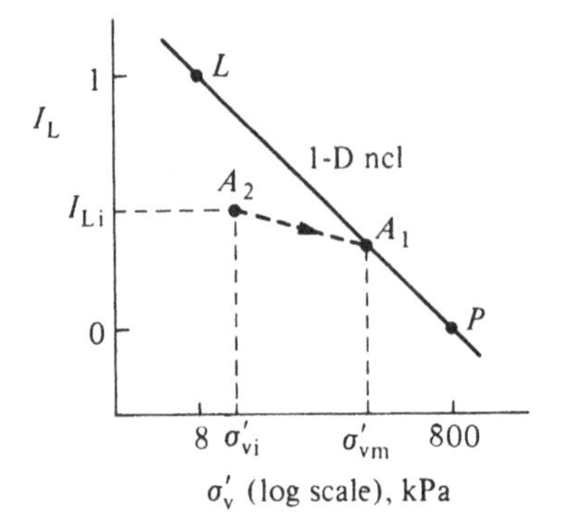

A_2 to intersect the normal compression line as shown in Fig. 9.17 to define past maximum stress σ'_{vm} and $n = \sigma'_{vm}/\sigma'_{vi}$ (a procedure suggested by Wroth, 1979). This procedure can be described analytically as

$$\frac{\ln n}{\ln 100} = \frac{(1 - I_{Li}) - \ln(\sigma'_{vi}/\sigma'_{vL})/\ln 100}{\Lambda^*} \qquad (9.26)$$

where

$$\Lambda^* = 1 - \frac{\kappa^*}{\lambda} \qquad (9.27)$$

and $\sigma'_{vL} = 8\,\text{kPa}$. Expression (9.26) can be conveniently rewritten

$$\log_{10} n = \frac{2(1 - I_{Li}) - \log_{10}(\sigma'_{vi}/8)}{\Lambda} \qquad (9.28)$$

with σ'_{vi} measured in kilopascals. This expression, with $\Lambda^* = 0.8$ (implying $\kappa^*/\lambda = 0.2$), has been used to estimate the variation of overconsolidation ratio with depth for the site on the Thistle field in the North Sea, for which strength estimates were shown in Fig. 9.8. Only very limited oedometer data are available to provide corroborative evidence for these estimates of overconsolidation ratio; but a plausible trend is indicated (Fig. 9.18), with a general fall of overconsolidation ratio towards 1 in the top 18 m and below this depth a higher value, where the strength data in Fig. 9.8 showed also a sharp increase.

More generally, charts of strength and liquidity and of overburden pressure (vertical effective stress) and liquidity can be used to make qualitative statements about the consolidation history of soil deposits (Wood, 1985b). Line X is expression (9.20) in Fig. 9.19a, a chart of liquidity

Fig. 9.18 Values of overconsolidation ratio for location in Thistle field, North Sea (•, estimated values, ×, values from oedometer tests).

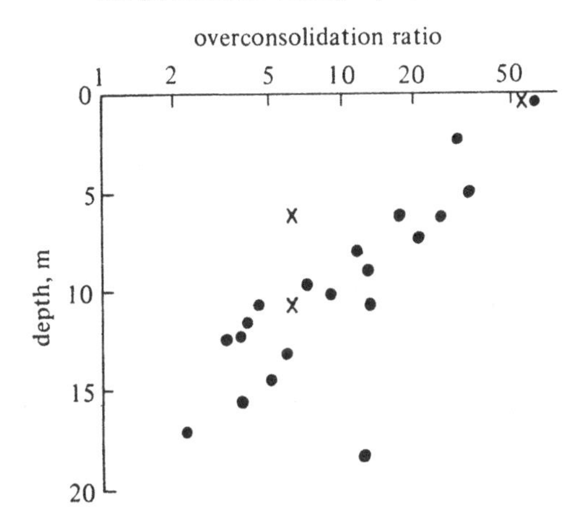

index and strength (cf. Fig. 9.9); in Fig. 9.19b, a chart of liquidity index and vertical effective stress, line Y is expression (9.24) (cf. Fig. 9.15). Four possibilities are distinguished in Figs. 9.19a, b.

Normally consolidated insensitive soils have strengths dependent on their liquidity which plot around line X (A in Fig. 9.19a). Such soils also have vertical effective stresses which plot around line Y (A in Fig. 9.19b). Since strength is a function primarily of water content, or liquidity index, overconsolidated soils also plot around line X (B in Fig. 9.19a). However, overconsolidated soils have lower liquidities than normally compressed soils and consequently have vertical effective stresses which plot below line Y (B in Fig. 9.19b).

Lines X and Y in Fig. 9.19 relate to soil which has been remoulded and disturbed and has lost its natural structure. Natural sensitive soils can have peak strengths considerably higher than their remoulded strengths and plot to the right of line X (C in Fig. 9.19a). When their structure is disturbed, the reliable strength for their in situ liquidity is considerably lower than this peak strength. Such soils contain too much water to be good for them (i.e. their structure is much more open than their present stress level implies); they therefore plot above line Y (C in Fig. 9.19b). When their structure is disturbed, they may reach eventual equilibrium at the low liquidity corresponding roughly to the same vertical effective stress but with the expulsion of a large volume of water from their collapsing pores.

Another consolidation possibility is so-called underconsolidation. Typically, in a saturated soil deposit with the water table at the surface,

Fig. 9.19 Plotting of field data of undrained strength c_u, vertical effective stress σ'_v, and liquidity index I_L in relation to proposed unique relationships.

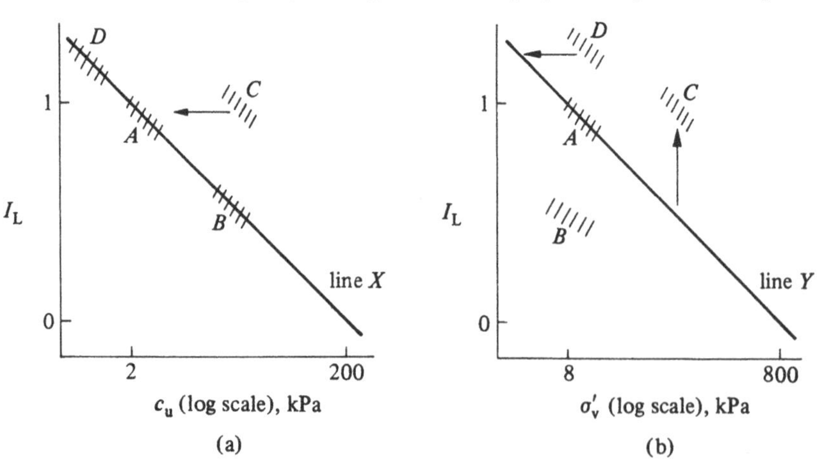

the effective vertical stress at depth z is

$$\sigma'_v = \gamma' z \tag{9.29}$$

However, if the rate of deposition of a soil is high, then part of the total overburden pressure at any depth

$$\sigma_v = \gamma z \tag{9.30}$$

is supported by excess pore pressures instead of by effective stresses. The pore pressure exceeds the hydrostatic pore pressure,

$$u > \gamma_w z \tag{9.31}$$

and the effective stresses are lower than expected from (9.29),

$$\sigma'_v < (\gamma - \gamma_w) z = \gamma' z \tag{9.32}$$

If values of vertical effective stress are calculated on the assumption that only hydrostatic pore pressures

$$u = \gamma_w z \tag{9.33}$$

are present, then the liquidity:vertical effective stress combination will plot to the right of line *Y* (*D* in Fig. 9.19b).

Underconsolidated soils are in fact normally compressed soils (i.e. the current vertical effective stress is the largest the soil has known and is the same as the preconsolidation pressure $\sigma'_{vc} = \sigma'_v$), but the vertical effective stresses are lower than expected. If the actual pore pressures u were known, then the effective stresses for these soils, calculated from

$$\sigma'_v = \sigma_v - u = \gamma z - u \tag{9.34}$$

should plot around line *Y*. The *strengths* of underconsolidated soils should still correspond with their actual liquidities since the in situ liquidities reflect the effective stresses of which the soil is actually aware. Thus, *D* in Fig. 9.19a lies around line *X*, and in terms of strength and liquidity, the soil looks just like any other young normally compressed deposit.

An indication that a particular soil is probably underconsolidated is of importance because large settlements are to be expected as the soil comes into equilibrium, with pore pressures falling to hydrostatic values, even before settlements occur under imposed surface loads.

9.4 Background to correlations

9.4.1 *Liquid limit*

Albert Atterberg (1911), a Swedish agricultural engineer, described a suite of tests that might be performed to determine the water contents at which the character of the mechanical behaviour of a clay went through various transitions. Of these, two have been absorbed into the soil mechanics canon: the liquid and plastic limit tests which were reckoned by Atterberg to define the range of water content for which a clay could be described as a plastic material.

Atterberg's test to determine the liquid limit (*flytgränsen*: literally, flowing limit) of soil involved mixing a pat of clay 'in a little round-bottomed porcelain bowl of 10–12 cm diameter'. A groove was cut through the pat of clay with a spatula, and the bowl was then struck many times against the palm of one hand. If after many blows the groove was only closed to an insignificant height at the base of the pat, then the clay was

Fig. 9.20 (a)–(c) Casagrande liquid limit apparatus (after British Standard BS1377:1975); (d) variation with water content of number of blows to close groove in Casagrande liquid limit test (after Casagrande, 1932); (e) section through groove in Casagrande liquid limit test.

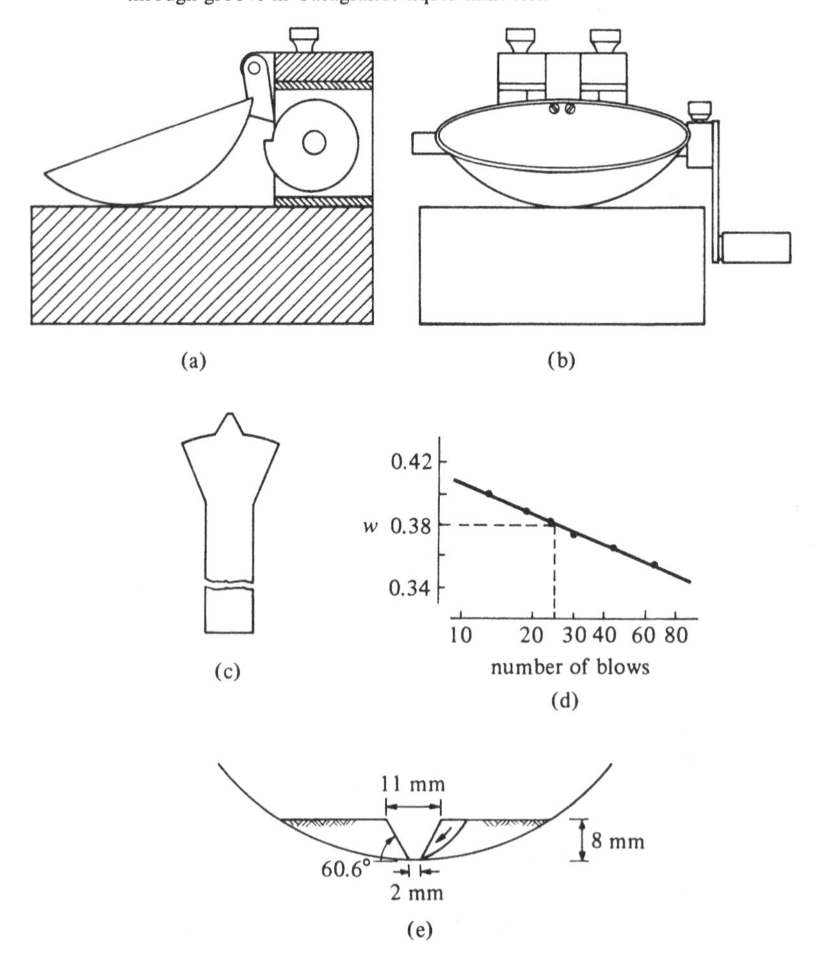

defined to be at its liquid limit: its water content was just not high enough for it to flow like a liquid given the opportunity to do so.

To make the measurement of the liquid limit more scientific, the test was standardised using the apparatus proposed by Casagrande (1932). This is a device (Figs. 9.20a, b) for performing mechanically the procedure described by Atterberg. The standard metal bowl containing the pat of soil is given repeated standard blows by being dropped through 10 mm onto a hard rubber base. The procedure is further tightened by requiring that, if a soil is at its liquid limit, the groove made in the pat of soil with a standard grooving tool (Fig. 9.20c) should close at its base over a distance of 13 mm in exactly 25 blows. In practice, liquid limit is usually obtained by interpolation from 'flow curves' (Casagrande, 1932) relating water content with number of blows (Fig. 9.20d).

In Section 9.2, the liquid limit of soils was associated with a particular undrained strength. The Casagrande liquid limit test requires the failure of miniature soil slopes (cross section, Fig. 9.20e): each slope is formed with a height of 8 mm and slope angle of 60.6°. Failure is induced by means of a series of shock decelerations as the bowl of the apparatus strikes the hard rubber base. The undrained stability of these slopes, of height H and formed in soil of strength c_u and total unit weight γ, can be studied in terms of the dimensionless group $c_u/\gamma H$. Since the unit weight γ depends on the value of the liquid limit, the strength c_u of a soil at its Casagrande liquid limit also depends on the value of that liquid limit.

Data of strengths of soils at water contents around their Casagrande liquid limits have been reported by Youssef, el Ramli, and el Demery

Fig. 9.21 Variation of undrained strength with water content at water contents close to liquid limit (after Youssef, el Ramli, and el Demery, 1965).

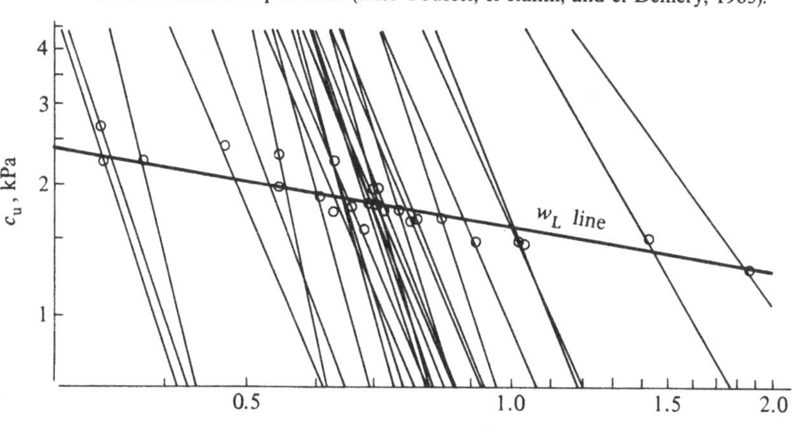

(1965) (Fig. 9.21). The strengths were measured by means of a laboratory miniature vane. There is a clear trend for the strength at the liquid limit c_L to fall as the liquid limit (determined with the Casagrande apparatus) increases, following the average line which has been drawn in the figure. The extreme values on this line are, for soil A: $w_L = 0.3$ $c_L = 2.4$ kPa; and for soil B: $w_L = 2.0$, $c_L = 1.3$ kPa. If we assume that the specific gravity of soil particles is the same for all the soils, $G_s = 2.7$, then water contents can be converted into unit weights using

$$\gamma = \frac{G_s(1 + w)\gamma_w}{1 + G_s w} \tag{9.35}$$

For soil $A, \gamma = 1.94\gamma_w$ and $c_u/\gamma H = 15.8$; for soil B, $\gamma = 1.27\gamma_w$ and $c_u/\gamma H = 13.0$. The closeness of the values of the dimensionless group $c_u/\gamma H$ shows the way in which the differences in strength and in unit weight for the soils balance each other.

The fall-cone test is now preferred to the Casagrande apparatus for determining the liquid limit, not only because it is more repeatable and consistent, and less operator sensitive but also because it provides a direct strength index; it is clear that the Casagrande apparatus cannot do this. Even though the combinations of cone parameters adopted in Scandinavia and Britain do appear to give liquid limits which correspond reasonably with Casagrande liquid limits, the very different procedures of the fall-cone test and the Casagrande device must lead to different values of liquid limits for extreme soil types.

9.4.2 *Plastic limit*

Atterberg's (1911) test to determine the plastic limit (*utrullgränsen*, literally, rolling-out limit) of soil involved rolling the clay on a piece of paper with the fingers until it formed 'fine threads'. The fine threads were put together and rolled again, and the process was repeated until the clay could no longer be rolled but broke into pieces. The clay was then defined to be at its plastic limit: its water content was just not high enough for it to be capable of plastic deformation and moulding.

The procedure for determining the plastic limit w_P has also been tightened. Casagrande (1932) notes that the water content at which a thread of soil crumbles depends on the diameter of the thread, and a diameter of 3 mm is now standard. British Standard test 3 specifies other details necessary for the correct performance of this test.

In Section 9.3, a strength of ~200 kPa was proposed for soils at their plastic limit. Schofield and Wroth (1968) have suggested that the plastic-limit test 'implies a tensile failure, rather like the split-cylinder or Brazil test of concrete cylinders'. In the split-cylinder test (Kong and Evans, 1975), a cylinder of concrete is loaded across a diameter (Fig. 9.22a); and when the load is increased, the cylinder eventually splits on this diameter. For an elastic cylinder of diameter D, loaded with point loads P per unit length, Timoshenko (1934) shows that there is a uniform tensile stress

$$\sigma_t = \frac{-2P}{\pi D} \tag{9.36}$$

acting across this diameter AA.* The normal compressive stress acting on the transverse diameter BB has the distribution shown in

Fig. 9.22 (a) Cylindrical specimen loaded diametrically in compression; (b) stresses on transverse diameter BB; (c) Mohr's circles of total (T) and effective (E) stress for element at centre of specimen (shear stress τ, total normal stress σ, effective normal stress σ').

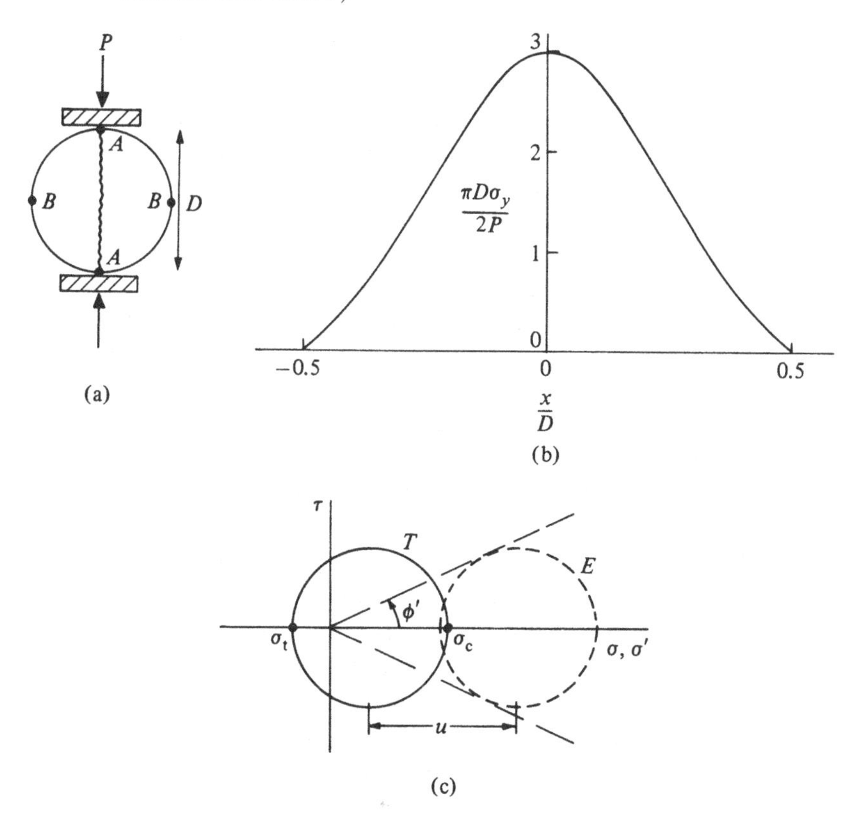

*For loads applied, practically, over a small part of the circumference, Wright (1955) shows that most of the diameter is still subjected to a tensile stress given by (9.36).

Fig. 9.22b, with a maximum at the centre,

$$\sigma_c = \frac{6P}{\pi D} \tag{9.37}$$

The vertical compressive stress has this magnitude at all points on the vertical diameter *AA*. The Mohr circle of total stress at all points on this diameter is therefore as shown in Fig. 9.22c. For soil which fails according to a frictional Mohr–Coulomb failure criterion, as discussed in Chapter 7, such a total stress state can be sustained only if there are negative pore pressures in the soil so that the Mohr circle of effective stress lies to the right of the total stress circle (Fig. 9.22c). The magnitude of this negative pore pressure depends, of course, on the effective angle of friction of the soil and on the size of the total stress circle, the undrained strength that is mobilised, which depends on the normal load *P* (Fig. 9.22a).*

If the strength at the plastic limit $c_P \sim 200\,\text{kPa}$, then $\sigma_c \sim 300\,\text{kPa}$ and $\sigma_t \sim 100\,\text{kPa}$. The total stress along the axis of the thread of clay being rolled is roughly zero, so the total mean stress is $p_P \sim 67\,\text{kPa}$. It was suggested in Section 9.3 that the mean effective stress at the critical state would be about twice the undrained strength at the same water content. That implies a mean effective stress $p'_P \sim 400\,\text{kPa}$ and a suction $u \sim -333\,\text{kPa}$. Suctions are able to exist in soils because of the action of surface tension at the air–water interface, where pore water menisci span between soil particles. Quite high suctions can exist without the soil becoming unsaturated, that is, without the menisci being drawn back from the surface of the soil. Suctions between 200 and 300 kPa are shown by Brady (1988) for London clay, with no external loading, at water contents near the plastic limit. The plastic limit test is probably a less consistent strength test than the Casagrande liquid limit test, but large suctions are certainly present in soils near their plastic limits, which can account for the mobilisation of significant undrained shear strengths.

9.4.3 *Plasticity and compressibility; liquidity and strength*

The simple correlations proposed in Section 9.3 between plasticity and compressibility, and between liquidity and strength, worked because of the assumption that there was a ratio of 100 between strengths of soils at their plastic and liquid limits. This figure was chosen as a convenient round number based on scattered data shown in Fig. 9.5. In the absence of any other information, this provides a good starting point for estimating

*The British Standard (BS 1377: 1975) says only that 'it is important to maintain uniform rolling pressure throughout the test'.

strengths and compressibilities from index properties. If other information is available, then it may be possible to improve these estimates.

It is proposed here that correlations are likely to work most satisfactorily for groups of related soils. Many different clay minerals contribute to the claylike properties of soils, such as plasticity. The plasticity of a soil depends on the type of clay mineral present as well as on the amount of clay present.* Soil samples taken over a limited geographical area are likely to have a common geological history, and though there may be variations in the proportion of clay present in individual samples, the nature of the clay mineral is likely to show much less variability.

Data of plasticity I_P and clay content C from a number of sites are shown in Fig. 9.23, adapted from Skempton (1953). Each set of points lies around a line through the origin, and Skempton defines the activity A of a soil as the slope of this line:

$$A = \frac{I_P}{C} \tag{9.38}$$

Fig. 9.23 Relationship between plasticity index I_P and clay fraction C for natural clays and clay minerals, with values of activity $A = I_P/C$ indicated (after Skempton, 1953).

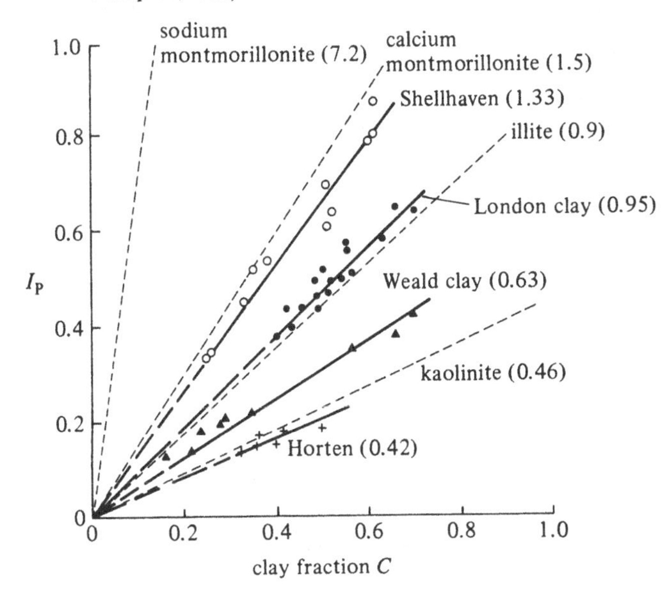

*A soil which consists only of sand particles does not exhibit any plasticity characteristics which can be observed in these standard tests for the liquid and plastic limits, and such soils are traditionally recorded as non-plastic.

Some activity values for soils containing pure clay minerals are also shown. Evidently, the activity of the soil is the plasticity of the pure clay fraction.

Variations of strength with liquidity index for two sets of artificial soils are shown in Fig. 9.24. The data are taken from tests on mixtures of kaolinite and sand, and mixtures of montmorillonite and sand, reported by Dumbleton and West (1970); the proportions of clay varied from 25 to 100 per cent. The authors report liquid limits determined using the Casagrande apparatus (Section 9.4.1). However, from their data of strengths (measured with a miniature laboratory vane), we could extract for each mixture the water contents corresponding to a strength of 1.7 kPa, to make the data consistent with the fall-cone index test of Section 9.2. Values of liquidity index could be calculated using these pseudo-fall-cone liquid limits.

The data for the montmorillonite and kaolinite mixtures diverge towards their plastic limits (as $I_L \rightarrow 0$), and though the spread within each band is reasonably small, the ratio of strengths at plastic and liquid limits is of the order of 100 for the montmorillonite mixtures but is much nearer to

Fig. 9.24 Variation of remoulded undrained strength c_u with liquidity index I_L for mixtures of (K) kaolinite and (M) montmorillonite with natural quartz sand (values of clay content C marked for each curve) (data from Dumbleton and West, 1970).

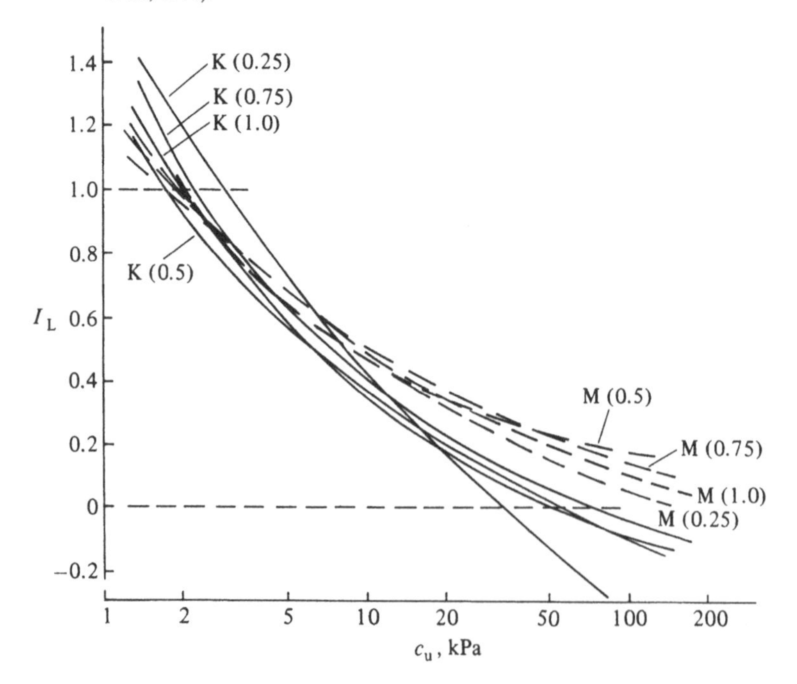

30 for the kaolinite mixtures. So the ratio 100 introduced in Section 9.3 can be replaced by a more general ratio $R = c_P/c_L$ which probably depends on the activity of the clay mineral present.

A more general form of (9.13) is then

$$\lambda = \frac{I_P G_s}{\ln R} \tag{9.39}$$

and the value of R can be estimated for a group of soils by inserting into (9.39) the value of compressibility λ measured in an oedometer test over a reasonable range of effective stress.

A corresponding more general form of (9.20) is then

$$c_u = c_L R^{(1 - I_L)} \tag{9.40}$$

where c_L is the strength of soils at their liquid limit, which was shown to be 1.7 kPa in Section 9.2 and was approximated to 2 kPa in Section 9.3.

9.4.4 *Liquidity and critical states*

The simple experimentally observed form of critical state lines

$$q = Mp' \qquad (9.41)(\text{cf. } 9.21)$$

$$V = \Gamma - \lambda \ln p' \qquad (9.42)(\text{cf. } 9.3)$$

requires three soil parameters for its definition: its slope in the compression plane λ (or C_c'), which has been shown in Section 9.3 to be related to plasticity index (9.13); its slope in the effective stress plane M (or ϕ'), which has been shown to be weakly related to plasticity index (Fig. 9.14); and its location in the compression plane, the parameter Γ, which is the value of specific volume at a mean effective stress $p' = 1$ unit of stress. Conventionally, the unit of stress is 1 kPa, but any units can be used without changing the position of the critical state line in the compression plane.

A link between Γ and w_L can now be deduced. Equation (9.42) can be written in terms of a reference pressure p_L', the mean effective stress at the critical state for a water content equal to the liquid limit (9.21). It then becomes

$$v = v_L - \lambda \ln \frac{p'}{p_L'} \tag{9.43}$$

where

$$v_L = 1 + G_s w_L \tag{9.44}$$

This is the specific volume for the soil at its liquid limit and has replaced Γ. If Γ is to be retained, then from expressions (9.42)–(9.44) (Figs. 9.25a, b),

$$\Gamma = 1 + G_s w_L + \lambda \ln p_L' \tag{9.45}$$

or with (9.39),

$$\Gamma = 1 + G_s\left(w_L + I_P \frac{\ln p'_L}{\ln R}\right) \tag{9.46}$$

For $p'_L = 4\,\text{kPa}$ and $R = 100$, (9.46) becomes

$$\Gamma = 1 + G_s(w_L + 0.3 I_P) \tag{9.47}$$

Although this may seem a little devious, it is an indication that the parameter Γ is essentially equivalent to liquid limit; a soil with a high liquid limit has a high value of Γ.

For the Weald clay (for which data of critical states were discussed in Section 6.3), Roscoe, Schofield, and Wroth (1958) quote liquid limit $w_L = 0.43$, plastic limit $w_P = 0.18$, and clay content $C = 0.4$. Then (9.47), using $G_s = 2.75$, leads to $\Gamma = 2.389$. In Section 6.3, a value of Γ of 2.072 was deduced from the experimental data, which were obtained over a range of mean effective stress of about 70–700 kPa. The value of Γ quoted there was itself an extrapolation from about 70 kPa to 1 kPa. Since it has

Fig. 9.25 (a,b) Relationship between Γ and liquid limit w_L ($I_L = 1$) for reference volumetric parameter on critical state line (csl); (c) linear approximations to curved critical state line.

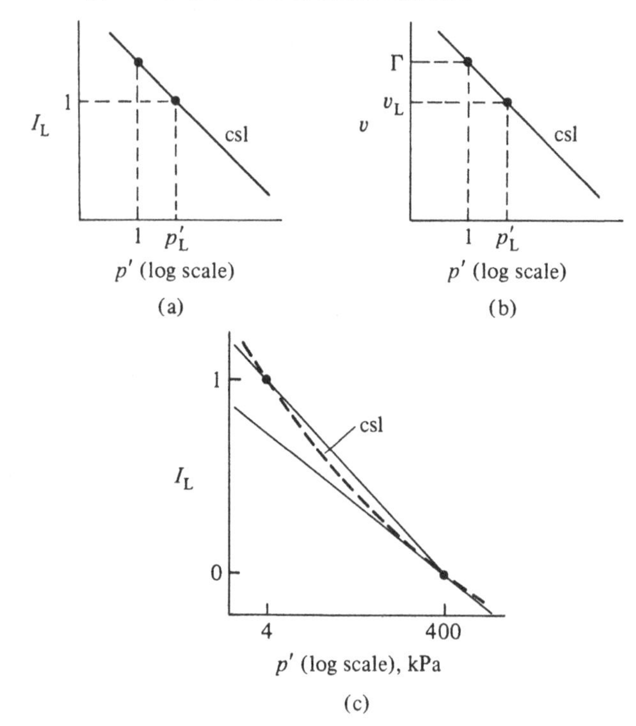

already been noted that the slope of critical state lines tends to fall as the mean stress increases, it is not surprising that this value of Γ is lower than that calculated from (9.47) (Fig. 9.25c). Of course, in practice, if a simple relationship such as (9.42) is to be used as an approximation for calculation purposes, then the values of λ and Γ should be chosen to give the best possible fit in the region of effective stresses of engineering interest in any particular problem.

Soils of different plasticities have different values of compressibility λ and produce different critical state lines when these are plotted in terms of water content or specific volume and mean effective stress (Fig. 9.26). The critical state lines for any two soils intersect at a point. Schofield and Wroth (1968) suggested that the critical state lines for *all* soils might pass through a single point in the $\ln p':v$ plane, which they called the Ω-point. This is probably too bold a generalisation. Here the proposal that critical state lines for related soils (soils perhaps of similar activity) pass through an Ω-point (Fig. 9.26) is explored.

Such an Ω-point constitutes a relationship between liquid limit and plasticity for soils. If p'_Ω, v_Ω, and w_Ω are values of mean effective stress, specific volume, and water content at the Ω-point, then for any one of the related soils,

$$\lambda = \frac{(w_L - w_\Omega)G_s}{\ln(p'_\Omega/p'_L)} \tag{9.48}$$

and with (9.39),

$$I_P = \frac{(w_L - w_\Omega)\ln R}{\ln(p'_\Omega/p'_L)} \tag{9.49}$$

The first expression, (9.48), is a relationship between compressibility λ and

Fig. 9.26 Critical state lines for related soils A, B, and C passing through single Ω-point in compression space.

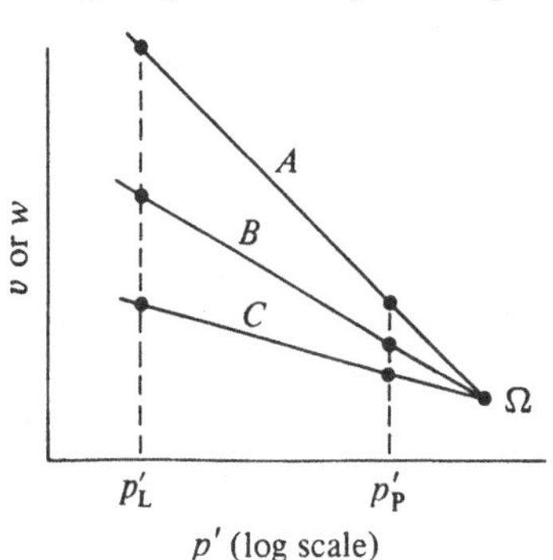

liquid limit w_L similar to the approximate relationship proposed by Terzaghi and Peck (1948) for remoulded clays, on the basis mainly of data collected by Skempton (1944) for a wide range of soils:

$$\lambda = 0.3(w_L - 0.1) \tag{9.50}$$

The second expression, (9.49), is a relationship of the same type as Casagrande's (1947) *A*-line (Fig. 9.27a), which is used for classification of soil types:

$$I_P = 0.73(w_L - 0.2) \tag{9.51}$$

This line is produced by Casagrande as an 'empirical boundary between typical inorganic clays which are generally above the *A*-line, and plastic soils containing organic colloids which are below it. Also located below the *A*-line are typical inorganic silts and silty clays...'. If the results of index tests on a number of samples of related soils are plotted on this plasticity chart, the points tend to lie on a straight line which is often approximately parallel to the *A*-line.

Given a group of related soils which plot on a line

$$I_P = A(w_L - B) \tag{9.52}$$

comparison with (9.49) shows that this is equivalent to defining an Ω-point given by

$$\ln \frac{p'_\Omega}{p'_L} = \frac{1}{A} \ln R \tag{9.53}$$

$$w_\Omega = B \tag{9.54}$$

or

$$v_\Omega = 1 + G_s B \tag{9.55}$$

Then if p'_L is fixed, for example at 4 kPa, and R is taken as 100, the position of each of the family of critical state lines in Fig. 9.26 immediately follows. Thus, $A = 0.73$ implies $p'_\Omega = 2197$ kPa, and $B = 0.2$ implies $w_\Omega = 0.2$ and $v_\Omega = 1.54$, for $G_s = 2.7$.

Plasticity data of Norwegian marine clays, from Bjerrum (1954), are plotted in Fig. 9.27b. These lie on a line nearly parallel to Casagrande's *A*-line

$$I_P = 0.79(w_L - 0.17) \tag{9.56}$$

which, with $G_s = 2.7$, $R = 100$, and $p'_L = 4$ kPa, implies $p'_\Omega = 1383$ kPa and $v_\Omega = 1.464$. These clays were deposited in a saline environment, but as a result of sea-level changes fresh water now flows through them. This fresh water leaches the salt from the clays and leaves them in a sensitive state (see Section 9.5). The effect of leaching the salt is to reduce the plasticity and liquid limit of the clay. Bjerrum (1954) quotes the results of a laboratory experiment in which two clay samples were sedimented in salt water; one was leached and the other left unleached, and index properties for the two samples were measured. The path of the leaching process is shown in Fig. 9.27b. The clay both before and after leaching is a Norwegian marine clay; the effect of leaching is merely to move the clay down the line (9.56) in the plasticity chart. In terms of critical state lines, this implies

Fig. 9.27 Relation between liquid limit w_L and plasticity index I_P. (a) Typical soils: 1, gumbo clays (Mississippi, Arkansas, Texas); 2, glacial clays (Boston, Detroit, Chicago, Canada); 3, clay (Venezuela); 4, organic silt and clay (Flushing Meadows, Long Island); 5, organic clay (New London, Connecticut); 6, kaolin (Mica, Washington); 7, organic silt and clay (Panama); 8, micaceous sandy silt (Cartersville, Georgia); 9, kaolin-type clays (Vera, Washington, and South Carolina) (after Casagrande, 1947). (b) Norwegian marine clays (data from Bjerrum, 1954, 1967).

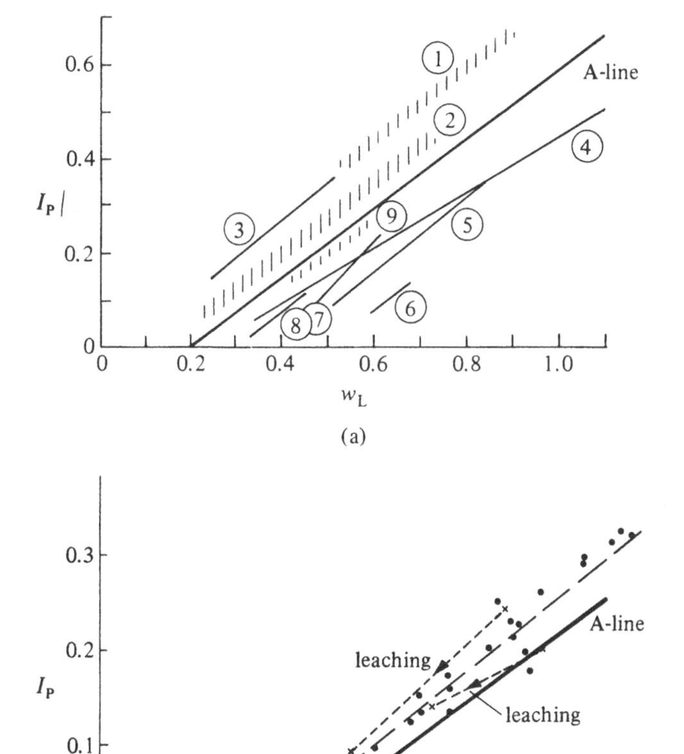

a reduction in compressibility without changing the Ω-point: leaching changes the critical state line from A to B to C in Fig. 9.26.

Thus, given some knowledge about a group of soils, the liquid limit w_L is sufficient to place a soil within this group and to obtain estimates for the location and slope of the critical state line for that soil.

9.4.5 *Liquidity and normal compression*

Elastic–plastic models constructed within the general framework discussed in Chapter 4 can be used to calculate the relative positions in the $\ln p':v$ compression plane of the critical state line and a normal compression line, for example, the one-dimensional normal compression line. This relative position is most conveniently described in terms of the ratio of stresses on the two lines, which in Chapter 4 have been assumed to be parallel, at any particular water content or specific volume.

In general, plastic potentials and yield loci in the $p':q$ effective stress plane are not identical. It was noted in Section 6.1 that the critical state

Fig. 9.28 Relation between values of mean effective stress p' on normal compression line (ncl) and critical state line (csl) at same specific volume (yl, yield locus; pp, plastic potential): (a) $p':q$ effective stress plane; (b) $v:p'$ compression plane; (c) $v:p'$ compression plane with p' plotted on logarithmic scale.

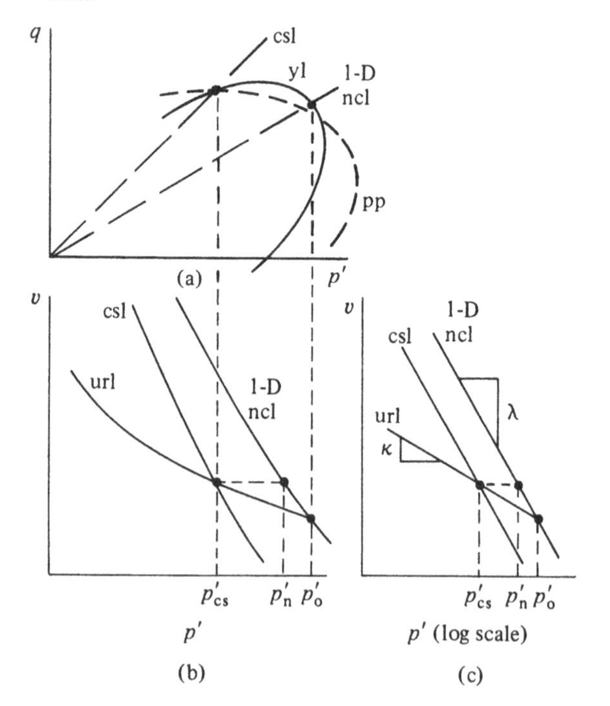

condition is actually associated with the point on the plastic potential curve where the tangent is horizontal. In Section 7.2 a parameter r was introduced which represented the ratio between the mean effective stress at the critical state and in normal compression for a particular yield locus (yl in Fig. 9.28a), that is, on a particular unloading–reloading line (url in Figs. 9.28b, c):

$$r = \frac{p'_o}{p'_{cs}} \tag{9.57)(7.22bis}$$

In the semi-logarithmic compression plane $\ln p':v$ (Fig. 9.28c), the normal compression and critical state lines have slope $-\lambda$, and the unloading–reloading line has slope $-\kappa$; the geometry of this diagram dictates that the ratio of mean effective stress on the normal compression and critical state lines at the same specific volume (or liquidity index) is

$$\frac{p'_n}{p'_{cs}} = r^{\Lambda} \tag{9.58}$$

where

$$\Lambda = 1 - \frac{\kappa}{\lambda} \tag{9.59)(6.19bis}$$

Normally compressed soil deposits often have a history of one-dimensional compression. In any soil deposit, it is easier to estimate the in situ vertical effective stress σ'_v than the in situ mean effective stress p', which depends also on the horizontal effective stress σ'_h:

$$p' = \frac{\sigma'_v + 2\sigma'_h}{3} \tag{9.60}$$

Stress paths followed by soils during one-dimensional compression (and unloading) are discussed in Section 10.3. Here it is appropriate only to note that the ratio K_0 of horizontal to vertical effective stresses for one-dimensionally compressed soil is

$$K_0 = \frac{\sigma'_h}{\sigma'_v} \tag{9.61}$$

and that for one-dimensionally normally compressed soil the value of $K_0 = K_{0nc}$ is given fairly accurately by the expression

$$K_{0nc} = 1 - \sin\phi' \tag{9.62}$$

The link between vertical effective stress and mean effective stress during one-dimensional normal compression is then

$$\frac{\sigma'_v}{p'} = \frac{3}{1 + 2K_{0nc}} = \frac{3}{3 - 2\sin\phi'} \tag{9.63}$$

At a given specific volume or liquidity index, if we combine (9.58) and (9.63), the ratio of vertical effective stress in one-dimensional compression to mean effective stress at the critical state is

$$\frac{\sigma'_{\mathrm{v}}}{p'_{\mathrm{cs}}} = \frac{3r^{\Lambda}}{1 + 2K_{0\mathrm{nc}}} \tag{9.64}$$

This relationship applies, for example, at the liquid limit $I_{\mathrm{L}} = 1$ where $p'_{\mathrm{cs}} = p'_{\mathrm{L}}$, and is related to the strength at the liquid limit c_{L} through (9.21). The vertical effective stress for one-dimensionally normally compressed soil with a water content equal to its liquid limit is σ'_{vL} which, from (9.64) and (9.21), is

$$\sigma'_{\mathrm{vL}} = \frac{6c_{\mathrm{L}}r^{\Lambda}}{M(1 + 2K_{0\mathrm{nc}})} \tag{9.65}$$

The parameters M and $K_{0\mathrm{nc}}$ are soil parameters linked to the angle of friction ϕ' through (9.22) and (9.62), respectively. Evaluation of this expression also requires values for r and Λ. The quantity r is a parameter of the soil model whereas Λ, from (9.59), is a parameter of the soil. We are deriving only approximate expressions in this chapter and it is hardly appropriate to rely on the Cam clay model of Chapter 5 for detailed numerical estimates of r, but it can be used to give general guidance.

According to the yield locus of the Cam clay model, the ratio of mean effective stress p' at a stress ratio η to the mean effective stress p'_{cs} at the critical state $\eta = M$ is

$$\frac{p'}{p'_{\mathrm{cs}}} = \frac{2M^2}{M^2 + \eta^2} \qquad \text{(9.66)(cf. 6.4 and 6.5)}$$

In terms of horizontal and vertical effective stresses σ'_{h} and σ'_{v}, stress ratio η is given by

$$\eta = \frac{3(\sigma'_{\mathrm{v}} - \sigma'_{\mathrm{h}})}{\sigma'_{\mathrm{v}} + 2\sigma'_{\mathrm{h}}} \tag{9.67}$$

and, hence, in one-dimensional normal compression, $\eta = \eta_{K\mathrm{nc}}$ where, from (9.61),

$$\eta_{K\mathrm{nc}} = \frac{3(1 - K_{0\mathrm{nc}})}{1 + 2K_{0\mathrm{nc}}} \tag{9.68}$$

and from (9.66),

$$r = \frac{2M^2}{M^2 + \eta_{K\mathrm{nc}}^2} \tag{9.69}$$

Using (9.22), (9.62), (9.68), and (9.69), we can calculate and plot the relationship between r and angle of friction ϕ' (Fig. 9.29). As ϕ' increases from 20° to 35°, r falls from about 1.5 to 1.4.

The value of Λ indicates the relative magnitudes of the slopes of the normal compression and unloading–reloading lines (Fig. 9.28c). From (9.59), $\kappa/\lambda = 1$ implies $\Lambda = 0$; $\kappa/\lambda = 0.5$ implies $\Lambda = 0.5$; $\kappa/\lambda = 0$ implies $\Lambda = 1$. Although the compressibility λ is well linked with plasticity of soils (Section 9.3), it is harder to make definitive statements about the elastic volumetric parameter κ. This is partly because κ, which represents volumetric changes occurring during *isotropic* unloading and reloading, is hardly ever measured and partly because a single unloading–reloading line is at best an idealisation of a real hysteretic response (Fig. 3.13a). Given such a response, the value of κ, determined perhaps as the average slope of the hysteresis loop, depends on the size of that loop and the extent to which the maximum mean stress is removed.

Much more often, a swelling index κ^* is measured which represents volume changes occurring during one-dimensional unloading (Fig. 9.30a) according to an expression of the form

$$v = v_s - \kappa^* \ln \sigma'_v \tag{9.25bis}$$

which involves only the vertical effective stress σ'_v applied in an oedometer test. It will be seen in Section 10.3.2 that changes in stress ratio η occur during one-dimensional unloading so that changes in mean effective stress and vertical effective stress are not in constant proportion, as implied by

Fig. 9.29 Ratio ($r = p'_o/p'_{cs}$) of mean effective stresses on one-dimensional normal compression line and critical state line for same yield locus related to angle of shearing resistance ϕ' according to Cam clay model.

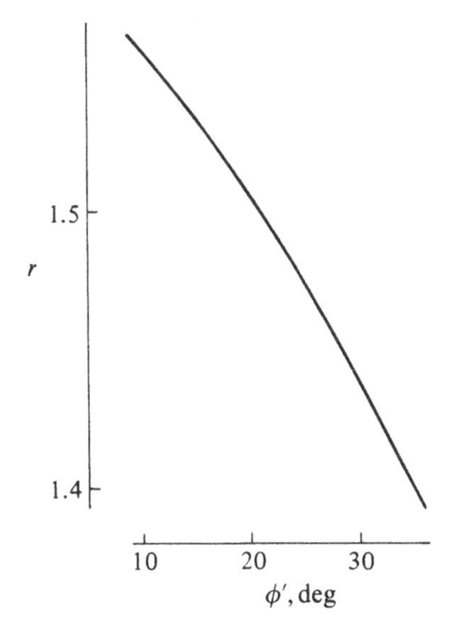

(9.63) during one-dimensional normal compression, and the connection between κ and κ^* is not straightforward.

Hysteresis loops seen in one-dimensional unloading and reloading tend to be much larger than those seen in unloading and reloading at constant stress ratio. A standard procedure for determining κ^* might involve

Fig. 9.30 (a) One-dimensional compression and unloading of clay from Gulf of Mexico (depth 0.82 m below seabed, water depth 3.7 km) (after Bryant, Cernock, and Morelock, 1967); (b) possible standard procedure for determining average slope of unloading–reloading cycle in oedometer by reducing vertical effective stress by factor of 4.

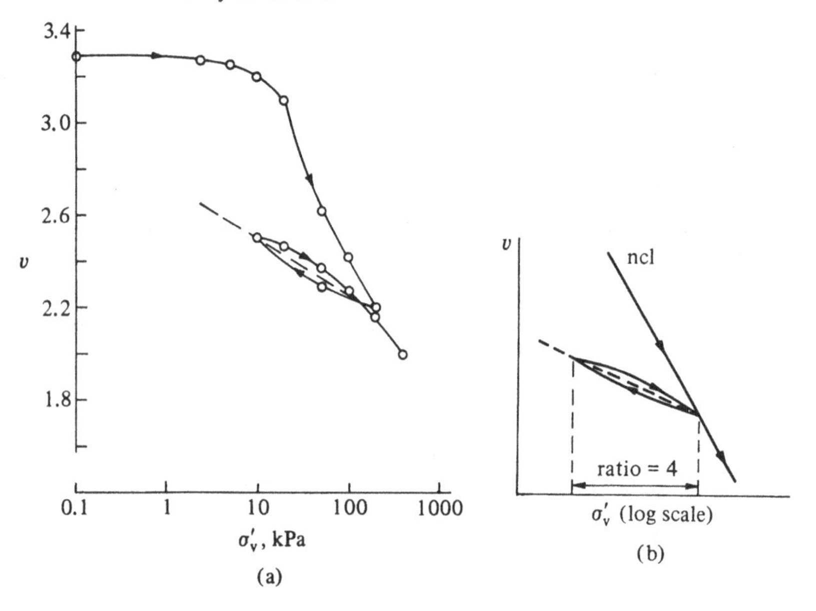

Fig. 9.31 Data of parameter Λ and plasticity index I_P (data collected by Mayne, 1980).

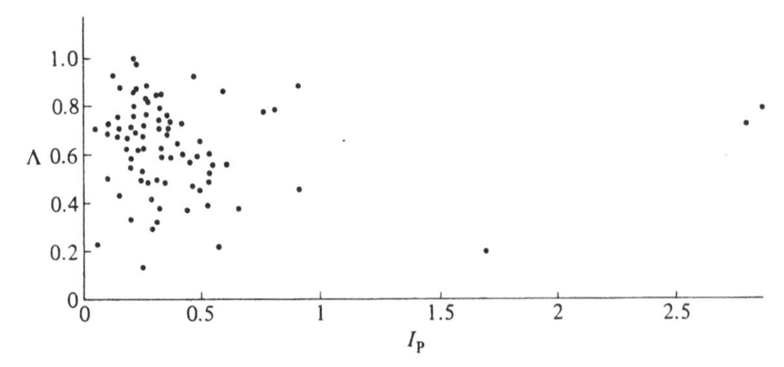

unloading the vertical effective stress by a definite factor of, for example, 2 or 4 (Fig. 9.30b). In the absence of such a standard procedure, the interpretation of published values of swelling index κ^* is difficult.

The parameter Λ was previously introduced in Section 7.2 as an exponent in an expression, (7.27), linking undrained strength and overconsolidation ratio. Mayne (1980) and Mayne and Swanson (1981) have fitted an expression of this form to published strength and overconsolidation data for many soils in order to deduce the value of Λ of which each soil seems to be aware. Their 105 values have been plotted against plasticity index I_p in Fig. 9.31. There is great scatter and no obvious trend, and some of the values seem implausibly low or high. There is possibly a fall of Λ with increasing plasticity. (The mean value of Λ is 0.63 with a standard deviation of 0.18.)

Leaving the choice of Λ open, we can evaluate the quantity $6r^{\Lambda}/[M(1+2K_{0nc})]$ from (9.65) for various values of angle of friction ϕ' (which controls M, K_{0nc}, and, through the Cam clay model, r) and different values of Λ. Curves are plotted in Fig. 9.32. For average soils with ϕ' in the range of 20° to 25° and Λ in the range of 0.6 to 0.8, a convenient value of this quantity is about 4. Thus, at a given specific volume or liquidity index, the vertical effective stress in one-dimensional normal compression is about four times the undrained strength and about twice the mean effective stress at the critical state. This is the value that was adopted in Section 9.3.

Fig. 9.32 Ratio of vertical effective stress in one-dimensional normal compression, and undrained strength at same specific volume as function of Λ and angle of shearing resistance ϕ' according to Cam clay model.

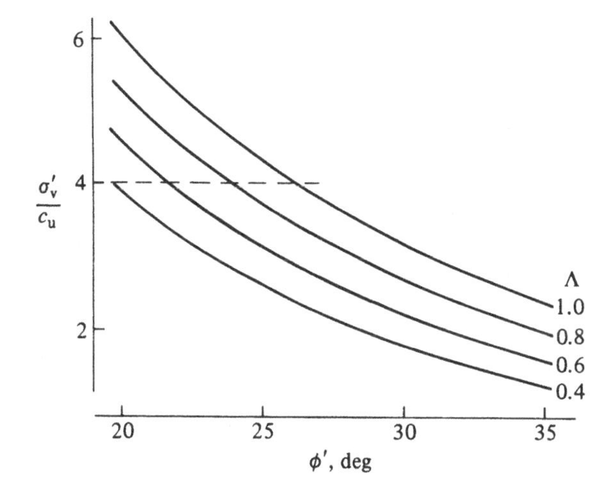

9.5 Sensitive soils

An example of a sensitive soil was shown in Fig. 9.11. Such soils have undisturbed strengths which may be considerably higher than their disturbed strengths. The ratio of undisturbed to remoulded strength is defined as the sensitivity S_t:

$$S_t = \frac{c_u}{c_{ur}} \tag{9.70}$$

Data of sensitivity and liquidity index for certain sensitive soils from various parts of the world are presented in Fig. 9.33. Related soils show a clear trend of increasing sensitivity with increasing liquidity index. For example, for the data of Norwegian marine clays in Fig. 9.33 (from Bjerrum, 1954), a relationship of the form

$$S_t = \exp(kI_L) \tag{9.71}$$

with $k \sim 2$ provides a reasonable fit. This implies a sensitivity $S_t \sim 7.4$ for a clay at its liquid limit ($w = w_L$, $I_L = 1$). A value $k = 0$ implies insensitive soil. Insensitive behaviour is usually observed for clays at or below their plastic limit, so (9.71) should be used only for $I_L > 0$.

Equations (9.70) and (9.71) can be combined with (9.20) to produce an expression with which undisturbed strengths can be estimated:

$$c_u = c_L R \exp[(k - \ln R)I_L] \qquad \text{for } I_L > 0 \tag{9.72}$$

This equation is best used for related soils with known values of k and R.

Fig. 9.33 Interrelationship between sensitivity and liquidity index for natural clays (△, Skempton and Northey, 1953; ○, Bjerrum, 1954; •, Bjerrum and Simons, 1960; +, Bjerrum, 1967).

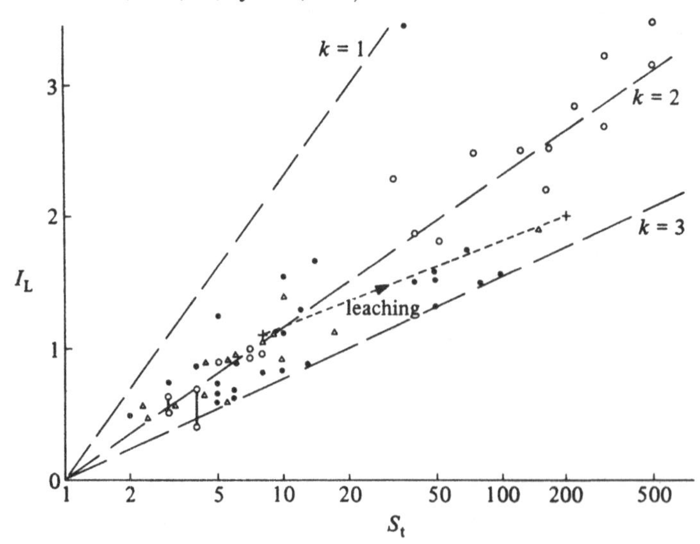

Bjerrum (1954) shows that the sensitivity of Norwegian clays is closely linked to the salt concentration in their pore water. It was noted in Section 9.4.4 that as the salt concentration changes, the index properties of the clay also change. As Fig. 9.34 shows, the plastic limit falls very slightly as the salt is leached out but the liquid limit falls markedly, and hence the plasticity drops considerably. The water content remains essentially constant because the structure of the clay does not change, so there is a marked rise in liquidity index. Data from adjacent borings in leached and unleached clay in Drammen are presented by Bjerrum (1967). The water content is more or less the same in both, but the liquidity index changes from about 1.1 to 2. An increase of sensitivity from 8 to 200–300 is quoted by Bjerrum (this leaching path is plotted on Fig. 9.33), whereas (9.71) with $k = 2$ suggests an increase from 9 to 55. High values of sensitivity are not likely to be accurate, and the precise value is of somewhat academic interest since it indicates primarily that it may be extremely dangerous to rely on the undisturbed strength for engineering purposes. Nevertheless, the link

Fig. 9.34 Changes in properties of a normally compressed Norwegian marine clay when salt concentration is reduced by leaching with fresh water: (a) sensitivity S_t; (b) undisturbed and remoulded strengths; (c) water content w, liquid limit w_L, and plastic limit w_P (after Bjerrum, 1954).

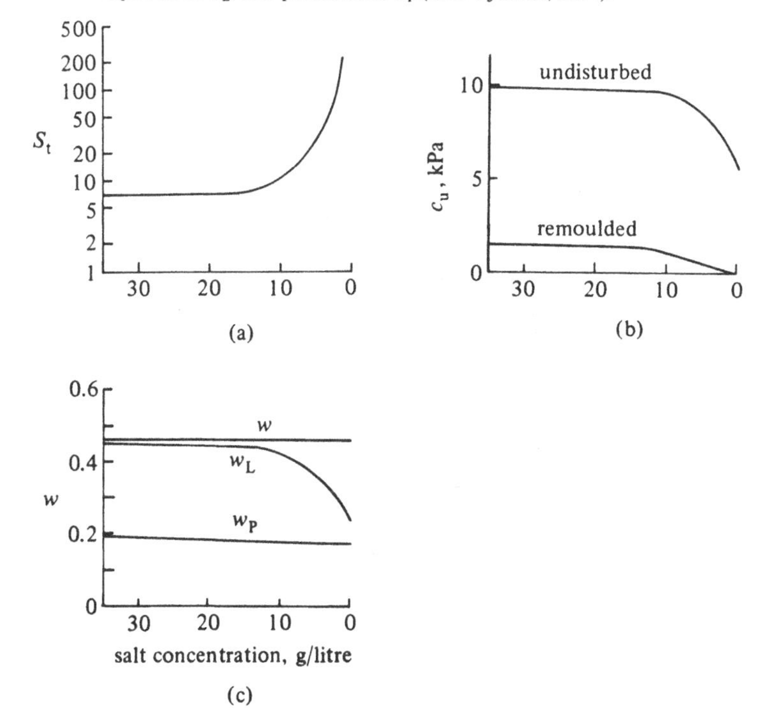

between sensitivity and liquidity at constant water content is clear. The effect of leaching is to move the state of the soil up line B (for $k = 2$) in Fig. 9.35: as the liquidity index increases, the separation of line B and line A ($k = 0$ for remoulded soil) increases, and hence the sensitivity increases.

A similar discussion can be applied to expression (9.24) relating vertical effective stress and liquidity index for normally compressed soils. It is evident from Fig. 9.34 that sensitive soils are being left with water contents that are too high to be good for them. The high liquidity and low in situ effective stress leaves a surplus of liquidity over that predicted by (9.24): this is an indication of the large amount of water that will be released

Fig. 9.35 Expected approximate relationships between liquidity index I_L and undrained shear strength c_u for insensitive or remoulded soil (line A) and for undisturbed, sensitive soil (line B).

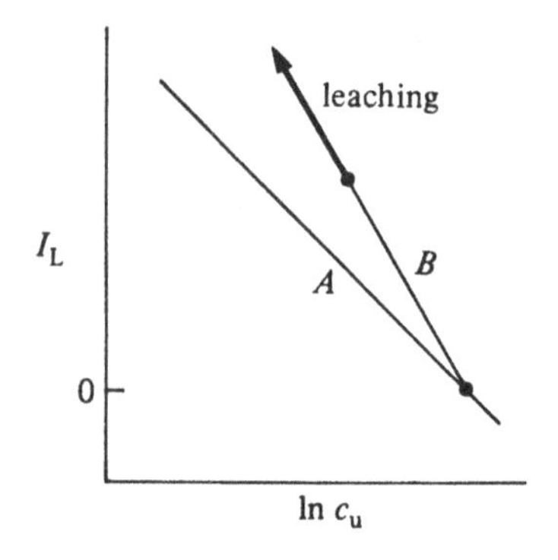

Fig. 9.36 Idealised relationship between liquidity index and effective overburden pressure, with contours of sensitivity (—) (after U.S. Dept. of the Navy, 1971) and expression (9.73) (– –).

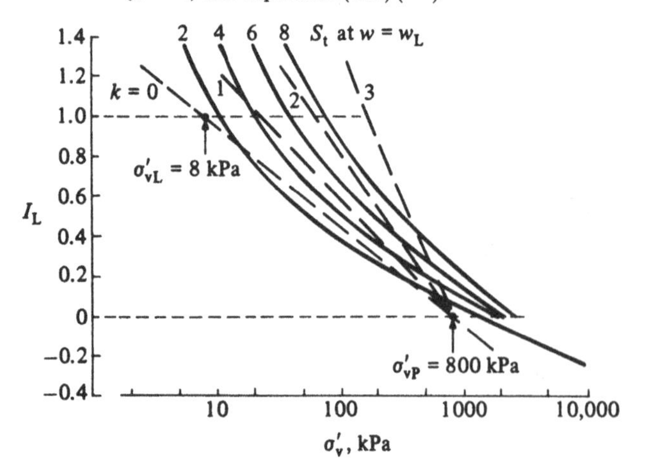

when the structure of the soil is disturbed and the soil tries to reach remoulded equilibrium under the in situ vertical effective stress. Slope failures in such soils can have devastating consequences.

The expression equivalent to (9.72) is

$$\sigma'_v = \sigma'_{vL} R \exp[(k - \ln R) I_L] \qquad \text{for } I_L > 0 \tag{9.73}$$

Plotting this with $R = 100$, $\sigma'_{vL} = 8\,\text{kPa}$, and various values of k gives a set of lines equivalent to the curves in Fig. 9.36, based on a chart from the design manual Navfac DM-7 (U.S. Department of the Navy, 1971).

Some field data of effective overburden pressure and liquidity index of undisturbed clays collected by Skempton and Northey (1953) are shown in Fig. 9.37. Expression (9.73) is plotted in this figure with values of k between 0 and 3. A value of $k = 3$ provides a reasonable upper limit to the data of sensitivity and liquidity of clays in Fig. 9.33 and also a reasonable outer limit to the data in Fig. 9.37.

Results of one-dimensional normal compression of slurried clays from Gosport and Horten are also shown in Fig. 9.37 for comparison with the

Fig. 9.37 Data of liquidity index and effective overburden pressure for European clays (after Skempton and Northey, 1953) (• and – field data; – – oedometer data for slurried clays).

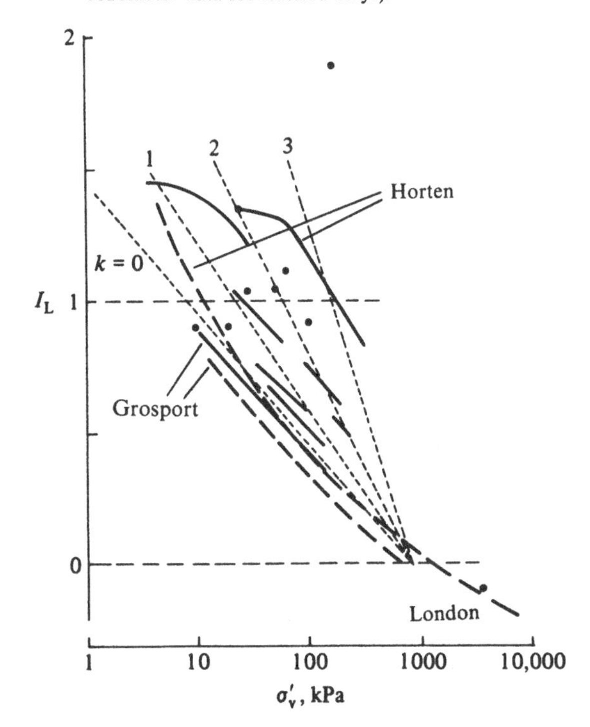

undisturbed compression curves. For both these clays, the slurried data lie close to the line for insensitive soil with $k = 0$. However, whereas the Gosport clay shows negligible in situ sensitivity, with the slurried and undisturbed data lying almost on top of each other, the in situ Horten clay plots close to the line for $k = 3$, indicating a high in situ sensitivity.

Expression (9.73) can be rewritten as

$$I_L = \left[\frac{\ln R}{\ln R - k}\right]\left[1 - \frac{\ln(\sigma'_v/\sigma'_{vL})}{\ln R}\right] \tag{9.74}$$

The ratio of the compressibilities of the undisturbed and remoulded soils is

$$\frac{\lambda_u}{\lambda_r} = \frac{\ln R}{\ln R - k} \tag{9.75}$$

Fig. 9.38 (a) Construction of relationship between specific volume v and vertical effective stress σ'_v expected for sensitive soil; (b) one-dimensional compression of (1) undisturbed and (2) remoulded Leda clay (after Quigley and Thompson, 1966).

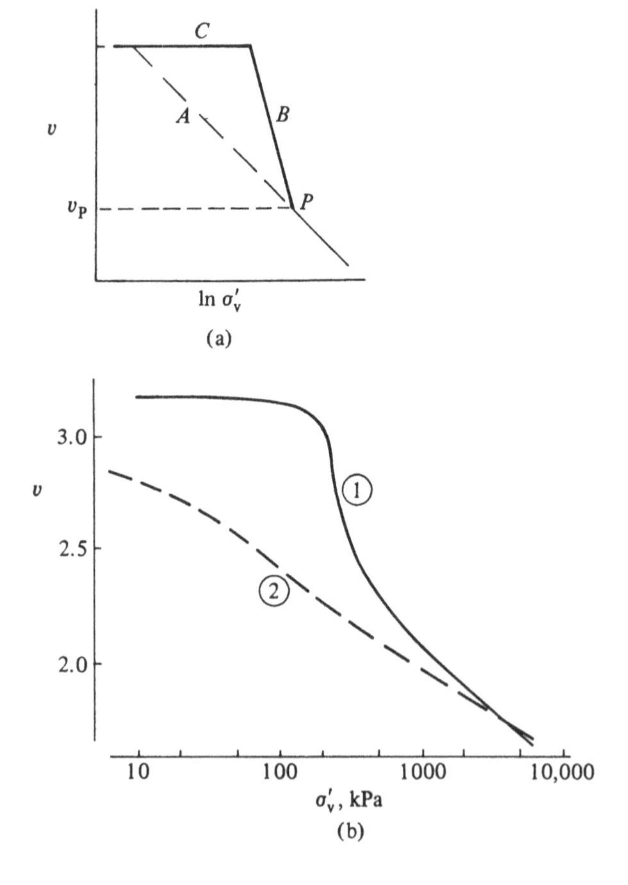

Terzaghi and Peck (1948) suggest that this ratio should be about 1.3 'for an ordinary clay of medium or low sensitivity'. With $R = 100$, $\lambda_u/\lambda_r = 1.3$ in (9.75) implies $k \sim 1$, which in Fig. 9.33 does indeed correspond to clay of medium or low sensitivity.

Yudhbir (1973) suggests a procedure by which field, in situ normal compression curves might be reconstructed or laboratory compression curves corrected for disturbance that occurred during sampling (as in Fig. 9.7). It is implicit in (9.74) that undisturbed and disturbed compression curves intersect at a water content equal to the plastic limit ($I_L = 0$); at this liquidity, sensitivity seems in general to be close to unity. With knowledge of index properties, (9.74) can be converted to a relationship between specific volume and vertical effective stress for completely disturbed soil ($k = 0$, line A in Fig. 9.38a), which can be expected to apply anyway for water contents below the plastic limit, beyond point P, with $v_P = 1 + G_s w_P$. If actual oedometer data on compression of disturbed samples were available, they could be used instead. From strength measurements (e.g. with a field vane or a fall-cone), some estimate of the sensitivity of the soil at its in situ water content can be obtained and used to estimate a value for k in (9.71). The normal compression line for in situ soil can then be generated from (9.74), using this value of k (line B in Fig. 9.38a). Recompression of sensitive soils at their in situ specific volume is often a very stiff process, particularly compared with subsequent normal compression (Figs. 9.30a and 9.38b) and so a third line can be drawn in Fig. 9.38a, a horizontal line C at the in situ specific volume. The resulting trilinear relationship, C–B–A, suitably smoothed to accommodate nature's abhorrence of sharp corners, is then a more realistic field compression curve (compare Fig. 9.38b).

9.6 Strength and overburden pressure

In Section 9.4.5, a convenient average value of 4 was chosen for the ratio of vertical effective stress during one-dimensional normal compression to the undrained strength of a soil at the same water content. This link between strength and overburden pressure should be examined further.

Many soil deposits are not normally compressed, so a general approximate relationship links strength with the equivalent consolidation pressure σ'_{ve} – the vertical effective stress which, in one-dimensional normal compression, would produce the current specific volume or water content (see Sections 6.2 and 7.4.1). The general relationship is

$$\frac{c_u}{\sigma'_{ve}} \simeq 0.25 \qquad (9.76)$$

The strength of a soil is linked with the size of the current, in situ yield locus. The Cam clay model was used to help justify (9.76), but Fig. 9.28 illustrated how a more general elastic–plastic model might be used. The size of an in situ yield locus can be characterised by the preconsolidation pressure σ'_{vc}, the yield point observed in an oedometer test (Section 3.3). For lightly overconsolidated clays, the difference between the preconsolidation pressure (point A_1 in Fig. 9.16b) and the equivalent consolidation pressure (point C in Fig. 9.16b) is small since the slope of the unloading line in this region is usually considerably flatter than the slope of the

Fig. 9.39 Data of ratio of undrained strength to vertical effective stress (c_u/σ'_v), and plasticity index I_P: (a) after Skempton (1957); (b) after Leroueil, Magnan, and Tavenas (1985), with relationships proposed by Skempton (1957) (line X) and Bjerrum (1973) (line Y) (•, organic content $< 3\%$; ○, organic content $> 3\%$).

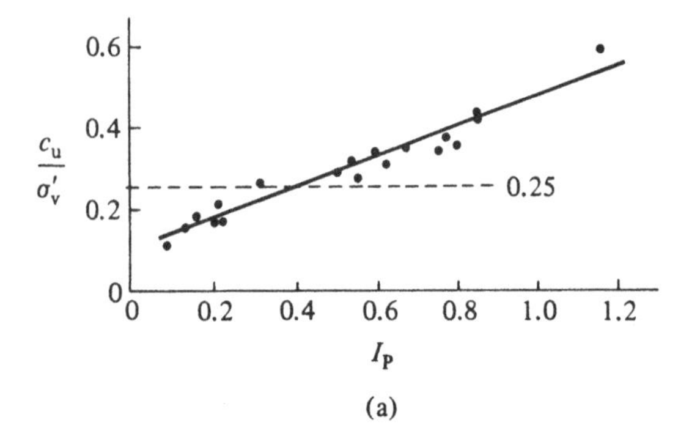

(a)

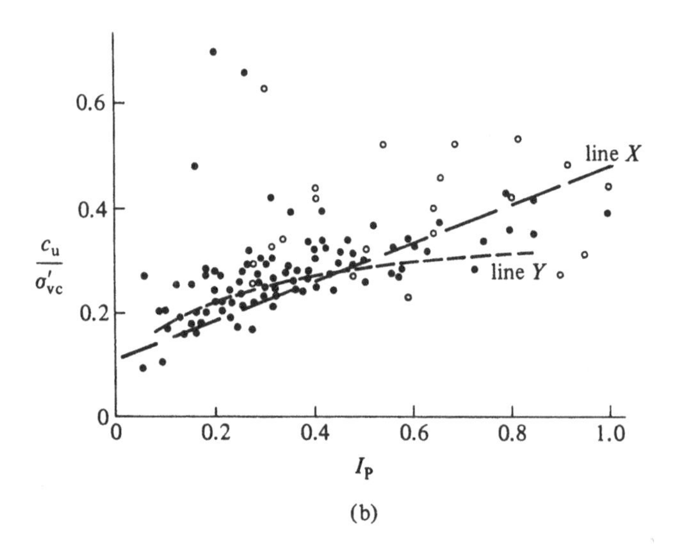

(b)

normal compression line. The consequences of assuming the relationship

$$\frac{c_u}{\sigma'_{vc}} = 0.25 \qquad (9.77)$$

to estimate strengths of clays are explored here.

An expression that is widely used in practice to assess strength data is that due to Skempton (1954b, 1957). Skempton showed data of c_u/σ'_v for a number of soils that were supposedly normally compressed (Fig. 9.39a) and deduced the relationship

$$\frac{c_u}{\sigma'_{vc}} = 0.11 + 0.37 I_P \qquad (9.78)$$

which links the ratio c_u/σ'_{vc} with plasticity I_P. The undrained strengths were peak values obtained from field vane tests. For normally compressed soils, the in situ vertical effective stress and the preconsolidation pressure are identical, $\sigma'_v = \sigma'_{vc}$, and in plotting the data for Fig. 9.39a, Skempton did not distinguish between them. With the gathering of more data, the scatter becomes greater and the nature of the dependence of c_u/σ'_{vc} on plasticity becomes less clear. Data collected by Leroueil, Magnan, and Tavenas (1985) are shown in Fig. 9.39b: here the undrained strengths are indeed divided by preconsolidation pressures. Expression (9.78) (line X) appears to provide a reasonable fit to the data.

Bjerrum (1972, 1973) analysed a large number of field records and proposed relationships between the strength ratio c_u/σ'_v and effective overconsolidation ratio σ'_{vc}/σ'_v which reduce to the curve shown in Fig. 9.39b (line Y) relating the strength ratio c_u/σ'_{vc} and plasticity index I_P. However, when many case histories of failures of embankments on soft clays are back-analysed, we find that the strengths estimated from Bjerrum's curve in Fig. 9.39b give a false estimate of stability: the factor of safety calculated at failure differs from unity by an amount dependent, roughly, on plasticity (Fig. 9.40a). Bjerrum (1972) drew an average straight line through his data of estimated factor-of-safety at failure (Fig. 9.40a) and then inverted this relationship to give a factor μ, which varied with plasticity (Fig. 9.40b), by which field vane strength data should be modified to give a better estimate of the strength which could be mobilised in the field. Bjerrum (1973) suggests that this μ factor may arise partly from effects of anisotropy of natural soils and partly from the dependence of strength on rate of shearing so the strength mobilised in a matter of seconds in vane tests is higher than the strength mobilised over a period of several weeks or months in embankment loading. Whatever the origin of this factor, when it is combined with Bjerrum's curve for variation of c_u/σ'_{vc} with plasticity (Fig. 9.39b), the resulting field strength ratio is almost

Fig. 9.40 (a) Variation of expected factor of safety at failure with plasticity index I_P, with relationship proposed by Bjerrum (1972) (line Z) (after Leroueil, Magnan, and Tavenas, 1985); (b) factor μ, for modifying measured vane shear strength, dependent on plasticity index I_P (after Bjerrum, 1972).

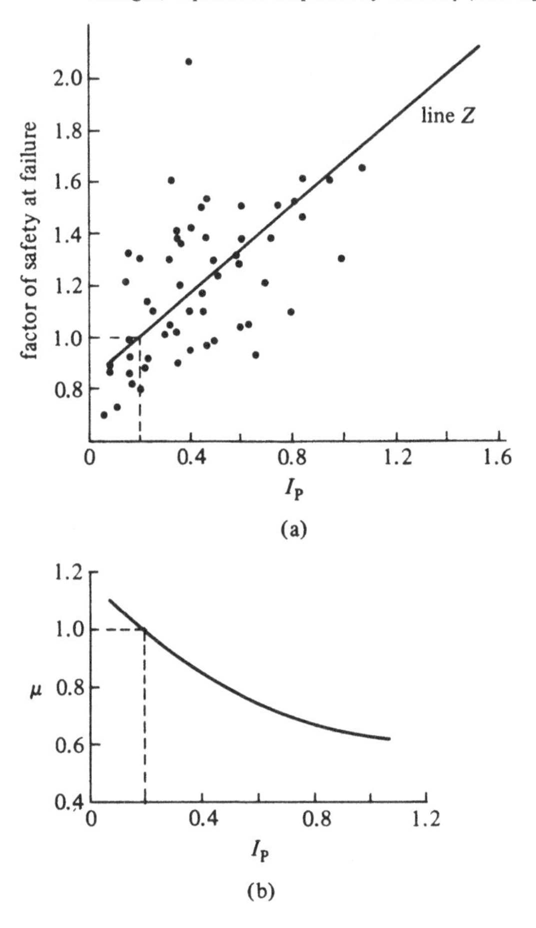

Fig. 9.41 Ratio of undrained strength to preconsolidation pressure (c_u/σ'_{vc}) obtained from combination of relationships proposed by Bjerrum (1973) (after Mesri, 1975).

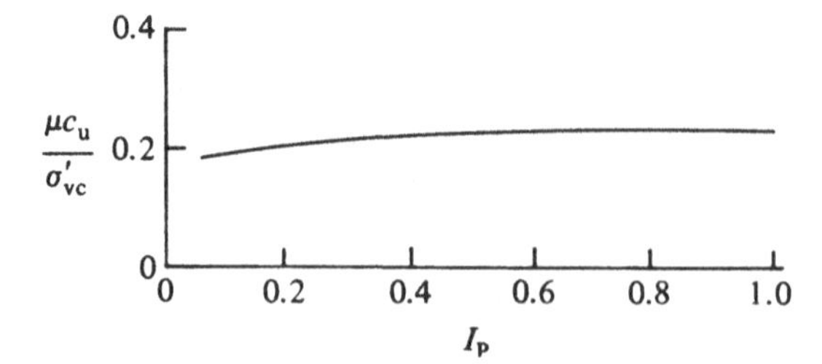

independent of plasticity (Fig. 9.41) with a roughly constant value of

$$\frac{c_u}{\sigma'_{vc}} \simeq 0.22 \tag{9.79}$$

as shown by Mesri (1975). Larsson's (1980) back analyses of field failures confirm that (9.79) provides a reasonable estimate of field strengths for inorganic clays. Practically, then, an expression similar to (9.77) is found to provide a better estimate of the strength which can be relied upon in the field than an expression such as (9.78), in which the strength ratio is allowed to increase with plasticity.

An obvious reason for the unjustified optimism of analyses based on peak vane strengths is precisely that these are peak strengths. The development of a failure beneath an embankment or foundation requires strains and deformations which take many elements of the soil well past their peak condition (Fig. 9.42), and an integrated peak strength along a failure surface is bound to give an unsafe result. A critical state strength measured at large deformation is more conservative. Limited data collected by Trak, LaRochelle, Tavenas, Leroueil, and Roy (1980) do indeed suggest (Fig. 9.43) that a strength ratio c_u/σ'_{vc} calculated using a critical state strength which they term the 'undrained strength at large strains' (USALS) has a value ~ 0.22 which is almost independent of plasticity.

The number 0.25 in (9.76) or (9.77) was an average number emerging from a calculation made using the Cam clay model. In fact, according to the Cam clay model, this number depends on the angle of friction of the soil. Figure 9.32 shows that as the angle of friction increases and the soil becomes stronger, the ratio σ'_v/c_u decreases and hence the inverse ratio

Fig. 9.42 Variation of shear strain γ and shear stress τ mobilised along failure surface.

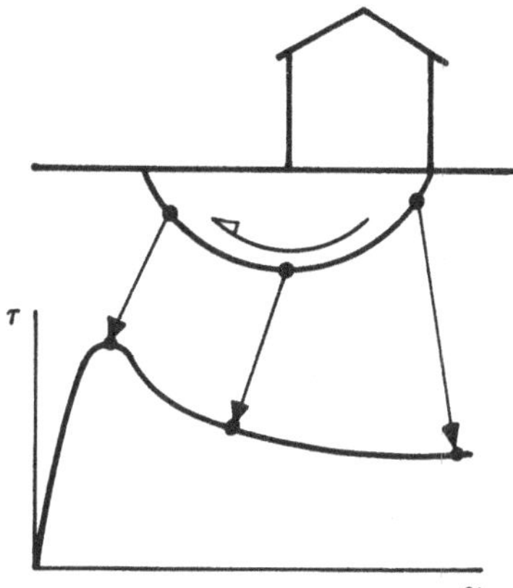

Fig. 9.43 Data of ratio of undrained strength at large strains to preconsolidation pressure, and plasticity index (after Trak, LaRochelle, Tavenas, Leroueil, and Roy, 1980).

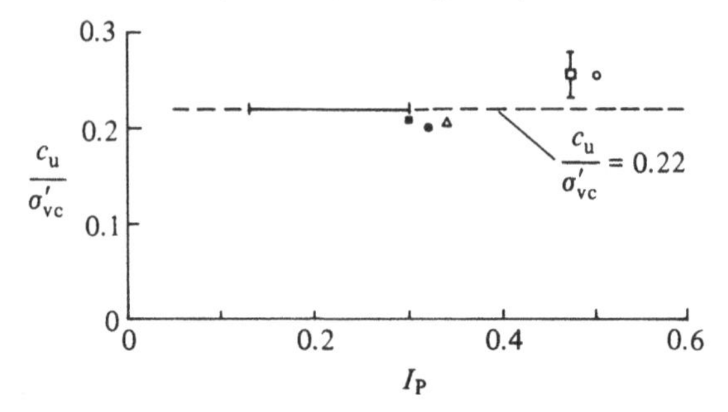

Fig. 9.44 Consequences of constant ratio of strength to preconsolidation pressure, on path followed during undrained shearing from one-dimensional normal compression line (1-D ncl) to critical state line (csl): (a) $\phi' = 20°$, (b) $\phi' = 25°$, (c) $\phi' = 30°$, and (d) $\phi' = 35°$.

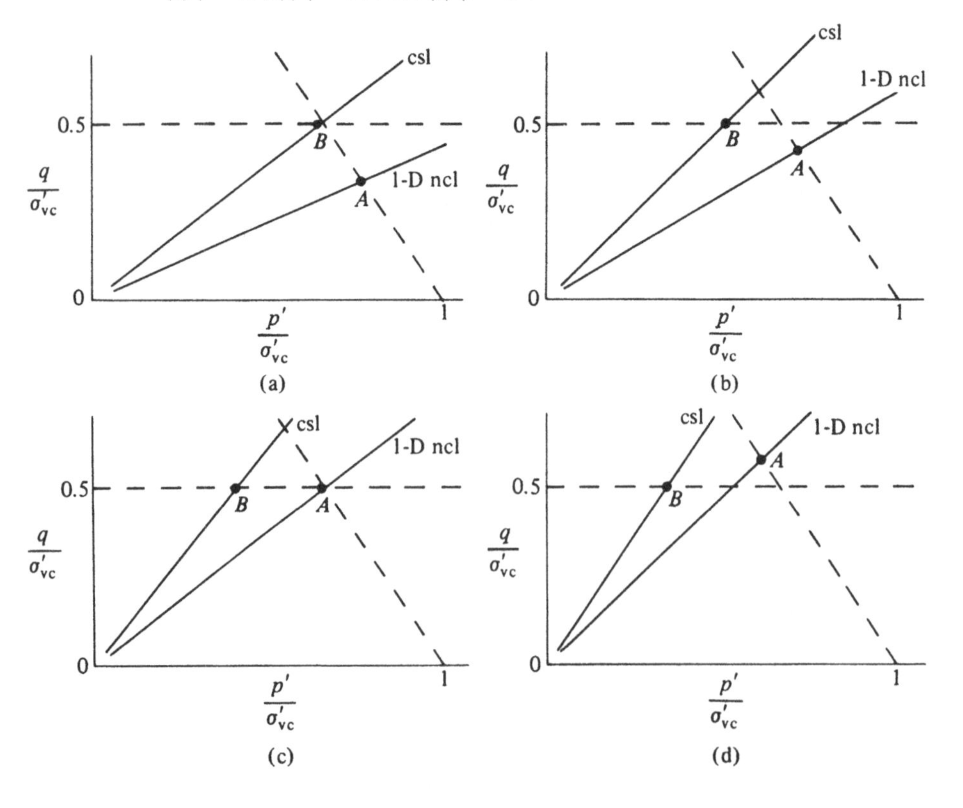

c_u/σ'_v increases. Angles of friction tend to fall with increasing plasticity (Fig. 9.14), so the Cam clay model on its own would produce a trend for variation of c_u/σ'_{vc} with plasticity I_P directly opposite to that described by expression (9.78) and to the general trend of the data in Fig. 9.39. The same applies to any other strength calculation based on friction (e.g. see Wroth, 1984).

If, in fact, field evidence supports a constant value of the ratio c_u/σ'_{vc} (calculated using critical state strengths) independent of plasticity and hence of angle of friction, then this has implications for the relative shapes of yield loci and plastic potentials to be used (Fig. 9.28) to describe the elastic–plastic behaviour of the soil. This can be demonstrated by considering stress states in a non-dimensional effective stress plane $p'/\sigma'_{vc}:q/\sigma'_{vc}$ (Fig. 9.44).

In one-dimensional normal compression, from (9.68) and (9.62),

$$\eta_{Knc} = \frac{3 \sin \phi'}{3 - 2 \sin \phi'} \tag{9.80}$$

Since the vertical effective stress is equal to the preconsolidation pressure $\sigma'_v = \sigma'_{vc}$, the initial in situ effective stress state must lie on the line

$$\frac{p'}{\sigma'_{vc}} + \frac{2}{3} \frac{q}{\sigma'_{vc}} = 1 \tag{9.81}$$

The initial effective stress state is fixed at the intersection of (9.80) and (9.81): points A in Fig. 9.44, plotted for $\phi' = 20°, 25°, 30°$, and $35°$.

Assume that failure is reached at large strains at a critical state,

$$\eta = M = \frac{6 \sin \phi'}{3 - \sin \phi'} \tag{9.22bis}$$

and also that from (9.77)

$$\frac{q}{\sigma'_{vc}} = \frac{2c_u}{\sigma'_{vc}} = 0.5 \tag{9.82}$$

This failure stress state is fixed at the intersection of (9.82) and (9.22): points B in Fig. 9.44. The path linking A and B cannot have the same shape for each of the four angles of friction, so it is an error to apply the Cam clay model without regard for soil type or plasticity and without regard for the marked difference between the shape of experimentally observed yield loci (in Chapter 3) and the assumed simple shape of the Cam clay yield loci (in Chapter 5).

In summary, the simple relationship

$$\frac{c_u}{\sigma'_{vc}} \simeq 0.25 \tag{9.77bis}$$

provides a reasonable indication of the way in which the undrained strength which can be usefully mobilised in the field varies with pre-consolidation pressure. Any attempt to refine this expression must, however, be based on a realistic numerical model of soil behaviour.

9.7 Conclusion

In this chapter a number of simple relationships between soil properties have been established against a background of critical state lines and the importance of volumetric parameters in controlling soil response. We have not intended to suggest that all sophisticated testing can be dispensed with but rather to show that simple predictive charts and formulae can be produced against which one can test the results of the apparently crude index tests and other data. Some of the assumptions that have been made and, in particular, some of the numerical values that have been chosen can be regarded with suspicion, but the patterns that have emerged are not really dependent on precise numerical values. If it is found that certain data in a collection of data from related soils appear to conflict with a pattern, then those data should either be rejected or be studied in more detail to explain the discrepancies.

Exercises

E9.1. A site investigation has been carried out for a proposed road bridge. The results of a conventional one-dimensional compression test performed on a sample taken from a depth of 3.6 m are given in Fig. 9.E1. For this sample the natural water content $w_i = 0.46$, the

Fig. 9.E1 Oedometer test on sample of silty clay.

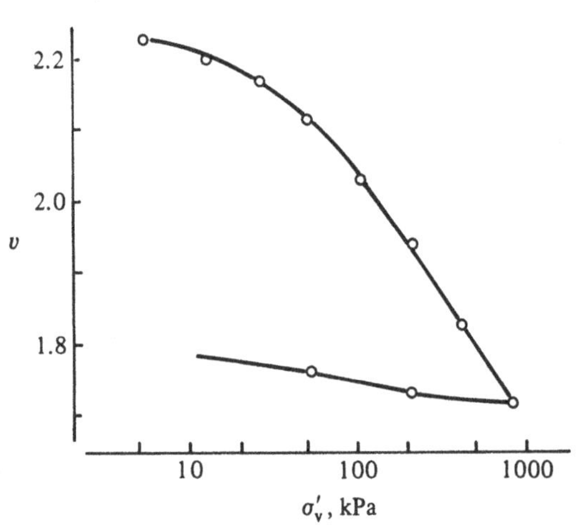

specific gravity of soil particles $G_s = 2.63$, the liquid limit $w_L = 0.46$, and plastic limit $w_P = 0.25$. The water table is at the ground surface. Estimate the in situ vertical effective stress at the depth of 3.6 m, and estimate the preconsolidation pressure and overconsolidation ratio from the results of the oedometer test.

The behaviour of the silty clay stratum from which the sample has been taken is to be represented by Cam clay. Adopting a value of $\phi' = 30°$, estimate suitable values of λ, κ, N, and Γ. Compare the value of λ with a value estimated from the given plasticity index.

Estimate the value of the undrained shear strength of the clay in triaxial compression. Compare this value with the measured strength of 9 kPa and with values estimated from knowledge of the plasticity of the soil and of the vertical effective stresses that it has experienced. Estimate the sensitivity of the clay and show that the liquidity index, vertical effective stress, and undrained strength data are broadly consistent.

E9.2. A bed of clay is being prepared for testing on a geotechnical centrifuge. The clay is prepared as a slurry at a water content twice the liquid limit and then compressed one-dimensionally under a uniform vertical effective stress which is increased in stages to 70 kPa. The block of clay is then 200 mm deep and is placed on the centrifuge and accelerated to 100 gravities with no surface loading but with the water table being maintained at the surface of the clay. The clay is allowed to reach equilibrium with zero excess pore pressure.

Estimate the variation of liquidity index, overconsolidation ratio, and undrained strength with depth through the clay.

E9.3. Obtain site investigation data from published papers or from files relating to a project with which you have been involved. Assess these data using the relationships between plasticity and compressibility, liquidity and strength, liquidity and vertical effective stress, and overconsolidation ratio that have been proposed in this chapter.

10

Stress paths and soil tests

10.1 Introduction

A complete description of the stress:strain behaviour of an elastic material can be embodied in a rather small number of parameters. For example, an isotropic linear elastic material (Section 2.1) requires only two independent parameters: Young's modulus and Poisson's ratio, or shear modulus and bulk modulus. A cross-anisotropic or transversely isotropic linear elastic material (Section 2.3) requires five parameters: which might be two Young's moduli, two Poisson's ratios, and a shear modulus. Once these elastic parameters have been determined, the response of the material to *any* changes in stress can be predicted. For the simple case of an isotropic elastic soil, the elastic parameters could be obtained from a uniaxial compression test, but the use of these parameters would in no way be confined to the prediction of the response in such tests. The testing programme required to determine the five elastic parameters for a cross-anisotropic elastic soil is rather more complex: simple compression tests in a triaxial apparatus are no longer sufficient, and tests which include rotation of principal axes are required (Graham and Houlsby, 1983). Nevertheless, once the values of the five parameters have been established, there is no limit to the range of stress paths to which the model can legitimately be applied.

A distinction has to be drawn between the description of the behaviour of inelastic soils and the description of the model which may be used to represent certain aspects of their behaviour. For it is certain that although simple elastic–plastic models of soil behaviour such as Cam clay may be very good at describing certain limited areas of soil response – those which are reasonably close to the data base of triaxial compression tests on which such models have typically been founded – their general success in matching the response of such a non-ideal material as soil is likely to be more doubtful.

Most of the data of soil behaviour that have been presented in previous chapters have been obtained from triaxial tests, primarily triaxial compression tests. In this chapter the stress paths that are actually followed by soil elements in certain typical geotechnical structures are considered. When such paths can be followed in a triaxial apparatus, then prediction of deformations, made with soil models based on triaxial test data, may be satisfactory. The further such paths deviate from the possibilities of the triaxial apparatus, the less satisfactory the predictions will be. The need to consider stress paths arises because real soils cannot be treated as isotropic or elastic materials: they must be treated as inelastic, anisotropic, and history-dependent materials whose behaviour cannot be described by a small number of elastic parameters. The general framework of elastic–plastic models of soil behaviour presented in Chapter 4 treats soils as just such non-ideal materials.

There are, of course, apparatus other than the triaxial apparatus and other tests which can be used to investigate soil response; some of these may match more closely the stress paths that are important in the ground. Each apparatus tests soil in its own way, and the relationships between the strengths of soils as measured by various tests are considered in Section 10.6.

The term *stress path* is familiar to many practising geotechnical engineers: It is linked with the stress path method of Lambe (1967). The stress path approach to predicting settlements of foundations requires triaxial tests on representative samples of soil which follow the stress changes estimated, usually from elastic analyses, to be relevant for selected elements of soil beneath the foundation. From the response observed in these tests, pseudo-elastic parameters can be deduced which are then applied to give numerical estimates of settlements.

This stress path method can in principle be extended to a wider range of geotechnical problems (Wood, 1984a). The necessary steps are

i. to identify critical soil elements around a structure,
ii. to estimate the stress paths followed by these elements as the structure is constructed or loaded,
iii. to perform laboratory tests on soil samples along these paths, and
iv. to estimate the deformations of the geotechnical structure from the results of these tests.

The problems of attempting to use this method are, however, that the actual stress paths followed are not independent of the very soil behaviour that is being investigated, that it may not be possible to follow the relevant stress paths in laboratory tests, and that the method of working from the results of the tests to the estimated performance of the structure may not be obvious. Nevertheless, consideration of stress paths does impose a discipline on engineers which should help them to identify the shortcomings of the tests that they are able to perform, and to assess the relevance of the numerical models of soil behaviour which they may wish to use to make their predictions.

10.2 Display of stress paths

Most of the data on soil behaviour that have been discussed here were obtained from triaxial tests in which conditions of axial symmetry were necessarily imposed. Axially symmetric stress paths have just two degrees of freedom: an axial principal stress σ_a and a radial principal stress σ_r (Fig. 10.1), and it has been convenient to display triaxial stress paths in the $p':q$ plane to emphasise the distinction between volumetric and distortional components of stress.

Conditions of axial symmetry occur infrequently in prototype situations; conditions of plane strain are more likely. As discussed in Section 1.5, plane strain effective stress paths can be displayed in terms of stress quantities s' and t,

$$s' = \frac{\sigma'_1 + \sigma'_3}{2} \qquad (10.1)(\text{cf. } 1.58)$$

$$t = \frac{\sigma'_1 - \sigma'_3}{2} \qquad (10.2)(\text{cf. } 1.60)$$

which are calculated in terms of the principal effective stresses σ'_1 and σ'_3 in the plane of shearing. The stress quantities s' and t are particularly useful for describing stress changes which are not associated with significant rotation of principal axes, and most of the stress paths presented in Section 10.4 fall into this category.

The stress quantities s' and t cannot convey information about the orientation of principal stresses σ'_1 and σ'_3 or about the magnitude of the

Fig. 10.1 Stresses on cylindrical triaxial sample.

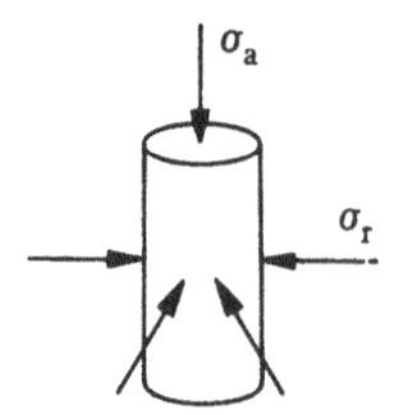

Fig. 10.2 General stress state specified by (a) normal and shear stresses referred to fixed axes and (b) principal stresses and principal axes.

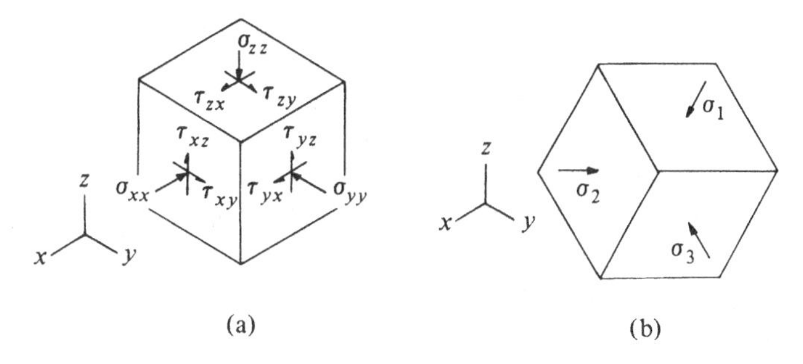

Fig. 10.3 (a) General mean stress p' and deviator stress q in principal effective stress space; (b) cube with sides defined by principal stress axes $\sigma'_1 : \sigma'_2 : \sigma'_3$; (c) assignment of major, intermediate, and minor principal stresses in sectors of deviatoric view of principal stress space.

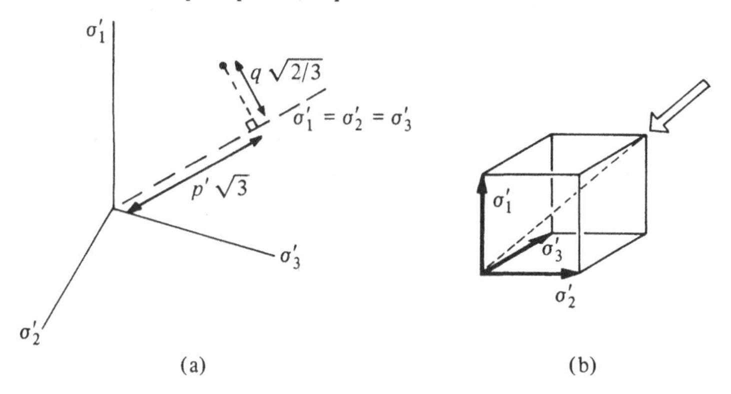

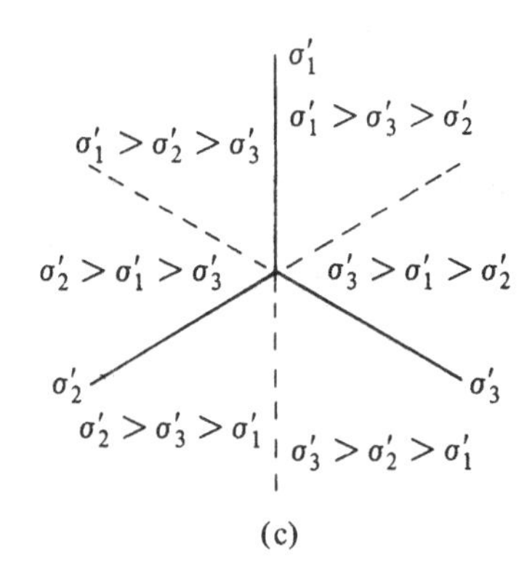

third principal stress σ'_2, orthogonal to the plane of shearing. Plane strain is a special case though it represents a situation which is met in practice rather more frequently than axial symmetry. A completely general stress state requires six quantities to describe it: these may be three normal stresses and three associated shear stresses (Fig. 10.2a) or three principal stresses and the directions of their three mutually orthogonal principal axes (Fig. 10.2b). In general, at a soil element in the ground, all six stress quantities are changing simultaneously. Six quantities are needed to describe the resulting strain increment, and there is no reason to suppose that principal axes of strain increment and principal axes of stress or of stress increment necessarily coincide.

Possibilities for controlled laboratory investigation of such general stress paths are, not surprisingly, limited. It is also unclear how such paths should best be displayed. This last problem is largely avoided here by restricting attention to paths in which no rotation of principal axes occurs.

The set of principal stresses $(\sigma'_1, \sigma'_2, \sigma'_3)$, where the subscripts (1, 2, 3,) do not signify any particular sequence of relative magnitudes, can be used as orthogonal cartesian axes to define a three-dimensional space (Fig. 10.3a) which can be viewed through two-dimensional projections. The most frequently used projection is the deviatoric or π-plane view (Fig. 10.3c), a view of principal stress space down the line $\sigma'_1 = \sigma'_2 = \sigma'_3$ (Fig. 10.3b). This view only shows differences of principal stresses, so it is important to remember that this is a view of *projected* information and that the paths shown do not in general lie in the planes depicted.

10.3 Axially symmetric stress paths

10.3.1 One-dimensional compression of soil

Some soils have been deposited rather uniformly over an area of large lateral extent, for example, in marine or lacustrine conditions. For such soils, symmetry dictates that soil particles can only have moved downwards during the process of deposition (Fig. 1.21); lateral movements would violate the symmetry. The deformation of such soils during deposition is entirely one-dimensional, and the effective stress state can be reproduced in a conventional triaxial apparatus.

The ratio of horizontal to vertical effective stresses in soils which have a history of one-dimensional deformation (which may include some overconsolidation) is called the *coefficient of earth pressure at rest* K_0:

$$\frac{\sigma'_h}{\sigma'_v} = K_0 \tag{10.3}$$

During monotonic one-dimensional normal compression, each state of deformation of a soil is essentially similar to all the preceding states, and the effective stress states have the same similarity. The value of K_0 is then found to be a constant $K_0 = K_{0nc}$ (Fig. 10.4a).

From the definitions of p' and q, the slope η_{Knc} of the one-dimensional normal compression line in the $p':q$ plane (Fig. 10.4b) is related to K_{0nc}:

$$\eta_{Knc} = \frac{3(1 - K_{0nc})}{1 + 2K_{0nc}} \tag{10.4a}$$

and

$$K_{0nc} = \frac{3 - \eta_{Knc}}{3 + 2\eta_{Knc}} \tag{10.4b}$$

The Cam clay model can be used to calculate values of η_{Knc}. The condition for one-dimensional compression is that

$$\frac{\delta\varepsilon_p}{\delta\varepsilon_q} = \frac{\delta\varepsilon_p^e + \delta\varepsilon_p^p}{\delta\varepsilon_q^e + \delta\varepsilon_q^p} = \frac{3}{2} \tag{10.5}$$

The ratio of all stress components remains constant during such one-dimensional normal compression, and hence

$$\frac{\delta q}{\delta p'} = \frac{q}{p'} = \eta_{Knc} \tag{10.6}$$

and

$$\frac{\delta q}{q} = \frac{\delta p'}{p'} \tag{10.7}$$

The elastic and plastic components of strain during one-dimensional

Fig. 10.4 One-dimensional loading in (a) $\sigma'_h:\sigma'_v$ and (b) $p':q$ effective stress planes

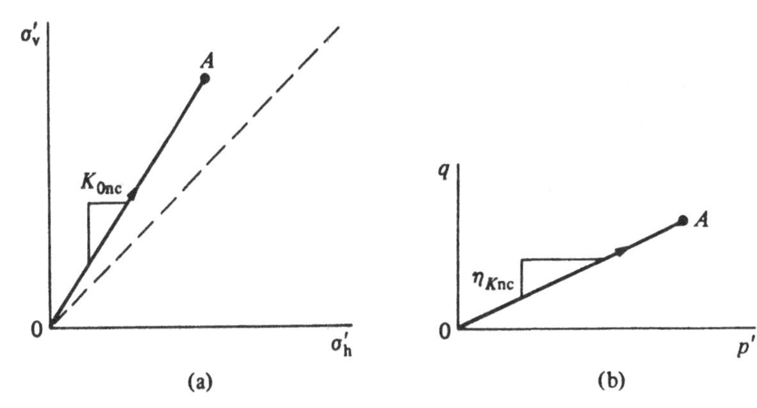

normal compression are, from the expressions in Section 5.2, with (10.7),

$$\delta\varepsilon_p^e = \frac{\kappa}{v}\frac{\delta p'}{p'} \tag{10.8}$$

$$\delta\varepsilon_q^e = \frac{2(1+\nu')}{9(1-2\nu')}\frac{\eta_{Knc}\kappa}{v}\frac{\delta p'}{p'} \tag{10.9}$$

$$\delta\varepsilon_p^p = \frac{\lambda-\kappa}{v}\frac{\delta p'}{p'} \tag{10.10}$$

$$\delta\varepsilon_q^p = \frac{2\eta_{Knc}}{M^2-\eta_{Knc}^2}\frac{\lambda-\kappa}{v}\frac{\delta p'}{p'} \tag{10.11}$$

Two elastic parameters were used in the Cam clay model in Section 5.2: the slope κ of the unloading–reloading line and the shear modulus G'. It is more convenient here to use Poisson's ratio ν' because this immediately expresses a ratio of components of elastic strains. The links between the various elastic parameters were given in Section 2.1.

Expression (10.5) is a condition on total strain increments into which (10.8)–(10.11) can be substituted to give the expression

$$\frac{\eta_{Knc}(1+\nu')(1-\Lambda)}{3(1-2\nu')} + \frac{3\eta_{Knc}\Lambda}{M^2-\eta_{Knc}^2} = 1 \tag{10.12}$$

from which values of η_{Knc} can be determined numerically for any set of values of ν', Λ $(=(\lambda-\kappa)/\lambda)$, and M. The first term on the left-hand side broadly expresses the elastic contribution, which is the entire contribution for $\Lambda = 0$. The second term expresses the plastic contribution.

Values of η_{Knc} can be converted to values of K_{0nc} using (10.4b), and the effect of different combinations of ν' and Λ on the variation of K_{0nc} with angle of shearing resistance ϕ' [which can be linked to M through (7.9)] is shown in Fig. 10.5a.

A soil that is composed of rigid, greatly interlocked particles can support its own weight without needing to push sideways very much to prevent lateral movement: such a soil would be expected to have a low value of K_{0nc}. Such a soil would also be able to mobilise a high angle of shearing resistance before shear deformations occurred. A material which has no frictional strength, such as water, produces a lateral push equal to the overburden pressure: in other words, $K_{0nc} = 1$. Some link between K_{0nc}

Fig. 10.5 Dependence of coefficient of earth pressure at rest for normally compressed soil, K_{0nc}, on angle of shearing resistance ϕ', (a) according to Cam clay model, (b) according to Jâky (1944), and (c) experimental data (•, clays; +, sands) (after Wroth, 1972; Ladd, Foott, Ishihara, Schlosser, and Poulos, 1977).

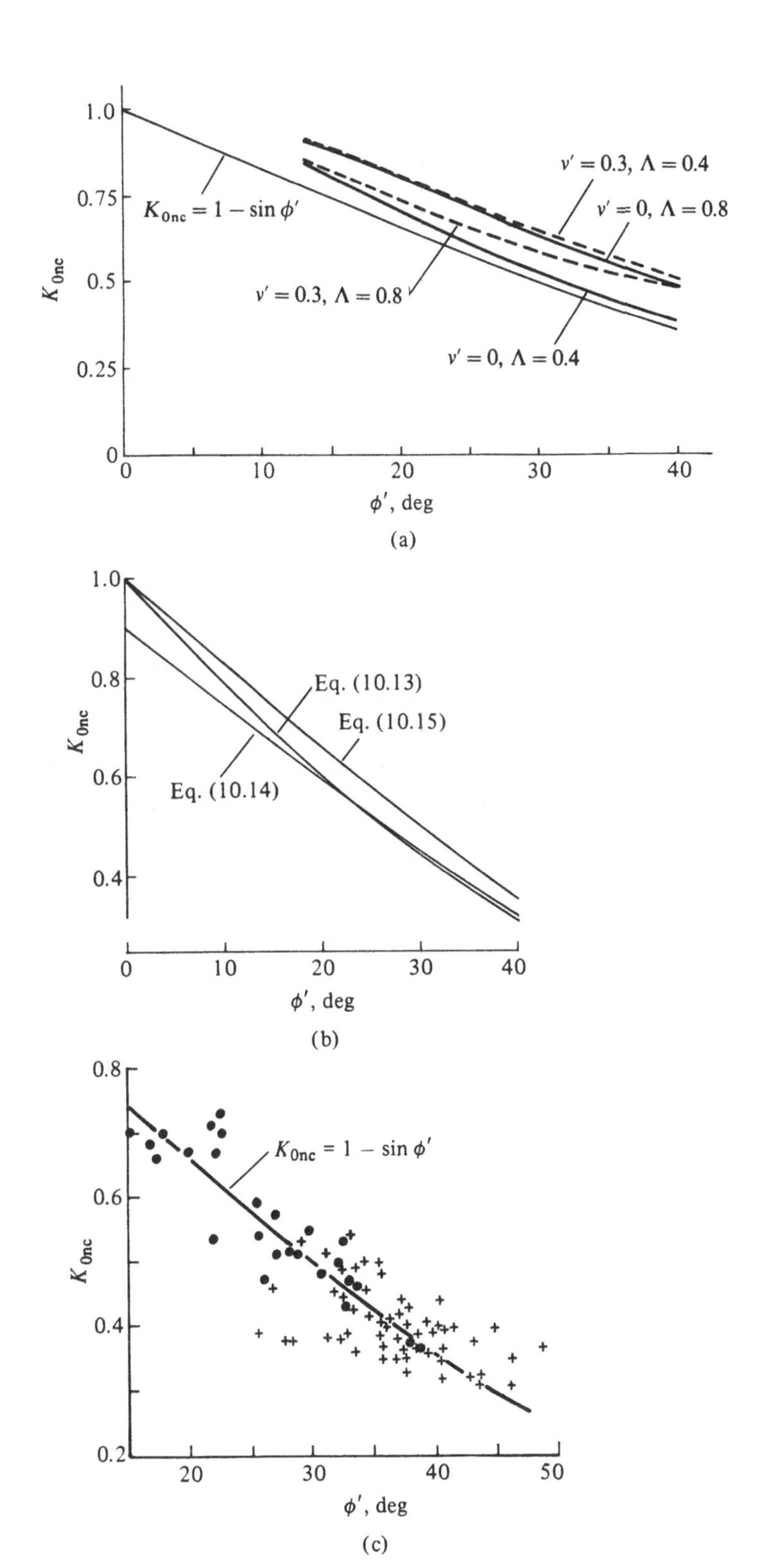

1.0
0.75
0.5
0.25
0
K_{0nc}
$K_{0nc} = 1 - \sin\phi'$
$\nu' = 0.3,\ \Lambda = 0.4$
$\nu' = 0,\ \Lambda = 0.8$
$\nu' = 0.3,\ \Lambda = 0.8$
$\nu' = 0,\ \Lambda = 0.4$
0
10
20
30
40
ϕ', deg
(a)
1.0
0.8
0.6
0.4
K_{0nc}
Eq. (10.13)
Eq. (10.15)
Eq. (10.14)
0
10
20
30
40
ϕ', deg
(b)
0.8
0.6
0.4
0.2
K_{0nc}
$K_{0nc} = 1 - \sin\phi'$
20
30
40
50
ϕ', deg
(c)

and angle of shearing resistance ϕ' might be anticipated. Jâky (1944) managed to arrive at the expression

$$K_{0nc} = \left(1 + \frac{2}{3}\sin\phi'\right)\left(\frac{1 - \sin\phi'}{1 + \sin\phi'}\right) \tag{10.13}$$

by thinking of conditions at the centre of the base of a heap of granular material. Jâky noted that this elaborate expression could be approximated satisfactorily by the expression

$$K_{0nc} \approx 0.9(1 - \sin\phi') \tag{10.14}$$

for values of ϕ' between about 20° and 45°, but also suggested that the expression

$$K_{0nc} = 1 - \sin\phi' \tag{10.15}$$

would be close enough for engineering purposes. This final expression has become enshrined in the soil mechanics canon with Jâky's name attached to it. The three expressions (10.13), (10.14), and (10.15) are plotted in Fig. 10.5b, and expression (10.15) is also plotted in Fig. 10.5a.

The experimental conditions necessary for the accurate measurement of K_0 are not easy to realise in practice (Bishop, 1958); even very small amounts of lateral movement can produce a significant change in the apparent value of K_0. Data of values of K_{0nc} and ϕ' for normally compressed clays and sands collected by Wroth (1972) and by Ladd et al. (1977) are shown in Fig. 10.5c, and (10.15) provides a reasonable fit to these points. Cam clay tends to predict values of K_{0nc} which lie above the empirical expression (Fig. 10.5a) unless simultaneous low values of both ν' and Λ are assumed (implying dominance of the deformation by low Poisson's ratio elastic response).

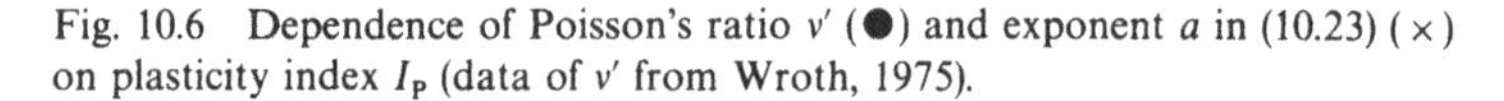

Fig. 10.6 Dependence of Poisson's ratio ν' (●) and exponent a in (10.23) (×) on plasticity index I_P (data of ν' from Wroth, 1975).

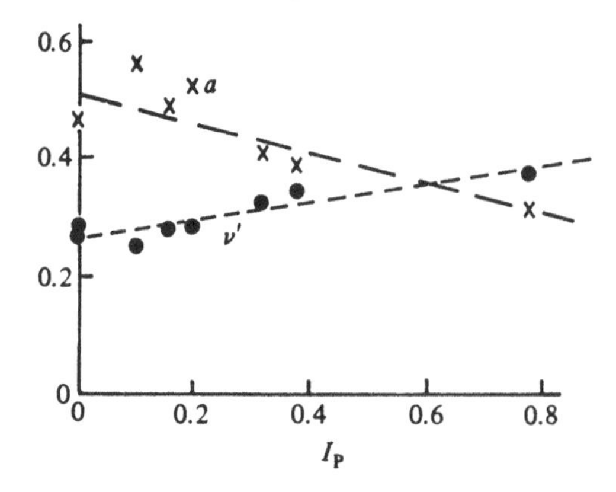

The nature of the angle ϕ' that should be used in (10.15) is somewhat uncertain. It is sometimes found that the angles of shearing resistance that can be mobilised in conditions of plane strain are rather higher than those that can be mobilised in triaxial compression. (Wroth, 1984, suggests that plane strain angles are typically 10 per cent higher.) Perhaps on grounds of the axial symmetry of one-dimensional deposition, a logical argument could be made in favour of triaxial compression. However, theoretical justifications of (10.15) cannot proceed very far, and it should rather be regarded as a convenient experimental finding. Angles of shearing resistance are most commonly measured in triaxial compression, and (10.15) with Fig. 10.5c can be regarded as establishing a simple correlation between these angles and K_{0nc}.

Many of the data in Fig. 10.5c relate to values of K_{0nc} and ϕ' for sands tested at a range of initial densities. It was noted in earlier chapters that it is difficult to change the structure of sand in proportional loading tests (in which all components of stress increase by the same proportion), of which one-dimensional compression is one example. When a sand is sheared, for example in triaxial compression, its void ratio changes until eventually the sand reaches a critical state condition, mobilising the critical state angle of shearing resistance. However, the value of the earth pressure coefficient at rest, K_{0nc}, depends on the in situ structure of the sand and can be expected to correlate with the peak angle of shearing resistance measured in triaxial compression, associated with the initial density and structure of the sand. It would not be appropriate to use the critical state angle in (10.15) to estimate the value of K_{0nc} for a dense sand.

In the elastic–plastic models of soil behaviour discussed in Chapters 4 and 5, volume changes occurring on a normal compression line were described by an expression

$$v = v_\lambda - \lambda \ln p' \qquad (10.16)(4.2\text{bis})$$

where v_λ is a reference value of specific volume v. During one-dimensional normal compression, mean effective stress p' and vertical effective stress σ'_v increase in a constant ratio:

$$\frac{p'}{\sigma'_v} = \frac{1 + 2K_{0nc}}{3} \qquad (10.17)$$

Consequently, changes in volume during one-dimensional normal compression can be described by

$$v = v'_\lambda - \lambda \ln \sigma'_v \qquad (10.18)$$

(where v'_λ is a different reference value of v). The compressibility λ that is determined is the same whether these normal compression data are plotted in terms of σ'_v or of p'.

10.3.2 One-dimensional unloading of soil

One-dimensional unloading of soil produces a more rapid drop of vertical effective stress than of horizontal effective stress. If it is supposed that soil behaves isotropically and elastically immediately on unloading, then the slope of this overconsolidation unloading stress path in the $p':q$ plane can be deduced from (10.12), setting $\Lambda = 0$ (which implies that there are no plastic deformations):

$$\frac{\delta q}{\delta p'} = \frac{3(1-2\nu')}{1+\nu'} \tag{10.19}$$

which implies

$$\frac{\delta \sigma'_{\text{h}}}{\delta \sigma'_{\text{v}}} = \frac{\nu'}{1-\nu'} \tag{10.20}$$

Wroth (1975) has deduced values of ν' from data of one-dimensional unloading of a number of soils; these are plotted against the plasticity of the soils in Fig. 10.6. The value of apparent Poisson's ratio is typically around 0.3 but rises slightly with increasing soil plasticity. A value $\nu' = 0.3$ in (10.20) implies a stress decrement ratio $\delta\sigma'_{\text{h}}/\delta\sigma'_{\text{v}} = 0.43$, whereas for most of the soils (particularly the plastic soils) shown in Fig. 10.5c, $K_{0\text{nc}}$ is above 0.5: the stress path followed on one-dimensional unloading does not retrace the stress path followed on one-dimensional normal compression. Typical stress paths are shown in Figs. 10.7a, b in $\sigma'_{\text{h}}:\sigma'_{\text{v}}$ and $p':q$ effective stress planes: OA for one-dimensional normal compression, AB for initial one-dimensional unloading.

Expression (10.20) can be converted into a relationship between K_0 and

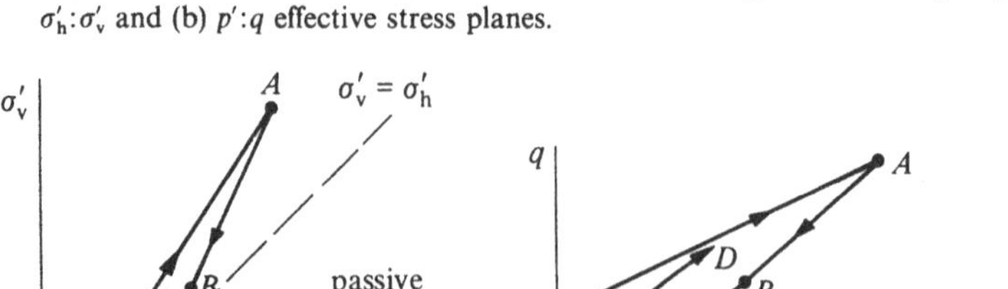
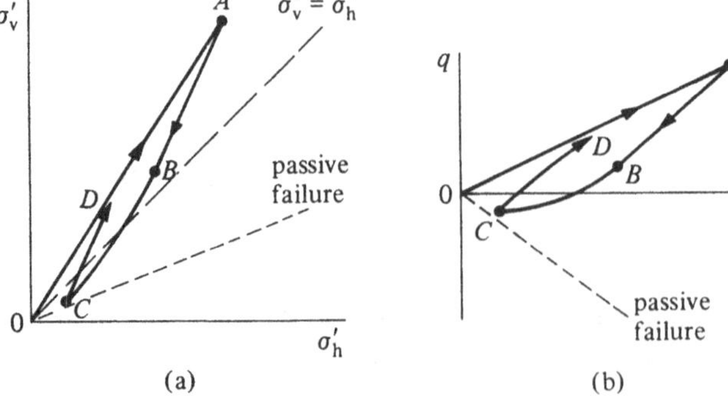

Fig. 10.7 Path followed on one-dimensional unloading and reloading in (a) $\sigma'_{\text{h}}:\sigma'_{\text{v}}$ and (b) $p':q$ effective stress planes.

overconsolidation ratio n:

$$K_0 = nK_{0nc} - (n-1)\frac{\nu'}{1-\nu'} \tag{10.21}$$

where

$$n = \frac{\sigma'_{vmax}}{\sigma'_v} \tag{10.22}$$

and σ'_{vmax} is the maximum value of σ'_v reached in one-dimensional normal compression.

The simple assumption of elastic unloading leading to the straight unloading stress path AB in Fig. 10.7 cannot be sustained beyond overconsolidation ratios of about 2; experimental measurements during one-dimensional unloading show significant curvature of the effective stress path BC (Fig. 10.7). A simple expression has been found for this unloading stress path. Schmidt (1966) plotted K_0 against overconsolidation ratio n using logarithmic axes for both. He found linear relationships such as that shown in Fig. 10.8, implying a relationship between K_0 and n of the form

$$K_0 = K_{0nc} n^a \tag{10.23}$$

The parameter a shows a small decrease with increasing plasticity (Fig. 10.6), but a value $a = 0.5$, as suggested by Meyerhof (1976), would be a reasonable round number to use for most soils, implying

$$\left(\frac{K_0}{K_{0nc}}\right)^2 = n \tag{10.24}$$

An example of the application of (10.15), (10.21), and (10.24) is shown

Fig. 10.8 Relationship between coefficient of earth pressure at rest K_o and overconsolidation ratio n for Weald clay (after Schmidt, 1966).

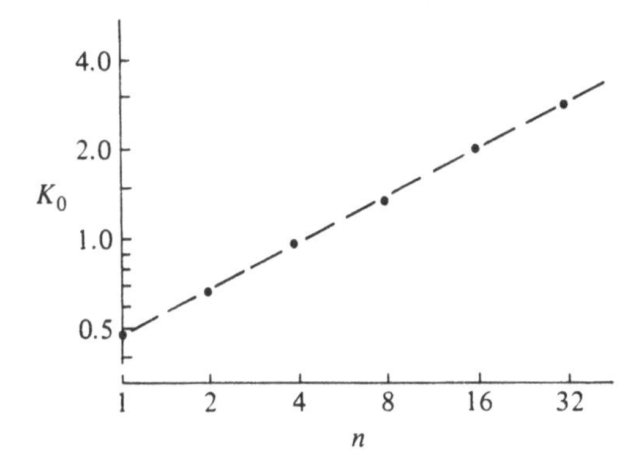

in Fig. 10.9. Ladd (1965) gives data of K_0 and overconsolidation ratio n for undisturbed Boston blue clay. This clay has plasticity $I_P = 0.15$ and an angle of shearing resistance (at maximum stress ratio) $\phi' = 32.75°$, which is the same value that would be predicted from expression (9.23). From (10.15), then, $K_{0nc} = 0.46$. For $I_P = 0.15$ in Fig. 10.6, $\nu' = 0.29$. With these values, the values calculated from expressions (10.21) and (10.24) are shown in Fig. 10.9 together with Ladd's data. Equation (10.24) provides a good fit, though of course with ν' set to an alternative value of 0.22 (deduced from the experimental observations for the Boston blue clay itself) a good initial fit could be obtained with expression (10.21), as also shown in Fig. 10.9.

An overconsolidation ratio n_p can be defined in terms of mean effective

Fig. 10.9 Observations and calculations of relationship between coefficient of earth pressure at rest K_0 and overconsolidation ratio n for Boston blue clay (data from Ladd, 1965).

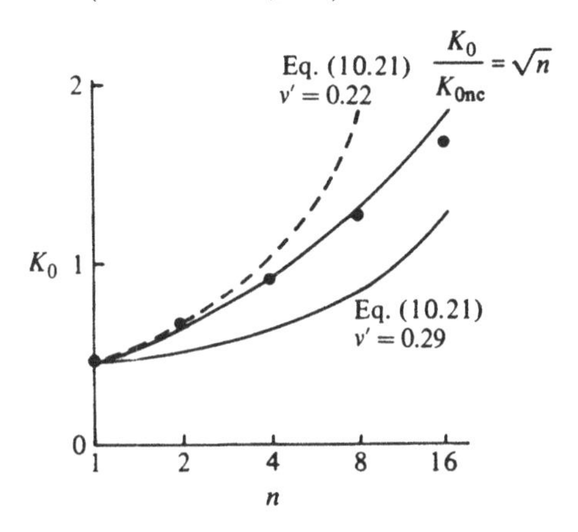

Fig. 10.10 Relationships between isotropic overconsolidation ratio n_p and overconsolidation ratio n.

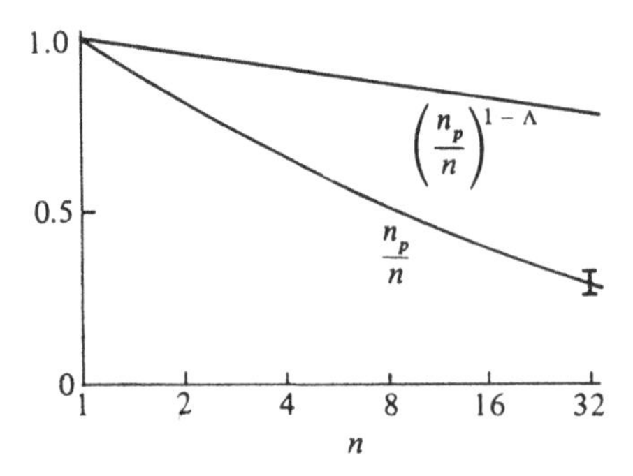

stress p':

$$n_p = \frac{p'_{max}}{p'} \tag{10.25}$$

where p'_{max} is the maximum value of p' reached in normal compression. Then, by comparison with (10.17) and (10.22),

$$\frac{n_p}{n} = \frac{1 + 2K_{0nc}}{1 + 2K_0} \tag{10.26}$$

Since K_0 is a function of n, this ratio n_p/n is not a constant, even for a particular soil. However, if the link between K_0 and n is provided by (10.24), then the variation of n_p/n with n can be deduced (Fig. 10.10). It is found to be rather insensitive to the value of K_{0nc}; the range for $n = 32$ in Fig. 10.10 corresponds to the range of values of K_{0nc} from 0.4 to 0.7 [$\phi' = 36.9°$ to $17.5°$ from (10.15)].

However, if it is assumed (as in the elastic–plastic models of Chapters 4 and 5) that recoverable volume changes occurring for stress paths lying within a current yield locus can be described by

$$v = v_\kappa - \kappa \ln p' \tag{10.27)(4.3bis}$$

then a simultaneous linear relationship between v and $\ln \sigma'_v$ with the same slope κ is not possible.

Expression (10.27) can be written as

$$v = v_c + \kappa \ln n_p \tag{10.28}$$

or

$$v = v_c + \kappa \ln n + \kappa \ln\left(\frac{n_p}{n}\right) \tag{10.29}$$

or

$$v = v_d - \kappa \ln \sigma'_v + \kappa \ln\left(\frac{n_p}{n}\right) \tag{10.30}$$

where

$$v_c = v_\kappa - \kappa \ln p'_{max} \tag{10.31}$$

and

$$v_d = v_c + \kappa \ln \sigma'_{vmax} \tag{10.32}$$

The effect of the changing ratio n_p/n on the volume changes occurring during one-dimensional unloading can best be illustrated by plotting (10.29) as $(v - v_c)/\kappa$ against n (Fig. 10.11). In (10.29), $(v - v_c)$ is the change in volume that has occurred since the vertical stress was reduced from its maximum normally compressed value σ'_{vmax}. The line

$$\frac{v - v_c}{\kappa} = \ln n \tag{10.33}$$

is shown in Fig. 10.11 for comparison. The curvature is not great, but because n increases more rapidly than n_p, as σ'_v falls more rapidly than p' (which depends also on σ'_h), the value of κ that would be deduced from one-dimensional unloading in an oedometer (in which only the vertical stress is usually measured) is only about 0.7 of the value of κ properly calculated in terms of change in mean effective stress p'. This must be taken into account if the only data of unloading that are available for a soil have been obtained from oedometer tests.

In Section 7.2, the form of experimentally observed critical state lines was used to predict the variation with overconsolidation of the ratio of undrained strength to in situ vertical effective stress. The expression that was deduced,

$$\frac{(c_u/\sigma'_{vi})}{(c_u/\sigma'_{vi})_{nc}} = \left(\frac{n}{n_p}\right) n_p^{\Lambda} \tag{10.34)(7.31bis}$$

involved both n and n_p. The ratio $(n_p/n)^{1-\Lambda}$ is plotted in Fig. 10.10 for $\Lambda = 0.8$; it falls gently from 1 to ~ 0.8 as n rises from 1 to 32. If it is assumed that

$$\left(\frac{n_p}{n}\right)^{1-\Lambda} \sim 1 \tag{10.35}$$

Fig. 10.11 Predicted volume change on one-dimensional unloading of soil.

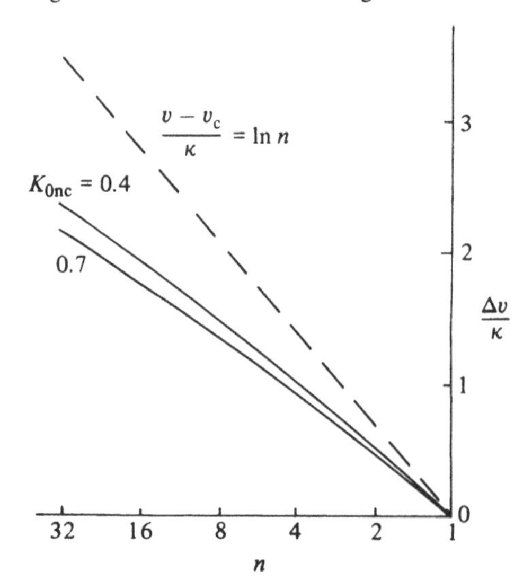

then, from (10.34),

$$\frac{(c_u/\sigma'_{vi})}{(c_u/\sigma'_{vi})_{nc}} \sim n^{\Lambda} \tag{10.36}$$

and, as noted in Section 7.2, this is the form of relationship that has been successfully fitted to published data.

The earth pressure coefficient at rest K_0 increases as a soil becomes progressively more overconsolidated. The horizontal effective stress falls more slowly than the vertical effective stress; and at some overconsolidation ratio, K_0 becomes greater than 1, and the horizontal stress becomes the major principal stress. The overconsolidation ratio n (or n_p) at which this occurs depends on the value of K_{0nc} according to (10.24); these values are plotted in Fig. 10.12.

The Mohr–Coulomb failure criterion discussed in Section 7.1 sets an upper limit to the ratio of major and minor principal effective stresses, σ'_{I} and σ'_{III}, respectively. For a soil with angle of shearing resistance ϕ', the maximum ratio occurs when the Mohr circle touches the failure line (Fig. 7.1a) and

$$\frac{\sigma'_{I}}{\sigma'_{III}} = \frac{1+\sin\phi'}{1-\sin\phi'} \tag{10.37}$$

In the one-dimensionally unloaded ground, this ratio is expected to set a limit to the value of K_0 that can be attained. When this limit is reached, then the soil is in a state of incipient passive failure, a mode of failure

Fig. 10.12 Values of overconsolidation ratio predicted to give earth pressure coefficient at rest K_0 equal to 1 and equal to passive pressure coefficient K_p as function of angle of shearing resistance ϕ'.

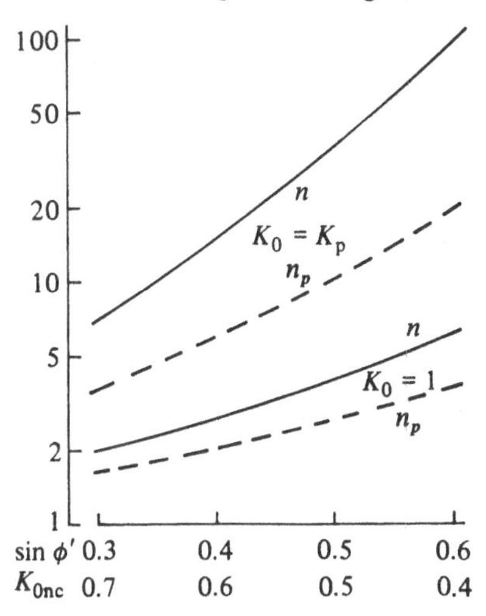

being driven by the high horizontal stresses against the weight of the soil. The values of n and n_p at which (10.37) and (10.24) coincide are also plotted in Fig. 10.12. For a soil with a high value of $K_{0nc} = 0.7$ [corresponding to $\phi' = 17.5°$ from (10.15)] and probable high plasticity, the required overconsolidation ratio n is only about 7. For a soil with a low value of $K_{0nc} = 0.4$ ($\phi' = 36.9°$) and probable low or zero plasticity, the required overconsolidation ratio is about 100.

Of course, the continuing one-dimensional symmetry of the loading condition in principle prevents a passive failure from taking place. However, in practice, if a heavily overconsolidated soil deposit is sitting in such a state of incipient passive failure, then any attempt to upset the symmetry releases the failure and large deformations may occur. For example, a deep excavation (Fig. 10.13a) may remove the vertical effective stress locally and lead to ground movements into the base of the hole. Nature, too, can upset the symmetry by eroding a river and valley. This again removes the vertical effective stress locally and allows the underlying strata to bulge in the bottom of the valley; a typical example in which such valley bottom bulge can be detected in the deformed strata is shown in Fig. 10.13b. Once such a passive failure has been released, there is a

Fig. 10.13 (a) Heave of base of excavation in heavily overconsolidated clay; (b) valley bottom bulge at site of Empingham Dam, Rutland (after Horswill and Horton, 1976).

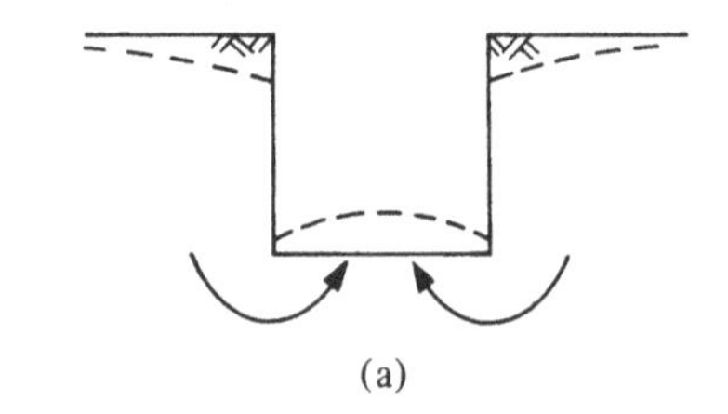
(a)

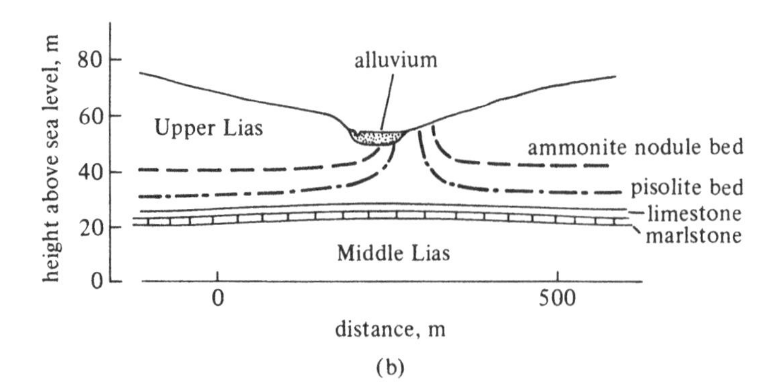

(b)

danger that the remaining soils may contain failure planes on which movements sufficient to reduce the strength of the soil to a residual value (Section 7.7) have occurred. Such a possibility must be taken into account in assessing the stability of any proposed structure, for example, an embankment dam at the site shown in Fig. 10.13b, which is to be founded on the soil.

The effective stress paths followed on one-dimensional unloading might be expected to become tangential to the passive failure line at high overconsolidation ratios (Fig. 10.7), though the curve plotted in Fig. 10.12 shows that these overconsolidation ratios are not unreasonably high for plastic soils. When soils are again re-loaded one-dimensionally, the value of K_0 again decreases, and the effective stress path tends towards the original stress path for the one-dimensionally normally compressed soil (*CD* in Fig. 10.7). Evidently, the value of K_0 for a soil deposit which is known to have been cyclically unloaded and reloaded can only be estimated within the bounds set by the normal compression line *OA* and the primary unloading curve *ABC*, starting from the preconsolidation state at *A*.

10.3.3 *Fluctuation of water table*

Parry (1970) notes that most natural deposits of soft clays are lightly overconsolidated, which is to say, with the background of Chapter 3, that the current effective stress state lies inside the current yield locus. Such overconsolidation may result from actual changes in effective

Fig. 10.14 Overconsolidation of soil resulting from fluctuation of water table: (a) variations of stress with depth with water table at ground surface; (b) variations of stress with depth with water table at depth d; (c) profile of overconsolidation ratio with depth.

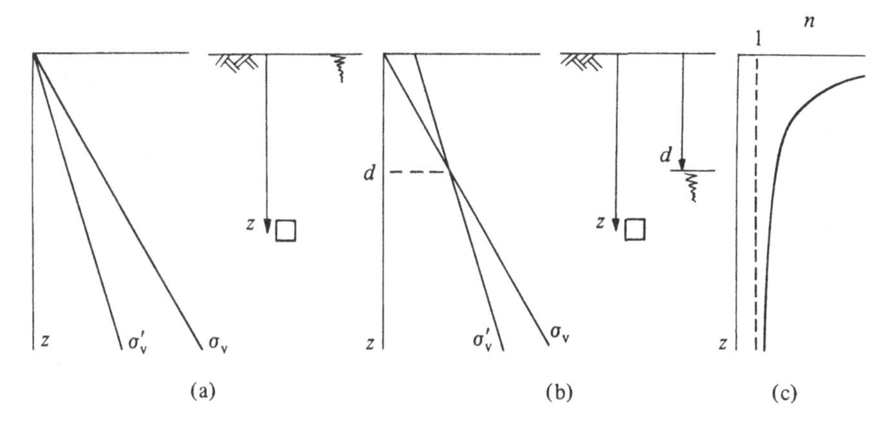

stresses, or it may appear as a result of other effects such as cementation between soil particles or time effects (which will be discussed in Section 12.2). Changes in effective stresses obviously occur where overburden has been eroded by ice or by water, but in soft soils it is very likely that effective stress changes have occurred as a result of fluctuations in water table, even without any erosion.

The effect of water table fluctuation is illustrated in Fig. 10.14. Suppose that the water table fluctuates between the ground surface (Fig. 10.14a) and a depth d (Fig. 10.14b). Suppose also that the soil remains saturated above the water table. Then with the water table at depth d, the effective vertical stress at depth z is

$$\sigma'_{v1} = \gamma z - \gamma_w(z - d) = \gamma' z + \gamma_w d \tag{10.38}$$

With the water table at the ground surface, the vertical effective stress at depth z is

$$\sigma'_{v2} = \gamma z - \gamma_w z = \gamma' z \tag{10.39}$$

and $\sigma'_{v2} < \sigma'_{v1}$ because more of the soil is buoyant in the groundwater. The overconsolidation ratio of the soil, supposing that the water table is currently at the ground surface, is

$$n = \frac{\sigma'_{v1}}{\sigma'_{v2}} = 1 + \frac{\gamma_w d}{\gamma' z} \tag{10.40}$$

and the soil becomes more nearly normally compressed ($n \rightarrow 1$) at depths much greater than d (Fig. 10.14c).

In discussing typical stress paths for geotechnical structures in subsequent sections, we assume that the soil starts from an initially lightly overconsolidated state – as a result, for example, of some past variation in the level of the water table.

10.3.4 Elements on centreline beneath circular load

Elements of soil within a soil deposit of large lateral extent and on the centreline beneath a circular load will be subjected to axially symmetric changes of stresses. This is the only engineering situation for which the stress path can be followed precisely in the conventional triaxial apparatus. Such situations are unusual but do occur; one can think of oil storage tanks as an example (Fig. 10.15a). Oil storage tanks are usually founded on soft soils near river estuaries or coastal sites, so an initial lightly overconsolidated effective stress state due to fluctuation of water table, as discussed in Section 10.3.3, is clearly appropriate.

The stress path followed by an element such as X beneath the oil tank in Fig. 10.15a is shown qualitatively in Figs. 10.15b, c. It has been assumed that the soil beneath the tank is initially one-dimensionally lightly overconsolidated with past maximum effective stress state A', present effective stress state B' and total stress state B. The separation BB' in the $p':q$ plane (Fig. 10.15b) is equal to the in situ pore pressure. The paths followed beneath a circular load will be considered in more detail in Section 11.2; here it suffices to note that the total stress changes BCD resulting from loading the tank are likely to involve an increase in total mean stress p and an increase in deviator stress q. If the tank is filled rapidly, then the response of the soil is likely to be undrained, and the effective stress path will show no change in mean effective stress p' until the current yield locus is reached ($B'C'$). The effective stress path for a lightly overconsolidated clay then turns to the left $C'D'$ as the soil yields and extra pore pressures are generated. If the tank is left and the excess pore pressures allowed to dissipate back to the in situ values existing at B and B', then the effective stress path will show an increase in mean effective stress $D'E'$, with perhaps little change in total stresses, so that dissipation can occur at approximately constant deviator stress q.

Fig. 10.15 (a) Element on centreline beneath circular tank on soft clay; (b) stress paths in $p':q$ and $p:q$ planes; (c) stress paths in deviatoric plane.

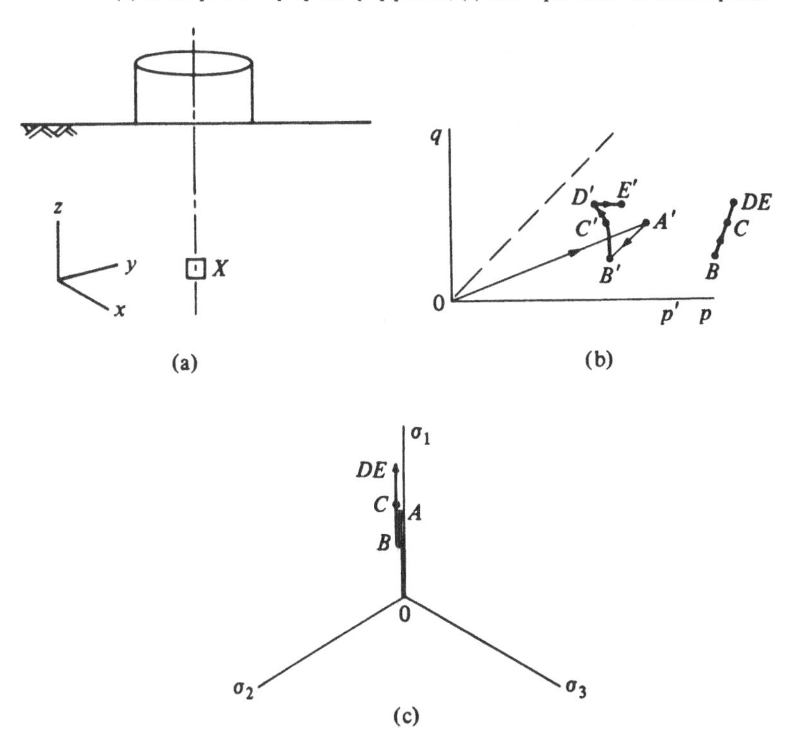

The stress changes are shown again in Fig. 10.15c in the deviatoric view of principal stress space with principal stresses σ'_1, σ'_2, and σ'_3 assigned to the z, x, and y directions, respectively. For these axially symmetric changes in stress, the stresses in the horizontal x and y directions (Fig. 10.15a) are always equal, $\sigma'_2 = \sigma'_3$. Hence the stress paths followed during one-dimensional compression and unloading (OAB), undrained loading beneath the tank (BCD), and subsequent dissipation of excess pore pressures, (DE), lie entirely on the vertical diameter of this view of stress space (on the σ_1 axis or its extension as seen in this view). The presence of pore pressure merely displaces the effective stress state from the total stress state in a direction parallel to the line $\sigma'_1 = \sigma'_2 = \sigma'_3$, and there is no distinction between total and effective stress paths in Fig. 10.15c.

10.4 Plane strain stress paths

10.4.1 One-dimensional compression and unloading

One-dimensional deformation is both an axially symmetric and a plane strain process, and the path followed during one-dimensional compression and unloading can be displayed in the $s':t$ effective stress diagram (Fig. 10.16). The principal effective stresses σ'_1 and σ'_3 are the vertical and horizontal effective stresses σ'_v and σ'_h, so combining (10.3) with (10.1) and (10.2) yields

$$\frac{t}{s'} = \frac{1 - K_0}{1 + K_0} \tag{10.41}$$

During one-dimensional normal compression,

$$K_0 = K_{0nc} = 1 - \sin\phi' \tag{10.15bis}$$

and

$$\frac{t}{s'} = \frac{1 - K_{0nc}}{1 + K_{0nc}} = \frac{\sin\phi'}{2 - \sin\phi'} \tag{10.42}$$

Fig. 10.16 One-dimensional loading and unloading in $s':t$ effective stress plane.

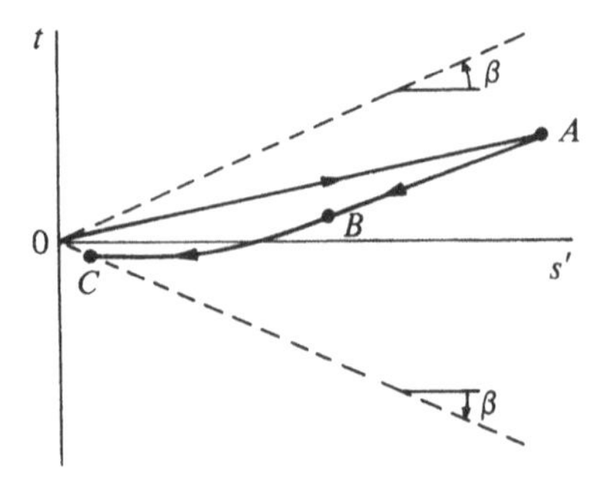

This is line OA in Fig. 10.16. A typical effective stress path ABC followed during one-dimensional unloading is also shown. The assignment of σ'_1 and σ'_3 to vertical and horizontal effective stresses does not imply any relative magnitude, so negative values of t are possible when the horizontal stress exceeds the vertical stress, $\sigma'_3 > \sigma'_1, K_0 > 1$.

The Mohr–Coulomb failure criterion (10.37) can also be converted to a relationship between s' and t. From (10.1) and (10.2),

$$\frac{t}{s'} = \pm \sin \phi' \tag{10.43}$$

where the positive sign refers to failure with the vertical stress greater than the horizontal stress: that is, *active* failure, with the weight of the soil assisting deformation. The negative sign refers to failure with the horizontal stress greater than the vertical stress: that is, *passive* failure, with the weight of the soil resisting deformation. These two failure lines are also plotted in Fig. 10.16; their slope β is given by

$$\tan \beta = \pm \sin \phi' \tag{10.44}$$

The definitions of s' and t, like the Mohr–Coulomb failure criterion, involve only the major and minor principal effective stresses and leave out of consideration the value of the intermediate principal effective stress σ'_2, which acts normal to the plane of shearing. Consequently, the Mohr–Coulomb failure criterion has a symmetry in the s':t plane (Fig. 10.16) which is absent in the p':q plane (see Section 7.1).

10.4.2 Elements beneath long embankment

The total and effective stress paths in the s:t and s':t stress planes for an element of soil X on the centreline beneath a long embankment (Fig. 10.17a) are qualitatively similar to the total and effective stress paths in the p:q and p':q stress planes for an element of soil on the centreline beneath a circular load (Section 10.3.4 and Fig. 10.15). Probable paths are shown in Fig. 10.17b, and the effect of dissipation of construction pore pressures again tends to move the effective stress state away from the active failure line ($D'E'$ in Fig. 10.17b).

The stress paths for this plane strain problem have been presented only in the s':t and s:t effective and total stress planes. It is instructive to consider, at least qualitatively, what the stress paths might look like in the deviatoric view of principal stress space that was introduced in Section 10.2 and used in Section 10.3.4 and Fig. 10.15c. The initial stress state B results from a one-dimensional history, so $\sigma'_2 = \sigma'_3$ and point B lies on the projection of the σ_1 axis (Fig. 10.18, where again principal stresses σ'_1, σ'_2, and σ'_3 are assigned to the z, x, and y directions, respectively). The precise details of path *BD*, as the difference between the vertical and horizontal stresses increases, depend on the details of the assumed soil model; the path shown in Fig. 10.18 is a qualitative estimate. As soon as plane strain deformation begins, the stresses σ'_2 and σ'_3 are no longer equal, and the stress path leaves the projection of the σ'_1 axis. This is just a reflection of the fact that plane strain stress paths cannot be followed in a conventional triaxial apparatus, in which two principal stresses are always equal, $\sigma'_2 = \sigma'_3$.

On the centreline, the directions of the principal stresses remain horizontal and vertical as the embankment is constructed. Away from the centreline (e.g. *Y* in Fig. 10.17a), there is no longer the same symmetry,

Fig. 10.17 (a) Elements beneath centre and edge of long embankment; (b) stress paths in $s':t$ and $s:t$ planes.

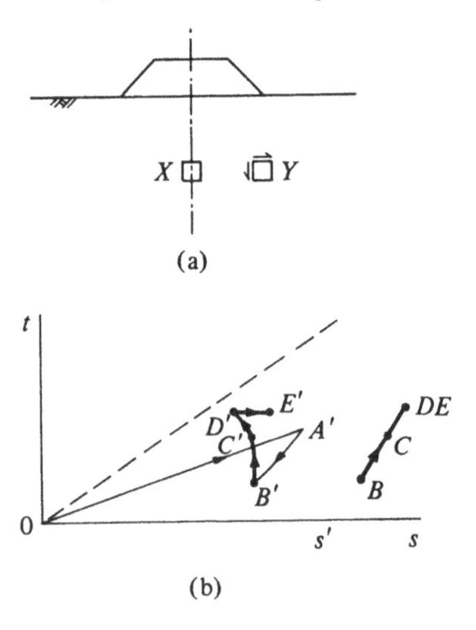

Fig. 10.18 Stress paths for elements deforming actively or passively in plane strain seen in deviatoric view of principal stress space.

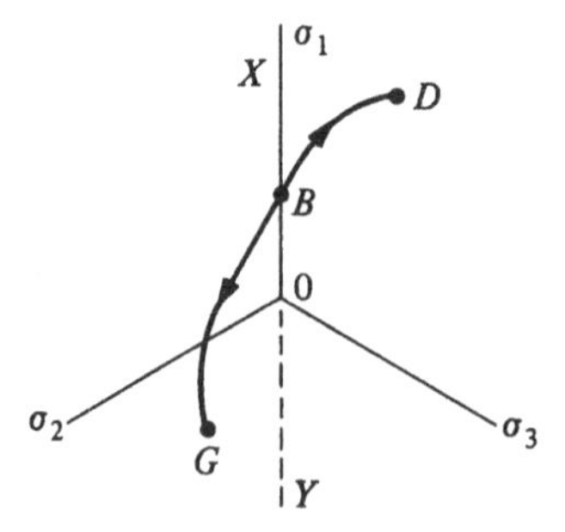

and vertical and horizontal planes experience a development of shear stresses as the embankment is built. The principal axes rotate, and the stress state can only be incompletely depicted in a stress plane whose axes are defined as functions of principal stresses, such as the $s':t$ plane of Fig. 10.17b or the deviatoric view of principal stress space of Fig. 10.18. There are still no shear stresses on planes normal to the length of the embankment, so there are still two principal effective stresses σ'_1 and σ'_3 in the xz plane. An $s':t$ or $s:t$ stress path could therefore still be generated for the xz plane, but the rotation of principal axes cannot be shown in such a diagram.

10.4.3 Elements adjacent to long excavation

A section through a long trench excavation is shown in Fig. 10.19a with elements of soil X and Y adjacent to the wall of the excavation and beneath the base of the excavation, respectively. The total and effective stress paths for element X (Fig. 10.19b) are essentially the same as those for an element behind a retaining wall which is moving forward, so the soil deforms actively with the vertical stress driving the deformation.

The directions of the principal stresses remain essentially unchanged during the movement of the wall, so the vertical and horizontal effective stresses σ'_v and σ'_h can be associated with the principal effective stresses σ'_1 and σ'_3. The vertical total stress at X remains essentially constant as

Fig. 10.19 (a) Elements of soil beside and beneath excavation; (b) stress paths in $s':t$ and $s:t$ planes.

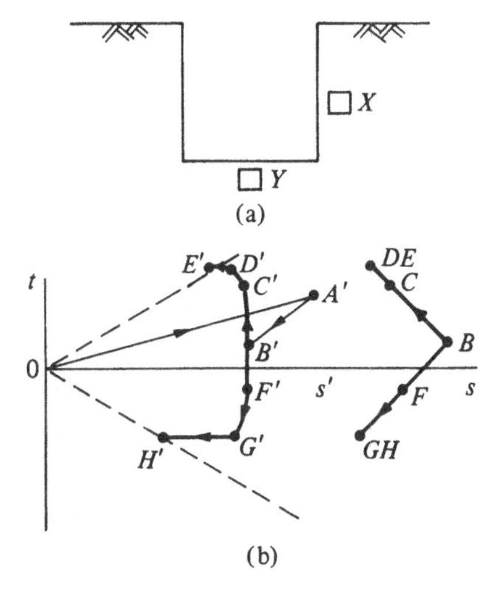

the wall deforms, so

$$\delta t = -\frac{\delta\sigma_h}{2} \tag{10.45a}$$

$$\delta s = \frac{\delta\sigma_h}{2} = -\delta t \tag{10.45b}$$

and the total stress path has a slope of -1 in the $s{:}t$ total stress plane.

As the soil moves forwards towards a state of active failure, $\delta\sigma_h < 0$ and, from (10.45a), $\delta t > 0$. The corresponding total stress path BCD is shown in Fig. 10.19b. If the deformation of the wall occurs rapidly, then the soil deformation is undrained and the effective stress path, again assuming an initially lightly overconsolidated condition, is $B'C'D'$ (Fig. 10.19b). The mean effective stress s' remains constant for plane strain constant volume deformation so long as the soil behaves elastically and isotropically ($B'C'$); but because the total and effective stress paths are converging, the pore pressure falls. When yielding occurs at C', the pore pressure may start to increase as the effective stress path turns towards the active failure line. What happens as pore pressure equilibrium is re-established in the soil behind the excavation depends on the drainage conditions that it imposes. If the water table is able to regain the position it occupied before the excavation was created, and if the total and effective stress states regain their in situ separation BB', then the effective stress state moves further towards the failure line ($D'E'$). In practice, however, it might be expected that the presence of the excavation will lead to a lower equilibrium water table in the adjacent soil, with lower pore pressures and improved stability.

The element Y beneath the base of the excavation is subjected to a large drop in total vertical stress with perhaps only a small change in total horizontal stress. Then

$$\delta t \sim \frac{\delta\sigma_v}{2} \tag{10.46a}$$

$$\delta s \sim \frac{\delta\sigma_v}{2} = \delta t \tag{10.46b}$$

and since $\delta\sigma_v < 0, \delta t < 0$. The total stress path BFG is shown in Fig. 10.19b. For undrained deformation, the effective stress path initially shows no change in mean effective stress s' until yielding occurs ($B'F'$), followed by reduction in effective stress towards the passive failure lines as the soil deforms plastically ($F'G'$). Because the total stress path shows a large decrease in mean stress s, the effect of dissipation of pore pressure may tend to move the effective stress state still closer to passive failure ($G'H'$), though, of course, details of this path depend on the ultimate ground water conditions.

In the deviatoric view (Fig. 10.18), the actively deforming element X again follows a stress path of the general shape BD, though in this case the deformation is being driven by a decrease in horizontal stress σ'_2 rather than an increase in vertical stress σ'_1. The passively deforming element Y follows a path such as BG. In both cases, the plane strain constraint causes σ'_2 and σ'_3 to differ from each other as soon as deformation begins, and both paths leave the projection of the σ'_1 axis.

10.4.4 Element in long slope

The analysis of the stability of a slope of soil which is long in its direction of dip was considered in Section 7.6. If the slope is also of large lateral extent (in its direction of strike), then deformations can be assumed to occur in plane strain. At an element X in such a slope (Fig. 10.20a), the principal axes of stress are certainly not vertical and horizontal, but the principal total stresses σ_1 and σ_3, and hence the values of s and t, can be calculated following the analysis of Section 7.6.

It was shown in Section 7.6 that the stability of the slope was severely reduced when pore pressures associated with steady seepage parallel to the slope and down the slope were present in the soil. The presence of any positive pore pressure in the soil makes the effective mean stress s' lower than the total mean stress s and hence impairs the stability of the slope. Conversely, the presence of any negative pore pressure in the soil makes the effective mean stress s' higher than the total mean stress s and hence enhances the stability of the slope.

In tropical areas such as Hong Kong, where rainfall and temperature changes are seasonal, high negative pore pressures may be present in near surface soils for much of the year as a result of low ground-water levels and less than complete saturation of soil voids. Slopes in such soils may be able to stand stably at angles considerably in excess of their effective

Fig. 10.20 (a) Element in long slope with infiltration of rainfall; (b) stress paths in s':t and s:t planes.

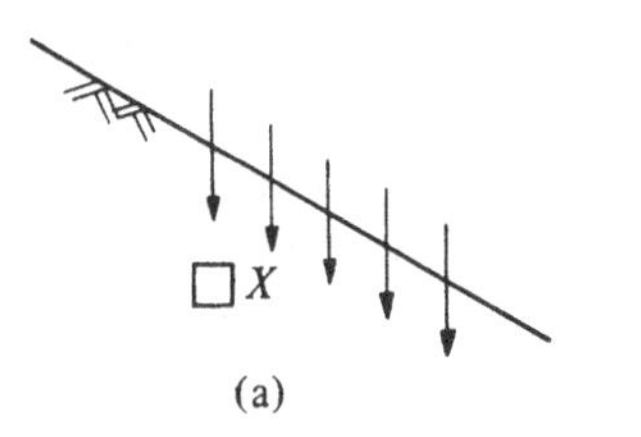

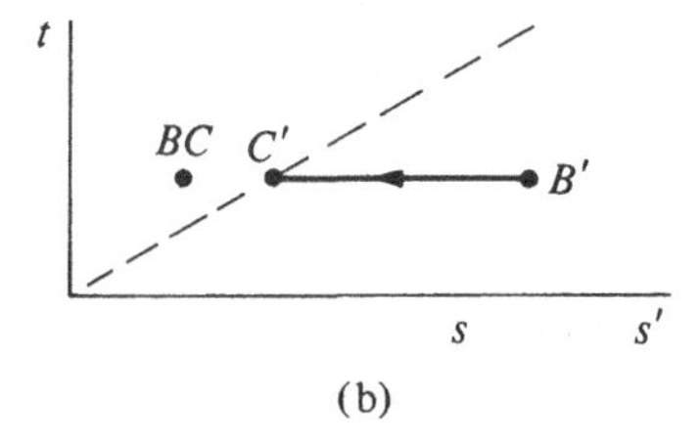

angle of friction (see Section 7.6) provided the presence of these negative pore pressures can be guaranteed. When the rain does come, water may infiltrate from the ground surface (Fig. 10.20a), fill the pores in the soil, and remove some or all of the negative pore pressure. Whether or not this leads to failure of the slope depends on the strength that the soil can mobilise on the approprite effective stress path. Brand (1981) suggests that the change in pore pressure occurs at essentially constant total stresses; in the s:t and s':t stress planes, the total and effective stress paths are as shown in Fig. 10.20b. The initial negative pore pressure puts the initial effective stress state B' to the right of the total stress state B ($s' > s$). As the pore pressure increases (becomes less negative), the effective stress path moves from B' to C' at constant shear stress t, and the total stress remains unchanged.

Previous examples of stress paths have demonstrated that plane strain conditions diverge from axially symmetric conditions, and this divergence is particularly evident when the deviatoric projection of Fig. 10.18 is considered. This example indicates that, even with the distinction between plane strain and axial symmetry temporarily set aside, the nature of the triaxial test which might be most relevant for assessing the performance of the soil in the slope is very different from the standard compression test that might automatically be performed. The standard drained test reaches failure after steady increase of stress ratio q/p', with deviator stress increasing at constant cell pressure. The test that is suggested by the effective stress path in Fig. 10.20b also induces failure by steady increase of stress ratio q/p' but with the deviator stress held constant as the mean effective stress is reduced.

10.5 General stress paths

It should be clear that the range of geotechnical situations for which stress paths can be qualitatively assessed with any confidence is rather limited. As the geotechnical structure becomes more complex, the stress paths also become more complex and more uncertain, and the possibility of reproducing them in a laboratory testing apparatus becomes more remote.

That is not intended as a cry of despair, however, because the whole object of developing numerical models for soil behaviour is precisely to provide a rational basis for the extrapolation from the known region of laboratory test data towards the unknown region of actual field response. Numerical analyses such as those presented in Section 11.3 will then give an indication of the stress paths expected by the computation and reveal the extent of the necessary extrapolation.

10.6 Undrained strength of soil in various tests

The behaviour of soils can be described in terms of effective stresses using numerical models of which Cam clay is only one example. There are many ways in which limiting conditions can be reached with these models, and each could be used to define a failure state and hence a strength for the soil. This section is concerned with a subset of these limiting conditions, failure states reached in undrained, constant volume tests which define values of undrained strength for the soil.

The undrained strength of soils can be measured in many different tests in the laboratory or in situ in the ground. Engineers often talk about *the* undrained strength of soil in a way which ignores not only the various modes of deformation to which soil elements can be subjected in these tests but also the differences between the stress and strain paths that may be followed in these tests, and the stress and strain paths that may be relevant for any particular problem of design or analysis. Stress paths in certain practical problems have been discussed in previous sections; here the differences arising from various testing configurations are considered.

Undrained strengths are typically used in traditional plastic collapse analyses for geotechnical structures which involve the rapid loading of clays. These analyses usually implicitly assume that the undrained failure of soils is governed by a Tresca failure criterion (Section 3.2), so that it is assumed that, irrespective of the mode of deformation, the maximum shear stress that the soil can support is the same. This makes it very easy to justify using undrained strengths obtained, for example, from triaxial tests for analysis of the collapse of plane structures but takes no account of the overall effective stress response of the soil, of which undrained testing explores only a small part. The behaviour of soils is fundamentally described in terms of effective stresses, and any other statement about soil response must be consistent with the underlying effective stress behaviour.

Various soil tests will be compared in Section 10.6.1 through the modes of constant volume deformation that they impose; this gives a clear visual impression of the differences. In Section 10.6.2, the Cam clay model will be used to calculate the undrained strengths, thought of as the values of maximum shear stress, that would be measured in different soil tests.

10.6.1 Modes of undrained deformation

A series of constant volume deformations of an initially cubical soil element are shown to an exaggerated scale in Fig. 10.21. The soil elements are assumed to have been taken from or to be in the ground with the z axis vertical.

Figure 10.21a shows the deformation that is imposed in a triaxial

Fig. 10.21 Modes of constant volume deformation: (a) triaxial compression, (b) triaxial extension, (c) plane strain compression, (d) plane strain extension, (e) pressuremeter cylindrical cavity expansion, (f) simple shear on vertical sample, (g) field vane, and (h) cone penetration test.

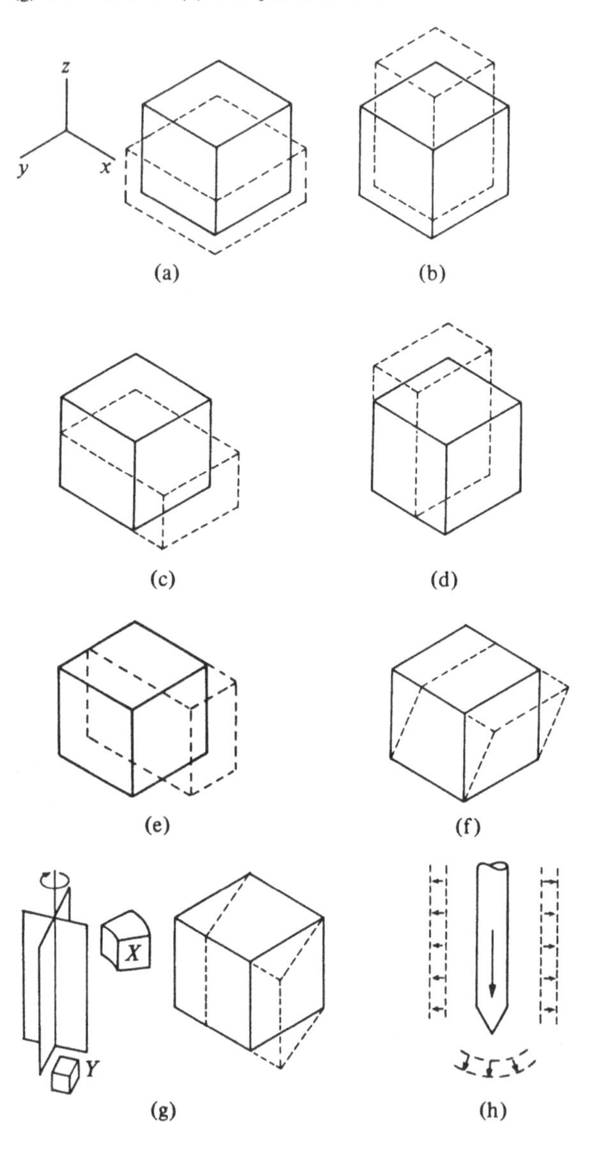

compression test. The radial tensile strain of the triaxial sample has half the magnitude of the axial compressive strain:

$$\varepsilon_z = -2\varepsilon_x = -2\varepsilon_y; \qquad \varepsilon_z > 0 \tag{10.47}$$

(with compressive strains positive). The deformation imposed in a triaxial extension test (Fig. 10.21b) is the opposite of this: whereas compression makes the element short and fat, extension makes it tall and thin. Thus,

$$\varepsilon_x = \varepsilon_y = \frac{-\varepsilon_z}{2}; \qquad \varepsilon_x > 0 \tag{10.48}$$

In plane strain tests on samples taken vertically from the ground, the strain in one of the originally horizontal directions, for example, the y direction, is zero. Plane strain compression (Fig. 10.21c) corresponds to active loading of the soil with the sample becoming shorter (ε_z compressive),

$$\varepsilon_z = -\varepsilon_x, \qquad \varepsilon_y = 0, \qquad \varepsilon_z > 0 \tag{10.49}$$

and plane strain extension (Fig. 10.21d) corresponds to passive loading of the soil with the sample becoming thinner (ε_x compressive),

$$\varepsilon_x = -\varepsilon_z, \qquad \varepsilon_y = 0, \qquad \varepsilon_x > 0 \tag{10.50}$$

Expansion of a long cylindrical cavity is used in the pressuremeter test to determine in situ properties of soils (Baguelin, Jézéquel, and Shields, 1978; Mair and Wood, 1987). A schematic diagram of a pressuremeter is shown in Fig. 10.22. A long cylindrical rubber membrane is expanded by internal pressure, and the radial or volumetric expansion of the cavity is measured. The soil is deformed in plane strain with the vertical z direction as the direction in which no strain occurs. Soil elements are compressed in the radial direction, and a balancing circumferential extension has to occur to maintain the constant volume condition. Converting this to a deformation mode for the cubical element (Fig. 10.21e), we find the strain path to be specified by

$$\varepsilon_y = -\varepsilon_x, \qquad \varepsilon_z = 0, \qquad \varepsilon_y > 0 \tag{10.51}$$

Fig. 10.22 Expansion of cylindrical cavity in pressuremeter test.

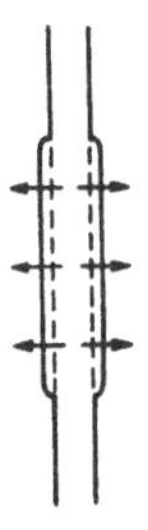

Such a deformation could be imposed in a plane strain apparatus in the laboratory, but it is more common for plane strain tests to be performed with no strain in one of the horizontal x or y directions, as shown in Figs. 10.21c, d.

The strain paths associated with undrained triaxial compression (TC), triaxial extension (TE), plane strain compression (PSC), plane strain extension (PSE), and pressuremeter expansion (PM) are shown in Fig. 10.23a in the deviatoric projection of principal strain space. Since these are all constant volume paths, the deviatoric projection provides a true view of each path, with no foreshortening.

Some results of simple shear tests on soils were presented briefly in Sections 6.5, 8.3, and 8.6. After conventional triaxial apparatus, simple shear apparatus are probably the most frequently used pieces of equipment for laboratory investigations of the stress:strain behaviour of soils; simple shear tests provide the only readily available opportunity for studying the effects of rotation of principal axes on the stress:strain behaviour of soils. The mode of deformation in a simple shear test is shown in Fig. 10.21f. The sample is deformed in plane strain so that $\varepsilon_y = 0$. In the xz plane, the deformation is such that the shape of the sample changes from a rectangle to a parallelogram. There is no direct strain in the x direction, and because the deformation is assumed to be at constant volume, there is no vertical strain either. The only non-zero component of strain is the shear strain γ_{xz}:

$$\varepsilon_x = \varepsilon_y = \varepsilon_z = 0; \qquad \gamma_{xz} \neq 0 \tag{10.52}$$

Fig. 10.23 (a) Deviatoric strain paths and (b) deviatoric stress paths according to Cam clay model with Mohr–Coulomb failure in constant volume triaxial compression TC, triaxial extension TE, plane-strain compression PSC, plane strain extension PSE, and pressuremeter cylindrical cavity expansion PM.

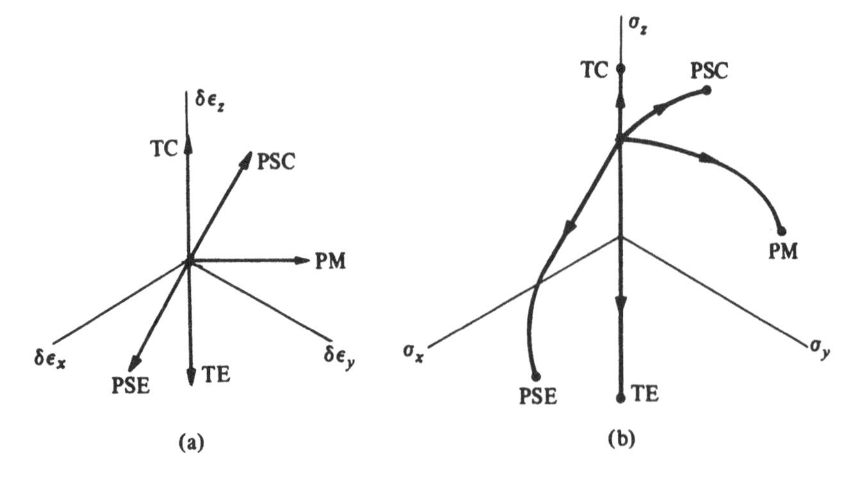

It is not easy to show the expected modes of deformation for in situ tests other than for the pressuremeter. Two tests commonly used for the in situ estimation of strengths of soil are the vane and the cone penetration test. The exaggerated deformation of an element X of soil around the outer radius of a cruciform vane being rotated in the ground, before a failure surface has developed, is shown, tentatively, in Fig. 10.21g. This element of soil is subjected to a deformation rather like the simple shear deformation of a sample with its vertical axis in the apparatus corresponding to a horizontal direction in the ground (compare Figs. 10.21f, g):

$$\varepsilon_x = \varepsilon_y = \varepsilon_z = 0; \qquad \gamma_{xy} \neq 0 \tag{10.53}$$

On the other hand, an element Y of soil which is going to form part of the top or bottom horizontal failure surface created by the rotating vane is, before the failure surface forms, subjected to a mode of deformation similar to the simple shear test on a sample with its vertical axis corresponding to the vertical direction in the ground (Fig. 10.21f).

A cone penetration test pushes elements of soil through a variety of modes of deformation. The penetration of the tip may subject soil elements ahead of the tip to deformations similar to those experienced in the expansion of a spherical cavity (Fig. 10.21h). Expansion of a spherical cavity in an isotropic soil subjects elements of soil to the constant volume deformation associated with conventional triaxial compression (Fig. 10.21a); there are now two circumferential directions which elongate equally as the radial deformation increases and radial compression occurs. However, only the elements of soil directly below the tip of the cone are correctly equivalent to vertical soil samples, that is, to conventional triaxial samples taken from the ground with their axes vertical, in the z direction (A in Fig. 10.24). Other elements are equivalent to inclined samples

Fig. 10.24 Inclined cylindrical samples taken out of the ground with their axes in a vertical plane.

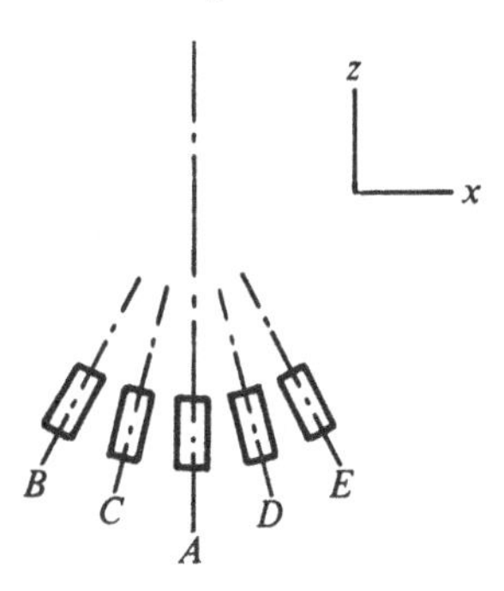

(B, C, D, and E in Fig. 10.24), and any anisotropy of the soil strongly influences their behaviour.

The penetration of the shaft of the cone is similar to the penetration of a driven pile which Randolph, Carter, and Wroth (1979) have suggested is equivalent to the undrained expansion (or rather creation) of a long cylindrical cavity. Soil elements away from the heavily disturbed zone adjacent to the cone might therefore be subjected to deformations similar to those imposed in the pressuremeter test (Fig. 10.21e). It is because the cone penetration test causes such a variety of modes of deformation that there is such an uncertainty concerning the correct factor to be used to convert cone penetration resistance to some shear strength parameter for a soil (e.g. see Meigh, 1987).

Each soil test subjects elements of soil to different modes of deformation. It is for this reason that it is inappropriate to talk about *the* undrained strength of a soil without stating by what means this particular strength was determined. Triaxial compression strengths and field vane strengths are perhaps the commonest strengths that are quoted in the literature. The various diagrams in Fig. 10.21 show that, quite apart from any other factors which may influence the strengths determined in any particular test, the modes of deformation are so different that similarity of values should be regarded as the exception rather than the rule.

10.6.2 Undrained strengths: Cam clay model

So far the discussion of the differences between various tests has been largely intuitive. It is possible to quantify these differences if assumptions are made about the way in which the soil behaves. When the strength of the soil was discussed in Chapter 7, it was really only the strength of soils in conventional triaxial compression (and extension) tests that was considered; stress and strain paths associated with common tests and with geotechnical constructions escape from the constraints imposed by the triaxial apparatus, and it is necessary to suggest what might be expected in other parts of the deviatoric projection of stress space, away from the diameter XOY in Fig. 10.18.

The Cam clay model described in Chapter 5 is a simple elastic–plastic model of soil behaviour which goes at least some way towards incorporating a rather more realistic description of the effective stress changes that occur in constant volume shearing. In Chapter 5 this model was described only in terms of stress changes that could be applied in the conventional triaxial apparatus. It is necessary to make some assumptions about how this observed response should be extrapolated to more general stress conditions. Experimental data for non-axially symmetric stress conditions are few (and to some extent contradictory). The simplest possible assumption is that the response of the soil depends on the mean effective stress p' and a general deviator stress q as defined in Section 1.4.1. For a set of principal stresses, $\sigma'_1, \sigma'_2, \sigma'_3$, the mean effective stress p' indicates the extent to which the stresses are all the same:

$$p' = \frac{\sigma'_1 + \sigma'_2 + \sigma'_3}{3} \qquad (10.54)(\text{cf. } 1.34)$$

The general deviator stress q is a function of principal stress differences and indicates the extent to which the stresses are not the same:

$$q = \sqrt{\frac{(\sigma'_2 - \sigma'_3)^2 + (\sigma'_3 - \sigma'_1)^2 + (\sigma'_1 - \sigma'_2)^2}{2}} \qquad (10.55)(\text{cf. } 1.35)$$

The values of p' and q represent (with factors of proportionality) the distance in principal stress space (Fig. 10.3a) along the mean stress axis $\sigma'_1 = \sigma'_2 = \sigma'_3$ and the orthogonal distance from the current effective stress state to this axis, respectively. It is convenient to assume that the stress:strain behaviour of soil depends only on p' and q: the Cam clay yield surface is assumed to be formed by rotation about the mean stress axis (or stress space diagonal $\sigma'_1 = \sigma'_2 = \sigma'_3$) (Fig. 10.25), and any section through the yield surface at constant mean effective stress p' is consequently a circle.

The Mohr–Coulomb failure criterion specifies a maximum value for the ratio between major and minor principal effective stresses:

$$\frac{\sigma'_{\text{I}}}{\sigma'_{\text{III}}} = \frac{1 + \sin\phi'}{1 - \sin\phi'} \qquad (10.37\text{bis})$$

This expression involves only two of the principal effective stresses and

Fig. 10.25 Ellipsoidal yield surface of Cam clay model in principal effective stress space (drawn for $M = 0.9$).

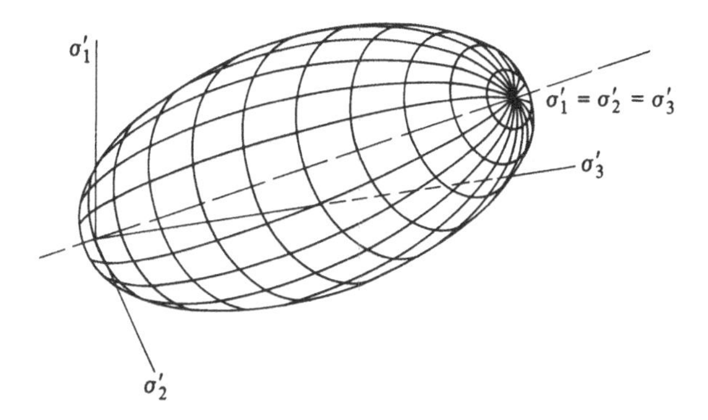

requires nothing of the remaining principal effective stress σ'_{II} except that it should be intermediate between the other two. In different parts of principal effective stress space $\sigma'_1:\sigma'_2:\sigma'_3$ (Fig. 10.3), the major and minor principal stresses are associated with different pairs selected from the three axes 1:2:3. The assignment of relative magnitudes of principal stresses can be seen in the deviatoric view (Fig. 10.3c). In terms of $\sigma'_1:\sigma'_2:\sigma'_3$, (10.37) becomes six separate expressions defining six planes through the origin of principal effective stress space which intersect to form an irregular hexagonal pyramid centred on the axis $\sigma'_1=\sigma'_2=\sigma'_3$ (Fig. 10.26a) and having an irregular hexagonal section in the deviatoric view (Fig. 10.26b). This can be compared with the shapes of the Tresca and von Mises yield criteria proposed for metals, which produce prismatic yield surfaces (Fig. 3.7) with regular hexagonal and circular sections in the deviatoric view (Fig. 3.8).

The experimental evidence from laboratory tests with general stress

Fig. 10.26 (a) Mohr–Coulomb failure criterion as hexagonal pyramid in principal effective stress space; (b) section through Mohr–Coulomb failure criterion at constant mean effective stress p' (drawn for $\phi' = 23°$).

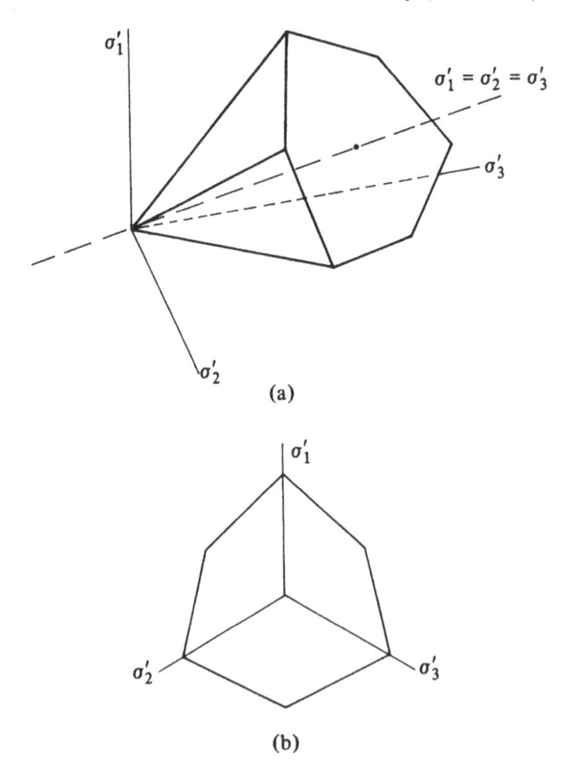

conditions (e.g. see Gens, 1982) suggests that the critical state failure of soils can be more accurately approximated by a Mohr–Coulomb failure criterion with a constant angle of shearing resistance, the irregular hexagonal pyramid of Fig. 10.26, than by the von Mises criterion that the generalisation of Cam clay with only p' and q implies (Fig. 10.25). It is assumed here, therefore, that although the prefailure stress:strain behaviour is described by Cam clay with expanding ellipsoidal yield surfaces as shown in Fig. 10.25, failure is governed by a Mohr–Coulomb failure criterion with angle of shearing resistance corresponding to the critical state reached with Cam clay in triaxial compression. The resulting model is then described by a sort of intersection of the Cam clay ellipsoid and the Mohr–Coulomb hexagonal pyramid (Fig. 10.27a). At any value of mean stress p', the value of general deviator stress q that can be

Fig. 10.27 (a) Ellipsoidal yield surface of Cam clay model combined with irregular hexagonal pyramid of Mohr–Coulomb failure criterion; (b) sections at constant mean effective stress p' (drawn for $M = 0.9$, $\phi' = 23°$).

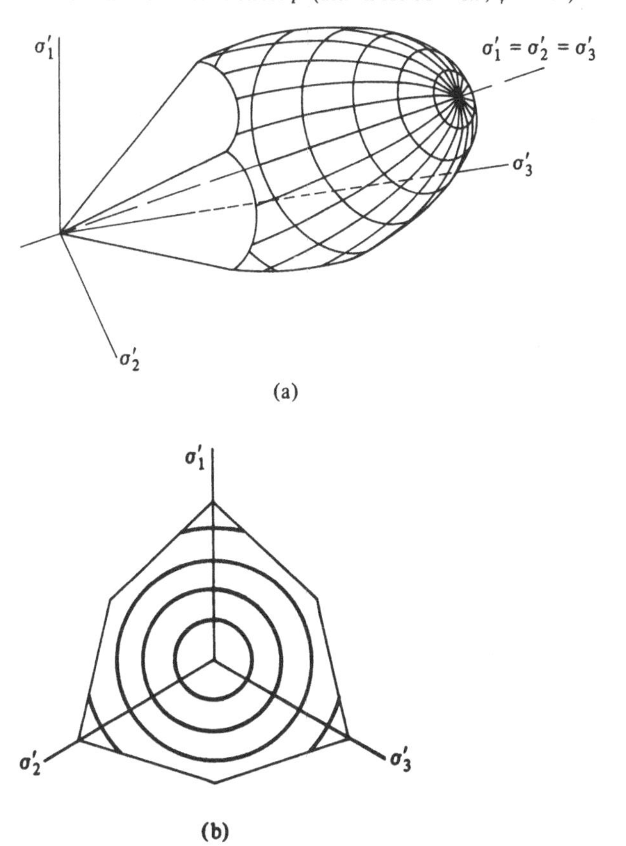

reached depends on the direction of loading in the deviatoric projection (Fig. 10.27b).

No attempt has been made to include the Hvorslev surface representation of peak strengths for heavily overconsolidated clays (Section 7.4.1) in this composite model, which is being applied here only to the estimation of undrained strengths of normally compressed clays. Tests on initially one-dimensionally normally compressed soil are considered. The initial effective stress state resulting from one-dimensional compression emerges from the model without further assumption, as described in Section 10.3.1.

Calculations of the undrained strengths expected in various tests have been made using $\nu' = 0.3$ and $\Lambda = 0.8$ as a typical set of parameters, and the effect of various values of M [related to angle of shearing resistance ϕ' in triaxial compression through (7.9)] has been studied. The resulting values of undrained strength c_u, which is usually the maximum shear stress measured in each particular test, are normalised with respect to the effective vertical consolidation pressure σ'_{zc} that exists before the test is started. The resulting strength ratios c_u/σ'_{zc} are shown in Fig. 10.28.

A typical set of stress paths in the deviatoric view of stress space is shown in Fig. 10.23b. So long as the soil is behaving isotropically and elastically inside the yield surface, constant volume deformation implies no change in mean effective stress p', and the deviatoric stress paths have the same directions as the deviatoric strain increment paths shown in

Fig. 10.28 Dependence of strength ratio c_u/σ'_{zc} on angle of shearing resistance ϕ' for normally compressed soil according to Cam clay model combined with Mohr–Coulomb failure criterion (TC, triaxial compression; TE, triaxial extension; PS, plane strain; PM pressuremeter; DSS, direct simple shear; FV field vane).

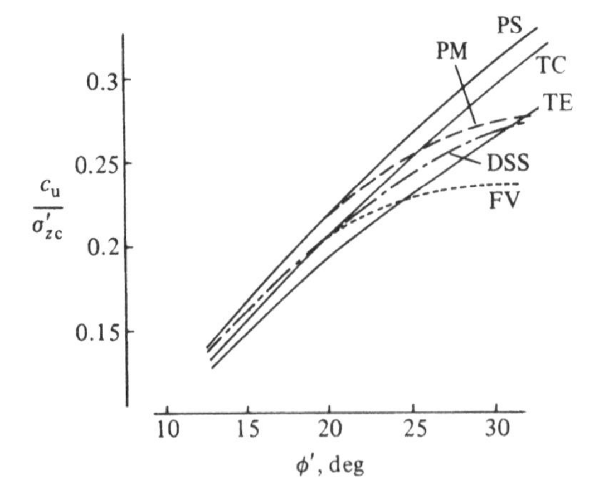

Fig. 10.23a. However, the stress paths bend round when yielding occurs. The occurrence of plastic volumetric strain in undrained deformation requires, following Section 4.3, changes in mean effective stress through development of pore pressures in order to produce equal and opposite elastic volumetric strains and an overall constant volume condition. Therefore, the stress paths shown in Fig. 10.23b do not actually lie in a single deviatoric plane at constant p'; the diagram represents projections of three-dimensional paths.

The symmetry contained in the model and in the Mohr–Coulomb failure criterion forces the strength ratios for plane strain compression and extension to be the same. According to this model, the strength ratio for plane strain is always higher than that for triaxial compression.

The pressuremeter tests run into an ambiguity. In the analysis of the pressuremeter test (Mair and Wood, 1987), the apparent undrained strength is calculated from the difference between the radial and circumferential effective stresses at failure, which implies, in terms of the stresses $\sigma'_x, \sigma'_y, \sigma'_z$,

$$c_u = \frac{\sigma'_y - \sigma'_x}{2} \tag{10.56}$$

There is a possibility, however, that for a low value of K_{0nc} (which from Fig. 10.5 implies a high value of angle of shearing resistance ϕ' and a high initial value of stress ratio $q/p' = \eta_{Knc}$), failure can occur with the vertical stress σ'_z still the major principal stress ($\sigma'_1 > \sigma'_3 > \sigma'_2$ in Fig. 10.3c, with principal stresses $\sigma'_1, \sigma'_2, \sigma'_3$ assigned to the z, x, y directions, respectively) instead of the intermediate stress as assumed in the analysis ($\sigma'_3 > \sigma'_1 > \sigma'_2$ in Fig. 10.3c, with the same assignment of stresses). The undrained strength which is calculated from (10.56) is not then the maximum shear stress sustained by the soil. For the particular set of soil parameters used here, the two results diverge only for values of ϕ' above about 31°.

For the simple shear test, further assumptions are required. De Josselin de Jong (1971) shows that there are two alternative sets of failure planes on which deformation can occur, both compatible with an overall simple shear deformation. The most obvious set of planes is the horizontal set shown in Fig. 10.29b, and the corresponding Mohr circle of effective stress (Fig. 10.29a) shows that the horizontal planes are planes on which the largest angle of friction in the soil is mobilised. Alternatively, deformation can take place on a vertical set of planes (Fig. 10.29d) with accompanying rotation to give the overall simple shear deformation. In this case, the maximum angle of friction is mobilised on these vertical planes (Fig. 10.29c), and the angle of friction mobilised on the horizontal boundary of the sample may be considerably lower. De Josselin de Jong maintains that this second mode of deformation is likely to be attained by many soils more easily than the first. Randolph and Wroth (1981) show some experimental evidence in support of this contention. In either case, however, the shear stress measured on the horizontal boundary of the sample is lower than the maximum shear stress experienced by the soil by a factor $\cos\phi'$ (see the Mohr circles in Figs. 10.29a, c). Simple shear deformation is a plane strain process, so it can be assumed in the Cam clay model that the changes of principal effective stresses that are experienced during the simple shear test are the same as those experienced during a plane strain compression test; the only difference, the rotation of the directions of the principal stresses, is assumed to have no effect per se. Then the shear strength in the simple shear test, which should more properly be called a measured maximum shear stress, is lower than the mobilised strength and can be calculated by multiplying the undrained

Fig. 10.29 Simple-shear deformation with sliding on (a), (b) horizontal planes and (c), (d) vertical planes combined with rotation (after de Josselin de Jong, 1971).

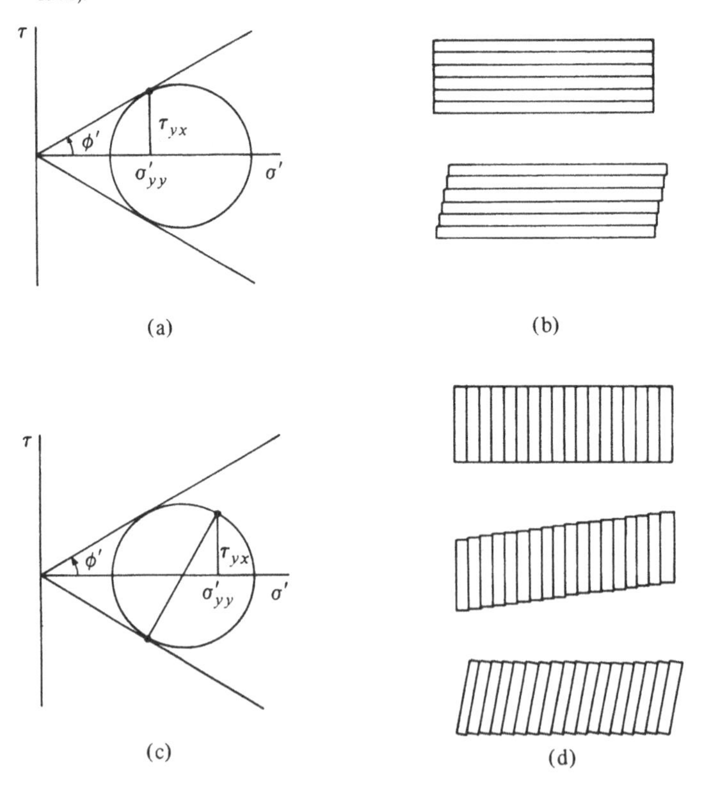

strength in plane strain compression by $\cos\phi'$. The factor $\cos\phi'$ decreases as ϕ' increases, and the curve for the apparent simple shear strength in Fig. 10.28 diverges steadily from the plane strain compression curve and actually reaches a peak at an angle of friction of about 40°.

It was suggested in Fig. 10.21g that elements of soil around the boundary of the cylinder of soil that is brought to failure in the ground by a rotating field vane would, before a failure surface actually forms, be subjected to a mode of deformation approximating to the simple shear of a sample taken out of the ground with a horizontal axis destined to become the vertical axis in the apparatus. By the same argument as in the previous paragraph, the shear stress in the soil on that vertical failure plane is lower than the undrained strength of the soil by a factor $\cos\phi'$. The undrained strength could be estimated from an appropriate plane strain test in which no rotation of principal axes occurs, which in this case is the pressuremeter test. Wroth (1984) presents experimental data and theoretical analyses which show that the contribution of the shear stress mobilised on the top and bottom surfaces of the failing cylinder of soil to the torque required to rotate the vane is often considerably smaller than the contribution from the vertical surface. The curve of vane strengths (FV in Fig. 10.28) is simply estimated from the pressuremeter strength (PM) multiplied by $\cos\phi'$. There is again the problem that for angles of shearing resistance greater than about 31° (a high angle for normally compressed clay) failure occurs according to the Mohr–Coulomb criterion, with the vertical effective stress still the major principal effective stress. If the contribution to the torque from the horizontal shear surfaces is reckoned to be non-negligible, then the shear stress on these surfaces is similar to that measured in the simple shear test, and the field vane curve (FV in Fig. 10.28) would move nearer to the simple shear curve (DSS).

It is not really worthwhile to labour the assumptions behind this particular model. It was noted anyway in Section 9.6 that the trend of major increase of strength ratio c_u/σ'_{zc} with increasing angle of shearing resistance ϕ' (and hence by implication with decreasing plasticity I_P) is not supported by experimental evidence. Nevertheless, similar conclusions about the dependence of the measured strength on the mode of testing have been drawn by, for example, Prévost (1979), Levadoux and Baligh (1980), Borsetto, Imperato, Nova, and Peano (1983), using other more elaborate models of soil behaviour.

For a typical value of ϕ' between 20° and 25°, the sequence of strength values shown in Fig. 10.28 is PS > PM > TC > DSS > FV > TE. Typical field data for an Italian soft clay site that are based on tests reported by Ghionna, Jamiolkowski, Lacasse, Ladd, Lancellotta, and Lunne (1983)

Table 10.1 *Undrained strength ratios in different tests*

Clay	I_P	c_u/σ'_{zc}					
		TC	TE	PSC	PSE	PM	DSS[a]
Bangkok (1)	0.85	0.72	0.37	—	—	—	0.60
Boston blue (2)	0.21	0.328	0.130	0.335	0.175	0.210	0.220
Drammen (1)	0.29	0.39	0.20	—	—	—	0.35
Haney (3)	0.18	0.268	0.168	0.296	0.211	—	—
AGS marine clay (4)	0.43	0.325	0.200	—	—	—	0.28
James Bay intact (5)	0.16	0.45	0.235	—	—	—	0.32
James Bay destructured (5)		0.335	0.20	—	—	—	0.275

Notes: (1) Data from Prévost (1979)
(2) Data from Levadoux and Baligh (1980)
(3) Data from Vaid and Campanella (1974)
(4) Atlantic Generating Station marine clay; data from Jamiolkowski, Ladd, Germaine, and Lancellotta (1985)
(5) Data from Jamiolkowski et al. (1985)

[a]Measured shear stress on horizontal surfaces, increased by 10% to allow for non-uniform boundary stresses (e.g. see Wood and Budhu, 1980).

Fig. 10.30 Field data of undrained strength measured with different tests at Porto Tolle, Italy (after Ghionna, Jamiolkowski, Lacasse, Ladd, Lancellotta, and Lunne, 1983) (TC, triaxial compression; TE, triaxial extension; PM, pressuremeter; DSS, direct simple shear; FV, field vane).

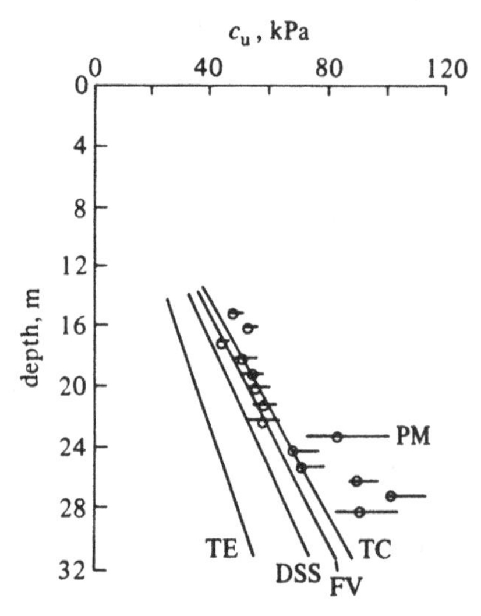

are shown in Fig. 10.30. The variation of deduced strength with depth differs from test to test; the sequence of magnitudes is similar to that deduced from Fig. 10.28 (except that DSS and FV are interchanged), but the spread of magnitudes is rather greater.

A series of values of the strength ratio c_u/σ'_{zc} determined from laboratory experiments on a number of one-dimensionally normally compressed clays are tabulated in Table 10.1. These data confirm the theoretical conclusion even though the numerical values may be different from those predicted with the Cam clay model. An additional moral can be drawn from the discussion about the simple shear and vane tests: in these the apparent undrained strength that is deduced is a shear stress mobilised on a particular plane or surface and is not necessarily the maximum shear stress experienced by the soil. There is even less reason to expect that this mobilised shear stress should be the same as an undrained strength measured as a maximum shear stress in another device.

10.7 Conclusion

In Sections 10.3, 10.4, and 10.5, qualitative stress paths for a number of field situations were discussed to show that different geotechnical constructions would load and deform soil elements in different ways. Lest it be thought that this comparison and distinction of stress paths was unnecessary, examination in Section 10.6 of the stress paths applied in different common soil tests showed that even something as apparently straightforward as undrained strength of soils is not independent of the stress path (or strain path) on which it is measured. In particular, undrained strengths (and other soil properties) determined under conditions of plane strain are not in general the same as undrained strengths (and other soil properties) determined under conditions of axial symmetry. However, most soil testing is actually done under conditions of axial symmetry, whereas plane strain conditions obtain more frequently in practice. It is necessary to be aware of the distinction when selecting soil parameters for design purposes.

Exercises

E10.1. Estimate the total and effective stress paths in s:t, s':t, and deviatoric stress planes for elements of soil behind a typical gravity wall retaining clay soil which moves rapidly towards a condition of (a) active, and (b) passive, failure.

Consider two long-term drainage conditions:

i. Drainage beneath the wall is so effective that the water table in the neighbourhood of the wall is drawn down to the level of the base of the wall.

ii. Drains through the wall become blocked, and the water table remains at the original ground surface which is level with the top of the wall.

E10.2. An element of Cam clay, with soil parameters $\lambda = 0.193$, $\kappa = 0.047$, $M = 0.97$, $N = 3.17$, and $G' = 2400\,\text{kPa}$ is in a normally compressed state subjected to the following stresses, referred to right-handed cartesian axes:

$\sigma'_{xx} = 110\,\text{kPa}$ $\quad \tau_{xy} = 25\,\text{kPa}$ $\quad \sigma'_{yy} = 54\,\text{kPa}$

$\sigma'_{zz} = 67\,\text{kPa}$ $\quad \tau_{yz} = 0$ $\quad \tau_{zx} = 0$

The element of soil is subjected to the following drained changes in stress, referred to the same axes:

$\delta\sigma'_{xx} = 2\,\text{kPa}$ $\quad \delta\tau_{xy} = 3\,\text{kPa}$ $\quad \delta\sigma'_{yy} = 1\,\text{kPa}$

$\delta\sigma'_{zz} = 4\,\text{kPa}$ $\quad \delta\tau_{yz} = 0$ $\quad \delta\tau_{zx} = 0$

Making appropriate assumptions, calculate the increments of volumetric and shear strain. Calculate direction cosines for the principal axes of stress increment and strain increment and for the principal axes of stress before and after the increment.

E10.3. A sample of Cam clay is set up in a true triaxial apparatus in a normally compressed condition with effective stress $\sigma'_1 = 200\,\text{kPa}$ and $\sigma'_2 = \sigma'_3 = 130\,\text{kPa}$. What should the principal effective stresses be at yield and at the critical state in the following undrained tests?

i. Axisymmetric compression $\quad \delta\varepsilon_1 > 0,\ \delta\varepsilon_2 = \delta\varepsilon_3$
ii. Axisymmetric extension $\quad \delta\varepsilon_1 < 0,\ \delta\varepsilon_2 = \delta\varepsilon_3$
iii. Plane strain active $\quad \delta\varepsilon_1 > 0,\ \delta\varepsilon_2 = 0$
iv. Plane strain passive $\quad \delta\varepsilon_2 = 0,\ \delta\varepsilon_3 > 0$
v. Pressuremeter expansion $\quad \delta\varepsilon_1 = 0,\ \delta\varepsilon_2 > 0$

Assume that the behaviour of the clay can be described in terms of general stress quantities p' and q, appropriately defined; that the mean effective stress at the critical state is independent of the preceding strain path (note that this is a different assumption from that in Section 10.6.2); and that the values of soil parameters for the Cam clay model are $M = 1.05$, $\lambda = 0.213$, $\kappa = 0.036$, $N = 3.58$, $G' = 1780\,\text{kPa}$.

E10.4. A thin-walled hollow cylindrical sample of soil of mean radius r and wall thickness t is compressed isotropically under an effective pressure P_1. It is then subjected to a drained test in which a torque αQ is applied, the axial stress is increased to $P_1 + \alpha a$, the internal pressure is reduced to $P_1 - \alpha b$, and the external pressure is increased to $P_1 - \alpha c$. Choosing suitable definitions of average stress components, find expressions for the principal stresses acting on a typical element of soil.

For a particular specimen, $20b = 20c = a = Q/(2\pi r^2 t)$ and $r = 10t$. The behaviour of the soil can be described by the Cam clay model with yield loci $p'/p'_o = M^2/(M^2 + \eta^2)$. If the sample has been previously isotropically normally compressed to a mean effective stress $2P_1$, at what value of α does the soil start to yield?

11

Applications of elastic–plastic models

11.1 Introduction

Some applications of elastic–plastic models to the evaluation or prediction of the behaviour of real geotechnical prototypes are now presented. The first application links back to the discussion of stress paths in the previous chapter: simple estimates of stress paths can be used to illustrate both the onset of plastic deformation in soil beneath a surface load and the influence that this plastic deformation has on the development of excess pore pressures and the settlement of foundations. This should be regarded as a pedagogic example to demonstrate the significance of elastic–plastic behaviour rather than as a recommended solution to a design problem. Once elastic–plastic descriptions of material non-linearity are introduced, it is rarely possible to avoid resort to computer-aided numerical analyses.

The following section summarises some more complete applications of elastic–plastic models in numerical analyses made using the finite element method. These examples have been chosen to illustrate the use of two different elastic–plastic models. Although it may appear that this book has placed undue emphasis on the Cam clay model, it is neither the only nor necessarily the best model which can be used for numerical predictions. There is an evident bias towards soft clay in these applications. The framework of volumetric hardening models developed in detail in Chapter 4 is particularly suited to the description of soft clays. Sands can also be described by elastic–plastic models, as has been hinted in Chapters 3 and 4, and stiff clays should in principle be described with the same models as soft clays. However, the models that have been successfully used for sand tend to be rather more complicated than those for clay, and stiff clays require rather careful consideration of the nature of the assumed elastic response. Some of the extra features that can usefully be incorporated are discussed in Chapter 12.

In any application of an elastic–plastic model, it is necessary to select values for the soil parameters on which the model is based. The way in which this can be done is indicated in each of the finite element analyses. In general, a model has to be 'tuned' by adjustment of the values of several parameters to produce the best possible match to the available data from laboratory (and possibly in situ) tests. There is a distinction between this selection of parameters to optimise the fit provided by a given model and the testing that might be required to verify the hypotheses on which that particular model is constructed. For example, the elastic–plastic soil models in Chapter 4 assume the existence of yield loci and plastic potentials for soils. However, it is not in general feasible to perform the extensive programme of probing tests necessary to confirm this assumption. The Cam clay model in Chapter 5 assumes in addition that the yield loci and plastic potentials are coincident and elliptical, but again it is not expected that the testing will have investigated these assumptions.

11.2 Circular load on soft clay foundation

11.2.1 Yielding and generation of pore pressure

On the centreline beneath a circular load, such as the tank in Fig. 11.1, conditions of axial symmetry obtain, and stress paths and strain increments can be described in terms of the triaxial variables p', q and $\delta\varepsilon_p, \delta\varepsilon_q$. A tank of radius 5 m is considered here. It is assumed that it is filled rapidly, so that the underlying deep bed of soft clay on which the tank is founded is initially loaded in an undrained manner. The soft clay is described using the Cam clay model of Chapter 5. However, in displaying effective stress paths, the Cam clay yield locus is indicated partly by a dashed line: it is shown that for this application there is little difference between statements made using the Cam clay model and

Fig. 11.1 Element X of soil at depth z on centreline beneath tank of radius a applying uniform circular load ζ.

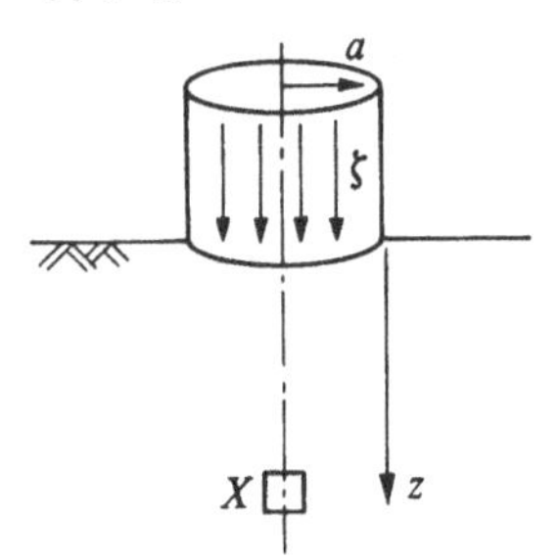

Table 11.1

Soil parameter	Symbol	Value
Saturated unit weight of soil	γ	16 kN/m^3
Unit weight of water	γ_w	9.81 kN/m^3
Angle of shearing resistance	ϕ'	23°
Slope of critical state line, from (7.9)	M	0.898
Coefficient of earth pressure at rest for clay in its normally compressed state, from (10.15)	K_{0nc}	0.61
Slope of normal compression line	λ	0.25
Slope of unloading–reloading line	κ	0.05
Hence from (5.20)	$(\lambda - \kappa)/\lambda = \Lambda$	0.8
Location of critical state line in v:ln p' plane	Γ	2.5
Which implies from (6.11)	$\Gamma + (\lambda - \kappa)\ln 2 = N$	2.64
Elastic shear modulus	G'	500 kPa

statements made using a volumetric hardening model which incorporates a shape of yield locus corresponding more closely to the shapes found from experimental probing of natural clays (see Chapter 3).

Values for a number of soil parameters are needed and are tabulated in Table 11.1.

In Section 10.3.3 we noted that most natural deposits of soft clay appear to be lightly overconsolidated in situ, and, in this example, we assume that the clay is lightly overconsolidated as a result of fluctuations of the water table, which is currently at the ground surface but has been down to a depth of 1 m. At each depth, the size of the current yield locus is controlled by the past maximum stresses which were experienced by the soil when the water table was at its lowest. In this normally compressed condition, the ratio of horizontal to vertical effective stresses was $\sigma'_{hc}/\sigma'_{vc} = K_{0nc}$. These preconsolidation stresses σ'_{vc} and σ'_{hc} are plotted in Fig. 11.2. From these, the corresponding values of p'_c and q_c and $\eta_c = q_c/p'_c$ can be calculated; hence, for each depth, the size of the current yield locus p'_o can be calculated from

$$\frac{p'_o}{p'_c} = 1 + \frac{\eta_c^2}{M^2} \tag{11.1}$$

For the combination of soil parameters used here, the values of p'_o are almost identical to the values of σ'_{vc}.

The current vertical effective stresses σ'_{vi} with the water table at the ground surface vary linearly with depth (Fig. 11.2). As a result of the fluctuation of the water table between the ground surface and depth

$d = 1$ m, the overconsolidation ratio n at depth z is

$$n = 1 + \left(\frac{\gamma_w}{\gamma'}\right)\left(\frac{d}{z}\right) \qquad (11.2)(10.40\text{bis})$$

where $\gamma' = \gamma - \gamma_w$ is the buoyant unit weight. The value of the earth pressure coefficient at rest K_0 for the overconsolidated clay is assumed to be given by

$$K_0 = K_{0nc} n^{1/2} \qquad (11.3)(\text{cf. } 10.24)$$

From this, values of the current horizontal effective stress σ'_{hi} can be calculated (Fig. 11.2) and hence values of mean effective stress p'_i and deviator stress q_i.

In terms of elastic–plastic soil models, the initially overconsolidated state of the clay implies that at all depths the initial effective stress state lies inside the initial yield locus (B' in Fig. 11.3). If an element of this soil is loaded without drainage in a conventional triaxial compression test, from its in situ stress state, the total stress changes follow the path BCD with slope $\Delta q/\Delta p = 3$, but the effective stress changes are fixed by the stress:strain behaviour of the soil and are largely independent of the total stress changes. For a soil which behaves isotropically and elastically prior to yield, the effective stress path shows no change in mean effective stress p' until the yield locus is reached (at C' in Fig. 11.3). Once the soil has yielded at C', the effective stress path turns towards the critical state at D', and pore pressures build up so that the recoverable, elastic expansion arising from the reduction in mean effective stress can balance the irrecoverable, plastic compression associated with the changing size of the yield locus.

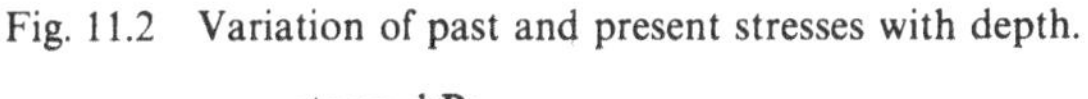
Fig. 11.2 Variation of past and present stresses with depth.

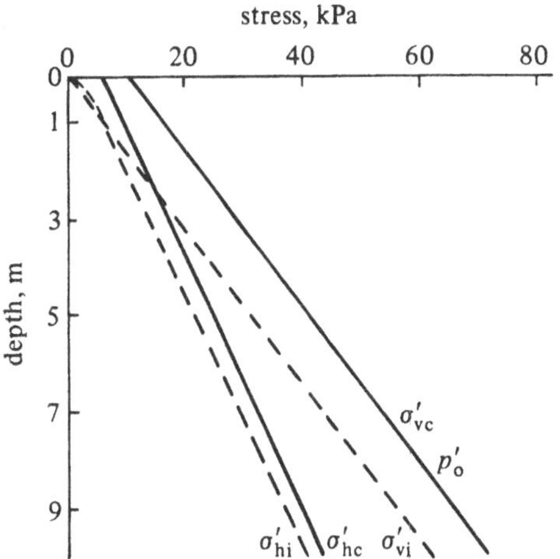

The effective stress path $B'C'D'$ is *the* undrained effective stress path which is followed whatever the total stress path, provided it involves an increase in deviator stress q. Elements on the centreline beneath the circular load (Fig. 11.1) experience axially symmetric changes of stress (Section 10.3.4) and, in undrained loading, follow the same effective stress changes $B'C'D'$ shown in Fig. 11.3, even though the total stress changes are quite different. While all the soil beneath the circular load is deforming elastically, the total stress changes can be deduced from straightforward

Fig. 11.3 Total and effective stress paths for undrained triaxial test on lightly overconsolidated soil.

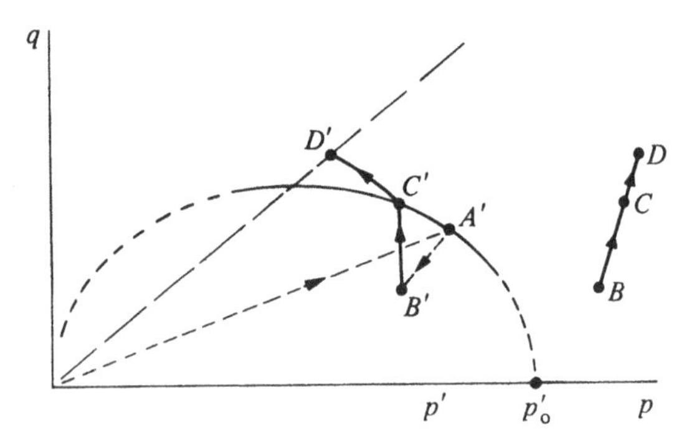

Fig. 11.4 Increases of total stresses on centreline beneath uniform circular load on elastic half-space (Poisson's ratio $\nu = 0.5$) (data from Poulos and Davis, 1974).

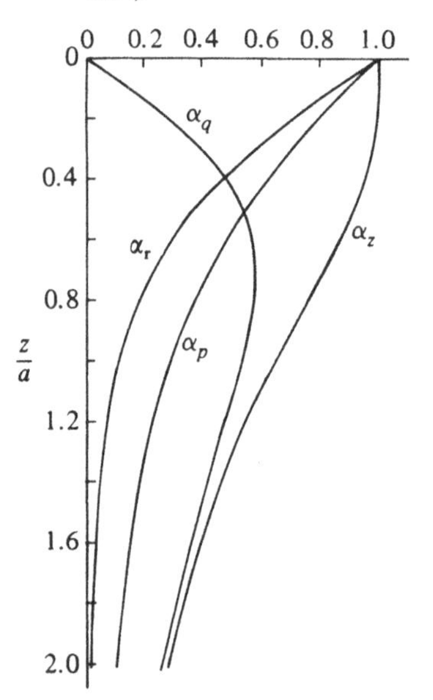

elastic analysis. Tables and charts of elastic stress distributions have been gathered by Poulos and Davis (1974) for a vast range of shapes and configurations of loading, including the stresses on the centreline beneath a flexible circular load of radius a, applying a uniform surface pressure ζ (Fig. 11.1). The variations of changes in dimensionless total vertical stress

$$\alpha_z = \frac{\Delta\sigma_z}{\Delta\zeta} \tag{11.4}$$

total horizontal or radial stress

$$\alpha_r = \frac{\Delta\sigma_r}{\Delta\zeta} \tag{11.5}$$

total mean stress

$$\alpha_p = \frac{\Delta p}{\Delta\zeta} \tag{11.6}$$

and deviator stress

$$\alpha_q = \frac{\Delta q}{\Delta\zeta} \tag{11.7}$$

with dimensionless depth z/a are shown in Fig. 11.4. The curves have been calculated for Poisson's ratio $\nu = 0.5$ as is appropriate for undrained deformation of isotropic elastic soil (Section 2.2). At the ground surface, $z/a = 0$, the changes in vertical stress and horizontal stress are both equal to the applied load, so the change in deviator stress is zero. The horizontal stress and mean stress decay fairly rapidly with depth, but the vertical stress falls off more gradually. The deviator stress rises to a maximum value at a depth equal to about three-quarters of the radius of the loaded area and is still more than half this maximum value at a depth equal to the diameter of the loaded area. It is seen in Section 11.2.2 that this is very significant for estimation of settlements.

The curves in Fig. 11.4 can be used to estimate the changes in total mean stress p and deviator stress q that occur at an element X (Fig. 11.1) beneath the load and hence to deduce the total stress path BC (see Fig. 11.5). As the tank is filled, the soil initially all responds elastically, and the excess pore pressure at any depth arises entirely from the change in total mean stress that has developed at that depth. So, for example, the pore pressure at X changes from its initial value because of the increase in total mean stress between B and C (Fig. 11.5). From (11.6), the ratio of change of pore pressure u to change of surface load ζ is

$$\frac{\delta u}{\delta\zeta} = \alpha_p \tag{11.8}$$

With a tank load of $\zeta = 10\,\text{kPa}$, for example, the variation of excess pore pressure with depth (see Fig. 11.6) matches the variation of α_p shown in Fig. 11.4.

Once the soil has yielded at C', the effective stress path is controlled by the simultaneous occurrence of plastic and elastic volumetric strains. According to the Cam clay model, the effective stress path is defined by

$$\frac{p'}{p'_y} = \left(\frac{M^2 + \eta_y^2}{M^2 + \eta^2}\right)^{\Lambda} \qquad (11.9)(\text{cf. } 5.19)$$

where p'_y and η_y are the values of mean effective stress and stress ratio at the moment during undrained loading at which plastic strains began, and p' and $\eta = q/p'$ are current values of mean effective stress and stress ratio.

Once the effective stress path has passed through the initial yield surface at C' (Fig. 11.5), elastic analyses are no longer strictly valid because yielding

Fig. 11.5 Total and effective stress paths for element on centreline beneath circular load.

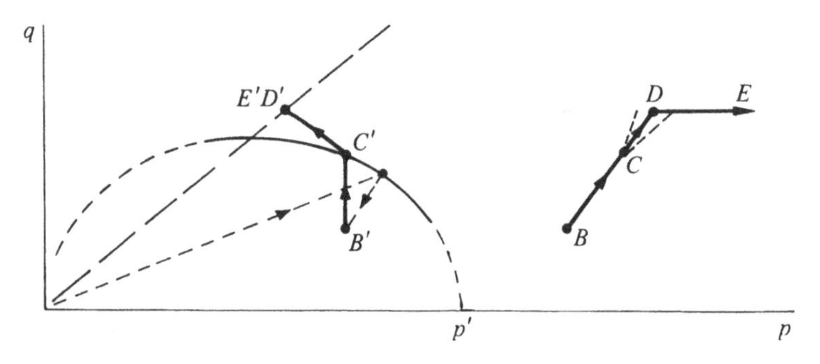

Fig. 11.6 Variation of pore pressure with depth for surface load $\zeta = 10$, 14.5, and 20 kPa.

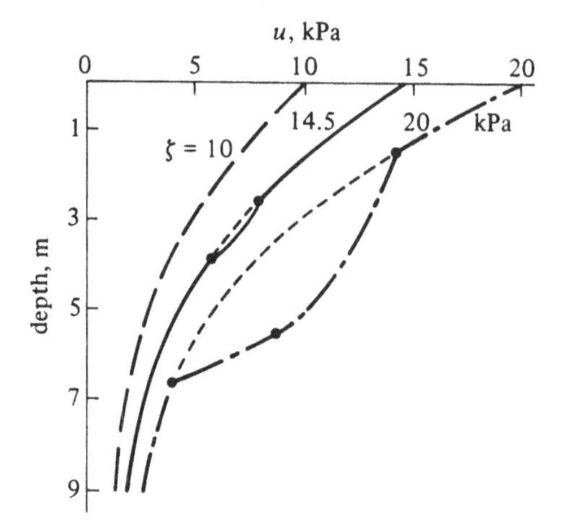

elements of soil have a lower stiffness for continued increase in shear stress than the surrounding still elastic soil. However, so long as only a small proportion of the soil beneath the load has reached yield, the changes in total stress calculated from the elastic analysis provide a reasonable initial estimate of the actual total stress path. The total stress path BCD is shown straight in Fig. 11.5, but the alternative dotted lines from C are added as a reminder of the uncertainty. It can be supposed anyway that the change in slope of the total stress path at C is less dramatic than the change in slope of the effective stress path at C' as it turns towards the critical state line $C'D'$.

Assuming that the total stress changes are indeed still given by the elastic analysis, we can calculate the excess pore pressures at any depth after some yielding has occurred from the changes of total and effective mean stresses. For a surface load ζ, the total mean stress at some depth can be determined from

$$p = p_{\mathrm{i}} + \alpha_p \zeta \tag{11.10}$$

and the deviator stress can be determined from

$$q = q_{\mathrm{i}} + \alpha_q \zeta \tag{11.11}$$

The effective mean stress p' can be deduced implicitly from (11.9) (but not explicitly since η is a function of mean effective stress) and then the current excess pore pressure:

$$\Delta u = (p - p_{\mathrm{i}}) - (p' - p'_{\mathrm{i}}) \tag{11.12}$$

This excess pore pressure is now made up of two parts: one due to the applied change in total mean stress and the other due to the change in effective mean stress which is the consequence of yielding having occurred.

For example, at a surface load $\zeta = 14.5\,\mathrm{kPa}$, the soil around a depth $z = 3\,\mathrm{m}$ $(z/a = 0.6)$ has yielded, and the curve of variation of excess pore pressure with depth (Fig. 11.6) shows a small blip at this depth. This depth at which yielding first began is the result of a combination of two effects: the past history of the soil, which leads to a steady increase in the size p'_{o} of the initial yield loci with depth (Fig. 11.2), and the applied loading, which (as is apparent from Fig. 11.4) produces the greatest shearing of the soil (greatest α_q and hence greatest increase in q) at a depth equal to about three-quarters of the radius of the uniformly loaded area.

Note that yielding and failure are in general quite distinct aspects of the mechanical behaviour of soils. The onset of yielding corresponds to a drop in stiffness of the soil, but in neither the triaxial test (Fig. 11.3) nor the element beneath the circular load (Fig. 11.5) is the passage of the effective stress path through the initial yield locus at C' significant for stability.

Whereas the laboratory test on a single element of soil terminates as the effective stress path approaches D' on the critical state line (Fig. 11.3), field loading of element X can continue. As the effective stress path for element X (Fig. 11.1) approaches point D' on the critical state line (Fig. 11.5) the tangent or incremental shear stiffness local to point X approaches zero, and the soil is unable to support any further shear stress. However, a distinction must be drawn between this local failure at a particular soil element and general failure of a geotechnical structure. Just as a steel structure requires several plastic hinges to form a mechanism of collapse (Baker and Heyman, 1969), so the geotechnical structure can fail only if a mechanism of failure exists through the soil beneath the structure (Fig. 11.7). This requires that many soil elements have reached a critical state; until that occurs, local attainment of critical states, or local failure, is contained and supported by surrounding unfailed soil.

The basic assumption of the critical state condition is that a state is reached at which deformation continues without further change in effective stresses: the effective stress path stops at point D' in Fig. 11.5. In particular, no further change in deviator stress q can occur:

$$\delta q = 0 \tag{11.13}$$

Changes in q arise from differences between changes in vertical total stress σ_z and radial total stress σ_r

$$\delta q = \delta\sigma_z - \delta\sigma_r \tag{11.14}$$

so (11.13) requires that, during this condition of contained failure,

$$\delta\sigma_z = \delta\sigma_r = \delta p \tag{11.15}$$

and changes of total vertical stress and total mean stress are identical. The change of mean effective stress p' is zero because the effective stress state has to remain at the critical state value; this implies that the pore pressure changes are equal to the change in total mean stress and hence to the change in total vertical stress:

$$\delta u = \delta\sigma_z \tag{11.16}$$

The effective stress state cannot change, so D' and E' (in Fig. 11.5) are at the same point. However, the total stress path can move sideways at constant q (DE in Fig. 11.5). An elastic analysis is not obviously relevant

Fig. 11.7 Mechanism of plastic collapse of long footing on clay.

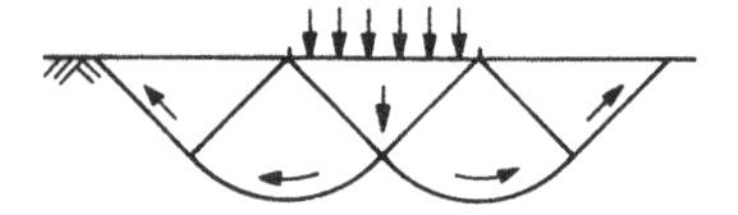

at this stage. However, finite element analyses show that the changes in vertical total stress beneath a surface load are rather independent of the soil model used, whereas the changes in horizontal total stress are much more dependent on the particular model – which controls how much resistance against lateral deformation the soil is able to provide (e.g. see Höeg, Christian, and Whitman, 1968; Burland, 1971). An illustration of this is provided by the fact that changes in total vertical stress calculated using an elastic analysis are independent of the value of Poisson's ratio ν that is used (see Poulos and Davis, 1974). It could be proposed that when contained failure is reached at a particular soil element, the total vertical stress is still given by elastic theory, whereas the total horizontal stress adjusts itself so that there is no nett change in deviator stress q. Equation (11.16) then implies that the rate at which pore pressure develops with continuing increase of surface load ζ is

$$\frac{\delta u}{\delta \zeta} = \alpha_z \tag{11.17}$$

At a surface load $\zeta = 20\,\mathrm{kPa}$, the soil has reached a critical state between depths of about 1.5 and 5.5 m and has yielded down to about 6.7 m. The variation of excess pore pressure with depth is shown in Fig. 11.6. Again, note that the pore pressures that are seen result from the combination of present stress path with past history: consideration only of changes in total stresses is insufficient to describe the response of the soil. Equally, it is not possible to say that a particular soil element will behave elastically or plastically from knowledge only of its past history; it is necessary also to know how it is to be loaded so that it is possible to estimate whether the transition from elastic to plastic response is likely to occur under the applied loads. This will be apparent again when simple calculations of settlements are considered in Section 11.2.2.

Another way of displaying the information of pore pressure generation and the onset of yielding at individual points is shown in Fig. 11.8. During the conventional undrained triaxial compression test shown in Fig. 11.3, changes in deviator stress Δq and axial total stress $\Delta\sigma_a$ are identical and the total stress path has slope $\delta q/\delta p = 3$. Initially, the change in pore pressure is equal to the change in total mean stress, and a plot of change in pore pressure against change in axial stress or deviator stress has slope $\delta u/\delta q = \delta u/\delta\sigma_a = \frac{1}{3}$ (*BC* in Fig. 11.8a). Once yielding begins, there is a sharp increase in the rate at which pore pressures develop with further increase in axial stress (*CD* in Fig. 11.8a). Even without any measurement of deformation during the test, the occurrence of yield could be identified from the effective stress path or the pore pressure changes.

If the pore pressure at element X (Fig. 11.1) is monitored with a piezometer as the surface load ζ is increased, then a plot of change of pore pressure against surface load has the form shown in Fig. 11.8b. From (11.8), the initial slope BC, before the soil at this position starts to yield, is α_p. From (11.17), the final slope DE, once the soil at this position has reached a critical state, is α_z. Between C and D, it was noted earlier that the total stress path may or may not have been significantly affected by the occurrence of yield; and it is not possible to make precise statements about the slope CD in Fig. 11.8b.

Two circular test fills were built by the Norwegian Geotechnical Institute on an instrumented soft clay site at Åsrum, about 100 km southwest of Oslo. Data from these field loading tests have been examined by Höeg,

Fig. 11.8 (a) Variation of pore pressure with applied deviator stress in undrained triaxial test on lightly overconsolidated soil; (b) variation of pore pressure with applied surface load for element on centreline beneath circular load; (c) variation of pore pressure with applied surface load for piezometers at depth 3 m at Åsrum (after Höeg, Andersland, and Rolfsen, 1969).

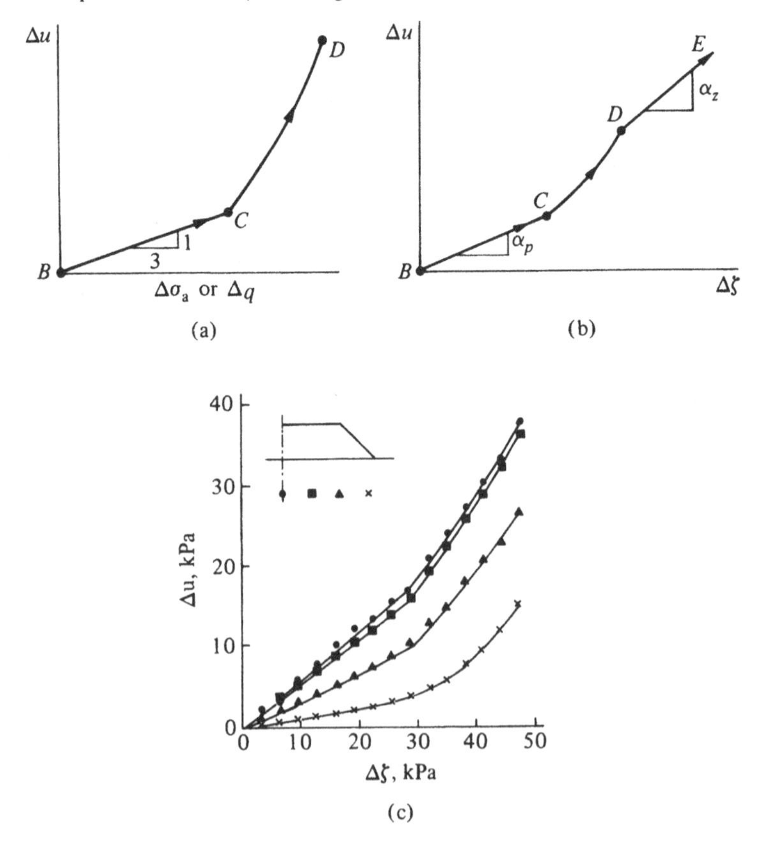

Andersland, and Rolfsen (1969) and by Parry and Wroth (1981). A typical section through one circular test fill is shown in Fig. 11.8c, together with the observed pore pressure changes measured at four points beneath the fill. These plots are the observed equivalent of Fig. 11.8b. The test fill had a top radius of 5 m and side slopes of 1:1. With a mean radius of 6.25 m for an equivalent uniform surface load (Fig. 11.1), the depth 3 m of the piezometers in Fig. 11.8c corresponds to a ratio $z/a = 0.48$. From the curves in Fig. 11.4, at this depth $\alpha_p = 0.567$, whereas the measured rate of initial increase in pore pressure with applied surface load for the piezometer on the centreline in Fig. 11.8c is 0.59.

The discussion has so far been restricted to elements on the centreline of circular loads where changes of stress are axially symmetric. At elements off the centreline, conditions are certainly not axially symmetric (compare Section 10.4.2). However, if the soil behaves isotropically and elastically before yield, then whatever the changes in total stress, even if they include rotation of principal axes, the effective stress path should show no change in mean effective stress (see Chapter 2), and the pore pressure changes will still be equal to the changes in total mean stress predicted by the elastic analysis. Further tables in Poulos and Davis (1974) have to be consulted for elements off the centreline, but Höeg et al. (1969) show that elastic calculations of the initial pore pressure development seem to provide a reasonable estimate of the observed response both on and off the centreline. A similar conclusion is drawn by Clausen (1972) and Clausen et al. (1984) for another field loading test at Mastemyr, Norway.

Equation (11.17) and Fig. 11.8b show that the ultimate slope of the plot of excess pore pressure against applied surface load for a particular point on the centreline should be equal to α_z, once the critical state has been reached at that point. For $z = 3$ m, on the centreline at Åsrum, $z/a = 0.48$ and, from Fig. 11.4, $\alpha_z = 0.919$, whereas the observed final slope in Fig. 11.8c is $\delta u/\delta\zeta = 1.1$. In this region, the calculation has certainly become too simplistic. In reality, a natural clay such as that at Åsrum will show some strain softening, associated with loss of structure, once a failure state has been reached (Leroueil, Magnan, and Tavenas, 1985).

11.2.2 Yielding and immediate settlement

Design of a foundation usually requires a check that the foundation will not reach an ultimate limit state; that is, it will not collapse under the applied loads. Such a calçulation, which requires knowledge of the strength of the soil, is typically based in the theory of plasticity, analysing mechanisms of collapse such as that in Fig. 11.7 (e.g. see Atkinson, 1981). However, it is also necessary to ensure that the foundation does not deform so much under working loads that, even without actually collapsing, it reaches a serviceability limit state and ceases to perform the functions for which it has been constructed. Traditionally, settlement calculations have divided the total settlement into two components: an immediate undrained settlement and long-term consolidation settlement associated with the dissipation of excess pore pressures. Tavenas and Leroueil (1980) suggest that loading of geotechnical structures is usually accompanied by significant pore pressure dissipation, so that the assumption of initial undrained conditions is not satisfactory. However, calculation of soil behaviour under such partially drained conditions cannot be achieved without numerical analyses such as those described in Section 11.3; for the simple calculations to be presented here, the traditional separation of immediate and long-term effects is retained.

Calculations of settlement are typically performed using an elastic analysis. Harr (1966), for example, has shown that the settlement at the centre of a uniformly loaded area on the surface of a semi-infinite mass of isotropic elastic soil of shear modulus G and Poisson's ratio ν can be written as:

$$\rho = (1 - \nu)\frac{\zeta a I_\rho}{G} \tag{11.18}$$

where ζ is the applied surface load, a a dimension of the loaded area, and I_ρ a dimensionless influence factor, whose value depends on the shape of

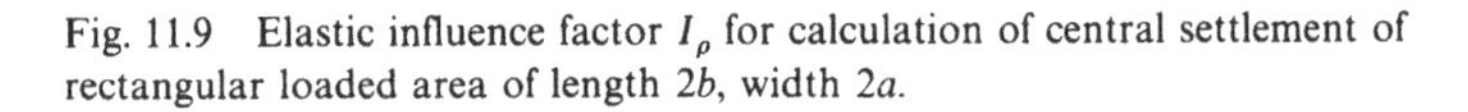

Fig. 11.9 Elastic influence factor I_ρ for calculation of central settlement of rectangular loaded area of length $2b$, width $2a$.

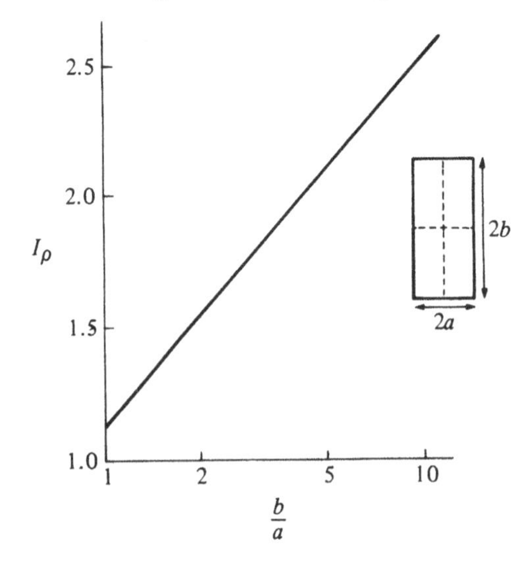

Table 11.2 *Vertical strains beneath circular load*

Depth (m)	Immediate vertical strains (%)		Consolidation vertical strains (%)	
	$\zeta = 10\,\mathrm{kPa}$	$\zeta = 14.5\,\mathrm{kPa}$	$\zeta = 10\,\mathrm{kPa}$	$\zeta = 14.5\,\mathrm{kPa}$
1	0.189	0.273	0.75	2.9
3	0.378	$0.549^{e} + 2.25^{pa}$	0.25	5.2
5	0.353	0.513	0.11	0.8
7	0.275	0.399	0.05	0.08
9	0.210	0.304	0.03	0.04

[a] e denotes elastic component; p denotes plastic component.

the loaded area. For a circular loaded area of radius $a, I_\rho = 1$; for a square loaded area of side $2a, I_\rho = 1.122$; and for a rectangular loaded area $2a \times 2b, I_\rho$ varies approximately linearly with $\ln b/a$ (Fig. 11.9). The settlement at the corner of a rectangular loaded area is half the settlement at the centre given by (11.18). The settlement at the edge of a circular loaded area is $2/\pi = 0.637$ times the settlement at the centre given by (11.18).

Consider the circular tank in Fig. 11.1. It was noted in the previous section that for an applied surface load $\zeta = 10\,\mathrm{kPa}$, the effective stress changes lie entirely within the initial yield loci and are consequently accompanied only by elastic deformations. The vertical strains occurring at each depth can be calculated from

$$\delta\varepsilon_z = \delta\varepsilon_q = \frac{\delta q}{3G} \tag{11.19}$$

(During axisymmetric constant volume deformation, axial strains and triaxial shear strains are identical.) For the purposes of calculation of settlements, the soil beneath the tank might be divided into a number of layers, each 2 m thick; the elastic strains are then estimated with (11.19) from the changes in deviator stress occurring at the mid-depth of each layer. These are tabulated in the second column of Table 11.2. Summing deformations over five layers, to a depth $z/a = 2$, and treating the values in Table 11.2 as representative of the average strain in each layer, we can estimate the elastic settlement of the centre of the circular tank to be 28.1 mm, using $G = 500\,\mathrm{kPa}$ in (11.19). However, using (11.18), with $\nu = 0.5$ for constant volume deformation, we calculate a settlement $\rho = 50\,\mathrm{mm}$. The estimate based on the deformation of the top 10 m of soil is thus about 44 per cent too low. The reason for this error is apparent from study of Fig. 11.4: the maximum value of $\alpha_q = \delta q/\delta\zeta$ is about 0.58 at a dimensionless depth $z/a \approx 0.75$, and α_q has only fallen to about half this value at $z/a = 2$, which for the tank of radius 5 m corresponds to a depth of 10 m. The change of q, the shear stress applied to the soil, is thus significant at great depths in the soil, and soil conditions at great depths may need to be taken into account when calculating settlements.

When the applied surface load is increased to 14.5 kPa, the soil layer around the depth 3 m starts to deform plastically (see Fig. 11.6). The strains at each depth are tabulated in Table 11.2; the elastic–plastic Cam clay model is now required to estimate the strains where yielding has occurred. The large plastic contribution to the strain at depth 3 m produces a settlement of 44.9 mm in the layer between depths of 2 and 4 m, which is of the same order as the total elastic settlement of 40.8 mm calculated over the top 10 m. From (11.18), a total elastic settlement of 72.5 mm is calculated, allowing for the soil at greater depth. The total settlement is then 117.4 mm, of which 73 per cent occurs in the top 10 m. The occurrence of yielding thus has a major effect in concentrating deformation near the surface. The occurrence of yielding also makes the elastic analysis increasingly irrelevant: the settlement of 117.4 mm could be attributed to a uniform elastic soil with shear modulus reduced from 500 kPa to 309 kPa; but this is not a helpful way of escaping from the realities of irrecoverable plastic deformations and would give a false impression about the distribution of strain with depth in the soil.

The surface load of $\zeta = 14.5$ kPa that has been reached in these simple settlement calculations is evidently pretty low. However, this calculation procedure, in which the estimation of total stress changes is uncoupled from the description of the stress:strain response of the soil, breaks down as soon as any soil element on the centreline is supposed to have reached

Fig. 11.10 Variation in vertical strain with applied surface load for settlement gauges on centreline at Mastemyr (after Clausen, Graham, and Wood, 1984).

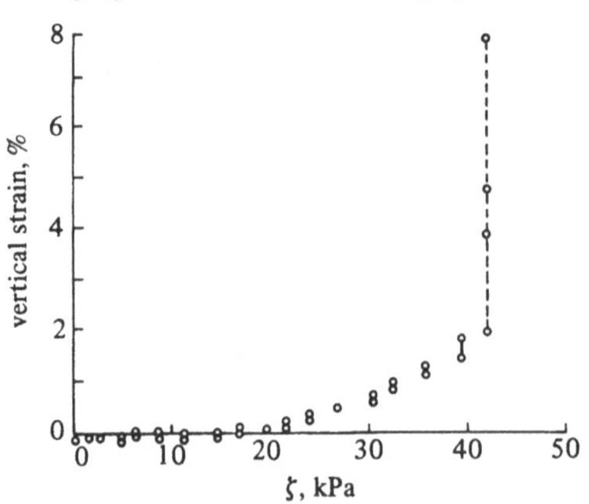

a critical state. For the soil actually to reach a critical state, infinite shear strains and correspondingly infinite settlements would have to occur. For higher loads, a more rigorous form of analysis cannot be avoided.

Data from another field loading test performed by the Norwegian Geotechnical Institute at Mastemyr, about 15 km southeast of Oslo have been presented by Clausen (1972) and examined by Clausen et al. (1984). A circular test fill of base diameter 20 m was constructed similar to that at Åsrum. The extensive instrumentation of the soil beneath the test fill included settlement gauges at various points, including two at depths 2.3 m and 6 m, more or less on the centreline of the test fill. The relative settlements of these two gauges have been converted to vertical strains (which for axisymmetric undrained conditions are identical to triaxial shear strains) and plotted against the applied surface load in Fig. 11.10. It might be deduced that yielding has occurred somewhere on this gauge length at a surface load of about 18 kPa. This gauge length which extends over a distance of 3.7 m does not provide a particularly precise pin-pointing of the occurrence of yield. However, this single example does illustrate the principle that yielding should be discernible from field observations of strains, as well as from field observations of excess pore pressures (Figs. 11.8b, c).

11.2.3 Yielding and coefficient of consolidation

The immediate undrained settlement arises as a result of the shearing of the soil at constant volume beneath the surface loading when this loading is applied rapidly. This shearing is accompanied by the generation of excess pore pressures. If the load is kept constant, these excess pore pressures dissipate, and further strains occur as the effective stresses in the soil increase. This is the process known as consolidation of soil, a time-dependent diffusion of pore pressures with flow of water down gradients of excess pore pressure. The rate at which water can flow through the soil under a given pressure gradient depends on the permeability k of the soil. Nett flow of water into or out of an element results in a change in volume of that element associated with changes in effective stresses. The complete link between volume changes and changes in effective stresses requires a model of the stress:strain behaviour of the soil such as the elastic–plastic models described here.

The original development of the theory of consolidation of soils by Terzaghi (1923) considered only the one-dimensional time-dependent deformation of soils that could be observed in an oedometer. In this simple one-dimensional consolidation theory, the rate of diffusion of pore pressure perturbations through the soil is characterised by a coefficient of consolidation c_v,

$$c_v = \frac{k}{m_v \gamma_w} \tag{11.20}$$

which is a ratio of permeability k to so-called coefficient of volume compressibility m_v. This coefficient m_v is a ratio of increment of vertical strain to increment of vertical effective stress in an oedometer test and is thus a one-dimensional or confined compliance. The results of oedometer tests are usually presented as plots of equilibrium sample height or specific volume v against vertical effective stress σ'_v (Fig. 3.12a). The height of the sample is at all times directly proportional to the specific volume v of the soil. The coefficient of volume compressibility is then

$$m_v = \frac{-\delta h}{h\,\delta\sigma'_v} = \frac{-\delta v}{v\,\delta\sigma'_v} \tag{11.21}$$

In one-dimensional deformation, vertical strain and volumetric strain are identical.

In Section 3.3, yielding in oedometer tests was associated with the increase of the vertical effective stress beyond the so-called preconsolidation pressure σ'_{vc}. This yield point seen in one-dimensional deformation represents just one point on the yield surface for the soil but marks the transition in the oedometer tests between stiff and less stiff response. The compressibility m_v is clearly not a fundamental constant soil parameter, since it must depend on consolidation history.

As noted in Section 3.3, detection of this yield point is often facilitated by plotting the results of oedometer tests with a logarithmic stress axis (Fig. 3.12b). In Section 4.2, we noted that one of the advantages of plotting compression data in such a semi-logarithmic compression plane was that the data of reloading, for stresses below the preconsolidation pressure $\sigma'_v < \sigma'_{vc}$, and the data of normal compression, for stresses greater than the preconsolidation pressure $\sigma'_v > \sigma'_{vc}$, are often found to lie on approximately straight lines. These lines were there described by the expressions

$$v = v_\lambda - \lambda \ln p' \qquad (11.22)\,(4.2\text{bis})$$

for normal compression, and

$$v = v_\kappa - \kappa \ln p' \qquad (11.23)\,(4.3\text{bis})$$

for unloading and reloading. If it is assumed that changes in mean effective stress p' and vertical effective stress σ'_v are always in constant proportion during one-dimensional compression (notwithstanding the discussion of Section 10.3.2), then (11.21) can be combined with the differential forms of (11.22) and (11.23) to relate the value of m_v to the more fundamental soil parameters λ and κ.

The differential form of (11.22) is

$$\delta v = -\lambda \frac{\delta p'}{p'} = -\lambda \frac{\delta \sigma'_v}{\sigma'_v} \tag{11.24}$$

so that during normal compression,

$$m_v = \frac{\lambda}{v\sigma'_v} \tag{11.25}$$

and from (11.20) the coefficient of consolidation is

$$c_v = \frac{kv\sigma'_v}{\lambda\gamma_w} \tag{11.26}$$

The differential form of (11.23) is

$$\delta v = -\kappa \frac{\delta p'}{p'} \approx -\kappa \frac{\delta \sigma'_v}{\sigma'_v} \tag{11.27}$$

so that during unloading and reloading,

$$m_v = \frac{\kappa}{v\sigma'_v} \tag{11.28}$$

and

$$c_v = \frac{kv\sigma'_v}{\kappa\gamma_w} \tag{11.29}$$

One consequence of these expressions is that for changes in vertical effective stress which lie either entirely below the preconsolidation pressure (inside the yield surface) or entirely above the preconsolidation pressure, the value of the compressibility m_v is roughly inversely proportional to

Fig. 11.11 Oedometer test on soft marine clay from Belawan, Sumatra, Indonesia (after Barry and Nicholls, 1982).

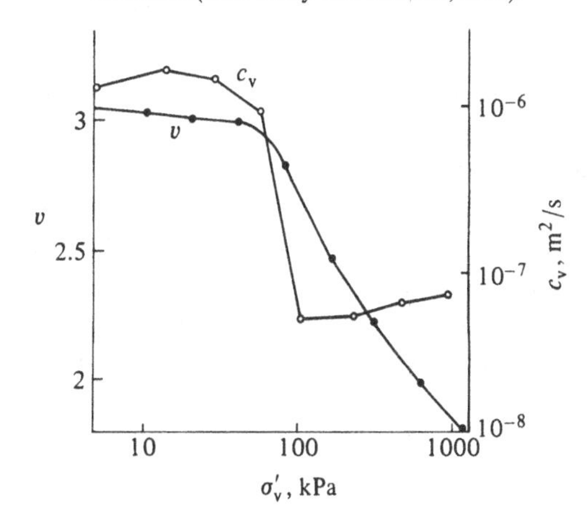

the effective stress. However, as the vertical effective stress passes through the preconsolidation pressure σ'_{vc}, though there is no sudden change in specific volume v or vertical effective stress σ'_v, there is a marked increase in m_v as the response of the soil changes from elastic, governed by κ (11.28), to elastic plus plastic, governed by λ (11.25).

The permeability of a soil is largely governed by the size of the pores through which the water flows and is consequently largely dependent on the void ratio or specific volume of the soil; the permeability therefore does not change as the preconsolidation pressure is passed. However, the rate of consolidation, as measured by the coefficient c_v [(11.26) and (11.29)], drops because of the increase in m_v as the soil passes through the yield surface. Such a drop in c_v is a familiar feature of the interpretation of conventional oedometer test results (Fig. 11.11).

This step change in coefficient of consolidation c_v as a yield point is passed arises not because there is a change in the rate at which pore water can flow to dissipate excess pore pressures – this is controlled by the permeability – but because there is an increase in the rate at which pore water is being expelled from the soil, for a given change in effective stress, as the soil tries to collapse under continued loading. The coefficient of consolidation c_v is not a satisfactory parameter for describing movement of pore water in general transient problems; a more fundamental approach is to make use of the permeability (which may well be permitted to change as the volume of the soil changes (Al-Tabbaa and Wood, 1987)) together with a complete effective stress model of the soil response which automatically indicates the volumetric stiffness of the soil under any change in stresses and hence produces a proper coupling of the effective stress changes and the flow of water in the soil. Of course, this more fundamental approach is likely to require numerical analyses such as those described in Section 11.3 and by Leroueil, Magnan, and Tavenas (1985).

11.2.4 Yielding and long-term settlement

Estimation of the settlement that accompanies the dissipation of excess pore pressures requires some measure of stiffness to associate with the changes in effective stress. A simple approach to the estimation of this long-term settlement might use an elastic analysis to estimate the change in *total* vertical stress on loading, equate this to the change in vertical *effective* stress that occurs on consolidation, and use data from oedometer tests to estimate the corresponding vertical strain. Such a procedure ignores the fact that, as shown in Section 11.2.1, the excess pore pressures that are generated during undrained loading are in general different from the increment in total vertical stress, and it is the transfer from excess pore pressure to effective stress that causes the long-term consolidation settlement.

A more plausible approach is that of Skempton and Bjerrum (1957). They propose, firstly, that the changes of total stresses should be calculated from an elastic analysis, and secondly, that the excess pore pressure that is generated should be calculated from these total stress changes using a pore pressure equation (Section 1.6),

$$\Delta u = b(\Delta p + a\,\Delta q) \qquad (11.30)(\text{cf. } 1.66)$$

which, for the circular load being considered here, can be written as

$$\Delta u = \Delta\zeta b(\alpha_p + a\alpha_q) \qquad (11.31)$$

where a and b are so-called pore pressure parameters, and $b = 1$ for saturated soil. Thirdly, they propose that the settlement ρ of a soil layer of total thickness H in which these pore pressures are dissipating can be calculated from the integral

$$\rho = \int_0^H m_v\,\Delta u\,dz \qquad (11.32)$$

where values of the compressibility m_v are taken from oedometer tests.

There are two flaws in the logic of this approach, however. On the one hand, the pore pressure parameter a in (11.30) and (11.31) is not a soil constant. It was demonstrated in Section 5.4 that its value depends on the stress history of the soil and the imposed stress changes; this was shown again in Section 11.2.1 and Fig. 11.6, where it emerged that the distribution of pore pressure with depth beneath a circular load depends on the extent to which yielding occurs in the soil beneath that load. On the other hand, the settlement that occurs during the dissipation of pore pressures is not the result of one-dimensional deformation of the soil, as the plucking of values of m_v in (11.32) from oedometer tests would seem to imply. It is to be expected that vertical deformations will be accompanied by lateral deformations.

It has already been shown in Section 11.2.1 that the Cam clay model (or any other elastic–plastic model) can be used in place of (11.30) and (11.31) to estimate the pore pressures that are generated when the soil is loaded without drainage. The same model can be used in place of a single value of m_v in (11.32) to estimate the vertical strains and hence the settlements that accompany the dissipation of these pore pressures.

The example of a circular tank on a deep layer of soft clay shown in Fig. 11.1 can be considered further. The immediate vertical strains occurring at the centre of the top five 2-m layers of soil on the centreline of the tank were tabulated in Table 11.2 for applied loads of 10 kPa and

14.5 kPa. At different depths the effective stress state will have progressed different distances along an undrained effective stress path, such as $B'C'D'$ in Fig. 11.5. Suppose, for simplicity, that the total stresses remain unchanged during the dissipation of the pore pressures generated by this undrained loading; there are three possibilities for the effective stress paths followed at any depth as the excess pore pressures dissipate. The simplest possibility is that neither the undrained loading nor the dissipation of excess pore pressures produces an effective stress state lying beyond the initial yield surface ($B'R'S'$ in Fig. 11.12a). The second possibility is that, though the undrained loading is entirely elastic, the process of dissipation of excess pore pressures takes the effective stress state through the initial yield locus (at T' on path $R'S'$ in Fig. 11.12b). The third possibility is that the undrained loading causes yielding of the soil (at C' in Fig. 11.12c) and expansion of the yield locus, and that the entire consolidation process is accompanied by plastic deformations ($R'S'$ in Fig. 11.12c).

The applied load of 10 kPa caused no yielding of the soil; the undrained shearing was purely elastic. It turns out that, with yield loci of the elliptical shape assumed in the Cam clay model, dissipation of the excess pore pressures set up by this load also causes no yielding of the soil (Fig. 11.12a). The vertical strains at each depth arise purely from elastic volumetric strains associated with increase in mean effective stress; these vertical strains are tabulated in Table 11.2. The distribution of pore pressure is the same as the elastic distribution of total mean stress (α_p in Fig. 11.4), which decays rapidly with depth. Although the deeper soil layers may contribute significantly to the vertical strains which produce the immediate settlement of the load, they do not provide any major contribution to the consolidation settlement. For the applied load of 10 kPa, the consolidation settlement is calculated to be 23.8 mm, compared with the immediate settlement of 50 mm calculated from (11.18). [The effective stress changes occurring under this load are entirely elastic; it would not be inconsistent to calculate the total settlement of 73.8 mm from (11.18) using an implied value of Poisson's ratio $\nu = 0.26$.]

The applied load of 14.5 kPa caused yielding of the soil only at depths of around 3 m. However, dissipation of the pore pressures generated by the application of this load produces changes in effective stresses which break through the initial yield loci at all depths up to about 6 m. At around 3 m depth, the effective stress path is of the type shown in Fig. 11.12c; at other depths above 6 m it is of the form shown in Fig. 11.12b, and below about 6 m it is once again of the form shown in Fig. 11.12a. As soon as the effective stress path meets the current yield locus, elastic volumetric strains associated with change in mean effective stress are accompanied by plastic volumetric strains *and plastic shear strains* associated with the expansion of the yield locus. The relative magnitudes of plastic volumetric and plastic shear strains are controlled by the direction of the normal to the plastic potential (yield locus) at the point of yielding. The resulting vertical strains at the various depths are shown in Table 11.2. The contribution of the deeper layers to the overall settlement is negligible. These layers, at depths roughly greater than the radius of the tank, are not brought to yield during the dissipation of excess pore

Fig. 11.12 Types of effective stress path during undrained loading and subsequent consolidation: (a) effective stress path lies entirely within initial yield surface; (b) yielding occurs during long-term dissipation of excess pore pressures; (c) yielding occurs during initial undrained loading.

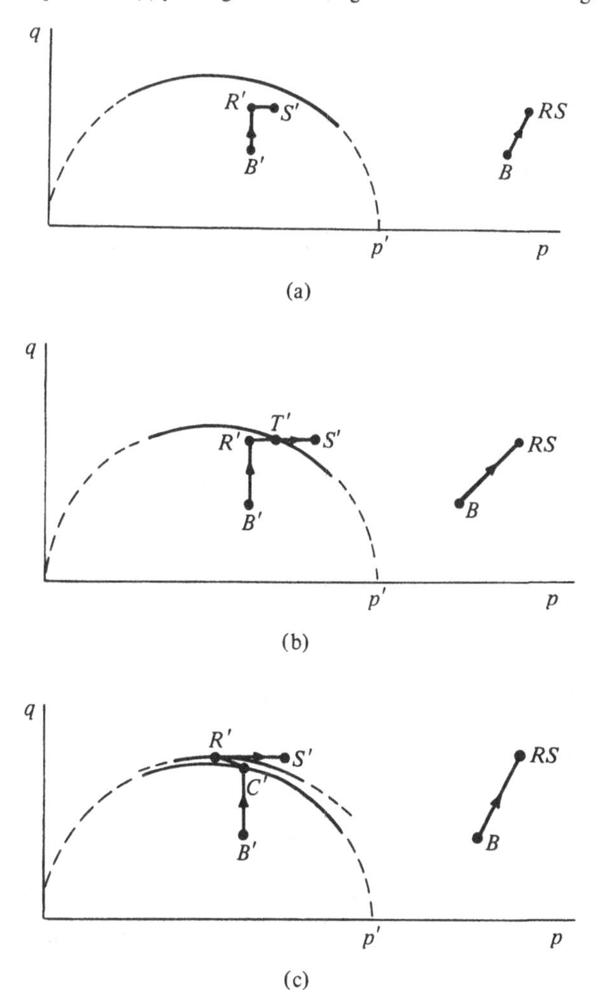

pressures. The consolidation settlement is calculated to be 180.4 mm and is dominated by the contribution of the plastic strains, compared with the immediate settlement under this load of 117.4 mm.

The Cam clay model that has been used for these settlement calculations is not a particularly sophisticated model, and the calculations are essentially hand calculations. Nevertheless, it does automatically build considerations of stress history and stress paths into the deduction of appropriate stiffnesses to be used for estimation of settlements. The deformation that occurs beneath the centre of a circular load may be axially symmetric, but it is certainly not one-dimensional, so data obtained from one-dimensional loading of soils in oedometer tests cannot be directly relevant. Selection of an appropriate value of volume compressibility m_v would be difficult because the degree of yielding to be expected under the actual stress paths would be uncertain unless guided by an elastic–plastic model.

One obvious criticism that can be levelled at the use of the Cam clay model for calculations such as these is that the shape of the Cam clay yield locus (Fig. 11.12) bears little resemblance to the shapes of yield loci that have been observed experimentally for natural clays (Figs. 3.15 and 3.22). However, the main divergence between the assumed and observed yield loci occurs for stress ratios below the one-dimensional normal compression line, whereas the effective stress paths that are generated by the tank loading of Fig. 11.1 and shown in a general way in Fig. 11.12 intersect the yield locus well above this line. Actually, it is only the shape of the part of the yield locus drawn with a solid line in Figs. 11.3 and 11.5 that has affected the calculation; the shape elsewhere is immaterial (hence the dashed lines at lower stress ratios in these figures).

11.3 Finite element analyses of geotechnical problems

There is a logical inconsistency in using an elastic analysis to estimate total stress paths and an elastic–plastic model to estimate the corresponding pore pressures or effective stresses and strains. This approach has the advantage of uncoupling the two parts of the calculation and can be justified on grounds of simplicity but not of rigour. Hand calculations, such as those described in the previous sections, using the Cam clay model, can be performed for elements that are undergoing simple modes of deformation such as axially symmetric deformation or perhaps plane strain deformation without rotation of principal axes. However, while *some* soil elements may be deforming in such special modes in any real geotechnical situation, many more elements are exercising much greater freedoms. Exact analyses of real problems using real soil models have to be performed numerically on computers, most commonly by a finite element technique.

The finite element method is described by Zienkiewicz (1977) as 'a general discretisation procedure of continuum problems posed by mathematically defined statements'. The mathematically defined statements for geotechnical problems are the equations of equilibrium and of flow (for problems involving movement of pore water), equations of compatibility of deformations, boundary conditions, and, of course, equations of stress:strain behaviour which link stresses and strains. In this section, finite element analyses of a number of geotechnical problems are presented briefly to show how elastic–plastic models of soil behaviour can be put to practical use. Further information about the finite element technique can be found in the books by Livesley (1983) and Britto and Gunn (1987).

11.3.1 Inhomogeneities within a triaxial test specimen

Although it is usually assumed that the triaxial test provides accurate information concerning the stress:strain behaviour of a single element of soil, a number of factors may lead to internal non-uniformities so that the external observations of stresses and deformations provide only an average description of the response of the soil. One cause of internal non-uniformity is the restraint provided by the top and bottom end platens (Fig. 11.13). Unless the soil sample wishes to deform one-dimensionally without lateral strain, any axial compression of the sample is accompanied by radial expansion; if the sample is to remain uniform, this radial expansion must occur over the full height of the sample including, in particular, the ends (Fig. 11.13b). If it is imagined that the soil is firmly stuck to the end platens, then radial movement at the ends is not possible, and the soil sample has to barrel (Fig. 11.13c) and become non-uniform, with deformation concentrated at the centre. Triaxial tests are sometimes performed with enlarged, lubricated ends (Fig. 11.13d) (Rowe and Barden, 1964) to encourage uniform deformation.

Even a sample provided with such smooth ends may still deform

Fig. 11.13 (a) Triaxial sample as prepared; (b) sample able to expand laterally; (c) sample bulging as result of end restraint; (d) sample tested with enlarged smooth end platens.

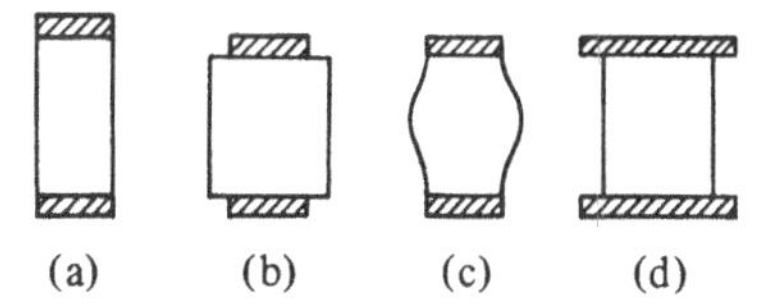

non-uniformly in a drained test if the rate at which the test is conducted is too fast for pore water to flow out of the sample and permit the soil to change its volume. This is the situation analysed by Carter (1982) using a finite element analysis with the Cam clay model to describe the stress:strain behaviour of the soil.

A diametral section through a triaxial sample is shown in Fig. 11.14. Considerations of symmetry indicate that it is only necessary to analyse one quarter of this section, and the division into finite elements used by Carter is shown in this figure. The end platens were assumed to be rigid and smooth, and the cylindrical side of the sample was subjected to a constant uniform normal total stress. Two drainage conditions were modelled: drainage only through the end platens, and drainage through all surfaces of the sample. The numerical experiments were then performed by increasing the axial displacement of the rigid end platen in small increments at a constant rate of strain. The sample was assumed to be initially uniform, isotropically normally compressed to a mean effective stress $p' = 207\,\text{kPa}$ ($30\,\text{lbf/in}^2$).

The soil was assumed to be Weald clay because experimental data were available with which some of the results of the finite element calculations could be compared. Soil parameters for the Cam clay model were deduced from published data concerning the behaviour of Weald clay. Data of isotropic compression and swelling reported by Henkel (1959) (Fig. 6.8) were used to deduce values of the slope of the normal compression line (5.8) $\lambda = 0.088$ and of the slope of the unloading–reloading line (4.3) $\kappa = 0.031$. The position of the isotropic normal compression line in the compression plane (Fig. 6.8) can be used to deduce the intercept Γ on the critical state line in the compression plane at a mean effective stress

Fig. 11.14 Finite element mesh for analysis of triaxial sample (after Carter, 1982).

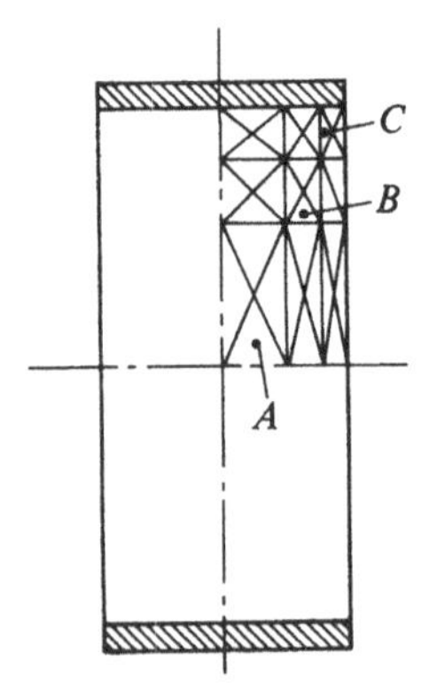

$p' = 1\,\text{kPa}$ [(6.10), (6.11)]. Carter quotes a value $\Gamma = 2.0575$. A value for the elastic shear modulus G' is required; but, as seen in the simple examples in Section 11.2, plastic strains dominate over elastic strains if the soil is yielding, and the precise value of G' has little influence on predicted response: Carter quotes a value $G' = 3000\,\text{kPa}$ deduced from the initial response of an undrained triaxial test on Weald clay. The slope M of the critical state line in the $p':q$ plane is required. On the basis of data presented by Bishop and Henkel (1957), Carter assumes a value $M = 0.882$ corresponding, from (7.10), to an angle of friction $\phi' = 22.6°$. Finally, a value for the permeability k of the soil, assumed isotropic, is required. Bishop and Henkel (1957) quote a value of coefficient of consolidation for normally compressed Weald clay $c_v = 6.67 \times 10^{-9}\,\text{m}^2/\text{s}$. This can be converted, through (11.26), to a value of permeability $k = 1.27 \times 10^{-12}\,\text{m/s}$ appropriate to Weald clay isotropically normally compressed to a mean stress $p' = 207\,\text{kPa}$. The soil parameters chosen by Carter are the following:

λ	0.088
κ	0.031
Γ	2.0575
M	0.882
ϕ'	22.6°
G'	3000 kPa
k	1.27×10^{-12} m/s

The externally observed deviator stress:axial strain curves for five axial strain rates are shown in Fig. 11.15 for the case of drainage only through the end platens. For comparison, experimental data from conventional undrained and conventional drained tests performed with the same drainage boundary conditions are also shown. The finite element analyses show that as the strain rate is increased, the apparent response changes from drained to undrained as less and less of the soil sample is able to achieve pore pressure equilibrium. Of course, since infinite time is theoretically required for complete dissipation of excess pore pressures and since experimental drainage path lengths must necessarily be finite, no test can ever be *fully* drained; but the slower the strain rate, the nearer the observed response approaches that ideal. On the other hand, in the absence of end restraints such as those shown in Fig. 11.13c, there is no reason why any non-uniformities should develop during an undrained test on an initially homogeneous soil. Undrained testing is about shearing at constant volume, which implies that although the pores may change in shape, they do not change in volume, and no flow of water in the soil sample is required.

Fig. 11.15 Stress:strain curves for drained triaxial compression of Weald clay with drainage only from both end platens; axial strain rates are (1) 4.2×10^{-8}/s, (2) 4.2×10^{-7}/s, (3) 1.7×10^{-6}/s, (4) 4.2×10^{-6}/s, and (5) 4.2×10^{-5}/s; with experimental results for drained test (A) and undrained test (B) (after Carter, 1982).

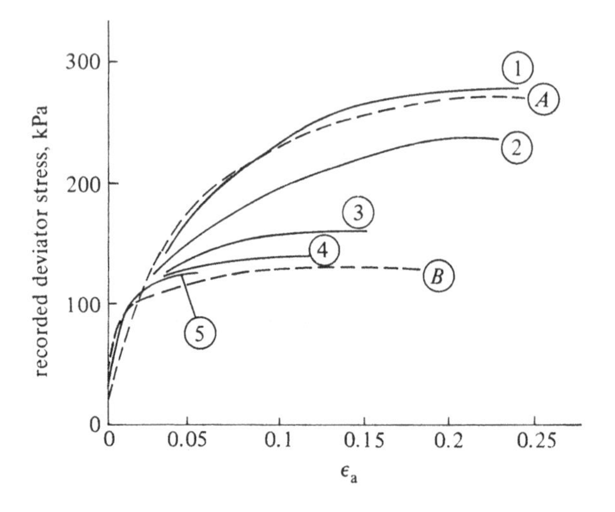

Fig. 11.16 Deviator stress q on soil sample at failure as function of axial strain rate in drained triaxial compression of Weald clay (○ computations, • measurements) (after Carter, 1982).

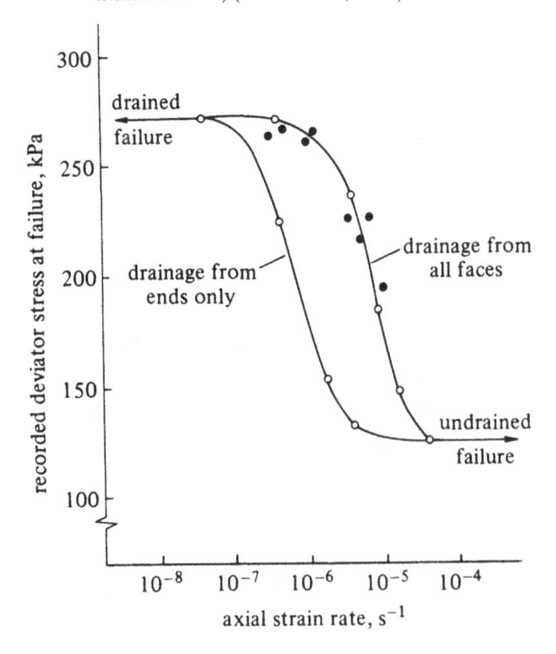

Fig. 11.17 Computed contours of stress and specific volume in triaxial specimen of Weald clay at axial strain $\varepsilon_a = 0.05$ during drained compression with axial strain rate 8.33×10^{-6}/s; drainage from all boundries (after Carter, 1982).

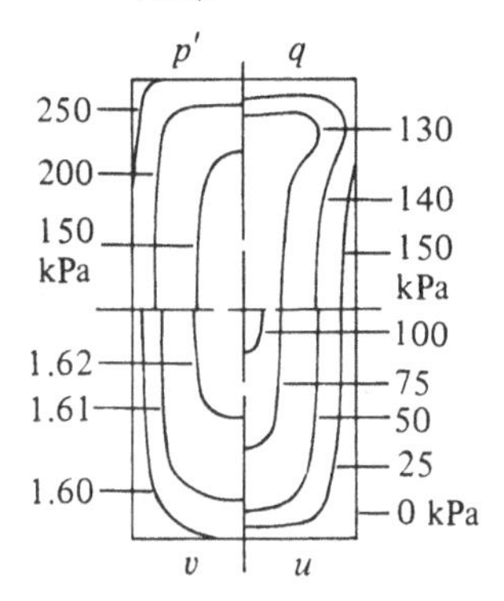

Fig. 11.18 Computed test paths for elements A, B, and C inside triaxial specimen of Weald clay during drained compression with axial strain rate 8.33×10^{-6}/s; drainage from all boundaries: (a) $p':q$ effective stress plane; (b) $v:p'$ compression plane; (c) generalised deviator stress q and shear strain ε_q; (d) specific volume v and shear strain ε_q.

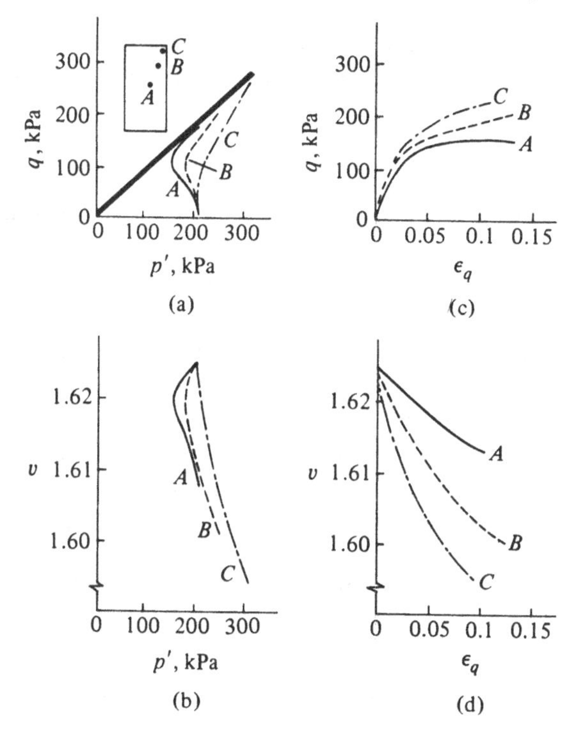

A consequence of the different stress:strain curves in Fig. 11.15 is that the apparent strength, calculated as the maximum value of deviator stress applied to the external boundaries of the sample, depends on the rate at which the sample is sheared. This is not a result of any viscous phenomena in the soil but simply an effect of increasing pore pressure disequilibrium as the rate of testing is increased. The apparent variation of strength with strain rate is shown in Fig. 11.16 with some experimental data reported by Gibson and Henkel (1954). The undrained and fully drained strengths provide bounds; of course, the more drainage that is provided, the shorter the average drainage path length and the faster a test can be performed without significantly affecting the observed strength.

The non-uniformity in the triaxial sample for a test conducted at an axial strain rate of $8.33 \times 10^{-6} \mathrm{s}^{-1}$, with drainage from the whole surface of the sample, is illustrated in Figs. 11.17 and 11.18. Figure 11.17 shows contours of mean effective stress p', a generalised deviator stress q (see Section 1.4.1, noting that, in general, local stress conditions are not axially symmetric), pore pressure u, and specific volume v at an imposed axial strain $\varepsilon_a = 0.05$. The paths followed during this test by the three elements A, B, and C in Fig. 11.14 are shown in Figs. 11.18a, b, c, d. The effective stress path followed at element C is close to that expected in a drained test (with $\delta q/\delta p' = 3$), whereas the effective stress path followed at element A looks much more like an undrained effective stress path. The stress:strain response measured externally is an average response; Carter suggests that the stress:strain curve for element B is close to this average.

11.3.2 Centrifuge model of embankment on soft clay

Finite element analyses are often used to predict the performance of prototype structures. Although field instrumentation can provide data with which these predictions can be compared, there are often many uncertainties about the precise soil conditions and boundary conditions at any particular site, and the quality of predictions can be masked by unknown natural variations. Model testing in the laboratory provides one route by which data which are relevant to field prototype loadings can be obtained under more carefully controlled conditions. In conventional model tests on the laboratory floor, the stresses at corresponding points in the model and prototype are scaled in the same ratio as the dimensions of the model and prototype. If the model is made with a length scale 1/100, then the stresses in the model are 1/100 times those in the prototype. If the behaviour of the soil were independent of stress level, then this would be of no consequence: strains in model and prototype would be the same and hence deformations would be scaled by the same factor,

1/100. However, it has been a recurrent theme in this book that the behaviour of most soils is extremely sensitive to stress level, and the success of elastic–plastic soil models is an indication that both stress history and stress increments must be correctly reproduced if the correct stress:strain response is to be obtained.

In centrifugal modelling (e.g. see Schofield, 1980), a small model constructed at, say, a length scale of 1/100 is rotated about a vertical axis (Fig. 11.19) in such a way that it experiences centripetal accelerations equivalent to 100 gravities (g) thus producing an increase in self weight by a factor of 100. Stresses and stress increments at corresponding points in the model and the prototype are now the same, and strains observed in the model should be directly relevant to the full-scale structure. Centrifuge model tests thus provide a convenient source of data at prototype stress levels and under controlled boundary conditions against which numerical predictions can be tested.

Almeida (1984) and Almeida, Britto, and Parry (1986) describe a series of centrifuge tests on models of embankments on soft clay. They also present a finite element analysis of one of the corresponding prototype

Fig. 11.19 Diagram of centrifuge model test.

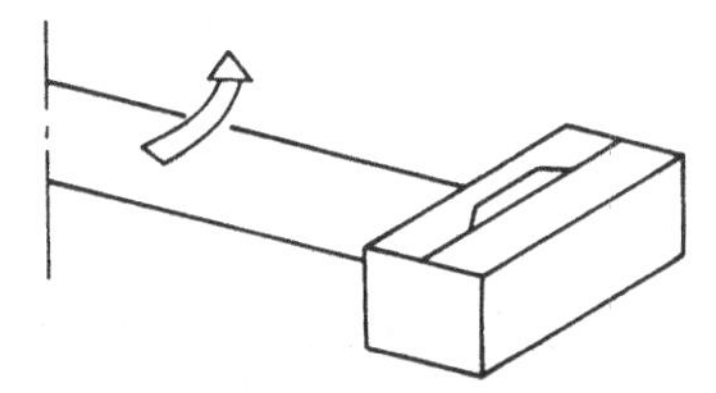

Fig. 11.20 Section through centrifuge model of embankment on soft clay, poured in five lifts (after Almeida, 1984).

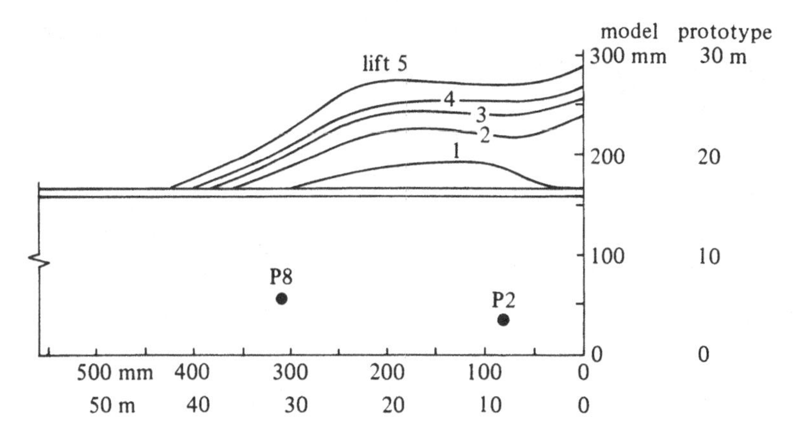

embankments. The soft clay was prepared by one-dimensional compression from a slurry in a strong box, first on the laboratory floor and then on the centrifuge at an acceleration of 100 *g*. The embankments were formed in layers by depositing sand from a series of hoppers while the centrifuge was running. A section through the embankment and soft clay for one test is shown in Fig. 11.20; a photograph of a model with the perspex side of the strong box removed, taken after a test in which failure of the soft clay had occurred, is shown in Fig. 11.21. The length of the model was 675 mm and its width 200 mm, corresponding to a prototype test bed 67.5 m by 20 m at full scale. Both model and prototype dimensions are indicated on Fig. 11.20.

To model typical prototype soft clay foundations in which a stiff crust overlies softer material, the soft clay was made in such a way that when it was ready for mounting on the centrifuge, it consisted of a layer of Cambridge gault clay about 40 mm thick over a layer of kaolin about 115 mm thick, corresponding at an acceleration of 100 *g* to a total prototype clay thickness of 15.5 m. When the sample was mounted on the centrifuge, a layer of sand 9 mm thick was placed on its surface to provide a small effective stress at the surface of the clay. The water table was maintained at the surface of this sand layer throughout the centrifuge model tests. The history of vertical effective stresses for the clay is shown

Fig. 11.21 Centrifuge model test of stage-constructed embankment on soft clay after failure (after Almeida, 1984).

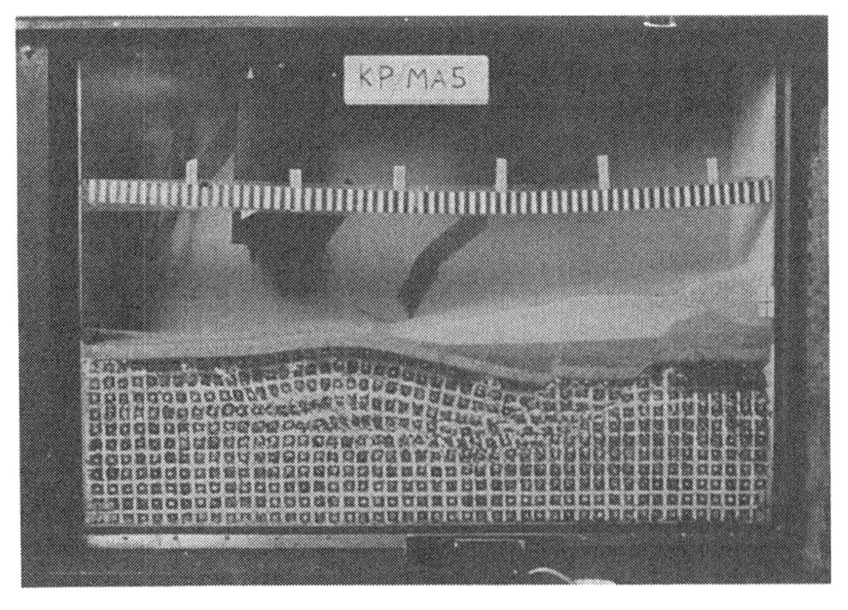

in Fig. 11.22a. First (*A* in the figure), the slurry was one-dimensionally compressed in stages to a vertical effective stress $\sigma'_v = 54$ kPa over a period of about 10 days. Then (*B*), with drainage prevented from the base of the clay, a surface pressure of 150 kPa was applied to its surface for about 2 hours, during which partial consolidation of the gault clay occurred and the profile of vertical effective stress could be deduced from measured pore pressures. Finally (*C*), with the strong box mounted on the centrifuge and an acceleration of 100 g applied for about 9 hours, the vertical effective stresses varied linearly with depth from 8 kPa at the base of the surface sand layer to 98 kPa at the base of the clay. The variation of overconsolidation ratio n with depth for the clay in this final condition before the construction of the embankment is shown in Fig. 11.22b.

The timing of the construction of the successive layers of the sand embankment on the soft clay is shown in Fig. 11.23. After each layer has been placed, the clay is allowed to consolidate and strengthen before the next layer is added. Flow of pore water occurs in both prototype and centrifuge model as a result of gradients of excess pore pressure. Pressures are correctly reproduced in the model tested at 100 g, but the linear dimensions are reduced by 1/100 and so the pore pressure gradients and flow rates are increased by a factor of 100. Flow-path lengths are also reduced by a factor 1/100 by comparison of model and prototype, and the nett result is that consolidation, which is controlled by rates of diffusion of pore pressures, occurs $100 \times 100 = 10^4$ times faster on the centrifuge at 100 g compared with the prototype tested at 1 g. The time scale for the

Fig. 11.22 (a) Vertical effective stresses and (b) overconsolidation ratio for clay in centrifuge model: (A) initial consolidation at 1 g; (B) partial consolidation at 1 g with top drainage only; (C) consolidation at 100 g (after Almeida, 1984).

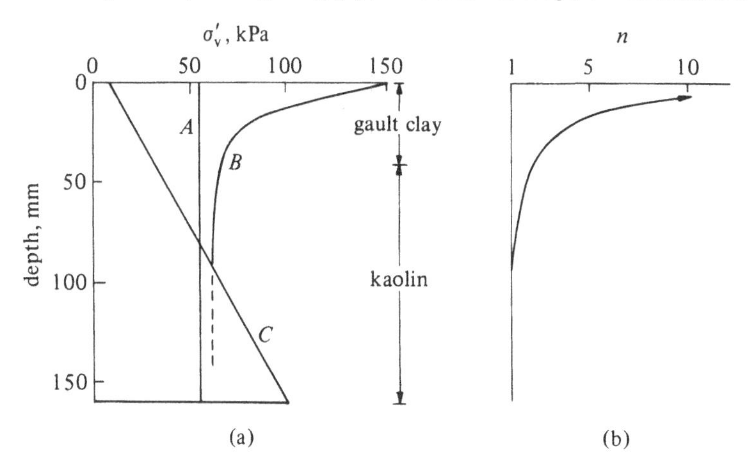

loading shown in Fig. 11.23 shows both model and prototype times, and one of the key benefits of centrifuge model testing is immediately apparent: the construction of the embankment on the centrifuge has taken about 5 hours, which corresponds to a period of about 7 years at full scale. The time saving is immense.

The visible face of the clay was prepared for observation of major discontinuities by sprinkling a grid of dark clay powder on the sample before it was mounted on the centrifuge; the grid can be seen, partially distorted by deformations of the model, in Fig. 11.21. The face of the clay was also prepared for observation of displacements by placing a grid of reflecting markers, also seen in Fig. 11.21. Measurements of the changing positions of these markers on photographs taken of the model while the centrifuge was running provide extensive information about the displacement field, and hence the strain field, within the clay foundation.

The finite element analysis was performed using the CRISP finite element program described by Britto and Gunn (1987). The finite element mesh is shown in Fig. 11.24; comparison with Fig. 11.20 reveals the correspondence between the elements in the embankment and the successive stages of construction. The fifth and final layer was modelled simply as a series of nodal surcharge loads. The clay foundation was divided into 85 linear strain triangular elements; the sand layer on top of the clay was divided into 12 linear strain quadrilateral elements; and the

Fig. 11.23 Loading history for centrifuge model (after Almeida, 1984).

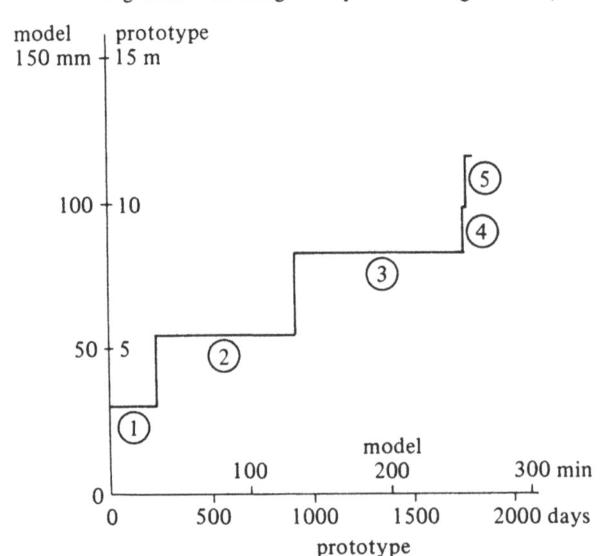

Table 11.3

	Kaolin	Gault clay
λ	0.25	0.219
κ	0.05	0.035
Γ	3.44	2.96
M	0.9	1.0
ϕ', deg	23.0	25.4
G'	$75p'$	2.25 MPa
k_{v_i}, m/s	$1.4\text{–}3.7 \times 10^{-9}$	0.937×10^{-9}
C_k	0.262	0.261

embankment was divided into 33 elements, a mixture of linear strain triangles and linear strain quadrilaterals as shown.

Soil properties were required for four material types. For want of any more satisfactory numerical soil model, the sand embankment and the surface sand layer were modelled as isotropic elastic materials with Young's modulus $E' = 3$ MPa and 2 MPa, respectively, and Poisson's ratio $\nu' = 0.3$ (implying shear moduli $G' = 1.15$ MPa and 0.77 MPa). The gault clay and kaolin were both modelled with the Cam clay model described in Chapter 5, with two sets of soil parameters which are tabulated in Table 11.3.

Values for kaolin of the slopes λ and κ of the normal compression and unloading–reloading lines were obtained from accumulated experience at Cambridge with oedometer and triaxial tests. The slope M of the critical state line in the effective stress plane and the parameter Γ describing the position of the critical state line in the compression plane were obtained from undrained triaxial tests with pore pressure measurement.

Values of λ and κ for gault clay were obtained from one-dimensional compression tests in oedometer and simple shear apparatus. The value of Γ was deduced, using the Cam clay model, from the position in the compression plane of the normal compression line found in the oedometer

Fig. 11.24 Finite element mesh for analysis of model embankment on soft clay (after Almeida, 1984).

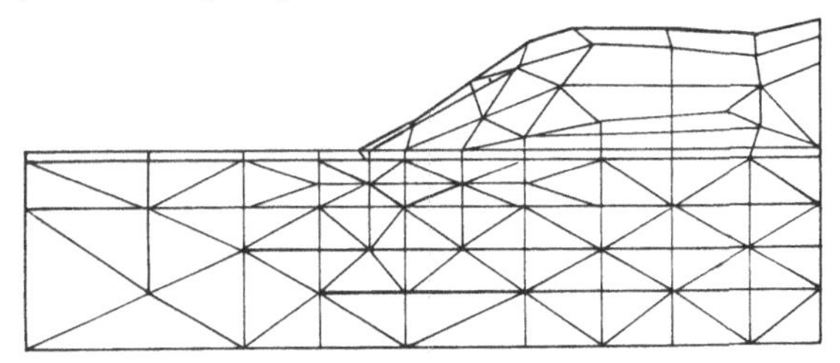

tests (compare Section 9.4.5). The value of M was estimated from an interpretation of strength data obtained in simple shear tests. In the absence of data from triaxial tests on gault clay, a constant value of shear modulus $G' = 2.25\,\text{MPa}$ was assumed. However, data from undrained triaxial tests on kaolin were used to support a relationship

$$G' = 75p' \tag{11.33}$$

and shear moduli for the kaolin were allowed to change as the mean effective stress changed. (This choice of elastic properties can result in curious values of Poisson's ratio for some stress histories.)

To be able to model the dissipation of excess pore pressures during the

Fig. 11.25 Profiles of (a) initial specific volume, (b) permeability, and (c) vertical and horizontal stress and size of initial yield locus for clay beneath embankment (after Almeida, 1984).

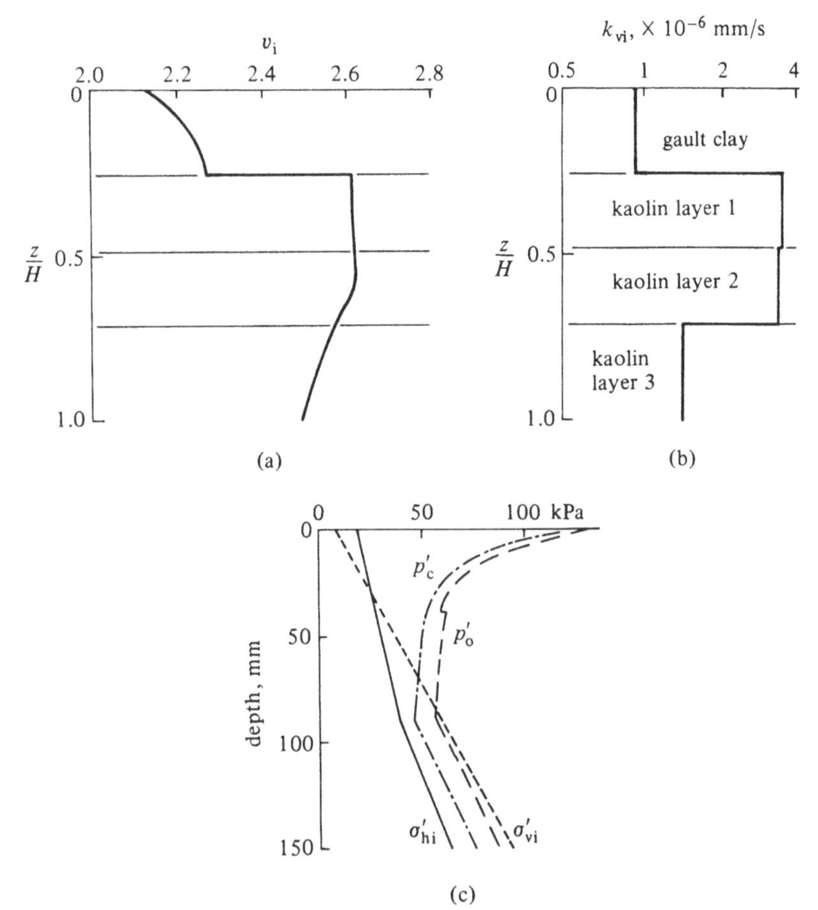

test, values were required for the permeabilities of the gault clay and the kaolin. For both soils it was assumed that the horizontal permeability k_h was 1.5 times the vertical permeability k_v. Values of k_v were calculated from rates of consolidation in oedometer tests on kaolin and gault clay and, for the kaolin only, from measurements of flow through a triaxial test specimen. For both soils, the permeability was allowed to vary with specific volume v according to the expression

$$k_v = k_{v_i} \exp\left(\frac{v - v_i}{C_k}\right) \tag{11.34}$$

where k_{v_i} and v_i are initial reference values and C_k a permeability coefficient. The values of C_k given in Table 11.3 were deduced from variations of permeability through oedometer tests. The reference, initial values of specific volume, and permeability assumed for this analysis are shown in Figs. 11.25a, b. The initial values of specific volume in Fig. 11.25a were calculated from the in situ mean effective stresses and the known history of overconsolidation. The in situ mean effective stresses were calculated from the in situ vertical effective stresses σ'_{vi} (C in Fig. 11.22) and the in situ horizontal effective stresses σ'_{hi} which were in turn calculated from the vertical effective stresses using an expression linking the coefficient of earth pressure at rest with the overconsolidation ratio of the soil,

$$K_0 = K_{0nc} n^a \qquad (11.35)(10.23\text{bis})$$

with $K_{0nc} = 0.69$ and $a = 0.401$ for both the kaolin and the gault clay. The initial size of the elliptical locus at any depth is specified by the intercept p'_o on the p' axis (Fig. 11.3) and is calculated from the maximum effective stresses experienced at each depth. The initial values of σ'_{vi}, σ'_{hi}, and p'_o are shown in Fig. 11.25c.

The centrifuge model test and the finite element analysis both provide data which can be used to compare numerical predictions and experimental observations. Here comparison is made in terms of the observations that might be made in a loading test of a real prototype trial embankment. The measured and predicted vertical displacements at the surface of the clay are shown in Fig. 11.26 for the start and end of load increment 3 (see Figs. 11.20 and 11.23). The general shape and magnitude of the deformation profile are quite well matched, though the observed settlements are considerably less than the predicted settlements at the right-hand edge of the model, which in the analysis is assumed to be a centreline and therefore, from symmetry, a smooth boundary. Perhaps the model container was in fact significantly rough on this surface and provided restraint against vertical movement of the soil. It also appears in Fig. 11.20 that the first layer of the embankment loads the clay foundation well away from the edge of the model container. It may be that the settlement profile produced by this first load increment is dominating the observed profile at later stages of the test.

The variation of settlement with time at one point near the surface of the clay beneath the embankment is shown in Fig. 11.27; agreement between the predicted and observed rates of settlement is extremely good. However, when the predicted and observed variations of vertical displacement with depth are compared (Fig. 11.28), it is clear that the deformations

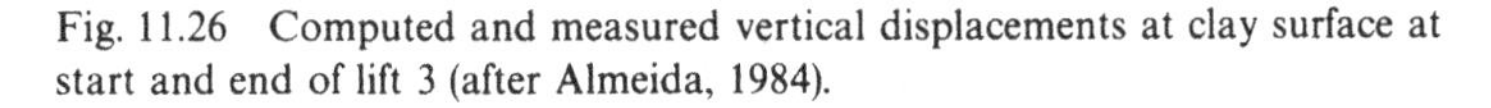
Fig. 11.26 Computed and measured vertical displacements at clay surface at start and end of lift 3 (after Almeida, 1984).

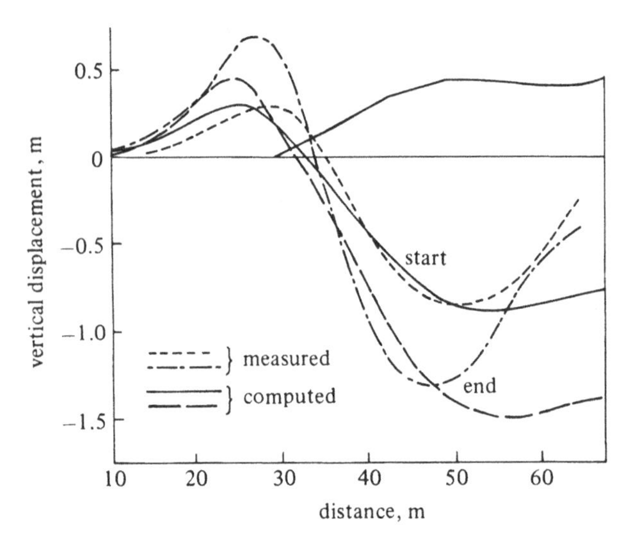

Fig. 11.27 Computed and measured development of settlement at clay surface with time (after Almeida, 1984).

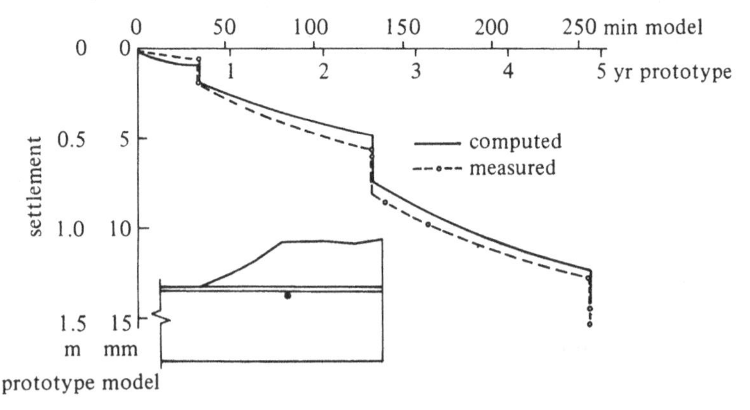

in the physical model are concentrated much more near the surface than those in the numerical calculation.

Horizontal displacements through the clay at the start and end of load increment 3 are compared in Fig. 11.29 for an imaginary inclinometer located near the toe of the embankment. Again, the deformation observed is concentrated much more towards the surface of the clay than expected from the numerical analysis.

Excess pore pressures measured and predicted at two piezometers (P2 and P8 in Fig. 11.20) are compared in Fig. 11.30. The agreements both of magnitude and of change with time are good. The effective stress paths

Fig. 11.28 Profile of computed and measured settlements with depth at start of lift 3 (location of profile shown in key diagram on right) (after Almeida, 1984).

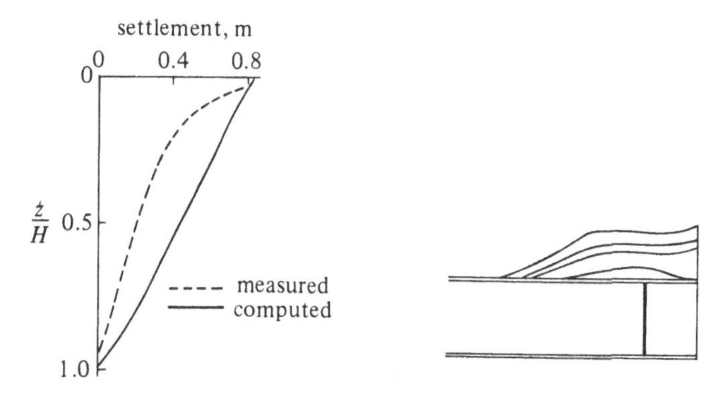

Fig. 11.29 Computed and measured equivalent inclinometer profile, at start and end of lift 3 (after Almeida, 1984).

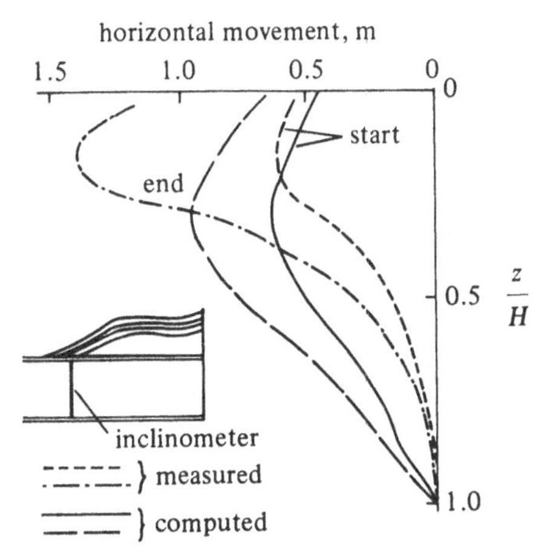

Fig. 11.30 Computed and measured development of pore pressure with time at piezometers (a) P2 and (b) P8 beneath embankment (after Almeida, 1984).

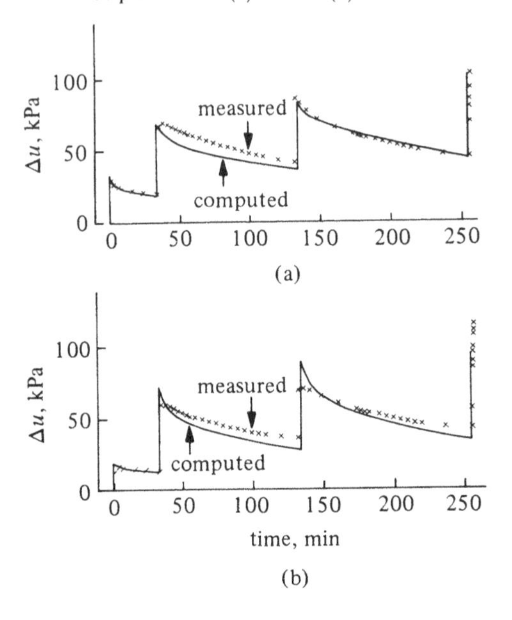

Fig. 11.31 Computed effective stress paths for elements of soil at positions of piezometers (a) P2 and (b) P8 (after Almeida, 1984).

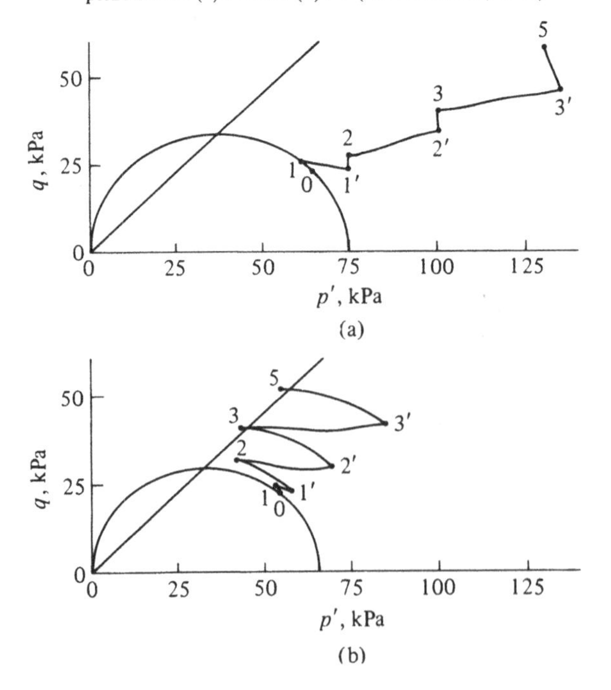

followed at these two piezometers, according to the finite element analysis, are shown in Fig. 11.31 in terms of mean effective stress p' and a generalised deviator stress q [Section 1.4.1, expression (1.35)]. (Of course, there is no possibility of verifying from measurements made in the centrifuge model test that these are in fact the actual effective stress paths followed.) These paths show that the soil around P2 near the assumed centre of the embankment is loaded with little increase in deviator stress but large increase in total mean stress with each successive load increment. Dissipation of excess pore pressures then permits substantial expansion of the yield surface to occur, and the resulting average effective stress path corresponds closely to compression with almost constant stress ratio $\eta = q/p'$. Much of the pore pressure at piezometer P8 results from drop of mean effective stress associated with plastic constant volume shearing with the application of each load increment. The expansion of the yield surface that occurs with dissipation of this pore pressure is much less substantial than at P2. The clay around P8 is being subjected to much more severe loading: lifts 2 and 3 bring the soil here to a state of contained failure, and lift 4/5 brings the soil to a state of failure which can no longer be contained.

No piezometers were installed in the soil to the left of the model where the clay was not subjected to any surface loading. However, the finite element analyses show that pore pressures are generated in this region with little change in total mean stress, so the effective mean stress is predicted to fall to a low value. A consequence of using (11.33) to estimate values of shear modulus is that in this region outside the embankment the shear stiffness of the kaolin is predicted to fall to a low value whereas the overlying gault clay is assumed to retain a constant shear modulus $G' = 2250\,\text{kPa}$. For the kaolin to have a shear modulus as high as this, the mean effective stress from (11.33) would have to be at least 30 kPa. It seems that the low shear stiffness of the kaolin to the side of the embankment probably accounts for the predicted much larger deformations in the lower parts of the clay (Figs. 11.28 and 11.29). It is also quite likely that the elastic behaviour of the sand assumed in the embankment and over the surface of the clay may tend to restrict horizontal movements near the clay surface. The tendency of elastic material to cling together is discussed in Section 12.3.

11.3.3 Experimental embankment on soft clay at Cubzac-les-Ponts

Centrifuge model tests can be performed relatively rapidly for the study of mechanisms of failure and for the study of time-dependent deformations. In a model at a length scale N (tested at an acceleration of N gravities), diffusion of pore water in the consolidation of soils is controlled by a time scale N^2; so at a length scale 100, 4.3 minutes in the centrifuge represent 30 days in the prototype. However, for modelling real prototypes – as opposed to classes of prototype structures – the small scale of a centrifugal model can make it difficult to represent accurately, within the confines of a small model, the natural variations of the vertical soil profile. At full scale, numerical predictions can be compared with the observations of carefully instrumented field trials, which can be conceived as an initial stage of construction to be subsequently incorporated in the complete project, or purely as a research trial. Although field trials produce data at much greater expense and at a much slower rate than centrifuge model tests, they do produce data from soil loaded to appropriate stress levels in its in situ state, and problems of sampling and disturbance are eliminated though knowledge of boundary conditions is less certain. Natural variations from site to site in a given locality always remain, but field trials provide the next step beyond model tests in the validation of numerical models of soil behaviour and methods of numerical analysis.

In 1972 the French Laboratoires des Ponts et Chaussées began a series of full-scale field tests to study the behaviour of embankments on soft clay. They deliberately chose to perform these tests on a site of their own to avoid the problems that inevitably arise when an instrumented field study is part of a real geotechnical construction for which external, non-scientific constraints may prevent the sequence of loading from being that which is desirable from the point of view of a research study and may make access to instrumentation during or after construction difficult or impossible. A site was chosen at Cubzac-les-Ponts about 30 km north of Bordeaux on the north bank of the Dordogne (Fig. 11.32). A section through the valley of the Dordogne at Cubzac-les-Ponts is shown in Fig. 11.33. The geotechnical profile at the site consists first of a thin layer of topsoil about 0.3 m thick, below which there is a layer of silty clay less than 2 m thick forming a surface desiccation crust, overconsolidated by seasonal variation of the water table between the ground surface and a depth of about 1.5 m. Then there is a layer of soft clay about 8 m thick of variable organic content. Water contents and index properties for the clay layer are shown in Fig. 11.34. These soft soils are underlain by a layer of sands and gravels some 5 m thick, and then calcareous rocks. It is the response of the soft clay to surface loading that was of primary interest in the field loading tests at Cubzac-les-Ponts.

Four test embankments have been constructed at Cubzac-les-Ponts. The locations of these are shown on a plan of the test site (Fig. 11.35). Embankment A was built to failure in 1974, and the failure height (4.5 m) of this embankment was used to guide the choice of heights of the subsequent embankments to give differing margins of safety against complete failure. Embankment B was built in 1975 to a height of 2.3 m with a factor of safety against immediate failure of 1.5 and was used to study the time-dependent consolidation of the soil under and adjacent to the embankment. Embankment C was built in 1978 to a height of 3 m with a factor of safety of 1.2 to study the continuing behaviour of the soil under conditions closer to rupture. Embankment D was built in 1981 to a height of 1 m with a factor of safety of 3.0 to study the behaviour of the soil under low added load.

The observation and analysis of embankment B are briefly described in this section. The detailed dimensions of this embankment are shown in Fig. 11.36a. The timing of its construction to its final height of 2.3 m over a period of about 7 days is shown in Fig. 11.36b. The embankment

Fig. 11.32 Location map of France.

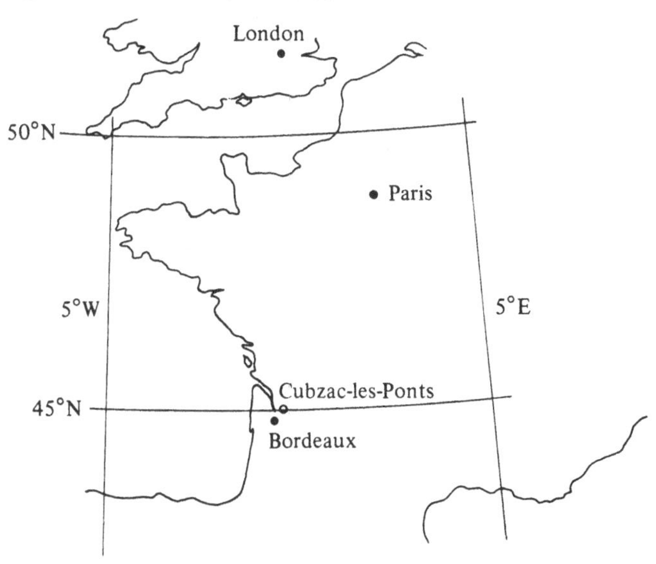

Fig. 11.33 Section through valley of the Dordogne at Cubzac-les-Ponts (after Magnan, Mieussens, and Queyroi, 1983).

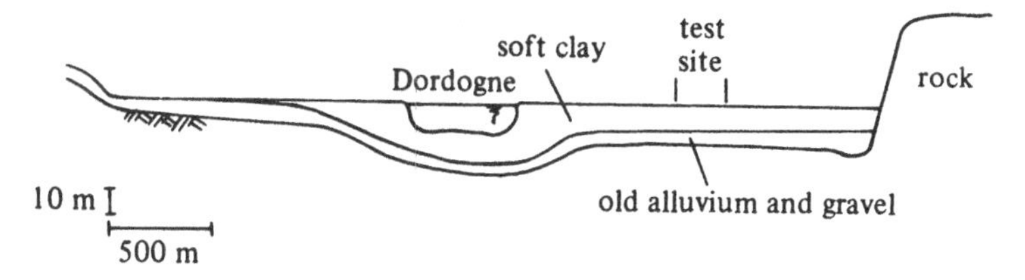

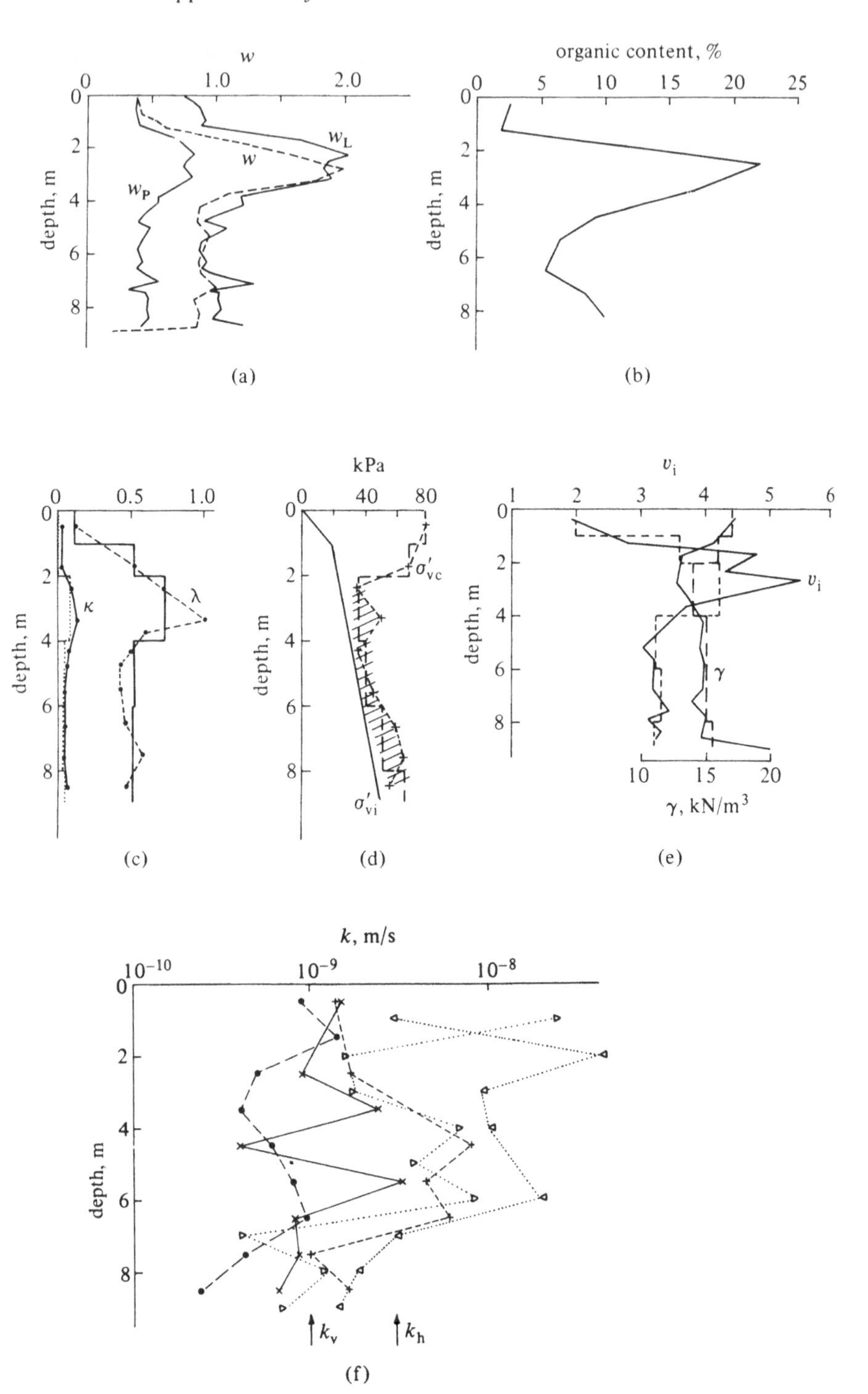
w
0
1.0
2.0
depth, m
w_L
w
w_P
(a)
organic content, %
0
5
10
15
20
25
depth, m
(b)
0
0.5
1.0
depth, m
κ
λ
(c)
kPa
0
40
80
depth, m
σ'_{vc}
σ'_{vi}
(d)
v_i
1
2
3
4
5
6
depth, m
v_i
γ
10
15
20
γ, kN/m^3
(e)
k, m/s
10^{-10}
10^{-9}
10^{-8}
depth, m
k_v
k_h
(f)

was constructed from coarse sand and gravel, with in situ unit weight 21 kN/m^3 and angle of repose 35°. Details of the field tests and analyses have been abstracted from Shahanguian (1981), Mouratidis and Magnan (1983), Magnan, Mieussens and Queyroi (1983), and Leroueil, Magnan, and Tavenas (1985).

The embankment and surrounding soil were heavily instrumented with 61 piezometers, 36 settlement gauges, 3 inclinometers, 3 horizontal strain gauges, 18 total pressure cells, and 114 surface markers. The instrumentation was placed predominantly in one transverse section of the embankment as indicated in Fig. 11.36a. The details of the instrumentation and a summary of the readings made over a period of about 2000 days after the construction of the embankment are given by Magnan et al. (1983).

Numerical analyses of the behaviour of this embankment over a period of about 6000 days from its construction are described by Mouratidis and Magnan (1983). They used a finite element analysis with the mesh of eight noded quadrilateral elements shown in Fig. 11.37. The mesh models just half the symmetrical embankment. Both vertical boundaries are assumed to be smooth, with vertical movement able to occur freely. The bottom boundary is assumed to be perfectly rough, so nodes on this boundary are fixed in space, and the presence of the gravel layer beneath the clay is allowed for by assuming that no excess pore pressures can develop at this level.

The numerical analysis was intended to be compared with the measurements obtained from the instrumented transverse section of the embankment, which was reckoned to be sufficiently remote from the ends to be deforming under conditions of plane strain. In the analysis of the plane centrifuge model test described in Section 11.3.2, the plane strain condition was included as a constraint on a general three-dimensional analysis. The analysis of the test embankment at Cubzac-les-Ponts was a two-dimensional analysis in which the stress acting on the transverse planes, in the direction of the length of the embankment, played no role and in which only stress changes in the plane of deformation were considered. Because of this, the elastic–plastic model that was used for

Fig. 11.34 Cubzac-les-Ponts: profiles of (a) water content w and index properties w_L, w_P; (b) organic content; (c) slopes of normal compression and unloading lines (λ, κ); (d) in situ vertical effective stress σ'_{vi} and preconsolidation pressure σ'_{vc}; (e) in situ specific volume v_i and unit weight γ; (f) vertical permeability k_v (•), horizontal permeability deduced from oedometer tests on horizontal samples with axial drainage k_{hh} (+), oedometer tests on vertical samples with radial drainage k_{hr} (x), and from in situ tests k_{hi} (▷ inward flow, ◁ outward flow) (after Magnan, Mieussens, and Queyroi, 1983).

the soil could be defined in terms of the plane strain effective stress quantities s', t and the corresponding work-conjugate strain increment quantities $\delta\varepsilon_s, \delta\varepsilon_t$, which were introduced in Section 1.5.

The anisotropic elastic–plastic model for the soft clay was constructed within the general framework of volumetric hardening elastic–plastic soil models discussed in Chapter 4 but differs from the Cam clay model used in the previous examples in three respects:

Fig. 11.35 Layout of test fills at Cubzac-les-Ponts.

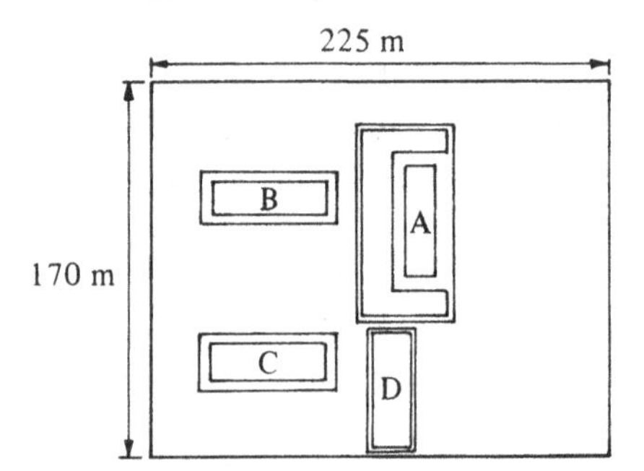

Fig. 11.36 Cubzac-les-Ponts: (a) section and plan of test fill B; (b) variation of height of embankment B with time (after Magnan, Mieussens, and Queyroi, 1983).

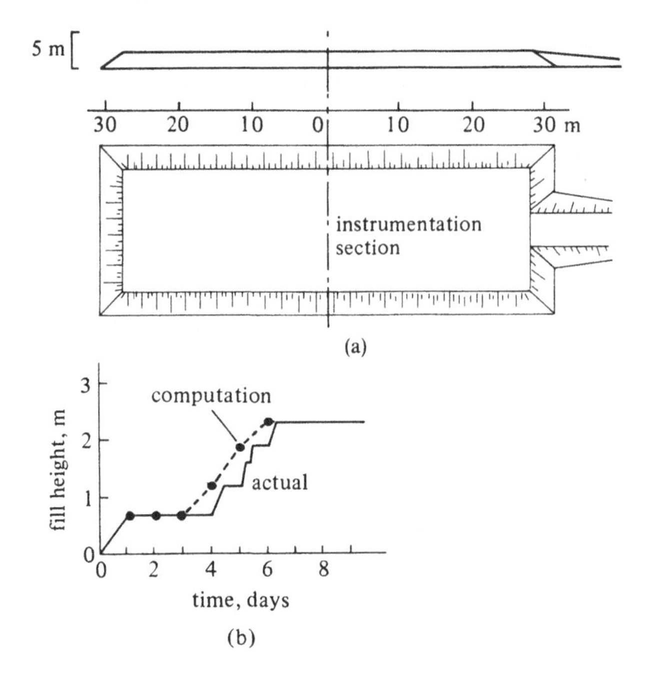

Table 11.4 *Parameters assumed in finite element analysis of test embankment B at Cubzac-les-Ponts*

Layer	v_i	γ (kN/m³)	σ'_{vc} (kPa)	κ	λ	E_h (kPa)	E_v (kPa)	G_{vh} (kPa)	k_v (m/s)	k_h (m/s)
Embankment	3.0	21	—	0.05	—	15000	15000	5250	11.6×10^{-6}	11.6×10^{-6}
0–1 m	2.0	17	78	0.017	0.12	1325	2650	930	1×10^{-9}	3×10^{-9}
1–2 m	3.6	16	68	0.022	0.53	2385	4770	1670	1×10^{-9}	3×10^{-9}
2–4 m	4.22	14	36	0.085	0.75	570	1140	400	1×10^{-9}	3×10^{-9}
4–6 m	3.24	15	41	0.048	0.53	955	1910	670	1×10^{-9}	3×10^{-9}
6–8 m	3.31	15	51	0.039	0.52	1500	3000	1050	1×10^{-9}	3×10^{-9}
8–9 m	3.2	15.5	65	0.048	0.52	1475	2950	1030	1×10^{-9}	3×10^{-9}

Note: For all layers, $\nu_{hh} = \nu_{vh} = 0.4$.

1. The elastic properties are assumed to be cross-anisotropic.
2. The yield locus is similar to those observed for natural clays and is not symmetrical about the mean effective stress p' axis.
3. The plastic potential is not the same as the yield locus, so the soil does not follow the postulate of normality or associated flow.

The necessary parameters for this model were obtained from the associated extensive programme of laboratory testing. The parameters chosen for the soil of the embankment and for the six layers into which the foundation soil was divided (Fig. 11.37) are tabulated in Table 11.4.

The sand of which the test embankment was constructed was modelled as an approximately isotropic elastic material with a high, isotropic, permeability $k_v = k_h = 11.6 \times 10^{-6}$ m/s (1 m/day).

The elastic response of the soft clay was assumed to be cross-anisotropic. Although, in principle, five independent parameters are required to specify such response (see Section 2.3), the nature of the anisotropy was specified in advance with the ratio of horizontal to vertical Young's moduli

Fig. 11.37 Finite element mesh for analysis of embankment B at Cubzac-les-Ponts (after Mouratidis and Magnan, 1983).

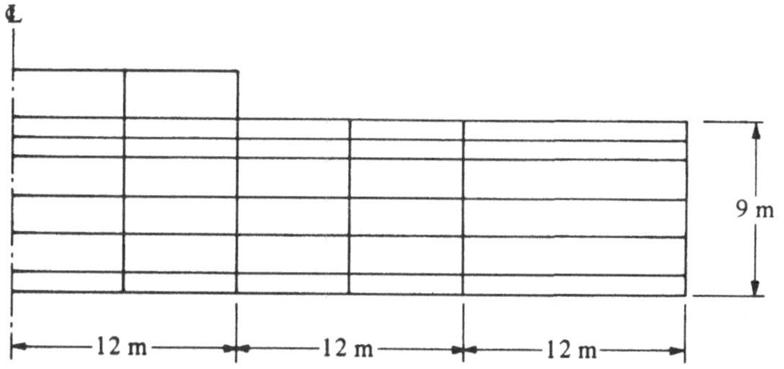

℄

9 m

12 m 12 m 12 m

$E_h/E_v = 0.5$, Poisson's ratios $\nu_{hh} = \nu_{vh} = 0.4$, and shear modulus for shearing in a vertical plane $G_{vh} = 0.35E_v$. The selection of five parameters was thus reduced to a problem of selection of one parameter, the vertical stiffness E_v, and this was deduced from the slope of the recompression line at the in situ effective vertical stress in conventional one-dimensional compression tests conducted in an oedometer.

For the general cross-anisotropic soil, Hooke's law (2.22) shows that the horizontal strain increment $\delta\varepsilon_h$ is given by

$$\delta\varepsilon_h = \frac{(1-\nu_{hh})\,\delta\sigma'_h}{E_h} - \frac{\nu_{vh}\,\delta\sigma'_v}{E_v} \tag{11.36}$$

For one-dimensional compression, $\delta\varepsilon_h = 0$, and hence

$$\frac{\delta\sigma'_h}{\delta\sigma'_v} = \left(\frac{E_h}{E_v}\right)\frac{\nu_{vh}}{1-\nu_{hh}} \tag{11.37}$$

The vertical strain increment $\delta\varepsilon_v$ is given by

$$\delta\varepsilon_v = -2\left(\frac{\nu_{vh}}{E_v}\right)\delta\sigma'_h + \frac{\delta\sigma'_v}{E_v} \tag{11.38}$$

and hence, with (11.37), the vertical stiffness is given by

$$E_v = \left[1 - \frac{2\nu_{vh}^2}{1-\nu_{hh}}\left(\frac{E_h}{E_v}\right)\right]\frac{\delta\sigma'_v}{\delta\varepsilon_v} \tag{11.39}$$

If it is assumed that unloading–reloading lines in oedometer tests are straight with slope κ in a semi-logarithmic $v:\ln\sigma'_v$ compression plane (compare Section 10.3.2),

$$v = v_1 - \kappa\ln\sigma'_v \tag{11.40}$$

then it follows that

$$\frac{\delta\sigma'_v}{\delta\varepsilon_v} = \frac{v\sigma'_v}{\kappa} \tag{11.41}$$

and hence, from (11.39),

$$E_v = \left(\frac{v\sigma'_v}{\kappa}\right)\left[1 - \frac{2\nu_{vh}^2}{1-\nu_{hh}}\cdot\frac{E_h}{E_v}\right] \tag{11.42}$$

or, with assumed values of E_h/E_v, ν_{vh}, and ν_{hh},

$$E_v = \frac{0.733v\sigma'_v}{\kappa} \tag{11.43}$$

All the cross-anisotropic elastic stiffnesses are related back to the slope κ of the unloading–reloading lines. The variation of κ with depth as determined from oedometer tests is shown in Fig. 11.34c, together with the distribution assumed in the numerical analysis (dotted line).

Triaxial tests to probe the shape of the initial yield locus for the clay at Cubzac-les-Ponts are described by Shahanguian (1981); his data are collected in a plot of the $p':q$ effective stress plane (Fig. 11.38), non-dimensionalised by dividing by the preconsolidation pressure σ'_{vc} at the appropriate depth. The shape is similar to that seen for other natural clays in Section 3.3. These are data from triaxial tests; but for the purpose of the plane strain analysis, the stress variables s' and t are sufficient to describe the principal effective stresses, and the data of Shahanguian in Fig. 11.38 were used to guide the choice of an elliptical yield locus in the $s':t$ effective stress plane (Fig. 11.39). This yield locus has a shape defined by the ratio μ of its minor and major axes and by the angle θ between the major axis and the mean stress s' axis. The angle θ was taken as $\theta = \tan^{-1}\frac{1}{3} = 18.4°$, so that the major axis of the ellipse corresponded to the stress path for one-dimensional normal compression $t/s' = (1 - K_{0nc})/(1 + K_{0nc})$ with $K_{0nc} \approx 0.5$ deduced from Shahanguian's data (solid curves in Fig. 11.38). The ratio μ was chosen so that the mean stress at the point where the ellipse cuts the mean stress s' axis was given by $s' = 0.6\sigma'_{vc}$ (Fig. 11.39); this requires $\mu = 0.534$.

A non-associated flow rule was chosen such that the direction of the plastic strain increment vector $(\delta\varepsilon^p_s:\delta\varepsilon^p_t)$ was given by the bisector of the normal to the yield locus at the point of yielding and the radial line to this yield point from the origin of the $s':t$ plane (Fig. 11.40). This implies

Fig. 11.38 Data from triaxial tests of yielding of clay at Cubzac-les-Ponts plotted in non-dimensional effective stress plane $p'/\sigma'_{vc}:q/\sigma'_{vc}$ (data from Shahanguian, 1981).

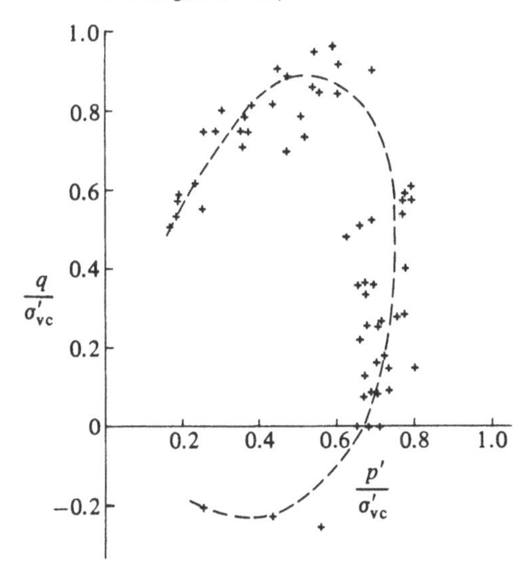

a shape of plastic potential in the $s':t$ plane as shown by the dashed curves in Fig. 11.39.

With the shape of the yield locus in the $s':t$ effective stress plane having the fixed elliptical form shown in Fig. 11.39, all that is required to complete the specification of the initial yield locus at any depth is the value of the preconsolidation pressure σ'_{vc} which, as implied in Fig. 11.38, controls the size of the initial yield locus for the clay at that depth. The spread of measured values of preconsolidation pressure obtained from one-dimensional loading in an oedometer is shown in Fig. 11.34d, together with the values assumed for each layer in the analysis (dashed line).

Change in size of the yield loci (with the shape remaining unchanged) was linked with plastic volumetric strains and hence, as described in Chapter 4, with the normal compression of the clay. This is controlled by the slope λ of the normal compression line in the $v:\ln s'$ or $v:\ln \sigma'_v$

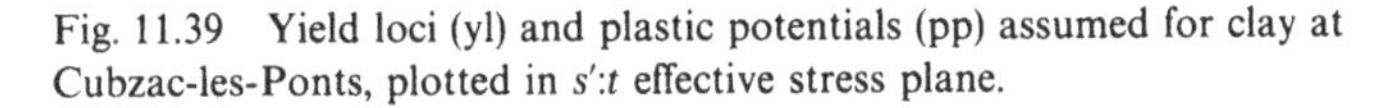

Fig. 11.39 Yield loci (yl) and plastic potentials (pp) assumed for clay at Cubzac-les-Ponts, plotted in $s':t$ effective stress plane.

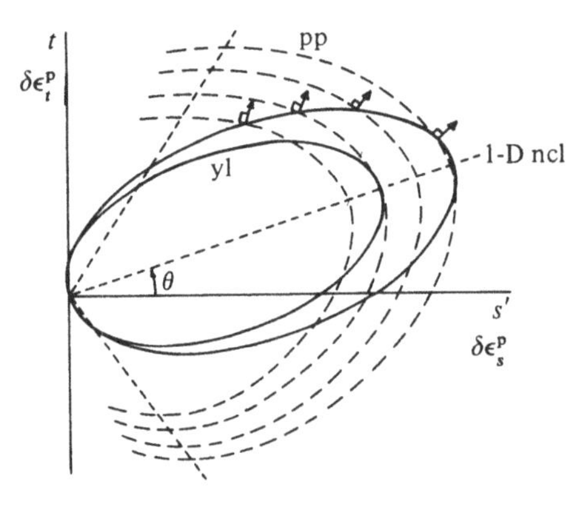

Fig. 11.40 Direction of plastic strain increment vector bisecting angle between stress vector and normal to yield locus.

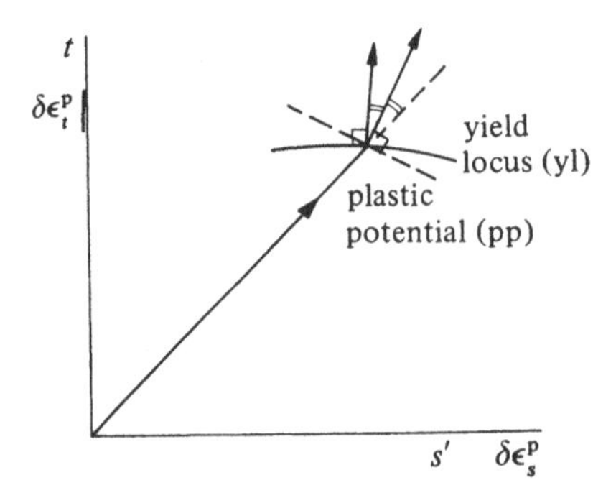

compression planes. Values of λ measured in oedometer tests are shown in Fig. 11.34c, together with the values assumed for each layer in the analysis (solid line).

Specification of the initial state of the soil required initial values of specific volume and of effective stresses. Values of initial specific volume were obtained from direct measurement (Fig. 11.34e) (assumed values shown by dashed line). Values of effective vertical stresses σ'_{vi} were calculated from knowledge of the in situ saturated unit weight γ of the soil (Fig. 11.34e) (assumed values shown by dashed line) and of the depth to the water table, which was assumed to be at a constant depth of 1 m for the purposes of the analysis. Values of effective horizontal stresses σ'_{hi} were calculated assuming a value of coefficient of earth pressure at rest $K_0 = \sigma'_{hi}/\sigma'_{vi} = 0.5$, irrespective of depth and of apparent overconsolidation ratio $\sigma'_{vc}/\sigma'_{vi}$, which had a maximum value of at least 3 in the desiccated clay near the ground surface (Fig. 11.34d).

Since it was intended to study the time-dependent deformations that occurred after the embankment was constructed, it was necessary to include values of vertical and horizontal permeabilities in the analysis. Vertical permeabilities were deduced from conventional oedometer tests on samples loaded in their original vertical direction with the flow of water, as excess pore pressures in the soil dissipated, also occurring in this vertical direction (Fig. 11.41a); values of vertical permeability k_v are shown in Fig. 11.34f. Horizontal permeabilities were deduced from two sorts of tests: (i) conventional oedometer tests on samples loaded in their original horizontal direction with flow of water also occurring in this horizontal direction (Fig. 11.41b), leading to values of horizontal permeability k_{hh} shown in Fig. 11.34f, and (ii) oedometer tests on samples loaded in their original vertical direction but with radial flow of water to a central sand drain (Fig. 11.41c), leading to values of horizontal permeability k_{hr} shown in Fig. 11.34f. In general, permeabilities measured

Fig. 11.41 Oedometer tests on (a) vertical sample with axial drainage, (b) horizontal sample with axial drainage, and (c) vertical sample with radial drainage.

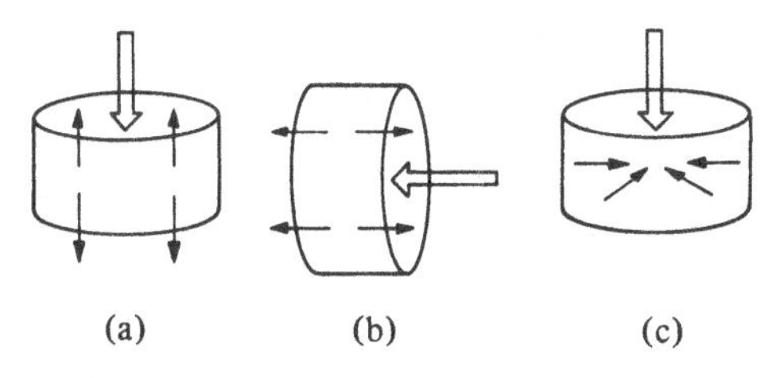

with radial flow in vertical samples (Fig. 11.41c) are lower than those measured with horizontal flow in horizontal samples (Fig. 11.41b).

Some in situ constant head flow tests were also performed, using a self-boring permeameter. The values of horizontal permeability k_{hi} deduced from these tests are also shown in Fig. 11.34f and are, in general, higher than the laboratory values. In fact, for the purposes of analysis, a uniform vertical permeability $k_v = 1 \times 10^{-9}$ m/s and a uniform horizontal permeability $k_h = 3 \times 10^{-9}$ m/s were assumed, as shown by the short arrows in Fig. 11.34f.

The embankment load was applied in the analysis by increasing the unit weight of the soil in the two embankment elements (Fig. 11.37) with time as shown in Fig. 11.36b. The numerical analysis proceeded with 25 time steps, chosen at intervals to follow the rapidly decreasing rate at which consolidation events develop: 6 steps of 1 day, 4 of 10 days, 10 of 100 days, and 5 of 1000 days.

Comparisons of observations of test embankment B at Cubzac-les-Ponts with numerical predictions made using the elastic–plastic soil model are shown in Figs. 11.42–11.45. The changing profile of the surface of the soil under the embankment is shown in Fig. 11.42. After about 2000 days, the measured and computed values on the centreline are similar; however, the

Fig. 11.42 Computed (--) and measured (—) settlements at surface of soil beneath embankment B at Cubzac-les-Ponts (after Magnan, Mieussens, and Queyroi, 1983; Mouratidis and Magnan, 1983).

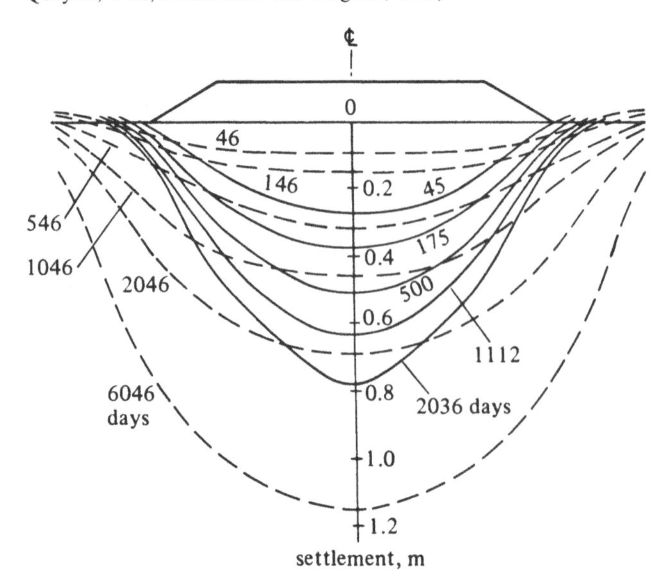

measured settlement initially occurs faster than is computed and is always more concentrated under the centre of the embankment.

When the settlements that develop with time in different layers of the soft clay beneath the embankment are compared (Fig. 11.43), it becomes clear that below a depth of 4 m the measurements and computations agree quite closely in terms of the rates and the magnitudes of settlements. In the upper layers, the measured settlement occurs much more rapidly than is computed, and in the top 2 m the measured long-term settlement is much greater than that computed.

Computed variations of horizontal movement with depth and time are compared with measurements made with an inclinometer installed beneath the toe of the embankment in Fig. 11.44. Although the magnitudes of horizontal movement are of the same order after about 2000 days, the

Fig. 11.43 Computed (– –) and measured (—) settlements in soil layers beneath embankment B at Cubzac-les-Ponts (a) layers 0–1 m, 1–2 m, 2–4 m; (b) layers 4–6 m, 6–9 m (after Magnan, Mieussens, and Queyroi, 1983; Mouratidis and Magnan, 1983).

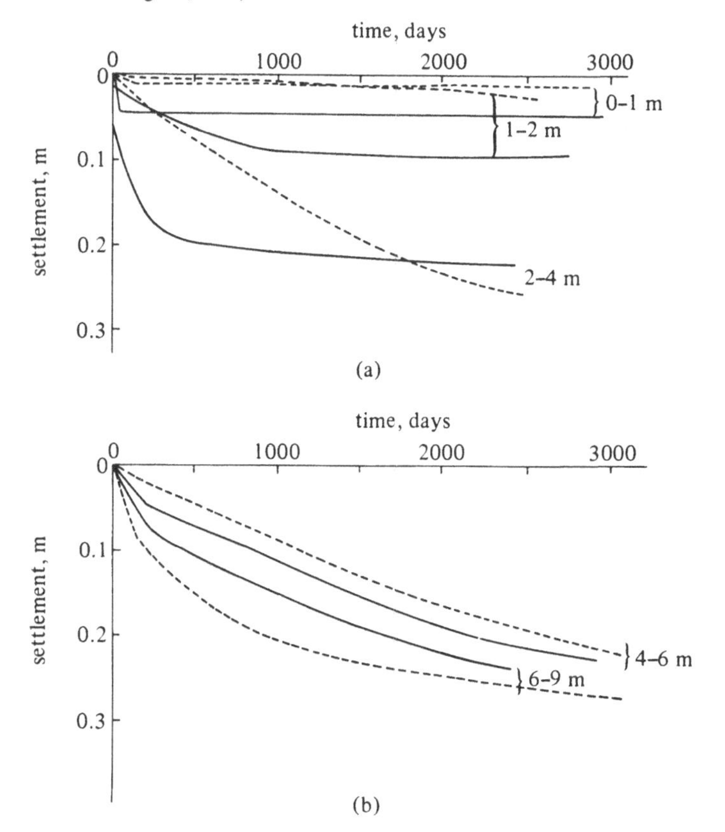

initial observed horizontal movement occurs much less rapidly than computed.

The measured and computed variations of excess pore pressure with depth beneath the centre of the embankment are shown in Fig. 11.45 for various times after construction of the embankment. The numerical analysis shows that an irregular distribution of pore pressures has developed during the rapid initial embankment loading. This results largely from the particular type of finite elements used to model the soft

Fig. 11.44 Computed (--) and measured (—) inclinometer profiles for toe of embankment B at Cubzac-les-Ponts (after Magnan, Mieussens, and Queyroi, 1983; Mouratidis and Magnan, 1983).

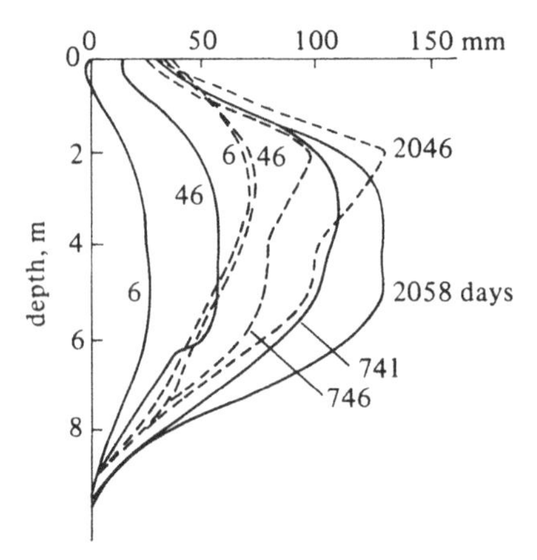

Fig. 11.45 Computed (--) and measured (—) pore pressures beneath centre of embankment B at Cubzac-les-Ponts (after Magnan, Mieussens, and Queyroi, 1983; Mouratidis and Magnan, 1983).

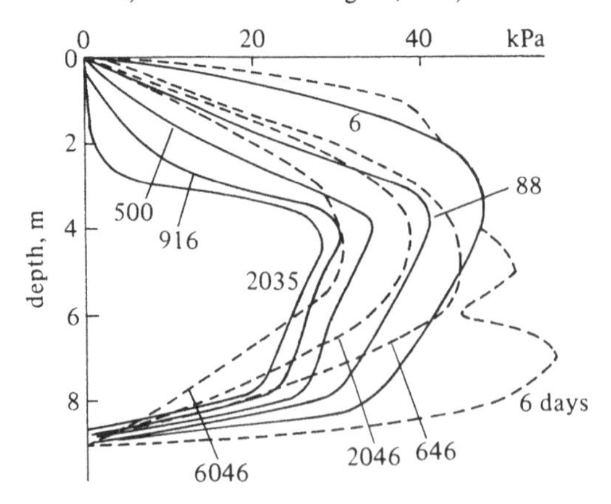

clay and from the rather coarse mesh used for the analysis (Fig. 11.37). However, these initial anomalies are rapidly damped out, and the first harmonic of pore pressure variations with depth quickly dominates the consolidation of the soil. Comparison with the measured distribution of pore pressure shows that dissipation occurs much more rapidly than is computed in the top 4 m of the soft clay and rather less rapidly towards the base of the clay layer.

Stress paths are shown in Fig. 11.46b for three soil elements: X at the top of the soft clay, Y in the middle of the soft clay, and Z at the bottom of the soft clay as shown in Fig. 11.46a. Elements X and Z are both on freely draining boundaries, and the pore pressure remains constant; so the changes in total stress are identical to the changes in effective stress. At both these locations, there is little change in effective stress after the load has been in place for about 40 days. At X, the assumed shape of yield loci (Fig. 11.39) allows rather high values of t/s' to be mobilised as

Fig. 11.46 (a) Location of elements X, Y, and Z; (b) computed effective stress paths for elements X, Y, and Z and total stress path for element Y, beneath embankment B at Cubzac-les-Ponts (after Mouratidis and Magnan, 1983).

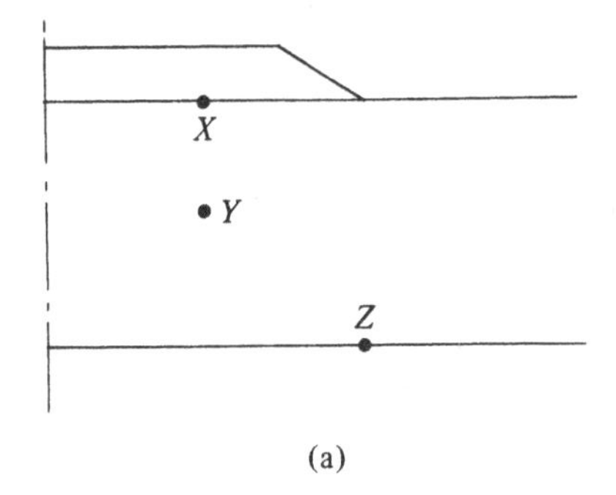

(a)

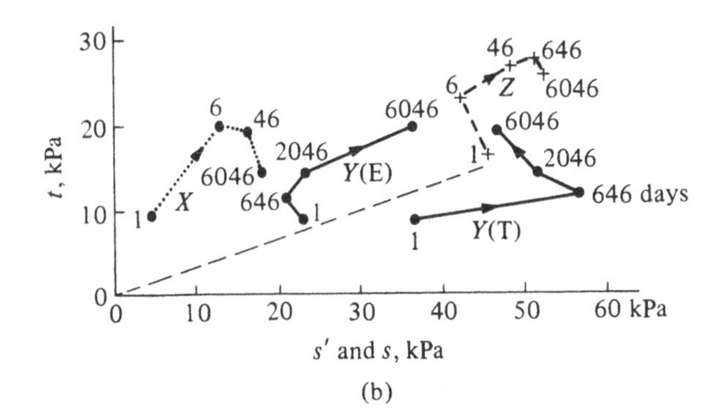

(b)

the embankment is constructed, and during this period the yield locus expands. After construction has ended, however, the effective stress path retreats inside the hardened yield locus, and the subsequent response is entirely elastic.

The response at Y is completely different. Here substantial excess pore pressures develop, and the total stress path is very different from the effective stress path. Substantial excess pore pressures remain for some 600 days, and the effective stress path suggests that during this period the soil at Y has hardly undergone any volumetric deformation. (To talk about *undrained* response during this period is probably not helpful: just because the pore pressure happens to remain approximately constant *at* some element, it cannot be inferred that no drainage is occurring *through* that element.) Significant changes in effective stress and total stress are continuing to occur at Y between 2000 and 6000 days after the placing of the load, and most of the hardening of the yield locus for the soil at Y occurs during this period.

Note that these effective stress paths have been shown in an effective stress plane $s':t$ calculated from values of principal effective stresses $\sigma'_1:\sigma'_3$, even though, at these soil elements off the centreline of the embankment, changes in magnitude of these principal stresses are accompanied by changes in their direction. These rotations of principal stresses cannot be displayed in an $s':t$ diagram.

The discrepancies between the computed and observed values of settlement, horizontal movement, and pore pressure in the upper region of the soft clay can probably be ascribed to the values of permeability assumed in the analysis. The actual clay is far from homogeneous, and it is likely, in particular, that the in situ horizontal permeability is much greater than the value assumed, especially in this upper region (Fig. 11.34f). The overall observed response (Figs. 11.43–45) implies that although dissipation of excess pore pressures occurs with horizontal flow of water from under the embankment, the soil here compresses with little initial lateral deformation. It is restrained by the undeforming soil outside the embankment.

11.4 Conclusion

This chapter began with discussion of the essentially qualitative use of the Cam clay model to illustrate the character of soil response that is to be expected around geotechnical structures and to emphasise the importance of dividing the response into elastic and plastic components. Latter sections have shown the application of elastic–plastic soil models, including Cam clay, to the calculation of pore pressures and deformations developing in real geotechnical problems, using finite element programs. The quality of these calculations depends on the specification and selection of soil parameters for the models.

It can be concluded that there is no difficulty in including the extra realism that elastic–plastic models (with or without coincident yield loci and plastic potentials) provide as compared with simpler, but less realistic, purely elastic analyses which may seriously misrepresent the manner in which the soil responds.

Exercises

E11.1. (a) A sample of soil is compressed one-dimensionally to a vertical effective stress of 100 kPa and then compressed further to a vertical stress of 200 kPa. During this one-dimensional normal compression, the earth pressure coefficient at rest $K_{0nc} = \sigma'_h/\sigma'_v = 0.6$. Plot the stress path for this compression in the $p':q$ plane.

(b) The water content of this saturated soil at a vertical stress 100 kPa is 0.45; at the end of the compression with a vertical stress 200 kPa, it is 0.40. The specific gravity of the soil particles is 2.7. Calculate the value of the compression parameter λ for the Cam clay model.

(c) The soil is then unloaded one-dimensionally to a vertical effective stress of 100 kPa, at which stage the value of $K_0 = 1.0$ and the water content is 0.41. Plot this new point in $p':q$, $p':v$, and $\ln p':v$ planes and calculate the value of the unloading–reloading parameter κ for the Cam clay model.

(d) The yield locus for the soil at the end of the compression is locally elliptical with the equation $p'/p'_o = M^2/(M^2 + \eta^2)$, where $\eta = q/p'$, $M = 1.0$, and $p'_o = 190.3$ kPa. Confirm that the stress state at the end of one-dimensional compression lies on this yield locus and sketch the yield locus in the $p':q$ plane.

(e) This yield locus is associated with the unloading–reloading line that has been plotted in the $p':v$ and $\ln p':v$ planes. At the critical state, $\eta = M$ in the expression for the yield surface. Plot the critical state line in the $p':q$, $p':v$, and $\ln p':v$ planes and deduce the values of the angle of shearing resistance ϕ' in triaxial compression and of the intercept Γ in the expression for the critical state line: $v = \Gamma - \lambda \ln p'$.

(f) The soil, unloaded to a vertical effective stress of 100 kPa, is now subjected to a conventional triaxial compression test with constant cell pressure. Sketch the total and effective stress paths that would be followed in drained and undrained tests and estimate the values of axial stress at which the yield locus and the critical state line are reached.

E11.2. A plane strain finite element analysis is to be performed of a long sand embankment overlying normally compressed clay. The clay is to be modelled as Cam clay with yield locus $q^2 = M^2 p'(p'_o - p')$, where $q^2 = \frac{1}{2}\{(\sigma'_2 - \sigma'_3)^2 + (\sigma'_3 - \sigma'_1)^2 + (\sigma'_1 - \sigma'_2)^2\}$. The clay is normally compressed to a vertical effective stress $\sigma'_v = 200\,\text{kPa}$, with $K_{0nc} = 0.65$, at which state $v = 2.3$.

What is the intermediate principal effective stress at failure in undrained plane strain?

With soil parameters $M = 0.9$, $\lambda = 0.25$, $\kappa = 0.05$, calculate the undrained strength in plane strain for this clay and compare this with the undrained strength in triaxial compression.

The finite element analysis approximates constant volume undrained behaviour by assuming no drainage and taking a bulk modulus for the pore fluid $K_w = 100\,\text{MPa}$. At a certain element beneath the embankment, the stress path is such that the vertical stress σ_z remains the major principal stress, and the change in minor total principal stress $\Delta\sigma_x = 0.7\,\Delta\sigma_z$. Show that the change in excess pore pressure when this element reaches the critical state is $\Delta u \simeq 172\,\text{kPa}$. Has the finite compressibility of the pore fluid significantly affected the undrained shear strength of this element of clay?

E11.3. A set of samples of Boston blue clay was taken from the site of a test embankment and subjected to conventional one-dimensional compression in an oedometer. The resulting data of

Fig. 11.E1 In situ stresses and preconsolidation pressures for Boston blue clay (• values of σ'_{vc} estimated from oedometer tests) (after D'Appolonia, Lambe, and Poulos, 1971).

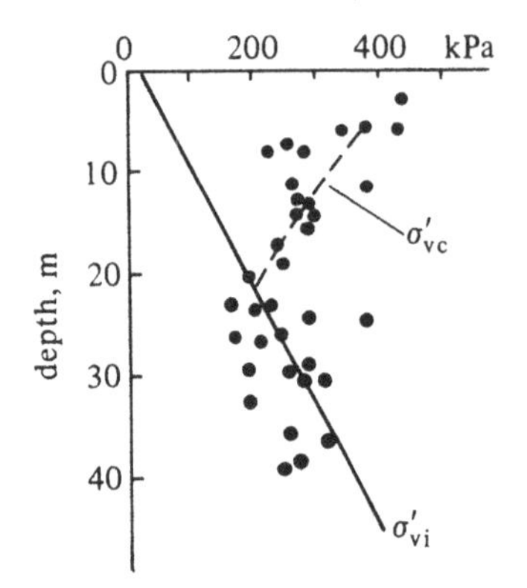

preconsolidation pressure σ'_{vc} and the estimated in situ effective overburden pressure σ'_{vi} are plotted against depth in Fig. 11.E1.

Estimate the profile with depth of undrained strength in triaxial compression, assuming that the Boston blue clay can be modelled by Cam clay with the following parameters: $M = 1.06$, $\lambda = 0.147$, $\kappa = 0.06$, $N = 2.808$, and $G' = 1250\,\text{kPa}$. Values of $K_0 = \sigma'_{hi}/\sigma'_{vi}$ should be estimated from expression (10.24) using values of overconsolidation ratio inferred from Fig. 11.E1.

E11.4. A road embankment is to be constructed on a site where the ground conditions consist of a deep stratum of clay. The performance of the underlying ground is to be monitored during construction of the embankment by taking readings from piezometers, settlement plates, and inclinometers.

One piezometer is installed on the centreline of the embankment at a depth of 10 m below the horizontal ground surface. The embankment is to be 80 m wide at its base, 15 m high, and made of granular fill of bulk unit weight $19\,\text{kN/m}^3$ with side slopes at 35° to the horizontal. The fill is to be placed sufficiently rapidly that negligible consolidation occurs in the clay during construction of the embankment.

From laboratory tests it is estimated that the clay has the following soil parameters: $\gamma = 19\,\text{kN/m}^3$, $\lambda = 0.147$, $\kappa = 0.06$, and $\phi' = 30°$ in plane strain. It is estimated that at a depth of 10 m the preconsolidation pressure is $\sigma'_{vc} = 162\,\text{kPa}$ and $K_0 = 0.72$. The ground water table is 1 m below the surface.

Estimate the response of the piezometer based on the following assumptions:

i. Plane strain conditions apply.
ii. The clay is to be modelled by a Cam clay-like model having an elliptical yield curve in the $(s':t)$ effective stress plane.
iii. After yield occurs, the distribution of total vertical stress in the ground can still be obtained by elastic theory.

By selecting two or three suitable stages of construction, plot the expected excess pore pressure in metres head of water against fill height.

E11.5. An oil storage tank 20 m in diameter and 12.5 m high is to be founded on the level surface of a thick stratum of soft clay. Laboratory tests show that the clay at a depth of 5 m, where the in situ vertical effective stress is $\sigma'_{vi} = 50\,\text{kPa}$, has an overconsolidation ratio of 2, for which the corresponding value of the coefficient of earth pressure at rest $K_0 = 1$. The clay has a value of $K_{0nc} = 0.55$

when in a normally compressed state and has a behaviour which can be modelled by Cam clay with a value of $M = 1$.

The tank is to be proof-loaded by rapidly filling it with water, and the performance of the ground is to be monitored by placing a piezometer at a point P, 5 m below the centre of the base of the tank. What head h of water in the tank would be just sufficient to cause yield in the soil element at P? Estimate the excess pore pressure u that would be recorded by the piezometer. Sketch the expected relationship between u and h during the whole operation of proof-loading the tank. Assume that after yield occurs, the distribution of total vertical stress increments is still given by elastic theory.

E11.6. A large expanse of soft clay has had 4 m of clay eroded from the surface. The water table has at all times been at the ground surface, and the saturated unit weight of the clay is $\gamma = 18\,\text{kN/m}^3$. The earth pressure coefficient at rest, K_0, varies linearly with overconsolidation ratio n from 0.6 at $n = 1$ through 1.0 at $n = 2.5$.

A long embankment of width 10 m is being constructed rapidly on this clay with fill of unit weight $16\,\text{kN/m}^3$. At what height of embankment will yield first occur in the clay at a depth of 5 m below the clay surface on the centreline of the embankment?

Use the Cam clay model, with a circular generalisation in the octahedral plane and with parameters $\lambda = 0.161$, $\kappa = 0.062$, $N = 2.828$ for $p' = 1$ kPa, and $M = 0.888$.

E11.7. A soft clay deposit has a level surface and a water table 0.5 m below ground level. A sample taken from a depth of 3 m is found to have an overconsolidation ratio $n = 1.75$. If overconsolidation is due only to past movements of ground water, deduce the value of overconsolidation ratio at a depth of 6 m. Take $\gamma = 15\,\text{kN/m}^3$ and $\gamma_w = 9.81\,\text{kN/m}^3$.

The yield envelope for the clay is given by the Cam clay model, and the earth pressure coefficient K_0 varies with overconsolidation ratio according to the expression $K_0 = 0.6\,n^{1/2}$. The critical state parameter $M = 1.0$.

An oil storage tank 18 m in diameter and 15 m high founded on this deposit is to be filled rapidly with oil of specific gravity 0.8. What depth of oil in the tank will cause a soil element at depth 6 m on the centreline to yield, and what excess pore pressure will then be recorded by a piezometer at this point?

E11.8. A long embankment, which can be considered as a strip load of width 25 m, is to be constructed rapidly on a deep deposit of lightly overconsolidated clay. It is known that the clay shows yield loci of the form $t/s' = m\ln(s'_o/s')$, with critical states $t = ms'$, where $m = 0.4$ is a soil constant. Describe qualitatively the relationship between the height of the embankment and the changes in pore pressure that would be registered by a piezometer placed in the clay on the centreline of the embankment.

At a depth of 10 m on the centreline, before construction, the past maximum and present effective stresses are estimated to be

past maximum	$\sigma'_{vc} = 120\,\text{kPa}$	$\sigma'_{hc} = 75\,\text{kPa}$
present	$\sigma'_{vi} = 90\,\text{kPa}$	$\sigma'_{hi} = 70\,\text{kPa}$

and the present pore pressure is $u_i = 80\,\text{kPa}$. Estimate the pore pressure that would be recorded by a piezometer at this depth when the embankment has reached a height of 7 m. The unit weight of the embankment fill material is $16\,\text{kN/m}^3$. It can be assumed that once the clay yields, the undrained effective stress path is given approximately by $\delta t = -m\,\delta s'/2$.

12

Beyond the simple models

12.1 Introduction: purpose of models

Models of the mechanical behaviour of soils have served two purposes in this book: they have been used primarily to illustrate facets of the observed behaviour of soils which might at first sight be considered extraordinary but which, with even a simple model of soil behaviour, can in fact be anticipated; however, Section 11.3 has shown how such models can be used in finite element analyses of geotechnical problems of practical importance. There are different requirements for models used for purely illustrative purposes and for models used for predicting the response of geotechnical structures. The illustrative model is expected to give a simplified but overall picture of soil behaviour, but the predictive model must be able to match rather closely the behaviour of the elements of real soil which are being deformed in a particular prototype. Some of the simplifying assumptions of the illustrative model are inappropriate for the predictive model if the predictions are to be useful. Many implicit or explicit assumptions have been made in this book in presenting the simple models; the effects of relaxing some of these assumptions are discussed briefly in this chapter.

12.2 Effects of time

Loading of a soil with no drainage, or with restricted drainage, in general leads to the generation of excess pore pressures as a result of prevented volume change in the soil. With time, these excess pore pressures tend to dissipate to equilibrium values. As pore pressures change, the effective stresses also change and the soil deforms. These deformations could be described as time-dependent deformations, but they are entirely explicable in terms of the changes in effective stress which the soil is experiencing. The time dependence of these deformations arises from the finite permeability of the soil, not from any extra constitutive, rheological properties of the soil skeleton. As has been seen in Section 11.3, the simple models introduced in this book are capable of reproducing this pattern of time-dependent behaviour.

However, some observations cannot be ascribed to changes in effective stress resulting from diffusion of pore water. Samples of many clays, when loaded one-dimensionally in an oedometer, show deformations continuing with time even when no measurable gradients of excess pore pressure remain. This sort of behaviour is usually called *creep* or *secondary consolidation*, and a coefficient of secondary consolidation c_α can be introduced to describe the further change of specific volume v with logarithm of time (Fig. 12.1) once pore pressures are assumed to have dissipated and so-called primary consolidation is complete:

$$v = v_1 - c_\alpha \ln\left(\frac{t}{t_1}\right) \tag{12.1}$$

where t_1 is a reference time and v_1 a reference value of specific volume. Expression (12.1) can be rewritten as

$$\frac{\delta v}{\delta t} = -\frac{c_\alpha}{t} \tag{12.2}$$

indicating that the volumetric strain rate decreases as time increases.

Experimental observations appear to show that for a given clay the value of c_α is low when creep is occurring at stresses lower than the preconsolidation pressure and that it increases as the preconsolidation pressure is exceeded (Figs. 12.2a, b). If a general variable compliance c is defined (Fig. 12.2a) as the current slope of the $\ln \sigma'_v : v$ relationship, whatever

Fig. 12.1 Coefficient of secondary consolidation c_α illustrated in one-dimensional compression of Mastemyr clay; load increment from 392 to 785 kPa (after Norwegian Geotechnical Institute, 1969).

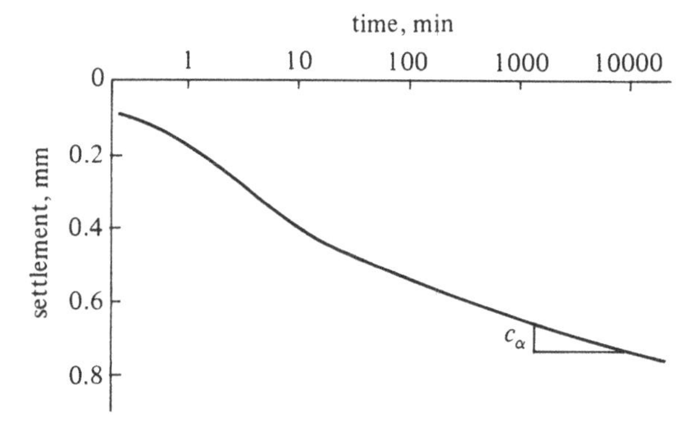

Fig. 12.2 (a) One-dimensional compression of Ottawa lacustrine clay; (b) variation of compliance c and coefficient of secondary consolidation c_α; (c) variation of ratio c_α/c (after Graham, Crooks, and Bell, 1983).

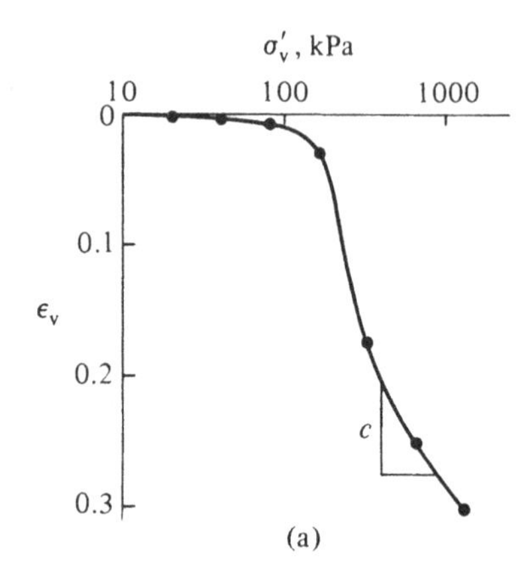

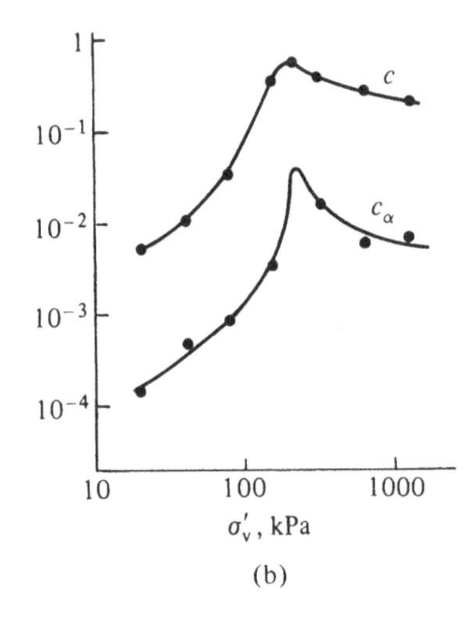

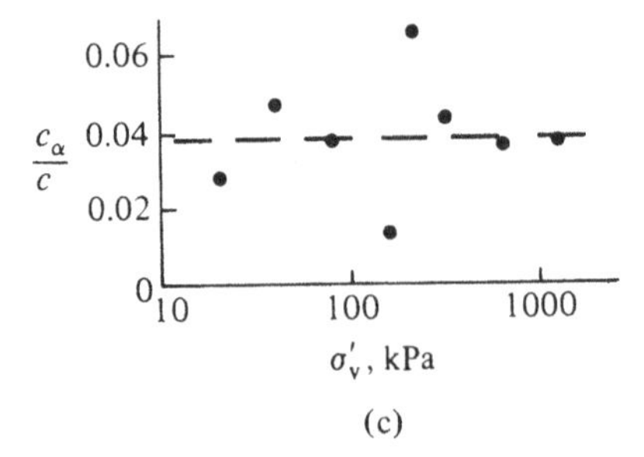

the value of σ'_v,

$$\frac{\delta v}{\delta \sigma'_v} = \frac{-c}{\sigma'_v} \tag{12.3}$$

then the variation of this with stress level is found to have a similar form to the variation of the creep coefficient c_α (Fig. 12.2b). The ratio c_α/c is found to be roughly constant for a particular soil (Fig. 12.2c), Mesri and Godlewski (1977) suggest $c_\alpha/c \sim 0.04$ with a hiatus around the preconsolidation pressure σ'_{vc}.

This creep or secondary compression is deformation resulting from readjustment of particle contacts at essentially constant effective stresses. The compliance c of the soil is an indication of how easily the structure of the soil can be made to collapse as the effective stresses are increased. When the soil is in a normally compressed condition, the structure is, in a sense, less stable and more ready to collapse than when the soil is in an overconsolidated condition. It is perhaps not surprising that creep deformations are also more significant when the soil is sitting in this 'unstable' normally compressed condition. Near the preconsolidation yield pressure, the structure is incipiently unstable, and the creep deformations represent a mixture of the elastic pre-yield and the plastic post-yield regimes. (Besides, in a sample being tested in an oedometer, the effective stress state is never uniform through the thickness of the sample: normally consolidated and overconsolidated states will certainly coexist.)

The term *creep* is usually used to describe deformations continuing with time at constant effective stresses. If a soil shows creep, then such creep deformations are occurring all the time, even though they can only be observed as creep when the stresses are held constant. In tests performed at different rates, at any particular effective stress state different amounts of creep deformation will have developed because of the different times that have elapsed, and different stress:strain relationships are observed. Typical results for one-dimensional compression tests and undrained triaxial compression tests performed at various rates of increase of load are shown in Fig. 12.3.

If the rate of loading is changed in the middle of a test, then it is typically found (e.g. Richardson and Whitman, 1963) that the state of the soil quickly jumps to the stress:strain curve appropriate to the new rate: the stress : strain curve that it would have followed if it had been loaded from the beginning at the new rate (Fig. 12.4). If an oedometer test performed with a fast rate of loading (*AB* in Fig. 12.5) is stopped, creep occurs (*BC*); and, as the rate of creep deformation gradually falls with time, the state of the soil moves to progressively lower $\ln \sigma'_v : v$ curves appropriate to the falling rates of deformation. When the loading is restarted (*CDE*), the

Fig. 12.3 (a) Oedometer tests on Batiscan clay with constant rate of strain: axial strain rates are (1) 1.43×10^{-5}/s, (2) 2.13×10^{-6}/s, (3) 1.07×10^{-7}/s (after Leroueil, Kabbaj, Tavenas, and Bouchard, 1985); (b) undrained triaxial compression tests on Drammen clay at various rates of strain: axial strain rates at peak are (1) 9.7×10^{-5}/s, (2) 1.6×10^{-6}/s, (3) 3.9×10^{-7}/s, (4) 6.4×10^{-8}/s, (5) 1.1×10^{-8}/s, (6) 3.9×10^{-9}/s (after Berre and Bjerrum, 1973).

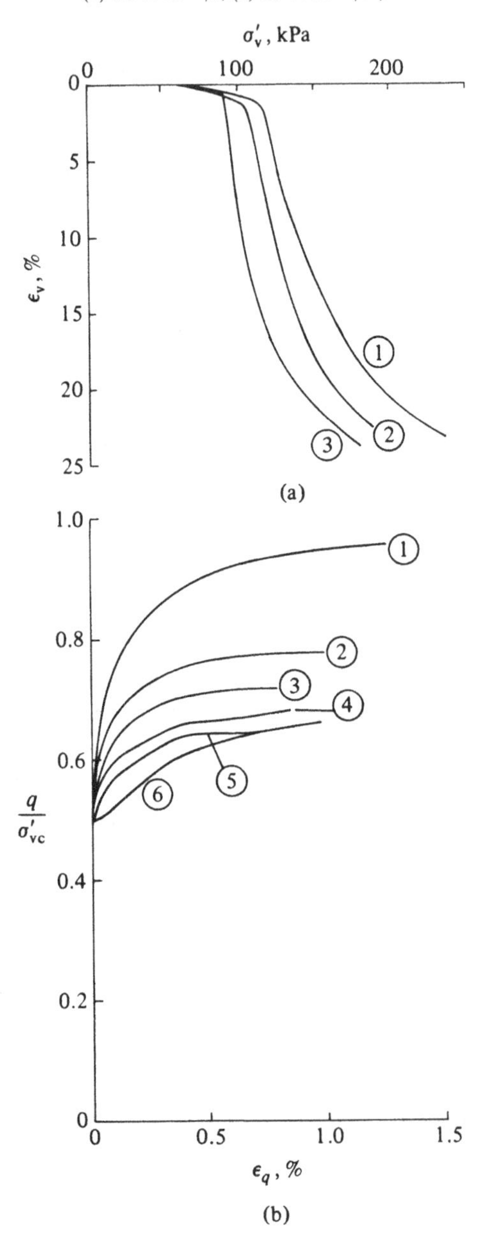

sample is found to have become apparently overconsolidated because the creep has taken the state of the soil to a position C below the equilibrium stress: deformation curve for the particular rate of loading. An apparent preconsolidation pressure would be assigned to point D even though the soil had never previously experienced effective stresses higher than those at B and C.

This is a process which occurs in nature, identified by Leonards and Ramiah (1960) and discussed in some detail by Bjerrum (1967). If a typical

Fig. 12.4 (a) Oedometer tests on Batiscan clay with step changes in axial strain rate: axial strain rates are (1) 1.05×10^{-7}/s, (2) 2.7×10^{-6}/s, (3) 1.34×10^{-5}/s (after Leroueil, Kabbaj, Tavenas, and Bouchard, 1985); (b) undrained triaxial compression test on Belfast clay with step changes in axial strain rate: axial strain rates are (1) 1.4×10^{-7}/s, (2) 1.4×10^{-6}/s, (3) 1.4×10^{-5}/s (after Graham, Crooks, and Bell, 1983).

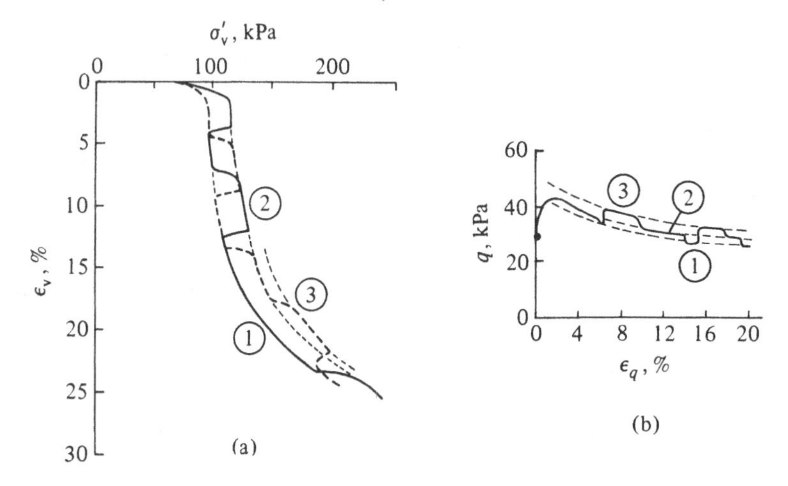

Fig. 12.5 Effect of creep or strain rate on one-dimensional compression of clay.

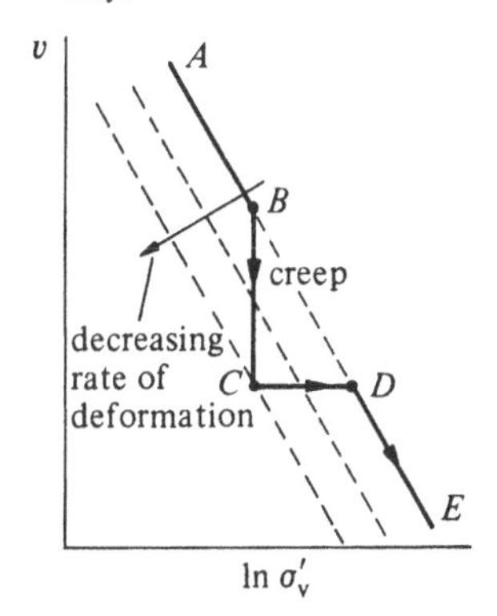

deposit of Norwegian marine clay is 3000 years old, then, in any laboratory or field loading applied over a time scale which is short by comparison with 3000 years, the clay appears to be overconsolidated. It is implicit in this qualitative model that as the vertical stresses are increased above the current values, the rate at which further creep deformations occur also increases and the benefit of increased stiffness may only be temporary.

The preconsolidation pressure observed in one-dimensional compression tests in an oedometer is, as emphasised in Section 3.3, related to just one point on the yield surface for the soil. If the value of the preconsolidation pressure depends on the rate at which the soil is being deformed, then it can be expected that the size of the whole yield surface also depends on the rate at which the soil is being deformed on the test paths which are used to probe for the occurrence of yielding. Experimental confirmation of this is provided by the data in Fig. 12.6 for the yielding of natural clay from St. Alban, Québec (Tavenas et al., 1978). These data are from tests having different durations of load increment; as the rate at which the soil is loaded decreases, the size of the yield locus decreases. Data of this type are few, but it seems not unreasonable to propose that the shape of the yield locus is independent of rate of loading, and only its size changes.

If soils were purely viscous materials (such as pitch, for example), then the surface of the earth would be flat. However, a natural way to extend the elastic–plastic models described in Chapter 4 to accommodate rate

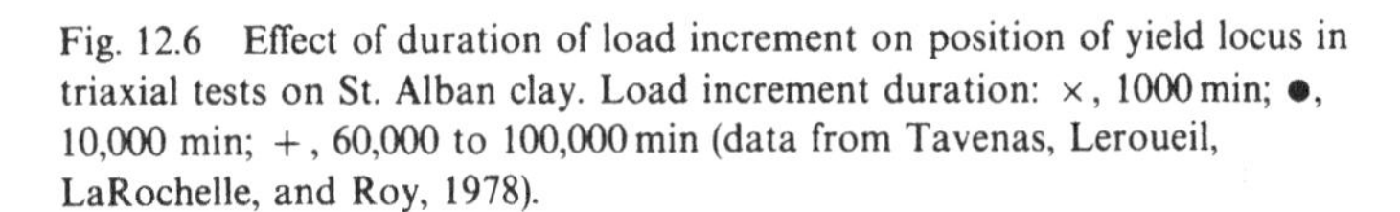

Fig. 12.6 Effect of duration of load increment on position of yield locus in triaxial tests on St. Alban clay. Load increment duration: ×, 1000 min; ●, 10,000 min; +, 60,000 to 100,000 min (data from Tavenas, Leroueil, LaRochelle, and Roy, 1978).

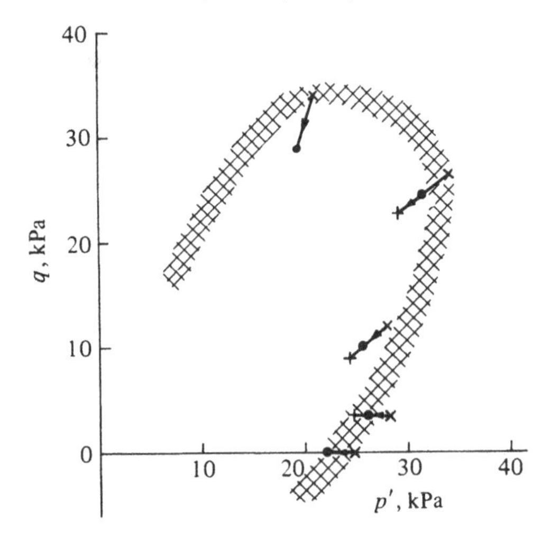

sensitivity and creep effects is to assume that there is a yield locus that limits elastically attainable states of stress in static, infinitely slow tests, but also to assume that at higher finite strain rates the size of the yield locus, indicated by the reference mean effective stress p'_o of Section 4.3, is increased by an amount depending on the strain rate. It is usually more convenient to make the inverse hypothesis that the rate at which plastic deformations occur (and these can now be called viscoplastic deformations) depends on the extent by which the present effective stress state lies outside the current static yield locus. The framework for such elastic–viscoplastic models was set up by Perzyna (1963) and Olszak and Perzyna (1966) and has been combined with a Cam clay model by Adachi and Oka (1982).

Three effective stress conditions are shown in Fig. 12.7. Point A lies inside the current static yield locus (syl). A change of effective stress to point B (Fig. 12.7a) does not pass beyond the static yield locus and is associated only with elastic time-independent deformations. Changes of effective stresses to points C and D, which lie outside the current static yield locus, lead to the onset of time-dependent viscoplastic deformations. At C (Fig. 12.7b) the stress ratio q/p' is low, and the mechanism of plastic deformation may involve plastic volumetric compression: that is, the static yield locus expands as the soil hardens and the rate of deformation decreases with time until, theoretically at infinite time, the current static yield locus has expanded to pass through C and plastic deformation ceases. At D (Fig. 12.7c) the stress ratio q/p' is high, and the mechanism of plastic deformation may involve plastic volumetric expansion: that is, the static yield locus contracts as the soil softens and the deformation of the soil accelerates to failure. In this simple elastic–viscoplastic model the possibility of creep occurring for stress states lying inside the yield locus or vertical effective stresses below the preconsolidation pressure has been ignored, but other variants are possible.

Fig. 12.7 Elastic–viscoplastic description of soil behaviour: (a) elastic stress change; (b) viscoplastic stress change with expansion of yield locus; (c) viscoplastic stress change with contraction of yield locus.

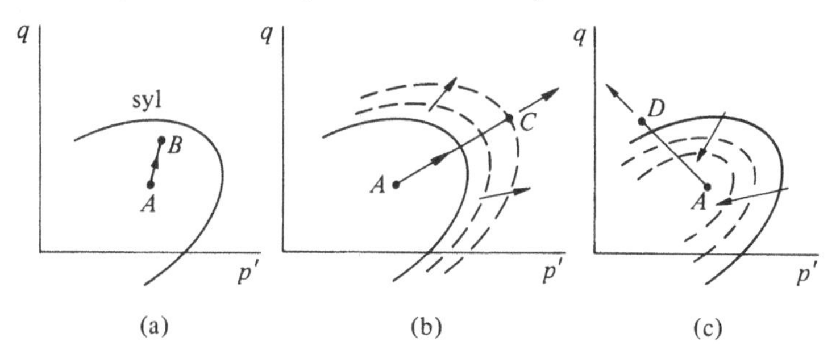

It is straightforward to set up equations relating strain rates to effective stresses and past histories. There are some effects, such as development of cementation between particles, which should perhaps be related to the length of time that has elapsed since a soil element was subjected to certain effective stresses and which lead to ageing or time-hardening phenomena which cannot be readily linked to rate of deformation. Time, however, knows no origin; so the point of onset of time hardening, particularly for soil elements in which the effective stresses are not constant, may be difficult to define.

It is worthwhile emphasising again that it is important, in looking at features of soil response, to distinguish between time effects due to creep or ageing and time effects due to dissipation of pore-water pressures. (Leroueil and Tavenas, 1981, have shown that the same field observations can sometimes be attributed to either of these effects.) On the whole, the dependence of soil behaviour on rate of testing – the viscous side of soil behaviour – has been left out of consideration in this book. There may, however, be many field situations in which rate effects must be incorporated as an important factor.

12.3 Inelastic elastic response

It has been assumed that yield surfaces exist which bound the regions of effective stress space which can be reached by a soil without incurring irrecoverable deformations. Within a yield surface, it has been assumed, deformations are entirely recoverable and elastic. Elastic response implies a one-to-one relationship between stress and strain. These assumptions are of course common to elastic–plastic models which might be used for any material, and, in this book, the idea of a yield surface was

Fig. 12.8 Annealed copper wire: (a) unloading and reloading in uniaxial tension (after Taylor and Quinney, 1931); (b) variation of tangent modulus with strain during reloading.

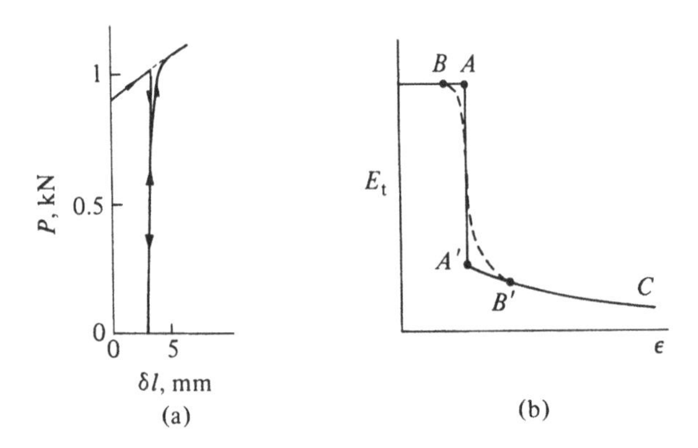

introduced with reference to tests on annealed copper wire and tubes. If the data for annealed copper are studied in detail (Fig. 12.8a), then it becomes apparent that there is not a sudden sharp drop in the tangent stiffness near the previous maximum load, or yield point, but rather a rapid reduction. The load:deflexion relationships do not actually have a discontinuous slope at this yield point. Nevertheless, the change in stiffness occurs over a sufficiently small range of loads or deformations for the assumption of a well-defined yield locus to be a good and credible hypothesis, with the simple note that nature can be expected to round off corners of material response. Until a yield point is neared, the response corresponds precisely to what one would expect of an ideal linear elastic material: the load:deflexion relationships are linear and identical on unloading and reloading, and a one-to-one relationship between stress and strain does exist in this region inside the yield surface. The ideal and actual changes of tangent stiffness (calculated from the local slope of the stress:strain or load:deformation relationship) around the yield point are shown in Fig. 12.8b. Ideally the stiffness shows a step change AA' at the yield point; actually the stiffness changes continuously from B to B'. The annealed copper (like soil) hardens after yielding, and the tangent stiffness continues to fall gently after yield (A' or B' to C).

By contrast, the yielding of soils is usually a much more gradual process, particularly if the soil is not one that has been provided with a significant

Fig. 12.9 Spestone kaolin: (a) unloading and reloading in undrained triaxial compression (after Roscoe and Burland, 1968); (b) variation of tangent modulus with strain during reloading.

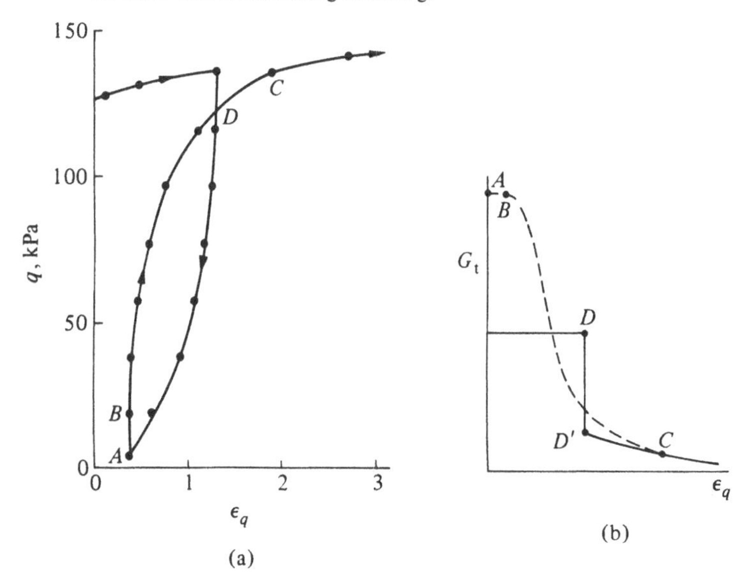

Fig. 12.10 Dependence of average secant modulus on size of unloading–reloading cycle.

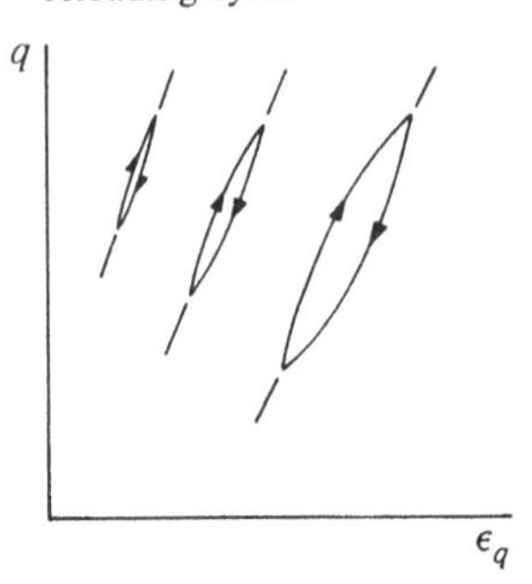

Fig. 12.11 Measurement of local axial strain and overall axial strain between end platens.

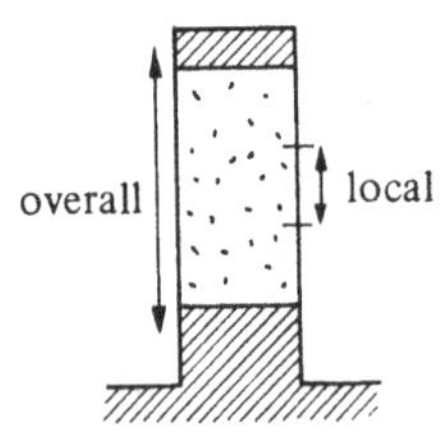

Fig. 12.12 Undrained triaxial compression of reconstituted North Sea clay (overconsolidation ratios 2, 4, 8): (a) variation of tangent modulus with strain; (b) contours of triaxial shear strain ε_q; (c) strain data interpreted as indication of changing current position of inner yield locus (data from Jardine, Symes, and Burland, 1984).

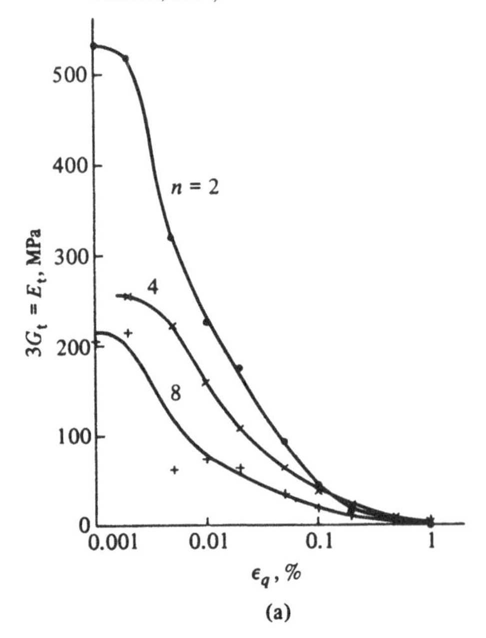

degree of chemical cementation bonding between the particles. Unloading–reloading cycles in undrained triaxial compression and isotropic compression were shown in Section 3.3; a typical cycle is illustrated in Fig. 12.9a. Not only is the change of stiffness much more gradual than for the annealed copper, but also the stress:strain response on unloading and reloading is hysteretic. This has a number of implications. There is no longer a one-to-one relationship between stress and strain in this supposedly elastic region; energy is dissipated on a closed stress cycle, which also implies inelastic response; and though an average secant modulus can be defined from the average slope of the unload–reload loop, the value of this average secant modulus depends on the amplitude of the unload–reload cycle (Fig. 12.10). The departure from the ideal relationship between tangent shear stiffness G_t and strain is shown in Fig. 12.9b. The strain range over which the response is truly elastic and non-dissipative

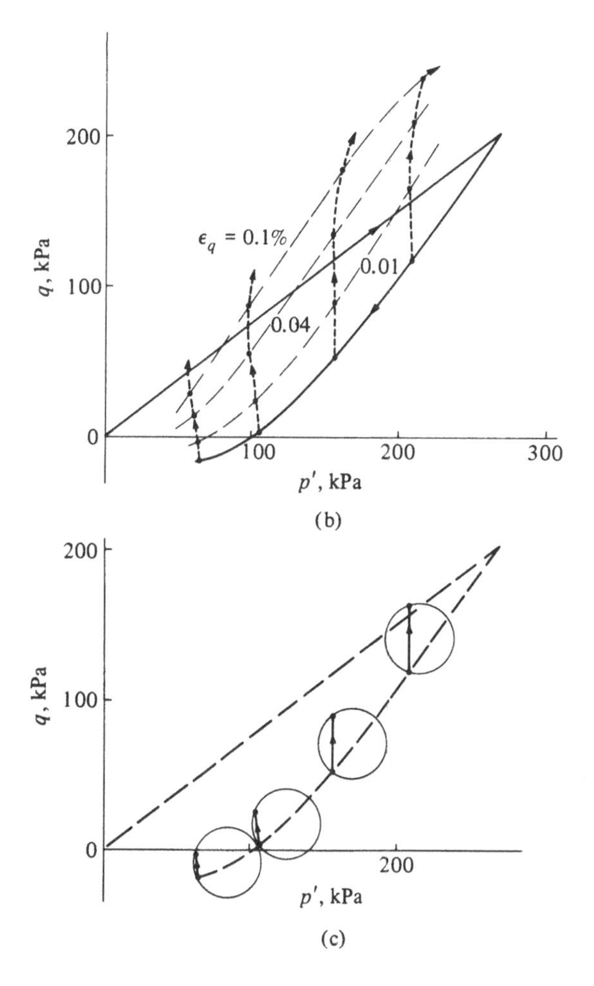

(AB) (Figs. 12.9a, b) may be very small by comparison with the strain required to bring the stiffness down to a clearly plastic, yielding level (around C), and the stiffness in the truly elastic region may be considerably higher than the average secant stiffness over the strain from A to D.

These are conclusions drawn also by Jardine, Symes, and Burland (1984), aided by accurate measurements of strain made locally on samples of stiff soils in the triaxial apparatus, over a gauge length rather than across the loading platens (Fig. 12.11); end effects, they argue, can easily obscure the observation of the initially extremely stiff response of the soil. Some of the variations of tangent stiffness with strain deduced from data obtained by Jardine et al. for a reconstituted clay obtained from the North Sea are shown in Fig. 12.12a. For the low plasticity soils they have studied, the region of truly elastic response rarely extends beyond a shear strain of about 0.01 per cent.

These high stiffnesses have been measured at the start of undrained

Fig. 12.13 (a) Stress path and (b) strain path for undrained compression of one-dimensionally overconsolidated soil.

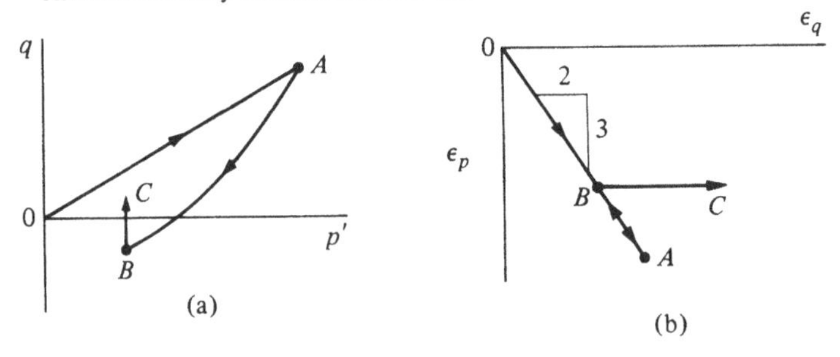

Fig. 12.14 (a) Inner yield locus (iyl) bounding region I of high stiffness, outer yield locus (oyl) bounding region II of intermediate stiffness, in effective stress plane; (b) deviator stress q and triaxial shear strain ε_q for stress path BCDEFG.

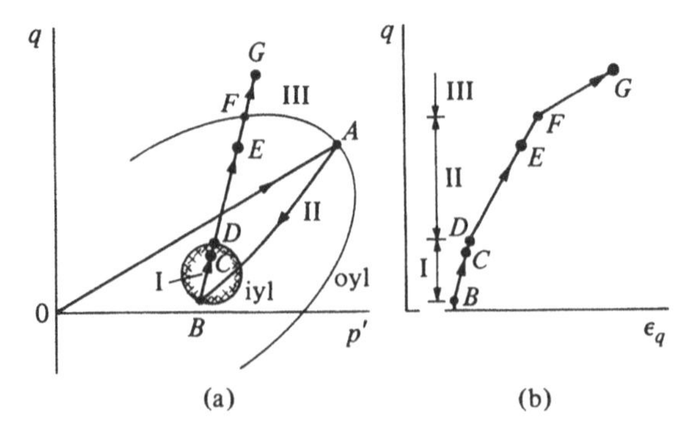

triaxial compression applied to samples of soil taken from the ground or samples of reconstituted soil which have been loaded and unloaded one-dimensionally in the laboratory. Either way, the undrained triaxial compression represents a change of loading direction in strain space or effective stress space (Fig. 12.13), and experimental evidence from triaxial, plane strain, and true triaxial tests (and more recent evidence from tests in which controlled rotation of principal axes of stress and strain can be imposed) suggests the more general proposition that high stiffnesses (and, by implication, truly elastic response) are only seen immediately after a major change in direction of an effective stress path or strain path. Appreciation of anticipated stress changes in relation to past history (as discussed in Chapter 10) is thus extremely important.

There are various ways in which the observed form of response can be incorporated into numerical models of soil behaviour. In so-called bounding surface models, for example, the stiffness is made dependent on the distance of the current effective stress state from the yield surface, and it decreases as the yield surface is neared, perhaps following the experimentally observed relationship of Fig. 12.12a. An alternative simple stratagem is to suppose that a small inner, or true, yield surface exists which is carried around by the current stress state and which bounds a small region of truly elastically attainable effective stress states. For example, after one-dimensional compression and unloading (OAB in Fig. 12.14a), this inner yield locus (iyl) in the $p':q$ plane might have been pushed to the position shown in Fig. 12.14a. A change of stress to a state C lying in zone I within this inner yield locus would be associated with high stiffness (Fig. 12.14b); a change of stress to a state E lying in zone II, between the inner yield locus (crossed at D) and the outer yield locus (oyl), would be associated with a lower stiffness; and a change of stress to a state G lying in zone III beyond the outer yield locus (crossed at F) would be associated with major plastic deformations. A change of direction of stress path at any stage requires the effective stress state to start by traversing the inner yield locus with consequent high stiffness and low strains. A model which combines this principle with the Cam clay model of Chapter 5 is described by Al-Tabbaa and Wood (1989).

Jardine, Symes, and Burland (1984) show (Fig. 12.12b) data of effective stress paths and strains during triaxial compression of samples which have been unloaded one-dimensionally to various overconsolidation ratios. If it is assumed that truly elastic response extends only up to a shear strain of 0.01 per cent, then the stiffness data of Fig. 12.12a and the contours of strain shown in Fig. 12.12b can be reinterpreted as indicating the changing position of the inner yield locus as the soil is unloaded

(Fig. 12.12c). In fact, even a shear strain of 0.01 per cent may be a considerable overestimate of the elastic strain region for many soils.

The effect of using such a translating or kinematic inner yield locus in finite element calculations of response of geotechnical structures is illustrated dramatically in analyses of the excavation in London clay for the underground car park at New Palace Yard beside the House of Commons, Westminster (reported by Simpson, O'Riordan, and Croft, 1979). A simplified plan and section of the excavation are shown in Fig. 12.15. The surrounding soil was supported by a diaphragm wall propped at successively lower depths as excavation proceeded. With adjacent important buildings very close to the exacavation, it was important to estimate expected movements of the surrounding ground, in particular the tilt that might be expected at the base of the clock tower Big Ben.

Fig. 12.15 (a) Plan and (b) section through excavation in London clay for car park at New Palace Yard, House of Commons, Westminster (after Simpson, O'Riordan, and Croft, 1979).

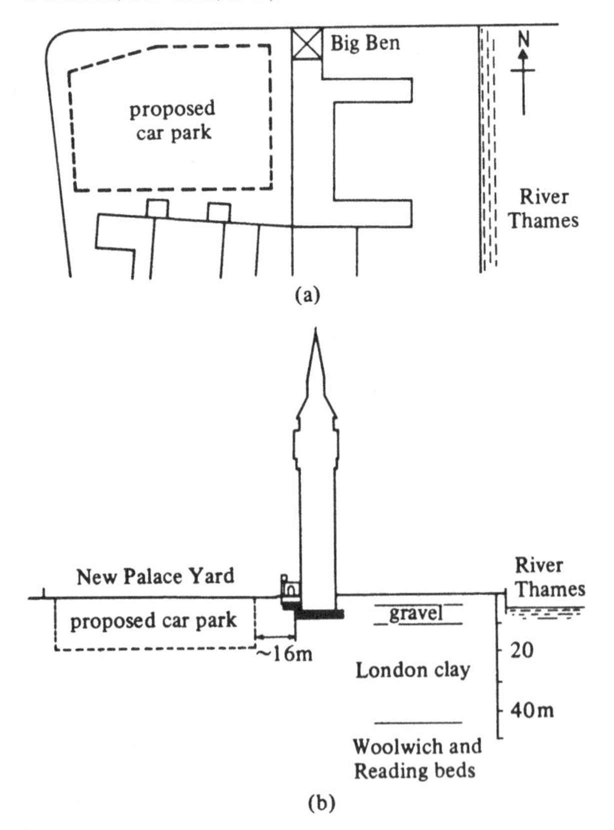

Early analyses by Ward and Burland (1973) and by St. John (1975) treated the soil as linearly elastic (but made some allowance for variation of modulus with depth). St. John compares results of calculations made assuming the deformation around the excavation to be either plane (with no deformation parallel to the wall) or axisymmetric (with all deformations in radial planes); only results of axisymmetric analyses are shown here. For the elastic material, the dominant effect of the excavation is the reduction in overburden pressure on the soil at the base of the excavation which consequently experiences vertical upward movement. The diaphragm wall is supported in this soil, and it too moves up carrying the soil behind the wall with it (Fig. 12.16). The constant volume condition of rapid undrained deformation then requires that this heave near the wall be balanced by a small settlement extending back to a large distance from the wall. Though the wall bulges somewhat towards the excavation, this does not affect the general shape of the profile of vertical movement behind the wall, and it is expected that Big Ben will tilt away from the excavation (Fig. 12.16). The calculated profile of vertical movement (according to the linear elastic axisymmetric analysis performed by Simpson et al.) is shown in Fig. 12.17a (curve *B*).

The elastic material clings together, and effects can be felt over great distances. In a material with a stiffness which decreases as the strain increases, the strains become disproportionately greater as the shear stress increases, and deformations become much more concentrated. (An extreme example of this is provided by the completely contained patterns of

Fig. 12.16 Tilt of Big Ben clock tower resulting from lifting of diaphragm wall with heave of excavation.

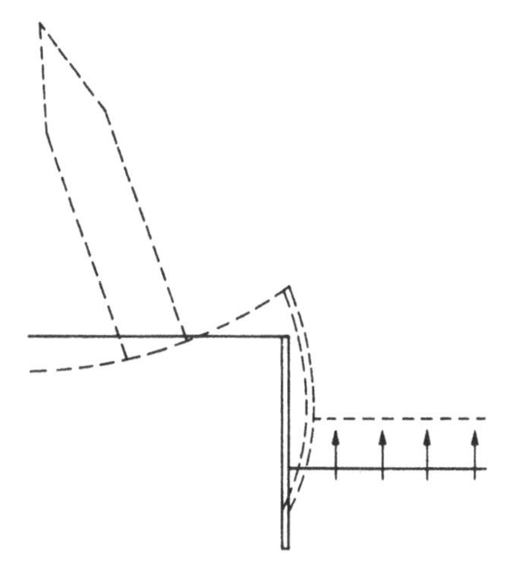

displacements associated with failure mechanisms in a perfectly plastic soil, as in Fig. 11.7.)

The profile of vertical movement calculated by Simpson et al., using a model incorporating an inner yield surface bounding a region of stress space associated with high stiffness, is also shown in Fig. 12.17a (curve *C*). In this model the tangent stiffness is assumed to fall by a factor of 10 when a small strain threshold is reached (Fig. 12.17b). Now the settlement further back from the wall is much smaller, and though there is some tendency to upward movement near the wall, the dominant effect is the conversion, at constant volume, of the bulge of the wall towards the excavation into a settlement trough behind the wall: Big Ben is expected to tilt towards the excavation. The observed profile of settlement is also

Fig. 12.17 (a) Observed and computed profiles of vertical movement adjacent to excavation for New Palace Yard car park: (*A*) measured, east wall; (*B*) computed, elastic soil; (*C*) computed, inner kinematic yield locus.
(b) Schematic diagram of variation of modulus with strain in soil model used for curve *C* (after Simpson, O'Riordan, and Croft, 1979).

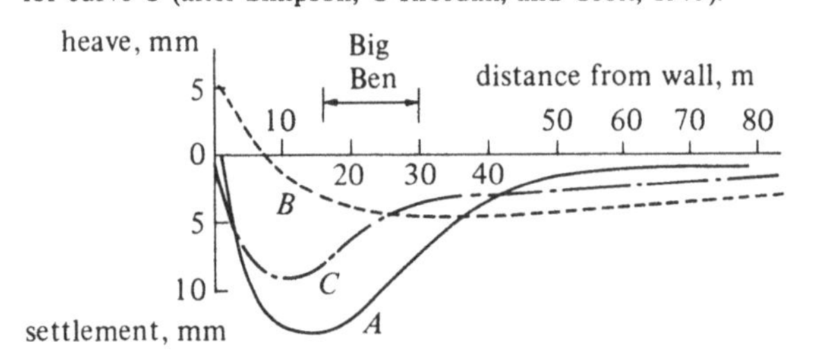

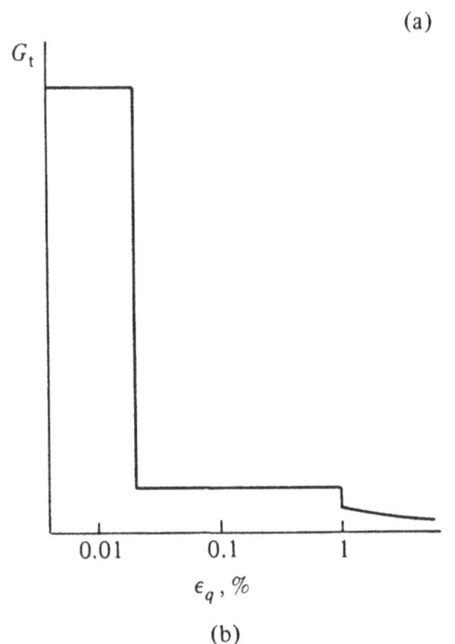

shown in Fig. 12.17a (curve *A*), and it corresponds quite nicely with the calculation using the inner kinematic yield surface. Big Ben did indeed tilt slightly towards the excavation, by about half a second (1 part in 7000) (Burland and Hancock, 1977).

In this case the movements involved are small, and the nett result of the various opposing effects is sensitive to the details of the model and to the values of the parameters chosen. However, the introduction of some irrecoverability into a region of stress space that might conventionally have been considered elastic has produced a marked change in the pattern of deformation calculated around the excavation.

Whether or not it is necessary to model the 'elastic' behaviour of a soil inelastically depends on the past history and future stress paths expected for that particular soil. If it is reckoned that the behaviour of a geotechnical structure will be dominated by elements of soil for which the stress changes are likely to involve major excursions beyond the current yield surface (according to the simple elastic–plastic model), then the resulting plastic deformations will dwarf the pre-yield deformations (Section 11.2), and a linear elastic description provided by a single average secant modulus will be quite adequate. This is usually the case for loading of soft soils: recent deposits and lightly overconsolidated soils. However, for stiff clays such as the London clay at New Palace Yard, the stress changes imposed by a geotechnical structure frequently lie entirely within the yield surface that one might have supposed to have been set up by

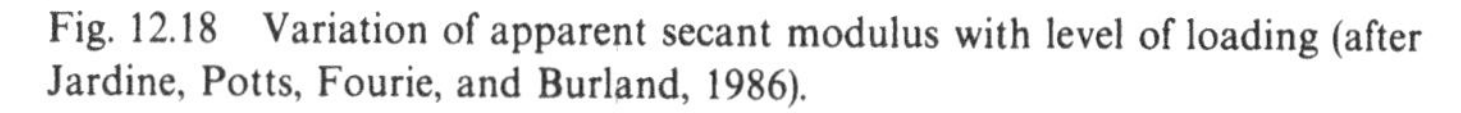

Fig. 12.18 Variation of apparent secant modulus with level of loading (after Jardine, Potts, Fourie, and Burland, 1986).

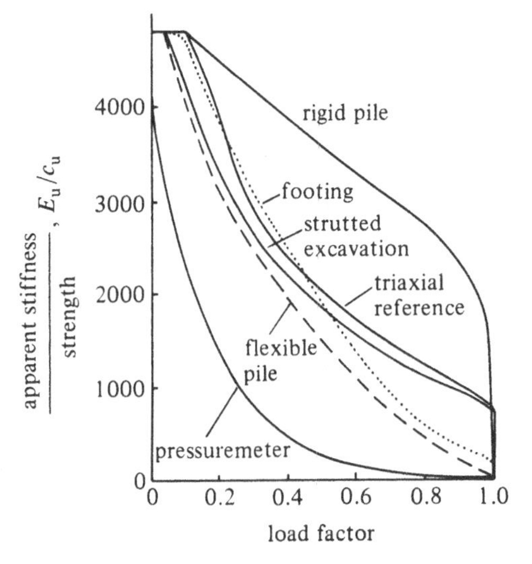

the past (geological) history of loading of the soil, and a more realistic description of the pre-yield behaviour is essential.

The kinematic hardening approach to soil modelling that has been outlined here makes it very clear that plastic strains start to be generated as soon as the effective stress state reaches the boundary of the 'inner' (true) yield locus (Fig. 12.14). However, if the changes in loading which are applied to the soil are essentially monotonic, then it may be numerically simpler to treat the soil response as non-linear 'elastic', with an elastic stiffness falling continuously with strain, as seen experimentally in Fig. 12.12a. Jardine, Potts, Fourie, and Burland (1986) performed finite element analyses for several common geotechnical situations to try to indicate how an equivalent single average stiffness should be chosen so that the settlement of a footing or pile or the movement of a wall, for example, might be calculated from a conventional linear elastic analysis – if an engineer insists on using such an analysis.

Results are presented in Fig. 12.18 as values of average secant modulus non-dimensionalised with undrained strength, plotted against load level non-dimensionalised with failure load. These make it very clear that the apparent average stiffness depends on both the type and the level of loading. The observed variation in stiffness in a triaxial test (shown in

Fig. 12.19 Response of soil to undrained cyclic loading according to Cam clay model: (a) effective stress path; (b) stress:strain response; (c) pore pressure:strain response.

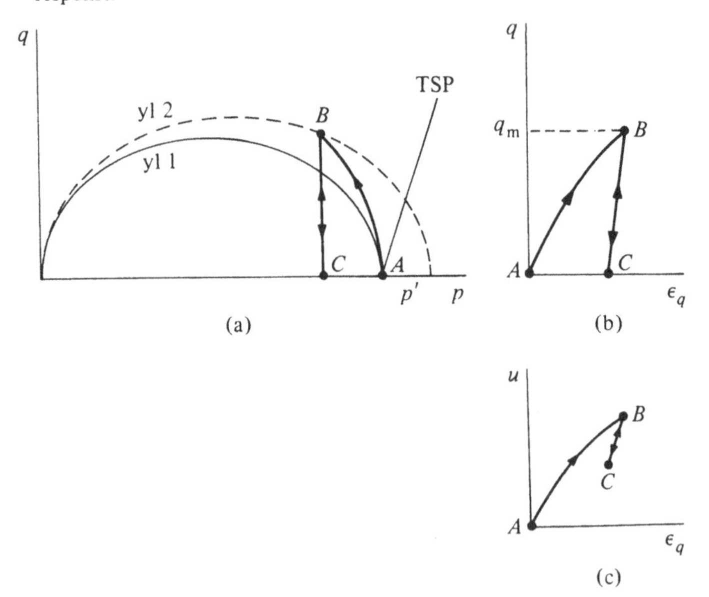

Fig. 12.18 as a reference curve and forming the input to the numerical analysis) may be of no direct help in choosing the single stiffness for a design calculation. Of course, the pattern of strains around the structure in a non-linear elastic soil is quite different from that in a linear elastic soil; so the curves in Fig. 12.18 relate only to particular points on the structures analysed.

Broadly, the simple elastic–plastic models of Chapters 4 and 5 are deficient where imposed effective stress paths lie entirely within the yield surface. Cyclic loading, even of soft soils, imposes such paths, and the simple models are again out of their depth. Consider undrained cycles of triaxial compression applied to normally compressed Cam clay (Fig. 12.19) with the deviator stress varying between zero and q_m. The first application of this stress (AB) causes yielding and plastic deformation (Fig. 12.19b), but its removal (BC) lies entirely within the expanded yield locus (yl 2) (Fig. 12.19a) and consequently produces purely elastic deformations (Fig. 12.19b). All subsequent applications and removals of the deviator stress merely retrace the path CBC in the stress plane (Fig. 12.19a) and

Fig. 12.20 Typical response observed in cyclic loading of clay: (a) effective stress path; (b) stress:strain response; (c) pore pressure:strain response (after Sangrey, Pollard, and Egan, 1978).

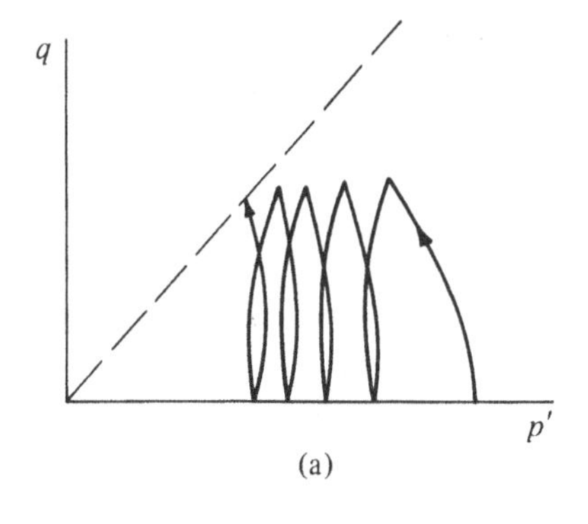

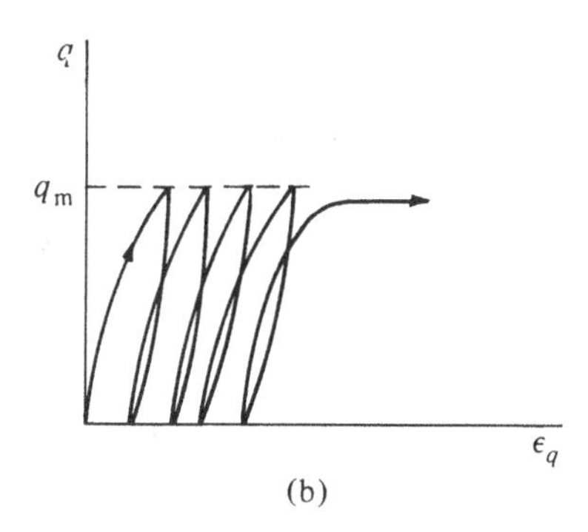

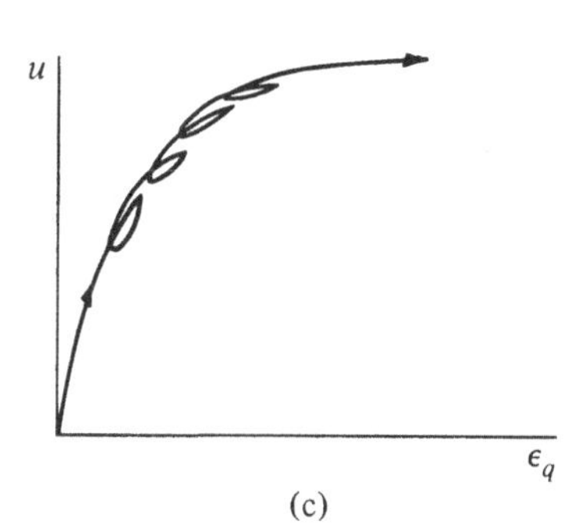

in the stress:strain relationship (Fig. 12.19b); the pore pressure changes that are observed (*CBC* in Fig. 12.19c) result only from the changes in total mean stress on the conventional (constant cell pressure) loading and unloading.

Experimentally (e.g. see Wood, 1982), however, continuing build up of pore pressure and deformation is usually observed (Fig. 12.20), with failure possibly being reached at a deviator stress well below the strength that could have been expected in static loading (Fig. 12.20a). Experimentally, it seems that data of the behaviour of soils under cyclic loading can be fitted into a coherent picture provided that changes in effective stresses (and changes in pore pressure) are properly taken into account. However, it is evident that if modelling of effects of cyclic or dynamic loading is important, then due attention must be paid to the description of the actually inelastic elastic deformations in the simple elastic–plastic model.

12.4 Evolution of yield loci

In the simple elastic–plastic models of soil response discussed in Chapter 4, it was assumed that yield loci always have the same shape in the $p':q$ effective stress plane and that they expand uniformly when the soil yields, this expansion being linked to the development of irrecoverable plastic volumetric strains. A typical experimentally observed yield locus for a natural soil might have the shape shown in Fig. 12.21a. A sample of this soil, starting from the isotropic effective stress state *A*, at the intersection of the current yield locus (yl) with the p' axis, would show very different responses in drained testing with constant mean effective stress p', according to whether the deviator stress was increased (*AB*, compression) or decreased (*AC*, extension). The apparent shear stiffness would be high in compression (Fig. 12.21b) because the effective stress path *AB* heads towards the interior of the yield locus – the elastic region – whereas the stiffness would be low in extension because the

Fig. 12.21 Different stiffnesses expected in triaxial compression (*AB*) and triaxial extension (*AC*) of soil with history of one-dimensional compression: (a) effective stress paths; (b) stress:strain responses.

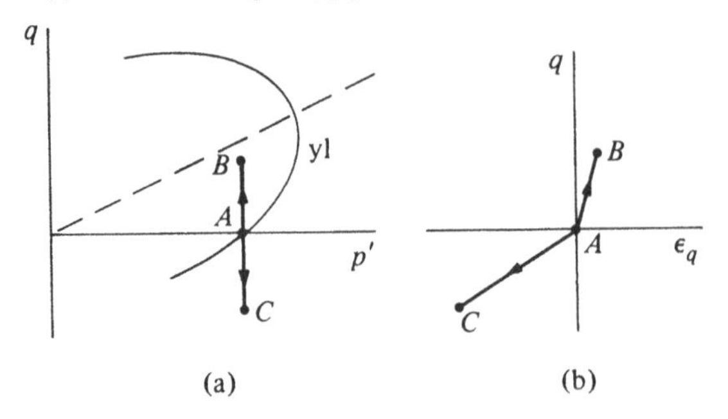

effective stress path AC is associated with yielding and irrecoverable deformation of the soil. The response of the soil in this isotropic stress state is thus anisotropic, and this anisotropy would be linked with the anisotropic past history of one-dimensional deformation of the soil in the ground, which has left the soil particles with certain preferred orientations and has resulted in the yield locus shown in Fig. 12.21a.

A sample of reconstituted soil which has only known isotropic stresses and deformations in its history would be expected to show yield loci which cross the p' axis at right angles (Fig. 12.22a) like the yield loci assumed in the Cam clay model, and the initial stiffness will be identical for small changes in deviator stress going into the compression or extension regions (AB and AC, respectively, in Figs. 12.22a, b).

There is no restriction on the effective stress changes that can be imposed on the soil. If the soil with the Cam clay yield locus of Fig. 12.22a were subjected to the stress path $OABC$ in Fig. 12.23a, where BC represents a large increase in mean effective stress at constant stress ratio q/p' (e.g. the stress ratio corresponding approximately to one-dimensional compression), then according to the assumptions that have been made here, no matter how long the section BC of the stress path, the current yield

Fig. 12.22 Identical stiffnesses expected in triaxial compression (AB) and triaxial extension (AC) of soil with history of isotropic compression: (a) effective stress paths; (b) stress:strain responses.

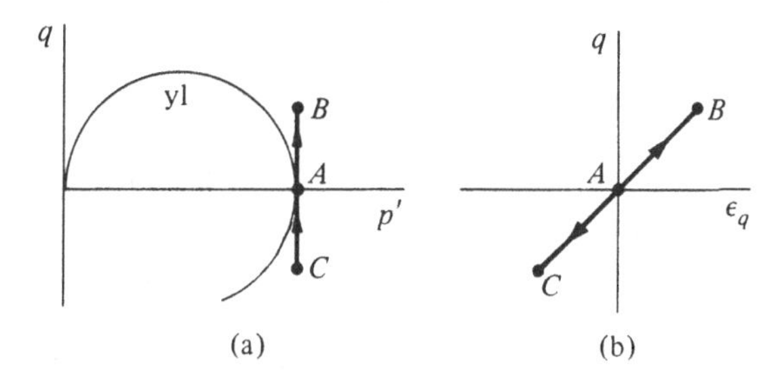

Fig. 12.23 Expected evolution of shape of yield locus with (a) major one-dimensional compression after initial isotropic compression and (b) major isotropic compression after inital one-dimensional compression.

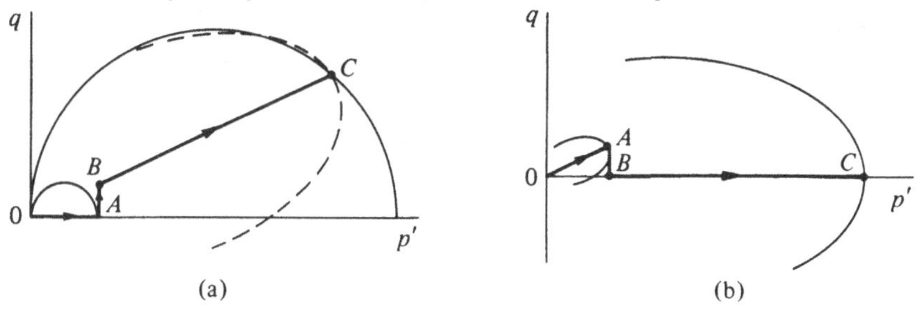

locus at C would still have the same elliptical shape centred on the p' axis. In reality, however, the yield locus at C would show some recognition of the almost entirely anisotropic stress history that the soil has experienced: it might resemble, rather, the yield locus for the natural soil, shown dotted in Fig. 12.23a. In other words, the hardening of the soil, the expansion of the elastic region, would have occurred preferentially in the direction of previous loading.

The procedure usually adopted for preparation of isotropically compressed samples of reconstituted soil for laboratory testing involves an initial one-dimensional compression in a mould to produce a sample which can be handled and placed in the triaxial apparatus for subsequent isotropic compression. It is hoped that the reverse transformation from anisotropic to isotropic yield locus will occur (Fig. 12.23b), provided the stresses reached during isotropic compression (BC) are sufficiently greater than those reached during the initial one-dimensional compression (OA). The different stress–dilatancy relationships (and if normality is supposed to apply, the different yield loci) deduced experimentally by Lewin (1973) for samples of clay with various compression histories were described in Section 8.6. He found that with a mean effective stress at C, about five times the mean effective stress at A, the initial anisotropy was virtually (but not completely) eliminated.

The other half of the assumption concerning the changing size of yield loci was that it was linked solely to the development of plastic volumetric strain. The yielding and plastic behaviour of soils was introduced in Chapters 3 and 4 following a discussion of some aspects of the yielding and plastic behaviour of annealed copper. Annealed copper, like other metals, does not change in volume as it yields and deforms plastically; the hardening of annealed copper is linked solely to the development of plastic shear strain. The way in which the mechanical behaviour of soils changes with the occurrence of irrecoverable deformations is linked with the changing arrangement of the soil particles, which is described in a limited way by the plastic volumetric strain and a little more by the plastic shear strain. One might postulate that part of the plastic hardening of soils is linked, like annealed copper, with the development of plastic shear strain. The size of the current yield locus in the $p':q$ effective stress plane is characterised by a reference pressure p'_o. In the Cam clay model and in the general volumetric hardening framework developed in Chapter 4, expansion of yield loci is linked only with plastic volumetric strain:

$$\delta p'_o = f(\delta\varepsilon^p_p) \qquad (12.4)(\text{cf. } 4.21)$$

However, the general analytical description of the plastic stress:strain relationship in Section 4.5 allowed hardening to be controlled also by shear strain:

$$\delta p'_{o} = f(\delta\varepsilon^{p}_{p}, \delta\varepsilon^{p}_{q}) \qquad (12.5)(\text{cf. } 4.39)$$

In the Cam clay model, p'_{o} represents the current yield stress in isotropic compression. An expression such as (12.5), admittedly only a general statement, looks as though it would place too great a coupling between development of plastic shear strain and hardening in isotropic compression. Consider the yielding of a sand. Irrecoverable volumetric deformation under isotropic or near isotropic stresses is largely an indication of particle crushing: that is, a change in the volumetric packing of the particles without change in their relative configuration (without major shifting and sliding). On the other hand, the yielding of sand (studied by Tatsuoka, 1972, and by Poorooshasb, Holubec, and Sherbourne, 1967, and discussed in Section 3.4), is very much concerned with the shearing of the sand: that is, major particle rearrangements without obvious particle crushing. Recall that the yield loci discussed in Section 3.4 showed no particular sign of closing on the p' axis. Yielding under near isotropic stresses was not studied by Tatsuoka or by Poorooshasb et al., but it seems unlikely that the irrecoverable shearing of the soil, which clearly strengthens the soil in shear, will produce major changes to the isotropic pressures needed to initiate particle crushing.

This suggests that two separate mechanisms of yielding might be

Fig. 12.24 (a) Two sets of yield loci separating predominantly shearing response (S) and predominantly volumetric response (V); (b) shear hardening relationship for set S; (c) volumetric hardening relationship for set V.

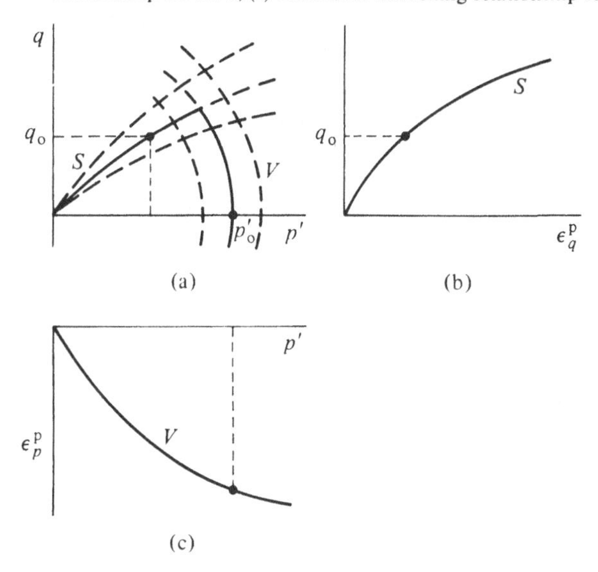

considered: one primarily concerned with the shearing of the soil and the other primarily concerned with the change in volume of the soil. Two separate families of yield loci can then be introduced (Fig. 12.24). One family (S) has a shape similar to the yield loci found experimentally (Section 3.4) and is defined by a reference deviator stress q_o (at some particular value of mean effective stress p') which changes only with irrecoverable shear strain ε_q^p according to some hardening relationship (Fig. 12.24b):

$$\delta q_o = f(\delta\varepsilon_q^p) \tag{12.6}$$

Another family (V) cuts the p' axis and is defined by a reference pressure p'_o which changes only with irrecoverable volumetric strain ε_p^p (Fig. 12.24c):

$$\delta p'_o = f(\delta\varepsilon_p^p) \tag{12.7}$$

These yield loci merely determine whether or not plastic deformations occur; the mechanism of plastic deformation is controlled by plastic potentials. Yielding on either family of curves is probably accompanied by both volumetric and shear strains even though hardening is linked with only one or the other component (just as yielding in Cam clay is accompanied by volumetric and shear strains even though the change in size of the yield loci is linked only with the volumetric component). This division of yielding into two families reflects the fact that shear strains are the result of the disturbance of the structure of the granular soil, and near isotropic compression does not produce any great disturbance to the structure. This arrangement of two sets of yield loci allows some evolution of the shape of the elastic region (the yield envelope) as the soil is loaded (Fig. 12.25), introducing some influence of the stress history.

A 'double hardening' model of this type has been used by Vermeer (1980) in a finite element analysis to estimate the boundary stresses on a deep, rigid, rough cylindrical die when a ferrite powder is compacted

Fig. 12.25 Evolution of shape of elastic region with two sets of yield loci: (a), (b), and (c) show successive stages of expansion of members of the two sets of loci for a non-monotonic stress path.

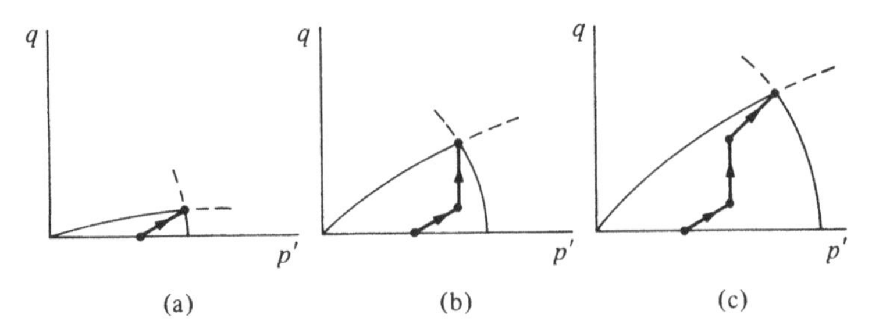

one-dimensionally (Fig. 12.26). The reasonable agreement of at least some aspects of this analysis with experimental observations illustrates that there are many granular materials other than soils to which numerical models developed for soils can be successfully applied. This application is a good one for this type of double hardening model because the stress paths in such a compaction problem involve significant increases in mean effective stress; much of the powder undergoes compression at rather low values of stress ratio q/p'.

A translating, kinematic, inner yield locus was introduced in Section 12.3 as a means of representing the high stiffnesses which are observed in soils

Fig. 12.26 (a) Section through cylindrical die filled with powder; (b) measured and computed load:displacement relationships; (c) measured and computed normal stresses on walls of die (after Vermeer, 1980).

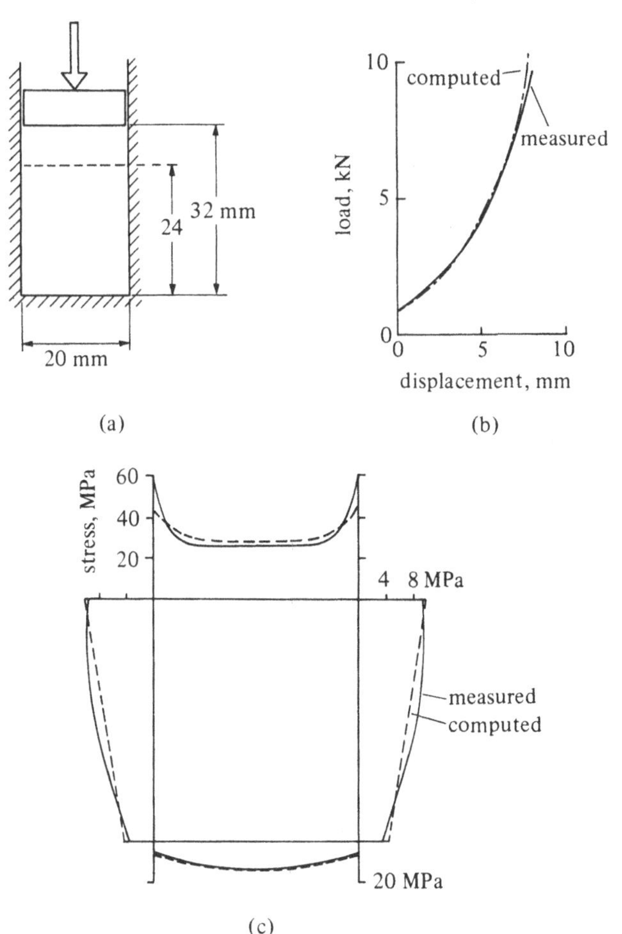

when effective stress paths change direction. The outer yield locus was retained, but it does not represent a boundary marking the onset of irrecoverable deformations since these start as soon as the inner yield locus is crossed; instead, it is a boundary between zones with different tangent stiffnesses. An attempt to investigate the shape of the current yield locus with stress paths such as *BCDEFG* in Fig. 12.14 would immediately run into a problem of choosing the magnitude of strain required for irrecoverable plastic behaviour to have been deemed to have occurred. Detection of the inner (true) yield locus requires a strain threshold less than the strain from *B* to *D*. Detection of the outer yield locus can be based on much less sensitive strain measurements.

Fig. 12.27 (a) Chain of parallel spring and slider elements; (b) load:displacement response of typical slider; (c) load:displacement response of chain of parallel spring and slider elements.

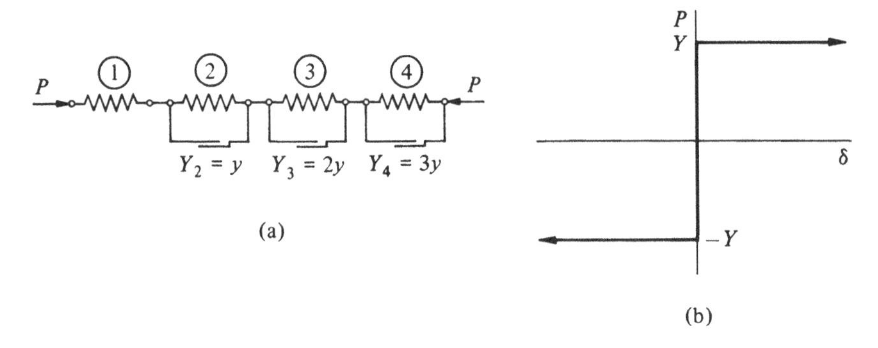

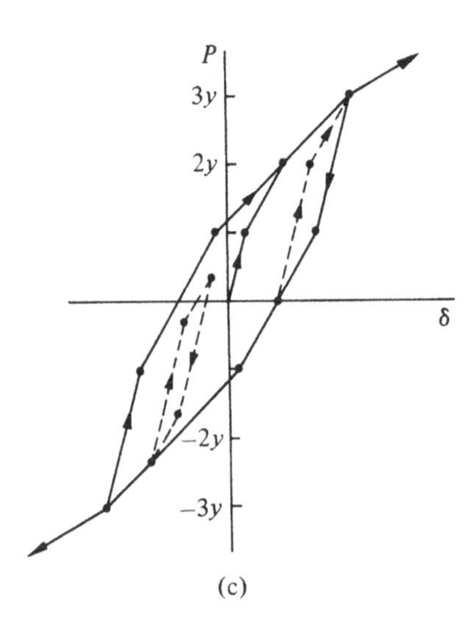

Strictly, of course, the inner yield locus is *the* yield locus; and as it is pushed around in stress space, the boundary of the elastic region certainly evolves with stress history, in position more than in size. An extension of this two-yield locus model to a model in which several yield loci are nested one inside the next can provide a convenient way of storing information about the history of the soil. A one-dimensional analogy may help to understand the pattern of response which such a model can reproduce.

Imagine a series of springs with parallel sliders connected as in Fig. 12.27a and subjected to an axial force P (compare Iwan, 1967). Each of the springs is assumed to have the same stiffness k, but the sliders have the rigid–perfectly plastic response shown in Fig. 12.27b with sliding loads Y, where $Y_2 = y$, $Y_3 = 2y$, and $Y_4 = 3y$. Initially, as the load P is increased from zero, only spring 1 deforms and the load:deflexion response has stiffness k (Fig. 12.27c). When $P = y$, however, slider 2 slides and spring 2 deforms; the tangent stiffness is now $k/2$ (Fig. 12.27c). When $P = 2y$, slider 3 slides and spring 3 comes into action; the stiffness drops to $k/3$. When $P = 3y$, slider 4 slides, and the stiffness drops to $k/4$. If the load is reduced again from $P = 3y$, all the sliders lock up, and only spring 1 is able to deform, with stiffness k, until the load has changed by $2y$ and slider 2 is able to slide in the opposite direction – and so on as the load is further reduced. The unloading response (Fig. 12.27c) is thus identical to the initial loading response but at twice the scale. If, when the load has reached $-3y$, the direction of loading is reversed again, a similar reloading response is observed (Fig. 12.27c), rejoining the original loading response at $P = 2y$. Whenever the direction of loading is changed, all the sliders lock up and only spring 1 is able to deform; two small cycles are shown within the larger cycle of Fig. 12.27c, and the tangent stiffness clearly depends on the direction of loading.

In the equivalent soil model, the sliders, whose sliding loads represent one-dimensional yield points, are replaced by yield surfaces of progressively larger sizes. The steadily increasing numbers of springs operating in series are replaced by steadily reducing tangent stiffnesses associated with the regions of effective stress space between the yield surfaces (fields of hardening moduli; compare Mróz, 1967); so the larger the yield surfaces that are moving, the lower the corresponding stiffness. A group of nested yield loci is shown in the $p':q$ plane in Fig. 12.28a, together with an initial stress point A. As the effective stress is changed from A to B (Fig. 12.28b), the successive yield loci are picked up and carried with the stress state, and the stiffness drops as successive yield loci are reached. Subsequent change of stresses to C (Fig. 12.28c) carries yield loci 1, 2, and 3 with the stress state but in such a way as to leave some record of the history of loading. Subsequent continuation to triaxial extension CD activates yield loci 1, 2, and 3 and is associated with a low stiffness (Fig. 12.28d). A retreat into triaxial compression towards B has initially to track across yield locus 1 at high stiffness (Fig. 12.28d). The anisotropic stress history is reflected in the current anisotropic stress:strain response.

Fig. 12.28 Set of nested yield loci translating under stress path ABC, for which (a) shows position of yield loci for stress state A, (b) shows position of yield loci for stress state B after stress path AB, and (c) shows position of yield loci for stress state C after stress path ABC; (d) stress:strain response in subsequent triaxial compression CB and extension CD.

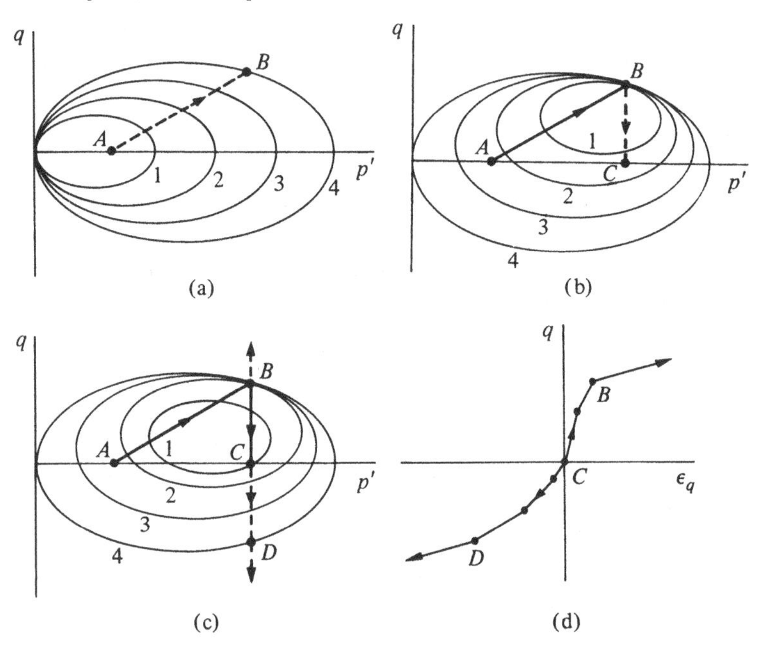

Fig. 12.29 (a) Set of nested yield loci subjected to isotropic compression AC; (b) stress:strain response in subsequent triaxial compression CB and extension CD.

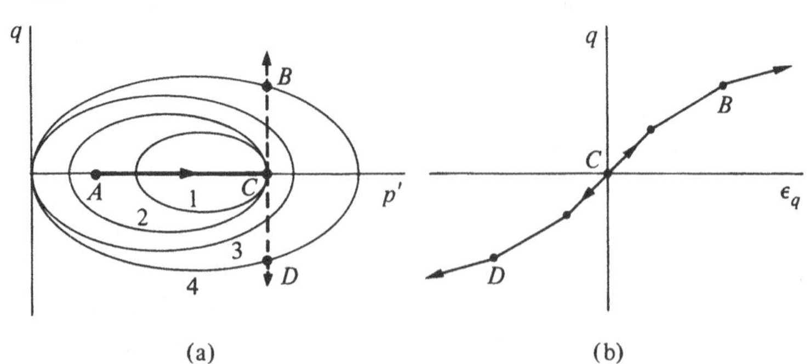

By contrast, a loading straight from A to C carries all the yield loci symmetrically along the p' axis (Fig. 12.29a), and the stiffness in both compression and extension is the same (Fig. 12.29b). The model thus shows isotropic effects for isotropic loading and anisotropic effects for anisotropic loading, and the location of the boundary of the region of effective stress space associated with any particular change in tangent stiffness evolves with the stress history. Some data obtained by Nadarajah (1973) in undrained triaxial compression and extension tests on samples of spestone kaolin are shown in Fig. 12.30 as a reminder of some of the experimental phenomena that such a model can reproduce. Figure 12.30 shows the

Fig. 12.30 (a) Stress paths and (b) stress:strain response for undrained triaxial compression and extension tests on isotropically normally compressed spestone kaolin (A) and one-dimensionally overconsolidated spestone kaolin (B) (after Nadarajah, 1973).

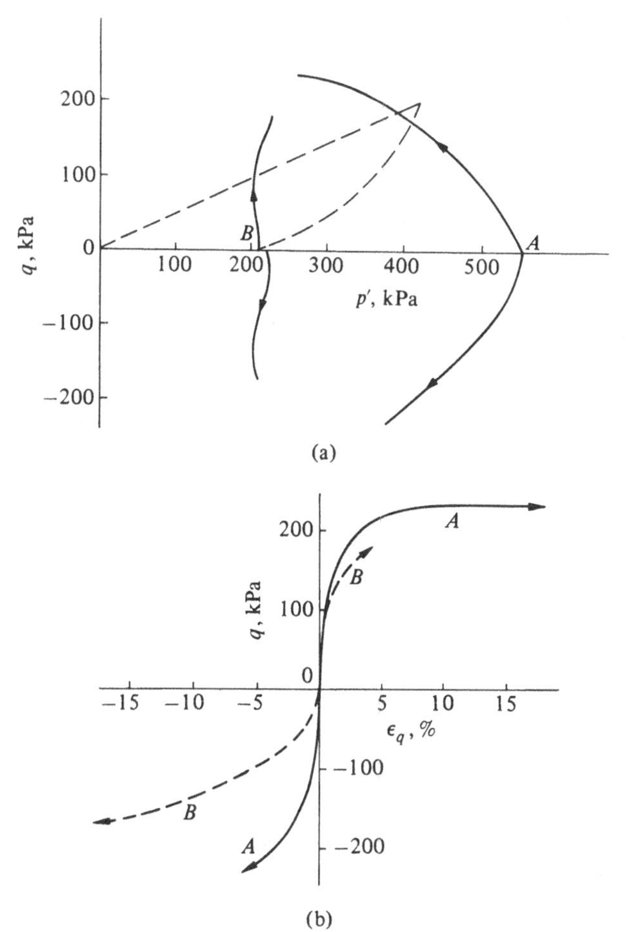

effective stress paths and stress:strain relationships for samples which have been isotropically compressed prior to undrained loading (curves *A*); the initial stiffness is the same in compression and extension. The figure also shows the effective stress paths and stress:strain relationships for samples which have been one-dimensionally compressed and then unloaded to an isotropic stress state before undrained loading was applied (curves *B*); the initial stiffness in compression is very much greater than the initial stiffness in extension.

This idea of nested translating yield loci has been discussed in a rather qualitative way: the yield loci shown in Figs. 12.28 and 12.29 have the Cam clay shape but have been shown only translating and not hardening though one might expect a Cam clay-like material to show some change in size of the yield loci as the soil is deformed. A more realistic model, such as those described by Mróz, Norris, and Zienkiewicz (1979) or Al-Tabbaa and Wood (1989), would have to balance translation and expansion of the yield loci according to some rule as a stress path is traversed; some assumption is also required about the mechanisms of plastic deformation appropriate for any particular stress state, in other words a plastic potential or flow rule for the soil is needed.

12.5 Concluding remarks: applicable models

Many, many numerical models for soil behaviour – constitutive relations – are described in the literature, and comprehensive surveys of available models can be found in the papers edited by Murayama (1985). In this chapter, we have given only the slightest hint of some of the possibilities, showing how they might be thought of as developments from the simple models discussed at greater length in other chapters and trying to take some of the mystery out of constitutive relations. The principles and hypotheses underlying constitutive models are often conceptually rather simple even though their development and presentation in terms of complete sets of equations may appear daunting.

Any geotechnical design calculation requires a conscious or unconscious choice of a model of soil behaviour. Settlement calculations typically assume that the soil is linear and elastic, and bearing capacity calculations typically assume that the soil is rigid and perfectly plastic. However, these basic models do not get very far; geotechnical constructions under working loads will certainly have proceeded well beyond a linear elastic range and yet are unlikely to have attained conditions of perfect plasticity. The analyses of the New Palace Yard excavation described in Section 12.3 indicate the poverty of an elastic analysis for indicating the pattern of deformations that might occur.

In choosing a model, an engineer must decide which effects are likely to be important enough to be included in the model. This decision must be based on information concerning the history of the soil and the future stress changes that the soil is likely to experience. But the choice of model also depends on other factors. The more effects there are to be built into the model, the more elaborate that model becomes, and the more soil parameters are required to specify the model. The more parameters that are required, the more complex the laboratory testing that is needed to determine their values becomes. Natural soils are heterogeneous, so some of their properties are likely to vary from place to place. Because extraction and testing of soil samples and in situ testing of soils cost money, a balance may have to be found between the *certain* cost of testing large numbers of samples to evaluate a large number of parameters for a heterogeneous soil and the *possible* benefit of using a complex rather than a simple model. The novelty of the project – location of structures and type of loadings – and the quality of the feasible site investigation may or may not justify the complexity of the model.

This book has been concerned with models in general and models in particular. The general model is the framework of critical state soil mechanics which links the compression and shearing of soils and which makes it clear that to understand soil behaviour, paths should always be studied both in the $p':q$ effective stress plane and in the $p':v$ compression plane; the vital effect of consolidation history or volumetric packing on subsequent response then becomes clear.

The particular model we have discussed is Cam clay, but it should have become clear that many models can be produced which, like Cam clay, (i) automatically generate a set of critical states at which plastic shearing can continue without change in effective stresses or volume and (ii) can be used to demonstrate that rather simple models founded in the theoretical framework of elasticity and plasticity can reproduce many of the important features of observed response of real soils, such as the existence of critical states, volume changes on shearing, and dependence of strength and stress:strain response on history.

Models such as Cam clay can be used successfully to match and predict soil behaviour in a wide range of laboratory (and in situ) tests and can be used in finite element computations to estimate the behaviour of prototype geotechnical structures. Particular success has been obtained with the Cam clay model in matching behaviour of prototypes which load softer deposits of soil.

Models such as Cam clay are firmly rooted in critical state soil mechanics and have the following particular properties:

i. They are described in terms of effective stresses and can include coupled effects of shearing and pore pressure generation or dissipation.
ii. They are based on simple physical notions of the way that the soil yields and deforms plastically.
iii. They automatically incorporate in a single model the compression and shearing of the soil.
iv. They are defined by a small number of parameters which can be determined from simple tests (Cam clay requires the slopes λ and κ of normal compression lines and unloading–reloading lines and the location Γ of the critical state line in the compression plane; the slope M of the critical state line in the effective stress plane; the shear modulus G' or Poisson's ratio ν'; and the permeability k).
v. They can make predictions for *all* stress and strain paths.

Simple models can give a good general impression (Fig. 12.31a); elaborate models can give a more detailed local picture (Fig. 12.31b).

Fig. 12.31 (a) *Galloway landscape*: George Henry (1889) (reproduced by permission of Glasgow Art Gallery and Museum); (b) *The bluidie tryste*: Joseph Noel Paton (1855) (reproduced by permission of Glasgow Art Gallery and Museum).

(a)

Hierarchies of models of increasing complexity might be produced to illustrate the various roles that different models can play (compare Vermeer and de Borst, 1984). The Cam clay model is a student's model, and its purpose is to describe and explain patterns of soil behaviour, enabling the student to proceed further in understanding soil than is possible with the simple, traditional, ideally elastic and perfectly plastic models (which have been called children's models to emphasise that soil mechanics today has advanced and matured). Developments beyond the student's models provide engineer's models applicable perhaps to particular classes of problems and introducing such extra features as are thought necessary for each problem. In the distance are the scientist's or philosopher's models which match comprehensively and precisely all aspects of soil response. Perhaps such models will, like the philosopher's stone, always remain elusive.

(b)

References

Adachi, T. and Oka, F. (1982), 'Constitutive equations for normally consolidated clays based on elasto-viscoplasticity', *Soils and Foundations* **22**(4), 57–70.

Airey, D.W., Budhu, M., and Wood, D.M. (1985), 'Some aspects of the behaviour of soils in simple shear', in P.K. Banerjee and R. Butterfield (eds.), *Developments in soil mechanics and geotechnical engineering – 2: Stress–strain modelling of soils* (Elsevier), pp. 185–213.

Almagor, G. (1967), 'Interpretation of strength and consolidation data from some bottom cores off Tel-Aviv, Polmakhim coast, Israel', in A.F. Richards (ed.), *Marine Geotechnique* (Urbana: Univ. of Illinois Press), pp. 131–53.

Almeida, M.S.S. (1984), *Stage constructed embankments on soft clays*, Ph.D. thesis, Cambridge University.

Almeida, M.S.S., Britto, A.M. and Parry, R.H.G. (1986), 'Numerical modelling of a centrifuged embankment on soft clay', *Canadian Geotechnical Journal* **23**, 103–14.

Al-Tabbaa, A. (1987) *Permeability and stress–strain response of speswhite kaolin*, Ph.D. thesis, Cambridge University.

Al-Tabbaa, A. and Wood, D.M. (1987), 'Some measurements of the permeability of kaolin', *Géotechnique* **37**(4), 499–503.

Al-Tabbaa, A. and Wood, D.M. (1989), 'An experimentally based 'bubble' model for clay', in S. Pietruszczak and G.N. Pande (eds.), *Numerical Models in Geomechanics NUMOG III* (London: Elsevier), pp. 91–9.

Andresen, A., Berre, T., Kleven, A., and Lunne, T. (1979), 'Procedures used to obtain soil parameters for foundation engineering in the North Sea', *Marine Geotechnology* **3**(3), 201–66.

Atkinson, J.H. (1981), *Foundations and slopes. An introduction to applications of critical state soil mechanics* (Maidenhead: McGraw-Hill).

Atkinson, J.H. and Bransby, P.L. (1978), *The mechanics of soils. An introduction to critical state soil mechanics* (Maidenhead: McGraw-Hill).

Atkinson, J.H., Evans, J.S., and Ho, E.W.L. (1985), 'Non-uniformity of triaxial samples due to consolidation with radial drainage', *Géotechnique* **35**(3), 353–5.

Atterberg, A. (1911), 'Lerornas forhållande till vatten, deras plasticitetsgränser och plasticitetsgrader', *Kungl. Lantbruks akademiens Handlingar och Tidskrift* **50**(2), 132–58.

ASCE = American Society of Civil Engineers, New York; ASME = American Society of Mechanical Engineers, New York; CIRIA = Construction Industry Research and Information Association, London; ICE = Institution of Civil Engineers, London; IUTAM = International Union of Theoretical and Applied Mechanics.

Baguelin, F., Jézéquel, J.F., and Shields, D.H. (1978), *The pressuremeter and foundation engineering* (Clausthal: Trans Tech Publications, Series on Rock and Soil Mechanics).
Baker, J., and Heyman, J. (1969), *Plastic design of frames. 1. Fundamentals* (Cambridge: Cambridge University Press).
Baran, P.A., and Sweezy, P.M. (1968), *Monopoly capital: An essay on the American economic and social order* (Harmondsworth: Penguin Books).
Barry, A.J., and Nicholls, R.A. (1982), 'Discussion', in *Vertical drains* (London: Thomas Telford), pp. 143–6.
Been, K., Crooks, J.H.A., Becker, D.E., and Jefferies, M.G. (1986), 'The cone penetration test in sands: Part I, state parameter interpretation', *Géotechnique* **36**(2), 239–49.
Been, K, and Jefferies, M.G. (1985), 'A state parameter for sands', *Géotechnique* **35**(2), 99–112.
Been, K., and Jefferies, M.G. (1986), 'Discussion: A state parameter for sands', *Géotechnique* **36**(1), 127–32.
Been, K., Jefferies, M.G., Crooks, J.H.A., and Rothenburg, L. (1987), 'The cone penetrometer test in sands: Part II, general inference of state', *Géotechnique* **37**(3), 285–99.
Bell, A.L. (1977), *A geotechnical investigation of post-glacial estuarine deposits at Kinnegar, Belfast Lough*, Ph.D. thesis, Queen's University, Belfast.
Berre, T. (1975), 'Bruk av triaksial- og direkte skjærforsøk til løsning av geotekniske problemer', in *Proc. Geoteknikermøde i København* (København: Polyteknisk Forlag), pp. 199–211.
Berre, T., and Bjerrum, L. (1973), 'Shear strength of normally consolidated clays', in *Proc. 8th Int. Conf. on Soil Mechs and Foundation Eng., Moscow* (Moscow: USSR National Society for Soil Mechanics and Foundation Engineering), vol. 1.1, 39–49.
Bishop, A.W. (1958), 'Test requirements for measuring the coefficient of earth pressure at rest', in *Proc. Brussels Conf. 58 on Earth Pressure Problems* (Brussels: Belgian Group of the International Society of Soil Mechanics and Foundation Engineering), vol. 1, pp. 2–14.
Bishop, A.W. (1959), 'The principle of effective stress', *Teknisk Ukeblad, Oslo* **39** (22 oktober), 859–63.
Bishop, A.W., and Henkel, D.J. (1957), *The measurement of soil properties in the triaxial test* (London: William Arnold).
Bjerrum, L. (1954), 'Geotechnical properties of Norwegian marine clays', *Géotechnique* **4**(2) 49–69.
Bjerrum, L. (1967), 'Engineering geology of Norwegian normally consolidated marine clays as related to settlements of buildings', 7th Rankine Lecture, *Géotechnique* **17**(2), 81–118.
Bjerrum, L. (1972), 'Embankments on soft ground', in *Proc. Specialty Conf. on Performance of Earth and Earth-Supported Structures, Purdue* (New York: ASCE), vol. 2, pp. 1–54.
Bjerrum, L. (1973), 'Problems of soil mechanics and construction of soft clays and structurally unstable soils', in *Proc. 8th Int. Conf. on Soil Mechs and Foundation Eng., Moscow* (Moscow: USSR National Society for Soil Mechanics and Foundation Engineering), vol. 3, pp. 111–59.
Bjerrum, L., and Flodin, N. (1960), 'The development of soil mechanics in Sweden, 1900–1925', *Géotechnique* **10**(1), 1–18.
Bjerrum, L., and Landva, A. (1966), 'Direct simple-shear tests on a Norwegian quick clay', *Géotechnique* **16**(1), 1–20.
Bjerrum, L., and Simons, N.E. (1960), 'Comparison of shear strength characteristics of normally consolidated clays', in *Proc. Research Conf. on Shear Strength of Cohesive Soils, Boulder, Colorado* (New York: ASCE), pp. 711–26.
Bolton, M.D. (1979), *A guide to soil mechanics*, (London: Macmillan Press).
Bolton, M.D. (1986), 'The strength and dilatancy of sands', *Géotechnique* **36**(1), 65–78.

Borsetto, M., Imperato, L., Nova, R., and Peano, A. (1983), 'Effects of pressuremeters of finite length in soft clay', in *Proc. Int. Symp. on Soil and Rock Investigations by In Situ Testing. Paris* (Organised by Comité Français de la Géologie de l'Ingénieur, Comité Français de la Mécanique des Sols, Comité Français de la Mécanique des Roches), vol. 2, pp. 211–15.
Brady, K.C. (1988), 'Soil suction and the critical state', *Géotechnique* **38**(1), 117–20.
Brand, E.W. (1981), 'Some thoughts on rain-induced slope failures', in *Proc. 10th Int. Conf. on Soil Mechs. and Foundation Eng., Stockholm* (Rotterdam: A.A. Balkema), vol. 3, pp. 373–6.
British Standards Institution (1975), *Methods of Test for Soils for Civil Engineering Purposes,* BS1377: 1975 (London: British Standards Institution).
Britto, A.M., and Gunn, M.J. (1987), *Critical state soil mechanics via finite elements* (Chichester: Ellis Horwood Ltd).
Brooker, E.W., and Ireland, H.O. (1965), 'Earth pressures at rest related to stress history', *Canadian Geotechnical Journal* **2**(1), 1–15.
Bryant, W.R., Cernock, P., and Morelock, J. (1967), 'Shear strength and consolidation characteristics of marine sediments from the western Gulf of Mexico', in A.F. Richards (ed.), *Marine Geotechnique* (Urbana: University of Illinois Press), pp. 41–62.
Burland, J.B. (1971), 'A method of estimating the pore pressures and displacements beneath embankments on soft, natural clay deposits', in R.H.G. Parry (ed.), *Stress–strain behaviour of soils (Proc. Rescoe Memorial Symp., Cambridge).* (Henley-on-Thames: G.T. Foulis & Co.), pp. 505–36.
Burland, J.B., and Hancock, R.J.R. (1977), 'Underground car park at the House of Commons, London: geotechnical aspects', *The Structural Engineer* **55**(2), 87–100.
Calladine, C.R. (1963), 'Correspondence: The yielding of clay', *Géotechnique* **13**(3), 250–5.
Calladine, C.R. (1985), *Plasticity for engineers* (Chichester: Ellis Horwood Ltd).
Carter, J.P. (1982), 'Predictions of the non-homogeneous behaviour of clay in the triaxial test', *Géotechnique* **32**(1), 55–8.
Casagrande, A. (1932), 'Research on the Atterberg limits of soils', *Public Roads* **13**(8) 121–30 and 136.
Casagrande, A. (1936), 'Characteristics of cohesionless soils affecting the stability of slopes and earth fills', *J. Boston Soc. Civil Engineers* **23**(1), 13–32.
Casagrande, A. (1947), 'Classification and identification of soils', *Proc. ASCE* **73**(6) part 1, 783–810.
Clausen, C–J.F. (1972), *Measurements of pore water pressure, settlements and lateral deformations at a test fill on soft clay brought to failure at Mastemyr, Oslo* (Oslo: Norwegian Geotechnical Institute), Technical Report 11.
Clausen, C–J.F., Graham, J., and Wood, D.M. (1984), 'Yielding in soft clay at Mastemyr, Norway', *Géotechnique* **34**(4), 581–600.
Coleman, J.D. (1962), 'Correspondence: Stress/strain relations for partly saturated soils', *Géotechnique* **12**(4), 348–50.
Collin, A. (1846), *Recherches expérimentales sur les glissements spontanés des terrains argileux* (Paris: Carilian-Goeurley et Dalmont); English translation by W.R. Schriever (1956), *Experimental investigation on sliding of clay slopes* (Toronto: University of Toronto Press).
Collins, K., and McGown, A. (1974), 'The form and function of microfabric features in a variety of natural soils', *Géotechnique* **24**(2), 223–54.
Crewdson, B.J., Ormond, A.L., and Nedderman, R.M. (1977), 'Air-impeded discharge of fine particles from a hopper', *Powder Technology* **16**, 197–207.
D'Appolonia, D.J., Lambe, T.W., and Poulos, H.G. (1971), 'Evaluation of pore pressures beneath an embankment', *Proc. ASCE, J. Soil Mechs and Foundations Div.* **97**(SM6), 881–98.

Davis, E.H. (1968), 'Theories of plasticity and the failure of soil masses', in I.K. Lee (ed.), *Soil mechanics – selected topics* (London: Butterworths), pp. 341–80.
de Josselin de Jong, G. (1971), 'Discussion, session 2'; in R.H.G. Parry (ed.), *Stress–strain behaviour of soils (Proc. Roscoe Memorial Symp., Cambridge).* (Henley-on-Thames: G.T. Foulis & Co.), pp. 258–61.
de Josselin de Jong, G. (1976), 'Rowe's stress-dilatancy relation based on friction', *Géotechnique* **26**(3), 527–34.
Donaghe, R.T., Chaney, R.C., and Silver, M.L. (eds.) (1988), *Advanced triaxial testing of soil and rock,* STP977 (Philadelphia: American Society for Testing and Materials).
Drucker, D.C. (1954), 'A definition of stable inelastic material', *J. Applied Mechanics, Trans. ASME* **26**, 101–6.
Drucker, D.C. (1966), 'Concepts of path independence and material stability for soils', in J. Kravtchenko and P.M. Sirieys (eds.), *Proc. IUTAM Symp. on Rheology and Soil Mechanics, Grenoble* (Berlin: Springer-Verlag), pp. 23–46.
Dumbleton, M.J., and West, G. (1970), *The suction and strength of remoulded soils as affected by composition* (Crowthorne: Road Research Laboratory), LR306.
El-Sohby, M.A. (1969), 'Deformation of sands under constant stress ratios', in *Proc. 7th Int. Conf. on Soil Mechs and Foundation Eng., Mexico* (Mexico City: Sociedad Mexicana de Mecánica de Suelos), vol. 1, pp. 111–19.
Fredlund, D.G. (1979), 'Appropriate concepts and technology for unsaturated soils', *Canadian Geotechnical Journal* **16**(1), 121–39.
Gens, A. (1982), *Stress–strain and strength characteristics of a low plasticity clay,* Ph.D. thesis, London University.
Ghionna, V., Jamiolkowski, M., Lacasse, S., Ladd, C.C., Lancellotta, R., and Lunne, T. (1983), 'Evaluation of self-boring pressuremeter, in *Proc. Int. Symp. on Soil and Rock Investigation by In Situ Testing, Paris* (Organised by Comité Français de la Géologie de l'Ingénieur, Comité Français de la Mécanique des Sols, Comité Français de la Mécanique des Roches), vol. 2, pp. 293–301.
Gibson, R.E., and Henkel, D.J. (1954), 'Influence of duration of tests at constant rate of strain on measured "drained" strength', *Géotechnique* **4**(1), 6–15.
Gibson, R.E., Knight, K., and Taylor, P.W. (1963), 'A critical experiment to examine theories of three-dimensional consolidation', in *Proc. European Conf. on Soil Mechs and Foundation Eng., Wiesbaden* (Essen: Deutsche Gesellschaft für Erd und Grundbau e.V.), vol. 1, pp. 69–76.
Graham, J., Crooks, J.H.A., and Bell, A.L. (1983), 'Time effects on the stress–strain behaviour of natural soft clays', *Géotechnique* **33**(3), 327–40.
Graham, J., and Houlsby, G.T. (1983), 'Elastic anisotropy of a natural clay', *Géotechnique* **33**(2), 165–80.
Graham, J., Noonan, M.L., and Lew, K.V., (1983), 'Yield states and stress–strain relationships in a natural plastic clay', *Canadian Geotechnical Journal* **20**(3), 502–16.
Hansbo, S. (1957), *A new approach to the determination of the shear strength of clay by the fall-cone test* (Stockholm: Royal Swedish Geotechnical Institute), *Proceedings* **14**.
Harr, M.E. (1966), *Foundations of theoretical soil mechanics.* (New York: McGraw-Hill).
Henkel, D.J. (1956), 'Discussion: Earth movement affecting LTE railway in deep cutting east of Uxbridge', in *Proc. ICE, Part II*, **5**, 320–3.
Henkel, D.J. (1959), 'The relationships between the strength, pore-water pressure, and volume-change characteristics of saturated clays', *Géotechnique* **9**(2), 119–35.
Henkel, D.J., and Skempton, A.W. (1955), 'A landslide at Jackfield, Shropshire, in a heavily overconsolidated clay', *Géotechnique* **5**(2), 131–7.
Heyman, J. (1972), *Coulomb's memoir on statics: an essay in the history of civil engineering* (Cambridge: Cambridge University Press).
Heyman, J. (1982), *Elements of stress analysis,* (Cambridge: Cambridge University Press).
Hill, R. (1950), *The mathematical theory of plasticity* (Oxford: Clarendon Press).

Hird, C.C., and Hassona, F. (1986), 'Discussion: A state parameter for sands', *Géotechnique* **36**(1), 124–7.

Höeg, K., Andersland, O.B., and Rolfsen, E.N. (1969), 'Undrained behaviour of quick clay under load tests at Åsrum', *Géotechnique* **19**(1), 101–15.

Höeg, K., Christian, J.T., and Whitman, R.V., (1968), 'Settlement of strip load on elastic–plastic soil', *Proc. ASCE, Journal of the Soil Mechanics and Foundations Division* **94**(SM2), 431–45.

Hooke, R. (1675), *A description of helioscopes, and some other instruments.* (London).

Horswill, P., and Horton, A. (1976), 'Cambering and valley bulging in the Gwash valley at Empingham, Rutland', *Phil. Trans Roy. Soc. London* **A283**, 427–51.

Houlsby, G.T. (1979), 'The work input to a granular material', *Géotechnique* **29**(3), 354–8.

Houlsby, G.T. (1982), 'Theoretical analysis of the fall cone test', *Géotechnique* **32**(2), 111–18.

Hvorslev, M.J. (1937), *Über die Festigkeitseigenschaften gestörter bindiger Böden* (København: Danmarks Naturvidenskabelige Samfund) Ingeniørvidenskabelige Skrifter A 45. English translation (1969), *Physical properties of remoulded cohesive soils* (Vicksburg, Miss.: U.S. Waterways Experimental Station), no 69–5.

Iwan, W.D. (1967), 'On a class of models for the yielding behavior of continuous and composite systems', *Trans. ASME, J. Appl. Mech.* **34**(E3), 612–17.

Jâky, J. (1944), 'A nyugalmi nyomâs tényezöje' ('The coefficient of earth pressure at rest'), *Magyar Mérnök és Épitész-Egylet Közlönye* (*J. of the Union of Hungarian Engineers and Architects*), 355–8.

Jamiolkowski, M., Ladd, C.C., Germaine, J.T., and Lancellotta, R. (1985), 'New developments in field and laboratory testing of soils', in *Proc. 11th Int. Conf. on Soil Mechs and Foundation Eng., San Francisco* (Rotterdam: A.A. Balkema), vol. 1, pp. 57–153.

Jardine, R.J., Potts, D.M., Fourie, A.B., and Burland, J.B. (1986), 'Studies of the influence of non-linear stress–strain characteristics in soil–structure interaction', *Géotechnique* **36**(3), 377–96.

Jardine, R.J., Symes, M.J., and Burland, J.B. (1984), 'The measurement of soil stiffness in the triaxial apparatus', *Géotechnique* **34**(3), 323–40.

Karlsson, R. (1977), *Consistency limits. A manual for the performance and interpretation of laboratory investigations, part 6* (Stockholm: Statens råd för byggnadsforskning).

Kolbuszewski, J.J. (1948), 'An experimental study of the maximum and minimum porosities of sands', in *Proc. 2nd Int. Conf. on Soil Mechs and Foundation Eng., Rotterdam* **1**, 158–65.

Kong, F.K., and Evans, R.H. (1975), *Reinforced and prestressed concrete.* (Walton-on-Thames: Nelson).

Ladd, C.C. (1965), 'Stress–strain behaviour of anisotropically consolidated clays during undrained shear', in *Proc. 6th Int. Conf. on Soil Mechs and Foundation Eng., Montreal* (Toronto: University of Toronto Press), vol. 1, pp. 282–90.

Ladd, C.C. (1981), 'Discussion on laboratory shear devices', in R.N. Yong and F.L. Townsend (eds.), *Laboratory shear strength of soil*, STP740 (Philadelphia: American Society for Testing and Materials), pp. 643–52.

Ladd, C.C., and Edgers, L. (1972), *Consolidated-undrained direct-simple shear tests on saturated clays* (Cambridge: Massachusetts Institute of Technology), Dept. of Civil Eng. research report R72–82.

Ladd, C.C., Foott, R., Ishihara, K., Schlosser, F., and Poulos, H.G. (1977), 'Stress-deformation and strength characteristics', in *Proc. 9th Int. Conf. on Soil Mechs and Foundation Eng., Tokyo* (Tokyo: Japanese Society of Soil Mechanics and Foundation Engineering), vol. 2, pp. 421–94.

Lade, P.V. (1977), 'Elasto-plastic stress–strain theory for cohesionless soil with curved yield surfaces', *Int. J. Solids and Structures* **13**(11), 1019–35.

Lambe, T.W. (1964), 'Methods of estimating settlement', *Proc. ASCE, Journal of the Soil Mechanics and Foundations Division* **90**(SM5), 43–67.
Lambe, T.W. (1967), 'Stress path method', *Proc. ASCE, Journal of the Soil Mechanics and Foundations Division* **93**(SM6), 309–31.
Larsson, R. (1980), 'Undrained shear strength in stability calculation of embankments and foundations on soft clays', *Canadian Geotechnical Journal* **17**(4), 591–602.
Larsson, R. (1981), *Drained behaviour of Swedish clays* (Linköping: Swedish Geotechnical Institute), Report 12.
Lee, K.L. and Seed, H.B., (1967), 'Drained strength characteristics of sands', in *Proc. ASCE, Journal of the Soil Mechanics and Foundations Division* **93**(SM6), 117–41.
Leonards, G.A., and Ramiah, B.K. (1960), 'Time effects in the consolidation of clays', in Papers on soils, 1959 meetings; *Symp. on Time Rates of Loading in Soil Testing*, STP254 (Philadelphia: American Society for Testing and Materials), pp. 116–30.
Leroueil, S., Kabbaj, M., Tavenas, F., and Bouchard, R. (1985), 'Stress–strain–strain rate relation for the compressibility of sensitive natural clays'. *Géotechnique* **35**(2), 159–80.
Leroueil S., Magnan, J–P., and Tavenas, F. (1985), *Remblais sur argiles molles* (Paris: Technique et Documentation, Lavoisier) English translation by D.M. Wood (1990) *Embankments on soft clays* (Chichester: Ellis Horwood Ltd.).
Leroueil, S., and Tavenas, F. (1981), 'Pitfalls of back-analyses', in *Proc. 10th Int. Conf. on Soil Mechs and Foundation Eng., Stockholm* (Rotterdam: A.A. Balkema), vol. 1, pp. 185–90.
Levadoux, J–N., and Baligh, M.M. (1980), *Pore pressures during cone penetration in clays* (Cambridge: Massachusetts Institute of Technology), Dept. of Civil Eng. research report R80-15.
Lewin, P.I. (1973), 'The influence of stress history on the plastic potential', in A.C. Palmer (ed.), *Proc. Symp. on Role of Plasticity in Soil Mechanics* (Cambridge: Cambridge University Engineering Department), pp. 96–105.
Livesley, R.K. (1983), *Finite elements: an introduction for engineers* (Cambridge: Cambridge University Press).
Love, A.E.H. (1927), *A treatise on the mathematical theory of elasticity*, 4th ed. (Cambridge: Cambridge University Press).
Luong, M.P. (1979), 'Les phénomènes cycliques dans les sables', *Journée de Rhéologie: Cycles dans les sols – rupture – instabilités.* (Vaulx-en-Velin: École Nationale des Travaux Publics de l'État), Publication 2.
Lupini, J.F., Skinner, A.E., and Vaughan, P.R. (1981), 'The drained residual strength of cohesive soils', *Géotechnique* **31**(2), 181–213.
McClelland, B. (1967), 'Progress of consolidation in delta front and prodelta clays of the Mississippi River', in A.F. Richards (ed.), *Marine Geotechnique* (Urbana: University of Illinois Press), pp. 22–40.
Magnan, J–P., Mieussens, C., and Queyroi, D. (1983), *Étude d'un remblai sur sols compressibles: Le remblai B du site expérimental de Cubzac-les-Ponts* (Paris: Laboratoire Central des Ponts et Chaussées), Rapport de recherche LPC 127.
Mair, R.J., and Wood., D.M. (1987), *Pressuremeter testing: Methods and interpretation*, CIRIA Ground Engineering Report: In-situ testing (London and Sevenoaks: CIRIA and Butterworths).
Marachi, N.D., Chan, C.K., and Seed, H.B. (1972), 'Evaluation of properties of rockfill materials', *Proc. ASCE, Journal of the Soil Mechanics and Foundations Division* **98**(SM1), 95–114.
Mayne, P.W. (1980), 'Cam-clay predictions of undrained strength', *Proc. ASCE, Journal of the Geotechnical Engineering Division* **106**(GT11), 1219–42.
Mayne, P.W., and Swanson, P.G. (1981), The critical-state pore pressure parameter from consolidated-undrained shear tests, in R.N. Yong and F.C. Townsend (eds.), *Laboratory shear strength of soil*, STP740 (Philadelphia: American Society for Testing and Materials), 410–30.

Meigh, A.C. (1987), *Cone penetration testing: Methods and interpretation*, CIRIA Ground Engineering Report: In-situ testing (London and Sevenoaks: CIRIA and Butterworths).
Mesri, G. (1975), 'Discussion: New design procedure for stability of soft clays', *Proc. ASCE, Journal of the Geotechnical Engineering Division* **101**(GT4), 409–12.
Mesri, G., and Godlewski, P.M. (1977), 'Time- and stress-compressibility interrelationship', *Proc. ASCE, Journal of the Geotechnical Engineering Division* **103**(GT5), 417–30.
Meyerhof, G.G. (1976), 'Bearing capacity and settlement of pile foundations', 11th Terzaghi Lecture, *Proc. ASCE, Journal of the Geotechnical Engineering Division* **102**(GT3), 197–228.
Mises, R. von (1913), 'Mechanik der festen Körper im plastisch-deformablen Zustand', *Nachrichten von der Königlichen Gesellschaft der Wissenschaften zu Göttingen, Mathematisch-physikalische Klasse*, 582–92.
Mitchell, J.K. (1976), *Fundamentals of soil behaviour* (New York: John Wiley & Sons).
Miura, N., Murata, H., and Yasufuku, N. (1984), 'Stress–strain characteristics of sand in a particle crushing region', *Soils and Foundations* **24**(1), 77–89.
Mouratidis, A., and Magnan, J–P. (1983), *Modèle élastoplastique anisotrope avec écrouissage pour le calcul des ouvrages sur sols compressibles* (Paris: Laboratoire Central des Ponts et Chaussées), Rapport de recherche LPC 121.
Mróz, Z. (1967), 'On the description of anisotropic work hardening', *J. Mech. Phys. Solids* **15**, 163–75.
Mróz, Z., Norris, V.A., and Zienkiewicz, O.C. (1979), 'Application of an anisotropic hardening model in the analysis of elasto-plastic deformation of soils', *Géotechnique* **29**(1), 1–34.
Murayama, S. (ed.) (1985), *Constitutive laws of soil*, Report of ISSMFE Subcommittee on Constitutive laws of soils and Proc. discussion session 1A, 11th Int. Conf. on Soil Mechanics and Foundation Engineering, San Francisco (Tokyo: Japanese Society of Soil Mechanics and Foundation Engineering).
Nadarajah, V. (1973), *Stress–strain properties of lightly overconsolidated clays*, Ph.D. thesis, Cambridge University.
Namy, D.L. (1970), *An investigation of certain aspects of stress–strain relationships for clay soils*, Ph.D. thesis, Cornell University, Ithaca.
Norwegian Geotechnical Institute (1969), *Results of direct shear, oedometer and triaxial tests on quick clay from Mastemyr* (Oslo: Norwegian Geotechnical Institute), internal report F 372–3.
Oda, M., Konishi, J., and Nemat–Nasser, S. (1980), 'Some experimentally based fundamental results on the mechanical behaviour of granular materials', *Géotechnique* **30**(4), 479–95.
Olsson, J. (1921), 'Metod för undersökning av lerors hållfasthetsegenskaper, tillämpad vid de geotekniska undersökningarna vid Statens Järnvägar', *Geologiska Förening Stockholm, Förhandlingar* **43**(5), 502–7.
Olszak, W., and Perzyna, P. (1966), 'On elastic/visco-plastic soils', in J. Kravtchenko and P.M. Sirieys (eds.) *Proc. IUTAM Symp. on Rheology and Soil Mechanics, Grenoble* (Berlin: Springer–Verlag), pp. 47–57.
Parry, R.H.G. (1956), *Strength and deformation of clay*, Ph.D. thesis, London University.
Parry, R.H.G. (1958), 'Correspondence: On the yielding of soils', *Géotechnique* **8**(4), 183–6.
Parry, R.H.G. (1970), 'Overconsolidation in soft clay deposits', *Géotechnique* **20**(4), 442–6.
Parry, R.H.G., and Wroth, C.P. (1981), 'Shear stress–strain properties of soft clay', in E.W. Brand and R.P. Brenner (eds.), *Soft clay engineering* (Amsterdam: Elsevier), pp. 309–64.
Perzyna, P. (1963), 'The constitutive equations for rate sensitive plastic materials', *Quarterly of Applied Maths* **20**(4), 321–32.
Poorooshasb, H.B., Holubec, I., and Sherbourne, A.N. (1966), 'Yielding and flow of sand in triaxial compression: PartI', *Canadian Geotechnical Journal* **3**(4), 179–90.

Poorooshasb, H.B., Holubec, I., and Sherbourne, A.N. (1967), 'Yielding and flow of sand in triaxial compression: Parts II and III', *Canadian Geotechnical Journal* **4**(4), 376–97.
Poulos, H.G., and Davis, E.H. (1974), *Elastic solutions for soil and rock mechanics* (New York: John Wiley & Sons).
Prévost, J–H. (1979), 'Undrained shear tests on clays', *Proc. ASCE, Journal of the Geotechnical Engineering Division* **105**(GT1), 49–64.
Quigley, R.M., and Thompson, C.D. (1966), 'The fabric of anisotropically consolidated sensitive marine clay', *Canadian Geotechnical Journal* **3**(2), 61–73.
Ramanatha Iyer, T.S. (1975), 'The behaviour of Drammen plastic clay under low effective stresses', *Canadian Geotechnical Journal* **12**(1), 70–83.
Randolph, M.F., Carter, J.P., and Wroth, C.P. (1979), 'Driven piles in clay – the effects of installation and subsequent consolidation', *Géotechnique* **29**(4), 361–93.
Randolph, M.F., and Wroth, C.P. (1981), 'Application of the failure state in undrained simple shear to the shaft capacity of driven piles', *Géotechnique* **31**(1), 143–57.
Reynolds, O. (1885), 'On the dilatancy of media composed of rigid particles in contact, with experimental illustrations', *Phil. Mag.* **20**, 469–81.
Reynolds, O. (1886), 'Experiments showing dilatancy, a property of granular material, possibly connected with gravitation', *Proc. Royal Inst. of Great Britain* **11**, 354–63.
Richardson, A.M., and Whitman, R.V. (1963), 'Effect of strain-rate upon drained shear resistance of a saturated remoulded fat clay', *Géotechnique* **13**(4), 310–24.
Roscoe, K.H. (1953), 'An apparatus for the application of simple shear to soil samples', *Proc. 3rd Int. Conf. on Soil Mechs and Foundation Eng., Zurich* (Zurich: Organising committee ICOSOMEF), vol. 1, pp. 186–91.
Roscoe, K.H., and Burland, J.B. (1968), 'On the generalised stress–strain behaviour of 'wet' clay', in J. Heyman and F.A. Leckie (eds.), *Engineering plasticity* (Cambridge: Cambridge University Press), pp. 535–609.
Roscoe, K.H., and Schofield, A.N. (1963), 'Mechanical behaviour of an idealised 'wet' clay', *Proc. European Conf. on Soil Mechanics and Foundation Engineering, Wiesbaden* (Essen: Deutsche Gesellschaft für Erd- und Grundbau e.V.), vol. 1, pp. 47–54.
Roscoe, K.H., Schofield, A.N., and Thurairajah, A. (1963), 'Yielding of clays in states wetter than critical', *Géotechnique* **13**(3), 211–40.
Roscoe, K.H., Schofield, A.N., and Wroth, C.P. (1958), 'On the yielding of soils', *Géotechnique* **8**(1), 22–52.
Rowe, P.W., (1962), 'The stress–dilatancy relation for static equilibrium of an assembly of particles in contact', *Proc. Roy. Soc. London* **A269**, 500–27.
Rowe, P.W. (1971), 'Theoretical meaning and observed values of deformation parameters for soil', in R.H.G. Parry (ed.), *Stress–strain behaviour of soils (Proc. Roscoe Memorial Symp., Cambridge)* (Henley-on-Thames: G.T. Foulis & Co.), pp. 143–94.
Rowe, P.W., and Barden, L. (1964), 'Importance of free ends in triaxial testing', *Proc. ASCE, Journal of the Soil Mechanics and Foundations Division* **90**(SM1), 1–27.
Saada, A.S., and Bianchini, G.F. (1975), 'Strength of one-dimensionally consolidated clays', *Proc. ASCE, Journal of the Geotechnical Engineering Division* **101**(GT11), 1151–64.
St. John, H.D. (1975), *Field and theoretical studies of the behaviour of ground around deep excavations in London clay*, Ph.D. thesis, Cambridge University.
Sangrey, D.A., Pollard, W.S., and Egan, J.A. (1978), 'Errors associated with rate of undrained cyclic testing of clay soils', *Dynamic geotechnical testing*, STP654 (Philadelphia: American Society for Testing and Materials), 280–94.
Schmertmann, J.H. (1955), 'The undisturbed consolidation behaviour of clay', *Trans. ASCE* **120**, 1201–33.
Schmidt, B. (1966), 'Discussion: Earth pressures at rest related to stress history', *Canadian Geotechnical Journal* **3**(4), 239–42.
Schofield, A.N. (1980), 'Cambridge Geotechnical Centrifuge operations', 20th Rankine Lecture, *Géotechnique* **30**(3), 227–68.

Schofield, A.N., and Wroth, C.P. (1968), *Critical state soil mechanics* (London: McGraw-Hill).
Shahanguian, S. (1981), *Détermination expérimentale des courbes d'état limite de l'argile organique de Cubzac-les-Ponts.* (Paris: Laboratoire Central des Ponts et Chaussées), Rapport de recherche LPC 106.
Sherwood, P.T., and Ryley, M.D. (1970), 'An investigation of a cone-penetrometer method for the determination of the liquid limit', *Géotechnique* **20**(2), 203–8.
Shibata, T. (1963), 'On the volume changes of normally-consolidated clays' (in Japanese), *Disaster Prevention Research Institute Annuals, Kyoto University* **6**, 128–34.
Simpson, B., O'Riordan, N.J., and Croft, D.D. (1979), 'A computer model for the analysis of ground movements in London clay', *Géotechnique* **29**(2), 149–75.
Skempton, A.W. (1944), 'Notes on the compressibility of clays', *Quarterly J. Geological Soc. of London* **100** (C: parts 1 & 2), 119–35.
Skempton, A.W. (1953), 'The colloidal 'activity' of clays', in *Proc. 3rd Int. Conf. on Soil Mechs and Foundation Eng., Zurich* (Zurich: Organising Committee ICOSOMEF), vol. 1, pp. 57–61.
Skempton, A.W. (1954a), 'The pore pressure coefficients A and B', *Géotechnique* **4**(4), 143–47.
Skempton, A.W. (1954b), 'Discussion of the structure of inorganic soil', *Proc. ASCE, Soil Mechanics and Foundations Division* **80** (Separate 478), 19–22.
Skempton, A.W. (1957), 'Discussion: The planning and design of the new Hong Kong airport', *Proc. ICE* **7**, 305–7.
Skempton, A.W. (1970a), 'The consolidation of clays by gravitational compaction', *Quarterly J. Geological Soc. of London* **125**(3), 373–411.
Skempton, A.W. (1970b), 'First-time slides in over-consolidated clays', *Géotechnique* **20**(3), 320–4.
Skempton, A.W. (1985), 'Residual strength of clays in landslides, folded strata and the laboratory,' *Géotechnique* **35**(1), 3–18.
Skempton, A.W., and Bjerrum, L. (1957), 'A contribution to the settlement analysis of foundations on clay', *Géotechnique* **7**(4), 168–78.
Skempton, A.W., and Henkel, D.J. (1957), 'Tests on London clay from deep borings at Paddington, Victoria and the South Bank', in *Proc. 4th Int. Conf. on Soil Mechs and Foundation Eng., London* (London: Butterworths Scientific Publications), vol. 1, pp. 100–6.
Skempton, A.W., and Northey, R.D. (1953), 'The sensitivity of clays', *Géotechnique* **3**(1), 30–53.
Spencer, A.J.M. (1980), *Continuum mechanics* (London: Longman).
Statens Järnvägars Geotekniska Kommission 1914–1922 (1922), *Slutbetänkande avgivet till Kungl. Järnvägsstyrelsen* (Stockholm: Statens Järnvägar), Geotekniska Meddelanden 2.
Stroud, M.A. (1971), *The behaviour of sand at low stress levels in the simple shear apparatus*, Ph.D. thesis, Cambridge University.
Tabor, D. (1951), *The hardness of metals* (Oxford: Clarendon Press).
Tatsuoka, F. (1972), *Shear tests in a triaxial apparatus – a fundamental study of the deformation of sand* (in Japanese), Ph.D. thesis, Tokyo University.
Tatsuoka, F. (1987). 'Discussion: The strength and dilatancy of sands', *Géotechnique* **37**(2), 219–25.
Tatsuoka, F., and Ishihara, K. (1974a), 'Yielding of sand in triaxial compression', *Soils and Foundations* **14**(2), 63–76.
Tatsuoka, F., and Ishihara, K. (1974b), 'Drained deformation of sand under cyclic stresses reversing direction', *Soils and Foundations* **14**(3), 51–65.
Tavenas, F., des Rosiers, J–P., Leroueil, S., LaRochelle, P., and Roy, M. (1979), 'The use of strain energy as a yield and creep criterion for lightly overconsolidated clays', *Géotechnique* **29**(3), 285–303.

Tavenas, F., and Leroueil, S. (1980), 'The behaviour of embankments on clay foundations', *Canadian Geotechnical Journal* **17**(2), 236–60.

Tavenas, F., Leroueil, S., LaRochelle, P., and Roy, M. (1978), 'Creep behaviour of an undisturbed lightly overconsolidated clay', *Canadian Geotechnical Journal* **15**(3), 402–23.

Taylor, D.W. (1948), *Fundamentals of soil mechanics* (New York: John Wiley).

Taylor, G.I., and Quinney, H. (1931), 'The plastic distortion of metals', *Phil. Trans. Roy. Soc.* **A230**, 323–62.

Terzaghi, K. von (1923), 'Die Berechnung der Durchlässigkeitsziffer des Tones aus dem Verlauf der hydrodynamischen Spannungserscheinungen', *Akademie der Wissenschaften in Wien, Sitzungsberichte, Mathematisch-naturwissenschaftliche Klasse, Part IIa* **132**(3/4), 125–38.

Terzaghi, K. von (1936), 'Stability of slopes of natural clay', in *Proc. 1st Int. Conf. on Soil Mechs and Foundation Eng., Harvard* (Cambridge, Mass: Harvard University Graduate School of Engineering), vol. 1, pp. 161–5.

Terzaghi, K., and Peck, R.B. (1948), *Soil mechanics in engineering practice* (New York: John Wiley).

Timoshenko, S. (1934), *Theory of elasticity* (New York: McGraw-Hill).

Trak, B., LaRochelle, P., Tavenas, F., Leroueil, S., and Roy, M. (1980), 'A new approach to the stability analysis of embankments on sensitive clays', *Canadian Geotechnical Journal* **17**(4), 526–44.

Tresca, H. (1869), 'Mémoire sur le poinçonnage et la théorie mécanique de la déformation des métaux', *Comptes rendus hebdomadaires des Séances de l'Académie des Sciences, Paris* **68**, 1197–201.

United States Department of the Navy (1971), *Design manual. Soil mechanics, foundations, and earth structures*, NAVFAC DM-7 (Alexandria: Department of the Navy, Naval Facilities Engineering Command).

Vaid, Y.P., and Campanella, R.G. (1974), 'Triaxial and plane strain behaviour of natural clay', *Proc. ASCE, Journal of the Geotechnical Engineering Division* **100**(GT3), 207–24.

Vardoulakis, I. (1978), 'Equilibrium bifurcation of granular earth bodies', in *Advances in analysis of geotechnical instabilities* (Waterloo, Ontario: University of Waterloo Press), SM study 13, Paper 3, pp. 65–119.

Vermeer, P.A. (1980), *Formulation and analysis of sand deformation problems* (Delft: Geotechnical Laboratory, Delft University of Technology), Report 195.

Vermeer, P.A. (1982), 'A simple shear-band analysis using compliances', in P.A. Vermeer and H.J. Luger (eds.), *Proc. IUTAM Symp. on Deformation and Failure of Granular Materials, Delft* (Rotterdam: A.A. Balkema), pp. 493–9.

Vermeer, P.A. (1984), 'A five-constant model unifying well-established concepts', in G. Gudehus, F. Darve, and I. Vardoulakis (eds.), *Constitutive relations for soils* (Rotterdam: A.A. Balkema), pp. 175–97.

Vermeer, P.A., and Borst, R. de (1984), 'Non-associated plasticity for soils, concrete and rock', *HERON* **29**(3), 1–64.

Vesić, A.S., and Clough, G.W. (1968), 'Behaviour of granular materials under high stresses', *Proc. ASCE, Journal of the Soil Mechanics and Foundations Division* **94**(SM3), 661–88.

Ward, W.H., and Burland, J.B. (1973), 'The use of ground strain measurements in civil engineering', *Phil. Trans. Roy. Soc.* **A274**, 421–8.

Watson, J.D. (1956), 'Earth movement affecting LTE railway in deep cutting east of Uxbridge', *Proc. ICE, Part II* **5**, 302–31.

Winterkorn, H.F., and Fang, H–Y. (1975), *Foundation engineering handbook* (New York: Van Nostrand-Reinhold).

Wong, P.K.K., and Mitchell, R.J. (1975), 'Yielding and plastic flow of sensitive cemented clay', *Géotechnique* **25**(4), 763–82.

Wood, D.M. (1974), *Some aspects of the mechanical behaviour of kaolin under truly triaxial conditions of stress and strain*, Ph.D. thesis, Cambridge University.

Wood, D.M. (1982), 'Laboratory investigations of the behaviour of soils under cyclic loading: a review', in G.N. Pande and O.C. Zienkiewicz (eds.), *Soil mechanics – transient and cyclic loads* (Chichester: John Wiley), pp. 513–82.

Wood, D.M. (1984a), 'Choice of models for geotechnical predictions', in C.S. Desai and R.H. Gallagher (eds.) *Mechanics of engineering materials* (Chichester: John Wiley & Sons), pp. 633–54.

Wood, D.M. (1984b), 'On stress parameters', *Géotechnique* **34**(2), 282–7.

Wood, D.M. (1985a), 'Some fall-cone tests', *Géotechnique* **35**(1), 64–8.

Wood, D.M. (1985b), 'Index properties and consolidation history', *Proc. 11th Int. Conf. on Soil Mechanics and Foundation Engineering, San Francisco* (Rotterdam: A.A. Balkema), vol. 2, pp. 703–6.

Wood, D.M., and Budhu, M. (1980), 'The behaviour of Leighton Buzzard sand in cyclic simple shear tests', in G.N. Pande and O.C. Zienkiewicz (eds.), *Proc. Int. Symp. on Soils under Cyclic and Transient Loading, Swansea* (Rotterdam: A.A. Balkema), vol. 1, pp. 9–21.

Wood, D.M., Drescher, A., and Budhu, M. (1979), 'On the determination of the stress state in the simple shear apparatus', *Geotechnical Testing Journal, American Society for Testing and Materials* **2**(4), 211–22.

Wood, D.M., and Wroth, C.P. (1978), 'The use of the cone penetrometer to determine the plastic limit of soils', *Ground Engineering* **11**(3), 37.

Wright, P.J.F. (1955), 'Comments on an indirect tensile test on concrete cylinders', *Magazine of Concrete Research* **7**(20), 87–96.

Wroth, C.P. (1958), 'Soil behaviour during shear – existence of critical voids ratios', *Engineering* **186**, 409–13.

Wroth, C.P. (1972), 'General theories of earth pressure and deformation', in *Proc. 5th European Conf. on Soil Mechs and Foundation Eng., Madrid* (Madrid: Sociedad Española de Mecanica del Suelo y Cimentaciones), vol. 2, pp. 33–52.

Wroth, C.P. (1975), 'In-situ measurement of initial stresses and deformation characteristics', in *Proc. Specialty Conf. on In-Situ Measurement of Soil Properties, Raleigh, North Carolina* (New York: ASCE), vol. 2, pp. 181–230.

Wroth, C.P. (1979), 'Correlations of some engineering properties of soils', in *Proc. 2nd Int. Conf. on Behaviour of Off-Shore Structures, London* (Cranfield: BHRA Fluid Engineering), vol. 1, pp. 121–32.

Wroth, C.P. (1984), 'The interpretation of in situ soil tests', 24th Rankine Lecture, *Géotechnique* **34**(4), 449–89.

Wroth, C.P., and Houlsby, G.T., (1985), 'Soil mechanics – property characterisation and analysis procedures', in *Proc. 11th Int. Conf. on Soil Mechs and Foundation Eng., San Francisco* (Rotterdam: A.A. Balkema), vol. 1, pp. 1–55.

Youssef, M.S., el Ramli, A.H., and el Demery, M. (1965), 'Relationships between shear strength, consolidation, liquid limit, and plastic limit for remoulded clays', in *Proc. 6th Int. Conf. on Soil Mechs and Foundation Eng., Montreal* (Toronto: Toronto University Press), vol. 1, pp. 126–9.

Yudhbir (1973), 'Field compressibility of soft sensitive normally consolidated clays', *Geotechnical Engineering* **4**(1), 31–40.

Yudhbir (1982), 'Collapsing behaviour of residual soils', in I. McFeat-Smith and P. Lumb (eds.), *Proc. 7th SE Asian Geotechnical Conf., Hong Kong* (Hong Kong: Hong Kong Institution of Engineers and Southeast Asian Geotechnical Society), vol. 1, pp. 915–30.

Zienkiewicz, O.C. (1977), *The finite element method (3rd ed.)* (Maidenhead: McGraw-Hill).

Zytynski, M., Randolph, M.F., Nova, R., and Wroth, C.P. (1978), 'On modelling the unloading–reloading behaviour of soils', *Int. J. for Numerical and Analytical Methods in Geomechanics* **2**, 87–94.

Index

9 780521 337823

Printed in the United States
By Bookmasters